THE MALAY ARCHIPELAGO

ORANG UTAN ATTACKED BY DYAKS.

THE

MALAY ARCHIPELAGO

THE LAND OF THE
ORANG-UTAN AND THE BIRD OF PARADISE

A NARRATIVE OF TRAVEL
WITH STUDIES OF MAN AND NATURE

BY

ALFRED RUSSEL WALLACE

AUTHOR OF "DARWINISM," "ISLAND LIFE," ETC.

London
MACMILLAN AND CO.
AND NEW YORK
1890

First Edition (2 vols. crown 8vo) February 1869.
Reprinted October 1869
New Edition (1 vol. crown 8vo) February 1872.
Reprinted October 1872, 1874, 1877, 1879, 1883, 1886.
New Edition (1 vol. extra crown 8vo) 1890.

This edition published by the Natural History Museum,
Cromwell Road, London SW7 5BD
Editorial additions © The Trustees of the
Natural History Museum, London 2023
ISBN 978 0 565 095390
A catalogue record for this book is available from
the British Library.
Printed by Toppan Leefung Printing Limited
10 9 8 7 6 5 4 3 2 1

TO

CHARLES DARWIN

AUTHOR OF "THE ORIGIN OF SPECIES"

I Dedicate this Book

NOT ONLY

AS A TOKEN OF PERSONAL ESTEEM AND FRIENDSHIP

BUT ALSO

TO EXPRESS MY DEEP ADMIRATION

For his Genius and his Works

FOREWORD

ALFRED RUSSEL WALLACE is celebrated for co-articulating the mechanism of evolutionary change with Charles Darwin. On 1st July 1858, two papers were read at the Linnean Society of London in Piccadilly – neither author was present. Darwin was at home in Down House, Kent and Wallace was recovering from a leg wound, acquired clambering over logs collecting insects, in Dorey on the island of New Guinea. Wallace had sent his essay entitled *On the Tendency of Varieties to Depart Indefinitely from the Original Type* to Darwin in February 1858. Darwin, upon reading it, saw the essay was a clear exposition of the same idea he had been gestating for many years. With generosity of spirit he insisted that Wallace's essay be presented alongside his own – and the rest, as they say, is history. Evolution by natural selection was an idea that forever changed how we see the world.

What was Wallace doing though in New Guinea when his paper was read in London? Wallace had spent several years collecting insects and birds in the Amazon. Not sponsored by a rich patron or the government, he planned to fund his travels through sale of the specimens he collected to an avid, natural history loving public in Victorian London. But disaster struck in 1852 when he was on the ship back – a fire broke out and all his collections were lost. Enough to make anyone want to stay away from travel for a while. Not Wallace. He immediately began to plan another collecting trip, again to be funded by the sale of collections. By 1854 he was off, this time to a part of the world known then as the Malay Archipelago – the complex network of islands straddling the equator north of Australia. He spent eight years criss-crossing the region in search of natural history specimens, mostly birds and insects. Once back in England, he spent another six years identifying and describing his collections – many of which were new to science. Only then did begin to write the story of his travels.

And what a story it is. The book is not organised chronologically, but rather by area – various visits to islands such as

Ceram (today's Seram in the Indonesian Maluku Islands) are pulled together into a single section in the book. This organisation mirrors Wallace's own focus on geography as a driving force for evolutionary change and provides a way for him to show its importance to the distribution of life on Earth. Here he describes what became known as the 'Wallace Line', where the fauna changes from one of Asian to one of Australian affinities, indicating profound changes on the Earth's surface.

Each chapter of the work begins with a discussion of the people of the region. While written in language that today sounds patronising and colonial, Wallace also reveals a deep respect for the peoples of the lands he visited; the last sentence of the book postulates they are more civilised than the culture of England of the time. He also details his collections from each area, one highlight of which was the fabulous birds of paradise from the island of New Guinea.

The tenth edition is reproduced here, and it is Wallace's personal copy with his handwritten annotations. It was published in 1890 (28 years after the first), and is Wallacean to the core. He generously adds information from subsequent collectors and admits where he was wrong, usually in footnotes. By the time this edition was published, all his collections were in the 'British Museum' (today's Natural History Museum, London) where they are still used by scientists from all over the world for study of the region, one of the world's tropical hotspots.

A book of a collector's travels could be dry and technical. Not this one. Wallace's curiosity and love of life shine through on every page. He nonchalantly talks of walking barefoot for several days after his shoes disintegrated, robberies of his money and many bouts of fever. But his focus on his goal never wavers. Sometimes, as in New Guinea waiting to leave a place where his collecting was frustrated by local monopolies, he expresses his disappointment, but what I have always loved about this book is its eternal optimism. The joy with which Wallace observed nature and the care he took for his companions is something we could all emulate.

SANDRA KNAPP, FLS FRS, Research Botanist
Natural History Museum, London

PREFACE TO THE TENTH EDITION

Since this work was first published, twenty-one years ago, several naturalists have visited the Archipelago ; and in order to give my readers the latest results of their researches I have added footnotes whenever my facts or conclusions have been modified by later discoveries. I have also made a few verbal alterations in the text to correct any small errors or obscurities. These corrections and additions are however not numerous, and the work remains substantially the same as in the early editions. I may add that my complete collections of birds and butterflies are now in the British Museum.

Parkstone, Dorset,
   October, 1890.

PREFACE TO THE FIRST EDITION

My readers will naturally ask why I have delayed writing this book for six years after my return; and I feel bound to give them full satisfaction on this point.

When I reached England in the spring of 1862, I found myself surrounded by a room full of packing-cases, containing the collections that I had from time to time sent home for my private use. These comprised nearly three thousand bird-skins, of about a thousand species; and at least twenty thousand beetles and butterflies, of about seven thousand species; besides some quadrupeds and land-shells. A large proportion of these I had not seen for years; and in my then weak state of health, the unpacking, sorting, and arranging of such a mass of specimens occupied a long time.

I very soon decided, that until I had done something towards naming and describing the most important groups in my collection, and had worked out some of the more interesting problems of variation and geographical distribution, of which I had had glimpses while collecting them, I would not attempt to publish my travels. I could, indeed, at once have printed my notes and journals, leaving all reference to questions of natural history for a future work; but I felt that this would be as unsatisfactory to myself as it would be disappointing to my friends, and uninstructive to the public.

Since my return, up to 1868, I have published eighteen papers, in the *Transactions* or *Proceedings* of the Linnæan Zoological and Entomological Societies, describing or cataloguing portions of my collections; besides twelve others

in various scientific periodicals, on more general subjects
connected with them.

Nearly two thousand of my Coleoptera, and many hundreds
of my butterflies, have been already described by various
eminent naturalists, British and foreign ; but a much larger
number remains undescribed. Among those to whom science
is most indebted for this laborious work, I must name Mr. F. P.
Pascoe, late President of the Entomological Society of London,
who has almost completed the classification and description of
my large collection of Longicorn beetles (now in his possession),
comprising more than a thousand species, of which at least nine
hundred were previously undescribed, and new to European
cabinets.

The remaining orders of insects, comprising probably more
than two thousand species, are in the collection of Mr. William
Wilson Saunders, who has caused the larger portion of them to
be described by good entomologists. The Hymenoptera alone
amounted to more than nine hundred species, among which were
two hundred and eighty different kinds of ants, of which two
hundred were new.

The six years' delay in publishing my travels thus enables
me to give what I hope may be an interesting and instructive
sketch of the main results yet arrived at by the study of my
collections ; and as the countries I have to describe are not
much visited or written about, and their social and physical
conditions are not liable to rapid change, I believe and hope
that my readers will gain much more than they will lose, by
not having read my book six years ago, and by this time
perhaps forgotten all about it.

I must now say a few words on the plan of my work.

My journeys to the various islands were regulated by the
seasons and the means of conveyance. I visited some islands
two or three times at distant intervals, and in some cases had to
make the same voyage four times over. A chronological arrangement would have puzzled my readers. They would never

have known where they were ; and my frequent references to
the groups of islands, classed in accordance with the peculiarities
of their animal productions and of their human inhabitants,
would have been hardly intelligible. I have adopted, therefore, a geographical, zoological, and ethnological arrangement,
passing from island to island in what seems the most natural
succession, while I transgress the order in which I myself visited
them as little as possible.

I divide the Archipelago into five groups of islands, as
follow :—

I. THE INDO-MALAY ISLANDS : comprising the Malay Peninsula and Singapore, Borneo, Java, and Sumatra.

II. THE TIMOR GROUP : comprising the islands of Timor,
Flores, Sumbawa, and Lombock, with several smaller
ones.

III. CELEBES : comprising also the Sula Islands and Bouton.

IV. THE MOLUCCAN GROUP :' comprising Bouru, Ceram,
Batchian, Gilolo, and Morty ; with the smaller islands
of Ternate, Tidore, Makian, Kaióa, Amboyna, Banda,
Goram, and Matabello.

V. THE PAPUAN GROUP : comprising the great island of
New Guinea, with the Aru Islands, Mysol, Salwatty,
Waigiou, and several others. The Ké Islands are described with this group on account of their ethnology,
though zoologically and geographically they belong to
the Moluccas.

The chapters relating to the separate islands of each of these
groups are followed by one on the Natural History of that
group ; and the work may thus be divided into five parts, each
treating of one of the natural divisions of the Archipelago.

The first chapter is an introductory one, on the Physical
Geography of the whole region ; and the last is a general
sketch of the Races of Man in the Archipelago and the surrounding countries. With this explanation, and a reference to
the Maps which illustrate the work, 1 trust that my readers

will always know where they are, and in what direction they are going.

I am well aware that my book is far too small for the extent of the subjects it touches upon. It is a mere sketch; but so far as it goes I have endeavoured to make it an accurate one. Almost the whole of the narrative and descriptive portions were written on the spot, and have had little more than verbal alterations. The chapters on Natural History, as well as many passages in other parts of the work, have been written in the hope of exciting an interest in the various questions connected with the origin of species and their geographical distribution. In some cases I have been able to explain my views in detail; while in others, owing to the greater complexity of the subject, I have thought it better to confine myself to a statement of the more interesting facts of the problem, whose solution is to be found in the principles developed by Mr. Darwin in his various works. The numerous illustrations will, it is believed, add much to the interest and value of the book. They have been made from my own sketches, from photographs, or from specimens; and such subjects only have been chosen as would really illustrate the narrative or the descriptions.

I have to thank Messrs. Walter and Henry Woodbury, whose acquaintance I had the pleasure of making in Java, for a number of photographs of scenery and of natives, which have been of the greatest assistance to me. Mr. William Wilson Saunders has kindly allowed me to figure the curious horned flies; and to Mr. Pascoe I am indebted for a loan of two of the very rare Longicorns which appear in the plate of Bornean beetles. All the other specimens figured are in my own collection.

As the main object of all my journeys was to obtain specimens of natural history, both for my private collection and to supply duplicates to museums and amateurs, I will give a general statement of the number of specimens I collected, and which reached home in good condition. I must premise that I

generally employed one or two, and sometimes three Malay
servants to assist me ; and for three years had the services of a
young Englishman, Mr. Charles Allen. I was just eight years
away from England, but as I travelled about fourteen thousand
miles within the Archipelago, and made sixty or seventy
separate journeys, each involving some preparation and loss
of time, I do not think that more than six years were
really occupied in collecting.

I find that my Eastern collections amounted to :

310	specimens of	Mammalia.
100	—	Reptiles.
8,050	—	Birds.
7,500	—	Shells.
13,100	—	Lepidoptera.
83,200	—	Coleoptera.
13,400	—	other Insects.

125,660 specimens of natural history.

It now only remains for me to thank all those friends to
whom I am indebted for assistance or information. My thanks
are more especially due to the Council of the Royal Geographical
Society, through whose valuable recommendations I obtained
important aid from our own Government and from that of
Holland ; and to Mr. William Wilson Saunders, whose kind and
liberal encouragement in the early portion of my journey was
of great service to me. I am also greatly indebted to Mr.
Samuel Stevens (who acted as my agent), both for the care he
took of my collections, and for the untiring assiduity with which
he kept me supplied, both with useful information, and with
whatever necessaries I required.

I trust that these, and all other friends who have been in any
way interested in my travels and collections, may derive from
the perusal of my book some faint reflexion of the pleasures I
myself enjoyed amid the scenes and objects it describes.

CONTENTS

THE MOLUCCAS

THE PAPUAN GROUP

LIST OF ILLUSTRATIONS

MAPS

THE MALAY ARCHIPELAGO

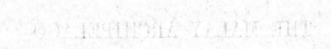

THE MALAY ARCHIPELAGO.

CHAPTER I.

PHYSICAL GEOGRAPHY.

IF we look at a globe or a map of the Eastern hemisphere, we shall perceive between Asia and Australia a number of large and small islands, forming a connected group distinct from those great masses of land, and having little connexion with either of them. Situated upon the Equator, and bathed by the tepid water of the great tropical oceans, this region enjoys a climate more uniformly hot and moist than almost any other part of the globe, and teems with natural productions which are elsewhere unknown. The richest of fruits and the most precious of spices are here indigenous. It produces the giant flowers of the Rafflesia, the great green-winged Ornithoptera (princes among the butterfly tribes), the man-like Orang-Utan, and the gorgeous Birds of Paradise. It is inhabited by a peculiar and interesting race of mankind—the Malay, found nowhere beyond the limits of this insular tract, which has hence been named the Malay Archipelago.

To the ordinary Englishman this is perhaps the least known part of the globe. Our possessions in it are few and scanty; scarcely any of our travellers go to explore it; and in many collections of maps it is almost ignored, being divided between Asia and the Pacific Islands.[1] It thus happens that few persons realize that, as a whole, it is comparable with the primary divisions of the globe, and that some of its separate islands are larger than France or the Austrian empire. The traveller, however, soon acquires different ideas. He sails for days, or even for weeks, along the shores of one of these great islands, often so great that its inhabitants believe it to be a vast continent. He finds that voyages among these islands are commonly

[1] Since the establishment of the British North Borneo Company the region is more known, but the Dutch Colonies are still rarely visited.

reckoned by weeks and months, and that their several inhabitants are often as little known to each other as are the
native races of the northern to those of the southern continent
of America. He soon comes to look upon this region as one apart
from the rest of the world with its own races of men and its own
aspects of nature; with its own ideas, feelings, customs, and
modes of speech, and with a climate, vegetation, and animated
life altogether peculiar to itself.

From many points of view these islands form one compact
geographical whole, and as such they have always been treated
by travellers and men of science; but a more careful and detailed study of them under various aspects, reveals the unexpected fact that they are divisible into two portions nearly
equal in extent, which widely differ in their natural products,
and really form parts of two of the primary divisions of the
earth. I have been able to prove this in considerable detail by
my observations on the natural history of the various parts of
the Archipelago; and as in the description of my travels and
residence in the several islands I shall have to refer continually
to this view, and adduce facts in support of it, I have thought
it advisable to commence with a general sketch of such of the
main features of the Malayan region as will render the facts
hereafter brought forward more interesting, and their bearing
on the general question more easily understood. I proceed,
therefore, to sketch the limits and extent of the Archipelago,
and to point out the more striking features of its geology,
physical geography, vegetation, and animal life.

Definition and Boundaries.—For reasons which depend mainly
on the distribution of animal life, I consider the Malay Archipelago to include the Malay Peninsula as far as Tenasserim, and
the Nicobar Islands on the west, the Philippines on the north,
and the Solomon Islands beyond New Guinea, on the east. All
the great islands included within these limits are connected
together by innumerable smaller ones, so that no one of them
seems to be distinctly separated from the rest. With but few
exceptions, all enjoy an uniform and very similar climate, and
are covered with a luxuriant forest vegetation. Whether we
study their form and distribution on maps, or actually travel
from island to island, our first impression will be that they form
a connected whole, all the parts of which are intimately related
to each other.

Extent of the Archipelago and Islands.—The Malay Archipelago
extends for more than 4,000 miles in length from east to west,
and is about 1,300 in breadth from north to south. It would
stretch over an expanse equal to that of all Europe from the
extreme west far into Central Asia, or would cover the widest
parts of South America, and extend far beyond the land into
the Pacific and Atlantic oceans. It includes three islands larger
than Great Britain; and in one of them, Borneo, the whole of
the British Isles might be set down, and would be surrounded

by a sea of forests. New Guinea, though less compact in shape, is probably larger than Borneo. Sumatra is about equal in extent to Great Britain; Java, Luzon, and Celebes are each about the size of Ireland. Eighteen more islands are, on the average, as large as Jamaica; more than a hundred are as large as the Isle of Wight; while the isles and islets of smaller size are innumerable.

The absolute extent of land in the Archipelago is not greater

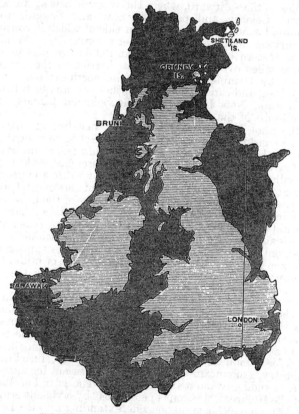

THE BRITISH ISLES AND BORNEO ON THE SAME SCALE.

than that contained by Western Europe from Hungary to Spain; but, owing to the manner in which the land is broken up and divided, the variety of its productions is rather in proportion to the immense surface over which the islands are spread, than to the quantity of land which they contain.

Geological Contrasts.—One of the chief volcanic belts upon the globe passes through the Archipelago, and produces a striking contrast in the scenery of the volcanic and non-volcanic islands. A curving line marked out by scores of active and hundreds of extinct volcanoes, may be traced through the whole length of Sumatra and Java, and thence by the islands of Bali, Lombock, Sumbawa, Flores, the Serwatty Islands, Banda, Amboyna, Batchian, Makian, Tidore, Ternate, and Gilolo, to Morty Island. Here there is a slight but well-marked break, or shift, of about 200 miles to the westward, where the volcanic belt again begins, in North Celebes, and passes by Siau and Sanguir to the Philippine Islands, along the eastern side of which it continues, in a curving line, to their northern extremity. From the extreme eastern bend of this belt at Banda, we pass onwards for 1,000 miles over a non-volcanic district to the volcanoes observed by Dampier, in 1699, on the north-eastern coast of New Guinea, and can there trace another volcanic belt, through New Britain, New Ireland, and the Solomon Islands, to the eastern limits of the Archipelago.

In the whole region occupied by this vast line of volcanoes, and for a considerable breadth on each side of it, earthquakes are of continual recurrence, slight shocks being felt at intervals of every few weeks or months, while more severe ones, shaking down whole villages, and doing more or less injury to life and property, are sure to happen, in one part or another of this district, almost every year. In many of the islands the years of the great earthquakes form the chronological epochs of the native inhabitants, by the aid of which the ages of their children are remembered, and the dates of many important events are determined.

I can only briefly allude to the many fearful eruptions that have taken place in this region. In the amount of injury to life and property, and in the magnitude of their effects, they have not been surpassed by any upon record. Forty villages were destroyed by the eruption of Papandayang in Java, in 1772 when the whole mountain was blown up by repeated explosions, and a large lake left in its place. By the great eruption of Tomboro in Sumbawa, in 1815, 12,000 people were destroyed, and the ashes darkened the air and fell thickly upon the earth and sea for 300 miles round. Even quite recently, since I quitted the country, a mountain which had been quiescent for more than 200 years suddenly burst into activity. The island of Makian, one of the Moluccas, was rent open in 1646 by a violent eruption, which left a huge chasm on one side, extending into the heart of the mountain. When I last visited it, in 1860, it was clothed with vegetation to the summit, and contained twelve populous Malay villages. On the 29th of December, 1862, after 215 years of perfect inaction, it again suddenly burst forth, blowing up and completely altering the appearance of the mountain, destroying the greater part of the inhabitants, and sending forth such volumes of ashes as to darken the air at Ternate, forty

miles off, and to almost entirely destroy the growing crops on
that and the surrounding islands.[1]

The island of Java contains more volcanoes, active and extinct,
than any other known district of equal extent. They are
about forty-five in number, and many of them exhibit most
beautiful examples of the volcanic cone on a large scale, single
or double, with entire or truncated summits, and averaging
10,000 feet high.

It is now well ascertained that almost all volcanoes have been
slowly built up by the accumulation of matter—mud, ashes, and
lava—ejected by themselves. The openings or craters, however,
frequently shift their position ; so that a country may be covered
with a more or less irregular series of hills in chains and masses,
only here and there rising into lofty cones, and yet the whole
may be produced by true volcanic action. In this manner the
greater part of Java has been formed. There has been some
elevation, especially on the south coast, where extensive cliffs of
coral limestone are found ; and there may be a substratum of
older stratified rocks ; but still essentially Java is volcanic ;
and that noble and fertile island—the very garden of the East,
and perhaps upon the whole the richest, the best cultivated, and
the best governed tropical island in the world—owes its very
existence to the same intense volcanic activity which still
occasionally devastates its surface.

The great island of Sumatra exhibits in proportion to its
extent a much smaller number of volcanoes, and a considerable
portion of it has probably a non-volcanic origin.

To the eastward, the long string of islands from Java, passing
by the north of Timor and away to Banda, are probably all due
to volcanic action. Timor itself consists of ancient stratified
rocks, but is said to have one volcano near its centre.

Going northward, Amboyna, a part of Bouru, and the west
end of Ceram, the north part of Gilolo, and all the small islands
around it, the northern extremity of Celebes, and the islands of
Siau and Sanguir, are wholly volcanic. The Philippine Archipelago contains many active and extinct volcanoes, and has
probably been reduced to its present fragmentary condition by
subsidences attending on volcanic action.

All along this great line of volcanoes are to be found more or
less palpable signs of upheaval and depression of land. The
range of islands south of Sumatra, a part of the south coast of
Java and of the islands east of it, the west and east end of

---

[1] More recently, in 1883, the volcanic island of Krakatoa was blown up in a terrific
eruption, the sound of the explosions being heard at Ceylon, New Guinea, Manilla, and
West Australia, while the ashes were spread over an area as large as the German Empire.
The chief destruction was effected by great sea waves, which entirely destroyed many
towns and villages on the coasts of Java and Sumatra, causing the death of between 30,000
and 40,000 persons. The atmospheric disturbance was so great that air-waves passed
three and a quarter times round the globe, and the finer particles floating in the higher
parts of the atmosphere produced remarkable colours in the sky at sunset for more than
two years afterwards and in all parts of the world.

Timor, portions of all the Moluccas, the Ké and Aru Islands, Waigiou, and the whole south and east of Gilolo, consist in a great measure of upraised coral-rock, exactly corresponding to that now forming in the adjacent seas. In many places I have observed the unaltered surfaces of the elevated reefs, with great masses of coral standing up in their natural position, and hundreds of shells so fresh-looking that it was hard to believe that they had been more than a few years out of the water ; and, in fact, it is very probable that such changes have occurred within a few centuries.

The united lengths of these volcanic belts is about ninety degrees, or one-fourth of the entire circumference of the globe. Their width is about fifty miles ; but, for a space of two hundred on each side of them, evidences of subterranean action are to be found in recently elevated coral-rock, or in barrier coral-reefs, indicating recent submergence. In the very centre or focus of the great curve of volcanoes is placed the large island of Borneo, in which no sign of recent volcanic action has yet been observed, and where earthquakes, so characteristic of the surrounding regions, are entirely unknown. The equally large island of New Guinea occupies another quiescent area, on which no sign of volcanic action has yet been discovered. With the exception of the eastern end of its northern peninsula, the large and curiously-shaped island of Celebes is also entirely free from volcanoes ; and there is some reascn to believe that the volcanic portion has once formed a separate island. The Malay Peninsula is also non-volcanic.

The first and most obvious division of the Archipelago would therefore be into quiescent and volcanic regions, and it might, perhaps, be expected that such a division would correspond to some differences in the character of the vegetation and the forms of life. This is the case, however, to a very limited extent ; and we shall presently see that, although this development of subterranean fires is on so vast a scale,—has piled up chains of mountains ten or twelve thousand feet high—has broken up continents and raised up islands from the ocean,—yet it has all the character of a recent action, which has not yet succeeded in obliterating the traces of a more ancient distribution of land and water.

Contrasts of Vegetation.—Placed immediately upon the Equator and surrounded by extensive oceans, it is not surprising that the various islands of the Archipelago should be almost always clothed with a forest vegetation from the level of the sea to the summits of the loftiest mountains. This is the general rule. Sumatra, New Guinea, Borneo, the Philippines and the Moluccas, and the uncultivated parts of Java and Celebes, are all forest countries, except a few small and unimportant tracts, due perhaps, in some cases, to ancient cultivation or accidental fires. To this, however, there is one important exception in the island of Timor and all the smaller islands around it, in which there is absolutely no forest such as exists in the other islands, and this character extends in a lesser degree to Flores, Sumbawa, Lombock, and Bali.

In Timor the most common trees are Eucalypti of several species, so characteristic of Australia, with sandal-wood, acacia, and other sorts in less abundance. These are scattered over the country more or less thickly, but never so as to deserve the name of a forest. Coarse and scanty grasses grow beneath them on the more barren hills, and a luxuriant herbage in the moister localities. In the islands between Timor and Java there is often a more thickly wooded country, abounding in thorny and prickly trees. These seldom reach any great height, and during the force of the dry season they almost completely lose their leaves, allowing the ground beneath them to be parched up, and contrasting strongly with the damp, gloomy, ever-verdant forests of the other islands. This peculiar character, which extends in a less degree to the southern peninsula of Celebes and the east end of Java, is most probably owing to the proximity of Australia. The south-east monsoon, which lasts for about two-thirds of the year (from March to November), blowing over the northern parts of that country, produces a degree of heat and dryness which assimilates the vegetation and physical aspect of the adjacent islands to its own. A little further eastward in Timor-laut and the Ké Islands, a moister climate prevails, the south-east winds blowing from the Pacific through Torres Straits and over the damp forests of New Guinea, and as a consequence every rocky islet is clothed with verdure to its very summit. Further west again, as the same dry winds blow over a wider and wider extent of ocean, they have time to absorb fresh moisture, and we accordingly find the island of Java possessing a less and less arid climate, till in the extreme west near Batavia rain occurs more or less all the year round, and the mountains are everywhere clothed with forests of unexampled luxuriance.

Contrasts in Depth of Sea.—It was first pointed out by Mr. George Windsor Earl, in a paper read before the Royal Geographical Society in 1845, and subsequently in a pamphlet *On the Physical Geography of South-Eastern Asia and Australia*, dated 1855, that a shallow sea connected the great islands of Sumatra, Java, and Borneo with the Asiatic continent, with which their natural productions generally agreed; while a similar shallow sea connected New Guinea and some of the islands adjacent to Australia, all being characterized by the presence of marsupials.

We have here a clue to the most radical contrast in the Archipelago, and by following it out in detail I have arrived at the conclusion that we can draw a line among the islands, which shall so divide them that one-half shall truly belong to Asia, while the other shall no less certainly be allied to Australia. I term these respectively the Indo-Malayan, and the

Austro-Malayan divisions of the Archipelago. (*See* Physical Map.)

In Mr. Earl's pamphlet, however, he argues in favour of the former land-connexion of Asia and Australia, whereas it appears to me that the evidence, taken as a whole, points to their long-continued separation. Notwithstanding this and other important differences between us, to him undoubtedly belongs the merit of first indicating the division of the Archipelago into an Australian and an Asiatic region, which it has been my good fortune to establish by more detailed observations.

Contrasts in Natural Productions.—To understand the importance of this class of facts, and its bearing upon the former distribution of land and sea, it is necessary to consider the results arrived at by geologists and naturalists in other parts of the world.

It is now generally admitted that the present distribution of living things on the surface of the earth is mainly the result of the last series of changes that it has undergone. Geology teaches us that the surface of the land and the distribution of land and water is everywhere slowly changing. It further teaches us that the forms of life which inhabit that surface have, during every period of which we possess any record, been also slowly changing.

It is not now necessary to say anything about *how* either of those changes took place ; as to that, opinions may differ ; but as to the fact that the changes themselves *have* occurred, from the earliest geological ages down to the present day, and are still going on, there is no difference of opinion. Every successive stratum of sedimentary rock, sand, or gravel, is a proof that changes of level have taken place ; and the different species of animals and plants, whose remains are found in these deposits, prove that corresponding changes did occur in the organic world.

Taking, therefore, these two series of changes for granted, most of the present peculiarities and anomalies in the distribution of species may be directly traced to them. In our own islands, with a very few trifling exceptions, every quadruped, bird, reptile, insect, and plant, is found also on the adjacent continent. In the small islands of Sardinia and Corsica, there are some quadrupeds and insects, and many plants, quite peculiar. In Ceylon, more closely connected to India than Britain is to Europe, many animals and plants are different from those found in India, and peculiar to the island. In the Galapagos Islands, almost every indigenous living thing is peculiar to them, though closely resembling other kinds found in the nearest parts of the American continent.

Most naturalists now admit that these facts can only be explained by the greater or less lapse of time since the islands were upraised from beneath the ocean, or were separated from the nearest land ; and this will be generally (though not always) indicated by the depth of the intervening sea. The enormous

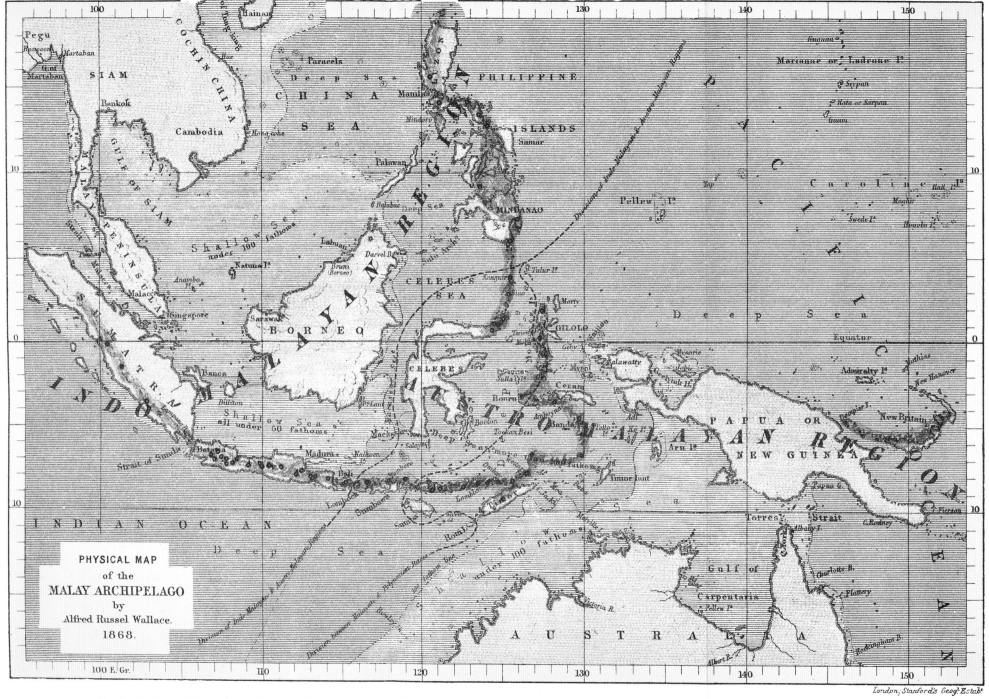

PHYSICAL MAP
of the
MALAY ARCHIPELAGO
by
Alfred Russel Wallace.
1868.

The Shallow Sea is lightly tinted   The active Volcanoes are shewn thus ●   London. Macmillan & Cº.   The Volcanic belts are colored Red

London, Stanford's Geogˡ Estabᵗ

thickness of many marine deposits through wide areas shows that subsidence has often continued (with intermitting periods of repose) during epochs of immense duration. The depth of sea produced by such subsidence will therefore generally be a measure of time ; and in like manner the change which organic forms have undergone is a measure of time. When we make proper allowance for the continued introduction of new animals and plants from surrounding countries, by those natural means of dispersal which have been so well explained by Sir Charles Lyell and Mr. Darwin, it is remarkable how closely these two measures correspond. Britain is separated from the continent by a very shallow sea, and only in a very few cases have our animals or plants begun to show a difference from the corresponding continental species. Corsica and Sardinia, divided from Italy by a much deeper sea, present a much greater difference in their organic forms. Cuba, separated from Yucatan by a wider and deeper strait, differs more markedly, so that most of its productions are of distinct and peculiar species ; while Madagascar, divided from Africa by a deep channel three hundred miles wide, possesses so many peculiar features as to indicate separation at a very remote antiquity, or even to render it doubtful whether the two countries have ever been absolutely united.

Returning now to the Malay Archipelago, we find that all the wide expanse of sea which divides the islands of Java, Sumatra, and Borneo from each other, and from Malacca and Siam, is so shallow that ships can anchor in any part of it, since it rarely exceeds forty fathoms in depth ; and if we go as far as the line of a hundred fathoms, we shall include the Philippine Islands and Bali, east of Java. If, therefore, these islands have been separated from each other and the continent by subsidence of the intervening tracts of land, we should conclude that the separation has been comparatively recent, since the depth to which the land has subsided is so small. It is also to be remarked, that the great chain of active volcanoes in Sumatra and Java furnishes us with a sufficient cause for such subsidence, since the enormous masses of matter they have thrown out would take away the foundations of the surrounding district ; and this may be the true explanation of the often-noticed fact, that volcanoes and volcanic chains are always near the sea. The subsidence they produce around them will, in time, make a sea, if one does not already exist.[1]

But it is when we examine the zoology of these countries that we find what we most require—evidence of a very striking character that these great islands must have once formed a part of the continent, and could only have been separated at a very recent geological epoch. The elephant and tapir of Sumatra and Borneo, the rhinoceros of Sumatra and the allied species of Java,

---

[1] It it now believed by most geologists that subsidence is produced by the weight of every fresh deposit of materials either in the sea or on the land. Accumulations of rock or ashes from volcanoes would, therefore, be itself a cause of subsidence.

the wild cattle of Borneo and the kind long supposed to be peculiar to Java, are now all known to inhabit some part or other of Southern Asia. None of these large animals could possibly have passed over the arms of the sea which now separate these countries, and their presence plainly indicates that a land communication must have existed since the origin of the species. Among the smaller mammals a considerable portion are common to each island and the continent ; but the vast physical changes that must have occurred during the breaking up and subsidence of such extensive regions have led to the extinction of some in one or more of the islands, and in some cases there seems also to have been time for a change of species to have taken place. Birds and insects illustrate the same view, for every family, and almost every genus of these groups found in any of the islands, occurs also on the Asiatic continent, and in a great number of cases the species are exactly identical. Birds offer us one of the best means of determining the law of distribution ; for though at first sight it would appear that the watery boundaries which keep out the land quadrupeds could be easily passed over by birds, yet practically it is not so ; for if we leave out the aquatic tribes which are pre-eminently wanderers, it is found that the others (and especially the Passeres, or true perching-birds, which form the vast majority) are often as strictly limited by straits and arms of the sea as are quadrupeds themselves. As an instance, among the islands of which I am now speaking, it is a remarkable fact that Java possesses numerous birds which never pass over to Sumatra, though they are separated by a strait only fifteen miles wide, and with islands in mid-channel. Java, in fact, possesses more birds and insects peculiar to itself than either Sumatra or Borneo, and this would indicate that it was earliest separated from the continent; next in organic individuality is Borneo, while Sumatra is so nearly identical in all its animal forms with the peninsula of Malacca, that we may safely conclude it to have been the most recently dismembered island.

The general result therefore at which we arrive is, that the great islands of Java, Sumatra, and Borneo, resemble in their natural productions the adjacent parts of the continent, almost as much as such widely-separated districts could be expected to do even if they still formed a part of Asia ; and this close resemblance, joined with the fact of the wide extent of sea which separates them being so uniformly and remarkably shallow, and lastly, the existence of the extensive range of volcanoes in Sumatra and Java, which have poured out vast quantities of subterranean matter and have built up extensive plateaux and lofty mountain ranges, thus furnishing a *vera causa* for a parallel line of subsidence—all lead irresistibly to the conclusion that at a very recent geological epoch the continent of Asia extended far beyond its present limits in a south-easterly direction including the islands of Java, Sumatra, and Borneo, and probably reaching as far as the present 100-fathom line of soundings.

The Philippine Islands agree in many respects with Asia and the other islands, but present some anomalies, which seem to indicate that they were separated at an earlier period, and have since been subject to many revolutions in their physical geography.

Turning our attention now to the remaining portion of the Archipelago, we shall find that all the islands from Celebes and Lombock eastward exhibit almost as close a resemblance to Australia and New Guinea as the Western Islands do to Asia. It is well known that the natural productions of Australia differ from those of Asia more than those of any of the four ancient quarters of the world differ from each other. Australia, in fact, stands alone : it possesses no apes or monkeys, no cats or tigers, wolves, bears, or hyenas ; no deer or antelopes, sheep or oxen ; no elephant, horse, squirrel, or rabbit ; none, in short, of those familiar types of quadruped which are met with in every other part of the world. Instead of these, it has Marsupials only, kangaroos and opossums, wombats and the duck-billed Platypus. In birds it is almost as peculiar. It has no woodpeckers and no pheasants, families which exist in every other part of the world ; but instead of them it has the mound-making brush-turkeys, the honeysuckers, the cockatoos, and the brush-tongued lories, which are found nowhere else upon the globe. All these striking peculiarities are found also in those islands which form the Austro-Malayan division of the Archipelago.

The great contrast between the two divisions of the Archipelago is nowhere so abruptly exhibited as on passing from the island of Bali to that of Lombock, where the two regions are in closest proximity. In Bali we have barbets, fruit-thrushes, and woodpeckers ; on passing over to Lombock these are seen no more, but we have abundance of cockatoos, honeysuckers, and brush-turkeys, which are equally unknown in Bali,[1] or any island further west. The strait is here fifteen miles wide, so that we may pass in two hours from one great division of the earth to another, differing as essentially in their animal life as Europe does from America. If we travel from Java or Borneo to Celebes or the Moluccas, the difference is still more striking. In the first, the forests abound in monkeys of many kinds, wild cats, deer, civets, and otters, and numerous varieties of squirrels are constantly met with. In the latter none of these occur ; but the prehensile-tailed cuscus is almost the only terrestrial mammal seen, except wild pigs, which are found in all the islands, and deer (which have probably been recently introduced) in Celebes and the Moluccas. The birds which are most abundant in the Western Islands are woodpeckers, barbets, trogons, fruit-

[1] I was informed, however, that there were a few cockatoos at one spot on the west of Bali, showing that the intermingling of the productions of these islands is now going on.

thrushes, and leaf-thrushes : they are seen daily, and form the great ornithological features of the country. In the Eastern Islands these are absolutely unknown, honeysuckers and small lories being the most common birds ; so that the naturalist feels himself in a new world, and can hardly realize that he has passed from the one region to the other in a few days, without ever being out of sight of land.

The inference that we must draw from these facts is undoubtedly, that the whole of the islands eastwards beyond Java and Borneo, with the exception, perhaps, of Celebes, do essentially form a part of a former Australian or Pacific continent, although some of them may never have been actually joined to it. This continent must have been broken up not only before the Western Islands were separated from Asia, but probably before the extreme south-eastern portion of Asia was raised above the waters of the ocean ; for a great part of the land of Borneo and Java is known to be geologically of quite recent formation, while the very great difference of species, and in many cases of genera also, between the productions of the Eastern Malay Islands and Australia, as well as the great depth of the sea now separating them, all point to a comparatively long period of isolation.

It is interesting to observe among the islands themselves, how a shallow sea always intimates a recent land-connexion. The Aru Islands, Mysol, and Waigiou, as well as Jobie, agree with New Guinea in their species of mammalia and birds much more closely than they do with the Moluccas, and we find that they are all united to New Guinea by a shallow sea. In fact, the 100-fathom line round New Guinea marks out accurately the range of the true Paradise birds.

It is further to be noted—and this is a very interesting point in connexion with theories of the dependence of special forms of life on external conditions—that this division of the Archipelago into two regions characterized by a striking diversity in their natural productions, does not in any way correspond to the main physical or climatal divisions of the surface. The great volcanic chain runs through both parts, and appears to produce no effect in assimilating their productions. Borneo closely resembles New Guinea, not only in its vast size and its freedom from volcanoes, but in its variety of geological structure, its uniformity of climate, and the general aspect of the forest vegetation that clothes its surface. The Moluccas are the counterpart of the Philippines in their volcanic structure, their extreme fertility, their luxuriant forests, and their frequent earthquakes ; and Bali with the east end of Java has a climate almost as dry and a soil almost as arid as that of Timor. Yet between these corresponding groups of islands, constructed as it were after the same pattern, subjected to the same climate, and bathed by the same oceans, there exists the greatest possible contrasts when we compare their animal productions. Nowhere does the ancient doctrine—that differences or similarities in the various forms of life that inhabit different countries are due to corresponding physical differences or similarities in the countries themselves—meet with so direct and palpable a contradiction. Borneo and New Guinea, as alike physically as two distinct countries can be, are zoologically wide as the poles asunder ; while Australia, with its dry winds, its open plains, its stony deserts, and its temperate climate, yet produces birds and quadrupeds which are closely related to those inhabiting the hot, damp, luxuriant forests which everywhere clothe the plains and mountains of New Guinea.

In order to illustrate more clearly the means by which I suppose this great contrast has been brought about, let us consider what would occur if two strongly contrasted divisions of the earth were, by natural means brought into proximity. No two parts of the world differ so radically in their productions as Asia and Australia, but the difference between Africa and South America is also very great, and these two regions will well serve to illustrate the question we are considering. On the one side we have baboons, lions, elephants, buffaloes, and giraffes ; on the other spider-monkeys, pumas, tapirs, ant-eaters, and sloths ; while among birds, the hornbills, turacos, orioles, and honeysuckers of Africa contrast strongly with the toucans, macaws, chatterers, and humming-birds of America.

Now let us endeavour to imagine that a slow upheaval of the bed of the Altantic should take place, while at the same time earthquake-shocks and volcanic action on the land should cause increased volumes of sediment to be poured down by the rivers, so that the two continents should gradually spread out by the addition of newly-formed lands, and thus reduce the Atlantic which now separates them to an arm of the sea a few hundred miles wide. At the same time we may suppose islands to be upheaved in mid-channel ; and, as the subterranean forces varied in intensity, and shifted their points of greatest action, these islands would sometimes become connected with the land on one side or other of the strait, and at other times again be separated from it. Several islands would at one time be joined together, at another would be broken up again, till at last, after many long ages of such intermittent action, we might have an irregular archipelago of islands filling up the ocean channel of the Atlantic, in whose appearance and arrangement we could discover nothing to tell us which had been connected with Africa and which with America. The animals and plants inhabiting these islands would, however, certainly reveal this portion of their former history. On those islands which had ever formed a part of the South American continent we should be sure to find such common birds as chatterers and toucans and humming-birds, and some of the peculiar American quadrupeds ; while on those which had been separated from Africa, hornbills, orioles, and honeysuckers would as certainly be found. Some portion of the upraised land might at different times have had a temporary connexion with both continents, and would then contain a certain amount of mixture in its living inhabitants. Such seems to have been the case with the islands of Celebes and the Philippines. Other islands, again, though in such close proximity as Bali and Lombock, might each exhibit an almost unmixed sample of the productions of the continents of which they had directly or indirectly once formed a part.

In the Malay Archipelago we have, I believe, a case exactly parallel to that which I have here supposed. We have indications of a vast continent, with a peculiar fauna and flora, having been gradually and irregularly broken up; the island of Celebes probably marking its furthest westward extension, beyond which was a wide ocean.[1] At the same time Asia appears to have been extending its limits in a south-east direction, first in an unbroken mass, then separated into islands as we now see it, and almost coming into actual contact with the scattered fragments of the great southern land.

From this outline of the subject, it will be evident how important an adjunct Natural History is to Geology; not only in interpreting the fragments of extinct animals found in the earth's crust, but in determining past changes in the surface which have left no geological record. It is certainly a wonderful and unexpected fact, that an accurate knowledge of the distribution of birds and insects should enable us to map out lands and continents which disappeared beneath the ocean long before the earliest traditions of the human race. Wherever the geologist can explore the earth's surface, he can read much of its past history, and can determine approximately its latest movements above and below the sea-level; but wherever oceans and seas now extend, he can do nothing but speculate on the very limited data afforded by the depth of the waters. Here the naturalist steps in, and enables him to fill up this great gap in the past history of the earth.

One of the chief objects of my travels was to obtain evidence of this nature; and my search after such evidence has been rewarded by great success, so that I have been enabled to trace out with some probability the past changes which one of the most interesting parts of the earth has undergone. It may be thought that the facts and generalizations here given, would have been more appropriately placed at the end rather than at the beginning of a narrative of the travels which supplied the facts. In some cases this might be so, but I have found it impossible to give such an account as I desire of the natural history of the numerous islands and groups of islands in the Archipelago, without constant reference to these generalizations which add

---

[1] Further study of the subject has led me to conclude that Celebes never formed part of the Austro-Malayan land, but that it more probably indicates the furthest eastward extension of the Asiatic continent at a very early period. (*See* the author's *Island Life*, p. 427.)

so much to their interest. Having given this general sketch of the subject, I shall be able to show how the same principles can be applied to the individual islands of a group as to the whole Archipelago ; and make my account of the many new and curious animals which inhabit them both more interesting and more instructive than if treated as mere isolated facts.

Contrasts of Races.—Before I had arrived at the conviction that the eastern and western halves of the Archipelago belonged to distinct primary regions of the earth, I had been led to group the natives of the Archipelago under two radically distinct races. In this I differed from most ethnologists who had before written on the subject ; for it had been the almost universal custom to follow William von Humboldt and Pritchard, in classing all the Oceanic races as modifications of one type. Observation soon showed me, however, that Malays and Papuans differed radically in every physical, mental, and moral character ; and more detailed research, continued for eight years, satisfied me that under these two forms, as types, the whole of the peoples of the Malay Archipelago and Polynesia could be classified. On drawing the line which separates these races, it is found to come near to that which divides the zoological regions, but somewhat eastward of it ; a circumstance which appears to me very significant of the same causes having influenced the distribution of mankind that have determined the range of other animal forms.

The reason why exactly the same line does not limit both is sufficiently intelligible. Man has means of traversing the sea which animals do not possess ; and a superior race has power to press out or assimilate an inferior one. The maritime enterprise and higher civilization of the Malay races have enabled them to overrun a portion of the adjacent region, in which they have entirely supplanted the indigenous inhabitants if it ever possessed any ; and to spread much of their language, their domestic animals, and their customs far over the Pacific, into islands where they have but slightly, or not at all, modified the physical or moral characteristics of the people.

I believe, therefore, that all the peoples of the various islands can be grouped either with the Malays or the Papuans ; and that these two have no traceable affinity to each other. I believe, further, that all the races east of the line I have drawn have more affinity for each other than they have for any of the races west of that line ;—that, in fact, the Asiatic races include the Malays, and all have a continental origin, while the races of Papuan type, including all to the east of the former, as far as the Fiji Islands, are derived, not from any existing continent, but from lands which now exist or have recently existed in the Pacific Ocean. These preliminary observations will enable the reader better to apprehend the importance I attach to the details of physical form or moral character, which I shall give in describing the inhabitants of many of the islands.

CHAPTER II.

SINGAPORE.

(A SKETCH OF THE TOWN AND ISLAND AS SEEN DURING SEVERAL VISITS FROM 1854 TO 1862.)

FEW places are more interesting to a traveller from Europe than the town and island of Singapore, furnishing, as it does, examples of a variety of Eastern races, and of many different religions and modes of life. The government, the garrison, and the chief merchants are English; but the great mass of the population is Chinese, including some of the wealthiest merchants, the agriculturists of the interior, and most of the mechanics and labourers. The native Malays are usually fishermen and boatmen, and they form the main body of the police. The Portuguese of Malacca supply a large number of the clerks and smaller merchants. The Klings of Western India are a numerous body of Mahometans, and, with many Arabs, are petty merchants and shopkeepers. The grooms and washermen are all Bengalees, and there is a small but highly respectable class of Parsee merchants. Besides these, there are numbers of Javanese sailors and domestic servants, as well as traders from Celebes, Bali, and many other islands of the Archipelago. The harbour is crowded with men-of-war and trading vessels of many European nations, and hundreds of Malay praus and Chinese junks, from vessels of several hundred tons burthen down to little fishing boats and passenger sampans; and the town comprises handsome public buildings and churches, Mahometan mosques Hindoo temples, Chinese joss-houses, good European houses massive warehouses, queer old Kling and China bazaars, and long suburbs of Chinese and Malay cottages.

By far the most conspicuous of the various kinds of people in Singapore, and those which most attract the stranger's attention, are the Chinese, whose numbers and incessant activity give the place very much the appearance of a town in China. The Chinese merchant is generally a fat round-faced man with an important and business-like look. He wears the same style of clothing (loose white smock, and blue or black trousers) as the meanest coolie, but of finer materials, and is always clean and neat; and his long tail tipped with red silk hangs down to his heels. He has a handsome warehouse or shop in town and a good house in the country. He keeps a fine horse and gig, and every evening may be seen taking a drive bareheaded to enjoy the cool breeze. He is rich, he owns several retail shops and trading schooners, he lends money at high interest and on good security, he makes hard bargains and gets fatter and richer every year.

In the Chinese bazaar are hundreds of small shops in which a

miscellaneous collection of hardware and dry goods are to be
found, and where many things are sold wonderfully cheap. You
may buy gimlets at a penny each, white cotton thread at four
balls for a halfpenny, and penknives, corkscrews, gunpowder,
writing-paper, and many other articles as cheap or cheaper than
you can purchase them in England. The shopkeeper is very
good-natured ; he will show you everything he has, and does
not seem to mind if you buy nothing. He bates a little, but not
so much as the Klings, who almost always ask twice what they
are willing to take. If you buy a few things of him, he will
speak to you afterwards every time you pass his shop, asking
you to walk in and sit down, or take a cup of tea, and you
wonder how he can get a living where so many sell the same
trifling articles. The tailors sit *at* a table, not *on* one ; and both
they and the shoemakers work well and cheaply. The barbers
have plenty to do, shaving heads and cleaning ears ; for which
latter operation they have a great array of little tweezers, picks,
and brushes. In the outskirts of the town are scores of carpenters and blacksmiths. The former seem chiefly to make
coffins and highly painted and decorated clothes-boxes. The
latter are mostly gun-makers, and bore the barrels of guns by
hand, out of solid bars of iron. At this tedious operation they
may be seen every day, and they manage to finish off a gun with
a flint lock very handsomely. All about the streets are sellers
of water, vegetables, fruit, soup, and agar-agar (a jelly made of
seaweed), who have many cries as unintelligible as those of
London. Others carry a portable cooking-apparatus on a pole
balanced by a table at the other end, and serve up a meal of
shell-fish, rice, and vegetables for two or three halfpence ; while
coolies and boatmen waiting to be hired are everywhere to be
met with.

In the interior of the island the Chinese cut down forest trees
in the jungle, and saw them up into planks ; they cultivate
vegetables, which they bring to market ; and they grow pepper
and gambir, which form important articles of export. The
French Jesuits have established missions among these inland
Chinese, which seem very successful. I lived for several weeks
at a time with the missionary at Bukit-tima, about the centre of
the island, where a pretty church has been built and there are
about 300 converts. While there, I met a missionary who had
just arrived from Tonquin, where he had been living for many
years. The Jesuits still do their work thoroughly as of old. In
Cochin China, Tonquin, and China, where all Christian teachers
are obliged to live in secret, and are liable to persecution, ex-
pulsion, and sometimes death,[1] every province, even those farthest
in the interior, has a permanent Jesuit mission establishment,
constantly kept up by fresh aspirants, who are taught the languages of the countries they are going to at Penang or Singapore. In China there are said to be near a million converts ; in

[1] Since the French settlement in Cochin China this is no longer the case.

Tonquin and Cochin China, more than half a million. One secret of the success of these missions is the rigid economy practised in the expenditure of the funds. A missionary is allowed about 30*l.* a year, on which he lives in whatever country he may be. This renders it possible to support a large number of missionaries with very limited means ; and the natives, seeing their teachers living in poverty and with none of the luxuries of life, are convinced that they are sincere in what they teach, and have really given up home and friends and ease and safety for the good of others. No wonder they make converts, for it must be a great blessing to the poor people among whom they labour to have a man among them to whom they can go in any trouble or distress, who will comfort and advise them, who visits them in sickness, who relieves them in want, and who devotes his whole life to their instruction and welfare.

My friend at Bukit-tima was truly a father to his flock. He preached to them in Chinese every Sunday, and had evenings for discussion and conversation on religion during the week. He had a school to teach their children. His house was open to them day and night. If a man came to him and said, "I have no rice for my family to eat to-day," he would give him half of what he had in the house, however little that might be. If another said, "I have no money to pay my debt," he would give him half the contents of his purse, were it his last dollar. So, when he was himself in want, he would send to some of the wealthiest among his flock, and say, " I have no rice in the house," or " I have given away my money, and.am in want of such and such articles." The result was that his flock trusted and loved him, for they felt sure that he was their true friend, and had no ulterior designs in living among them.

The island of Singapore consists of a multitude of small hills, three or four hundred feet high, the summits of many of which are still covered with virgin forest. The mission-house at Bukit-tima was surrounded by several of these wood-topped hills, which were much frequented by wood-cutters and sawyers, and offered me an excellent collecting ground for insects. Here and there, too, were tiger pits, carefully covered over with sticks and leaves, and so well concealed, that in several cases I had a narrow escape from falling into them. They are shaped like an iron furnace, wider at the bottom than the top, and are perhaps fifteen or twenty feet deep, so that it would be almost impossible for a person unassisted to get out of one. Formerly a sharp stake was stuck erect in the bottom ; but after an unfortunate traveller had been killed by falling on one, its use was forbidden. There are always a few tigers roaming about Singapore, and they kill on an average a Chinaman every day, principally those who work in the gambir plantations, which are always made in newly-cleared jungle. We heard a tiger roar once or twice in the evening, and it was rather nervous work hunting for insects among the fallen trunks and old sawpits, when one of these savage animals might be lurking close by, waiting an opportunity to spring upon us.

Several hours in the middle of every fine day were spent in these patches of forest, which were delightfully cool and shady by contrast with the bare open country we had to walk over to reach them. The vegetation was most luxuriant, comprising enormous forest trees, as well as a variety of ferns, caladiums, and other undergrowth, and abundance of climbing rattan palms. Insects were exceedingly abundant and very interesting, and every day furnished scores of new and curious forms. In about two months I obtained no less than 700 species of beetles, a large proportion of which were quite new, and among them were 130 distinct kinds of the elegant Longicorns (Cerambycidæ), so much esteemed by collectors. Almost all these were collected in one patch of jungle, not more than a square mile in extent, and in all my subsequent travels in the East I rarely if ever met with so productive a spot. This exceeding productiveness was due in part no doubt to some favourable conditions in the soil, climate, and vegetation, and to the season being very bright and sunny, with sufficient showers to keep everything fresh. But it was also in a great measure dependent, I feel sure, on the labours of the Chinese wood-cutters. They had been at work here for several years, and during all that time had furnished a continual supply of dry and dead and decaying leaves and bark, together with abundance of wood and sawdust, for the nourishment of insects and their larvæ. This had led to the assemblage of a great variety of species in a limited space, and I was the first naturalist who had come to reap the harvest they had prepared. In the same place, and during my walks in other directions, I obtained a fair collection of butterflies and of other orders of insects, so that on the whole I was quite satisfied with these my first attempts to gain a knowledge of the Natural History of the Malay Archipelago.

CHAPTER III.

MALACCA AND MOUNT OPHIR.

(JULY TO SEPTEMBER, 1854.)

BIRDS and most other kinds of animals being scarce at Singapore, I left it in July for Malacca, where I spent more than two months in the interior, and made an excursion to Mount Ophir. The old and picturesque town of Malacca is crowded along the banks of the small river, and consists of narrow streets of shops and dwelling-houses, occupied by the descendants of the Portuguese, and by Chinamen. In the suburbs are the houses of the

English officials and of a few Portuguese merchants, embedded in groves of palms and fruit-trees, whose varied and beautiful foliage furnishes a pleasing relief to the eye, as well as most grateful shade.

The old fort, the large Government House, and the ruins of a cathedral, attest the former wealth and importance of this place, which was once as much the centre of Eastern trade as Singapore is now. The following description of it by Linschott, who wrote two hundred and seventy years ago, strikingly exhibits the change it has undergone :—

"Malacca is inhabited by the Portuguese and by natives of the country, called Malays. The Portuguese have here a fortress, as at Mozambique, and there is no fortress in all the Indies, after those of Mozambique and Ormuz, where the captains perform their duty better than in this one. This place is the market of all India, of China, of the Moluccas, and of other islands round about, from all which places, as well as from Banda, Java, Sumatra, Siam, Pegu, Bengal, Coromandel, and India, arrive ships, which come and go incessantly, charged with an infinity of merchandises. There would be in this place a much greater number of Portuguese if it were not for the inconvenience and unhealthiness of the air, which is hurtful not only to strangers, but also to natives of the country. Thence it is that all who live in the country pay tribute of their health, suffering from a certain disease, which makes them lose either their skin or their hair. And those who escape consider it a miracle, which occasions many to leave the country, while the ardent desire of gain induces others to risk their health, and endeavour to endure such an atmosphere. The origin of this town, as the natives say, was very small, only having at the beginning, by reason of the unhealthiness of the air, but six or seven fishermen who inhabited it. But the number was increased by the meeting of fishermen from Siam, Pegu, and Bengal, who came and built a city, and established a peculiar language, drawn from the most elegant modes of speaking of other nations, so that in fact the language of the Malays is at present the most refined, exact, and celebrated of all the East. The name of Malacca was given to this town, which, by the convenience of its situation, in a short time grew to such wealth, that it does not yield to the most powerful towns and regions round about. The natives, both men and women, are very courteous, and are reckoned the most skilful in the world in compliments, and study much to compose and repeat verses and love-songs. Their language is in vogue through the Indies, as the French is here."

At present, a vessel over a hundred tons hardly ever enters its port, and the trade is entirely confined to a few petty products of the forests, and to the fruit, which the trees planted by the old Portuguese now produce for the enjoyment of the inhabitants of Singapore. Although rather subject to fevers, it is not at present considered very unhealthy.

The population of Malacca consists of several races. The ubiquitous Chinese are perhaps the most numerous, keeping up their manners, customs, and language; the indigenous Malays are next in point of numbers, and their language is the Lingua-franca of the place. Next come the descendants of the Portuguese—a mixed, degraded, and degenerate race, but who still keep up the use of their mother tongue, though ruefully mutilated in grammar; and then there are the English rulers, and the descendants of the Dutch, who all speak English. The Portuguese spoken at Malacca is a useful philological phenomenon. The verbs have mostly lost their inflections, and one form does for all moods, tenses, numbers, and persons. *Eu vai*, serves for "I go," "I went," or, "I will go." Adjectives, too, have been deprived of their feminine and plural terminations, so that the language is reduced to a marvellous simplicity, and, with the admixture of a few Malay words, becomes rather puzzling to one who has heard only the pure Lusitanian.

In costume these several peoples are as varied as in their speech. The English preserve the tight-fitting coat, waistcoat, and trousers, and the abominable hat and cravat; the Portuguese patronize a light jacket, or, more frequently, shirt and trousers only; the Malays wear their national jacket and sarong (a kind of kilt), with loose drawers; while the Chinese never depart in the least from their national dress, which, indeed, it is impossible to improve for a tropical climate, whether as regards comfort or appearance. The loosely-hanging trousers, and neat white half-shirt half-jacket, are exactly what a dress should be in this low latitude.

I engaged two Portuguese to accompany me into the interior; one as a cook, the other to shoot and skin birds, which is quite a trade in Malacca. I first stayed a fortnight at a village called Gading, where I was accommodated in the house of some Chinese converts, to whom I was recommended by the Jesuit missionaries. The house was a mere shed, but it was kept clean, and I made myself sufficiently comfortable. My hosts were forming a pepper and gambir plantation, and in the immediate neighbourhood were extensive tin-washings, employing over a thousand Chinese. The tin is obtained in the form of black grains from beds of quartzose sand, and is melted into ingots in rude clay furnaces. The soil seemed poor, and the forest was very dense with undergrowth, and not at all productive of insects; but, on the other hand, birds were abundant, and I was at once introduced to the rich ornithological treasures of the Malayan region.

The very first time I fired my gun I brought down one of the most curious and beautiful of the Malacca birds, the blue-billed gaper (Cymbirhynchus macrorhynchus), called by the Malays the "Rain-bird." It is about the size of a starling, black and rich claret colour with white shoulder stripes, and a very large and broad bill of the most pure cobalt blue above and orange below, while the iris is emerald green. As the skins dry the bill turns dull black, but even then the bird is handsome. When fresh killed, the contrast of the vivid blue with the rich colours of the plumage is remarkably striking and beautiful. The lovely Eastern trogons, with their rich brown backs, beautifully pencilled wings, and crimson breasts, were also soon obtained, as well as the large green barbets (Megalæma versicolor)—fruit-eating birds, something like small toucans, with a short, straight bristly bill, and whose head and neck are variegated with patches of the most vivid blue and crimson. A day or two after, my hunter brought me a specimen of the green gaper (Calyptomena viridis), which is like a small cock-of-the-rock, but entirely of the most vivid green, delicately marked on the wings with black bars. Handsome woodpeckers and gay king-fishers, green and brown cuckoos with velvety red faces and green beaks, red-breasted doves and metallic honeysuckers, were brought in day after day, and kept me in a continual state of pleasurable excitement. After a fortnight one of my servants was seized with fever, and on returning to Malacca, the same disease attacked the other as well as myself. By a liberal use of quinine, I soon recovered, and obtaining other men, went to stay at the Government bungalow of Ayer-panas, accompanied by a young gentleman, a native of the place, who had a taste for natural history.

At Ayer-panas we had a comfortable house to stay in, and plenty of room to dry and preserve our specimens ; but, owing to there being no industrious Chinese to cut down timber, insects were comparatively scarce, with the exception of butter-flies, of which I formed a very fine collection. The manner in which I obtained one fine insect was curious, and indicates how fragmentary and imperfect a traveller's collection must neces-sarily be. I was one afternoon walking along a favourite road through the forest, with my gun, when I saw a butterfly on the ground. It was large, handsome, and quite new to me, and I got close to it before it flew away. I then observed that it had been settling on the dung of some carnivorous animal. Thinking it might return to the same spot, I next day after breakfast took my net, and as I approached the place was delighted to see the same butterfly sitting on the same piece of dung, and suc-ceeded in capturing it. It was an entirely new species of great beauty, and has been named by Mr. Hewitson Nymphalis calydonia. I never saw another specimen of it, and it ; was only after twelve years had elapsed that a second individual reached this country from the north-western part of Borneo.

Having determined to visit Mount Ophir, which is situated in the middle of the peninsula about fifty miles east of Malacca, we engaged six Malays to accompany us and carry our baggage. As we meant to stay at least a week at the mountain, we took with us a good supply of rice, a little biscuit, butter, and coffee, some dried fish and a little brandy, with blankets, a change of clothes, insect and bird boxes, nets, guns, and ammunition. The distance from Ayer-panas was supposed to be about thirty miles.

Our first day's march lay through patches of forest clearings and Malay villages, and was pleasant enough. At night we slept at the house of a Malay chief, who lent us a ve-randah, and gave us a fowl and some eggs. The next day the country got wilder and more hilly. We passed through extensive forests, along paths often up to our knees in mud, and were much annoyed by the leeches for which this district is famous. These little creatures infest the leaves and herbage by the side of the paths, and when a passenger comes along they stretch themselves out at full length, and if they touch any part of his dress or body, quit their leaf and adhere to it. They then creep on to his feet, legs, or other part of his body and suck their fill, the first puncture being rarely felt during the excitement of walking. On bathing in the evening we generally found half a dozen or a dozen on each of us, most frequently on our legs, but sometimes on our bodies, and I had one who sucked his fill from the side of my neck, but who luckily missed the jugular vein. There are many species of these forest leeches. All are small, but some are beautifully marked with

RARE FERNS ON MOUNT OPHIR.

stripes of bright yellow. They probably attach themselves to deer or other animals which frequent the forest paths, and have thus acquired the singular habit of stretching themselves out at the sound of a footstep or of rustling foliage. Early in the afternoon we reached the foot of the mountain, and encamped by the side of a fine stream, whose rocky banks were overgrown with ferns. Our oldest Malay had been accustomed to shoot birds in this neighbourhood for the Malacca dealers, and had been to the top of the mountain, and while we amused ourselves shooting and insect hunting, he went with two others to clear the path for our ascent the next day.

Early next morning we started after breakfast, carrying blankets and provisions, as we intended to sleep upon the mountain. After passing a little tangled jungle and swampy thickets through which our men had cleared a path, we emerged into a fine lofty forest pretty clear of undergrowth, and in which we could walk freely. We ascended steadily up a moderate slope for several miles, having a deep ravine on our left. We then had a level plateau or shoulder to cross, after which the ascent was steeper and the forest denser till we came out upon the "Padang-batu," or stone field, a place of which we had heard much, but could never get any one to describe intelligibly. We found it to be a steep slope of even rock, extending along the mountain side farther than we could see. Parts of it were quite bare, but where it was cracked and fissured there grew a most luxuriant vegetation, among which the pitcher plants were the most remarkable. These wonderful plants never seem to succeed well in our hot-houses, and are there seen to little advantage. Here they grew up into half-climbing shrubs, their curious pitchers of various sizes and forms hanging abundantly from their leaves, and continually exciting our admiration by their size and beauty. A few coniferæ of the genus Dacrydium here first appeared, and in the thickets just above the rocky surface we walked through groves of those splendid ferns Dipteris Horsfieldii and Matonia pectinata, which bear large spreading palmate fronds on slender stems six or eight feet high. The Matonia is the tallest and most elegant, and is known only from this mountain, and neither of them is yet introduced into our hot-houses.

It was very striking to come out from the dark, cool, and shady forest in which we had been ascending since we started, on to this hot, open rocky slope where we seemed to have entered at one step from a lowland to an alpine vegetation. The height, as measured by a sympiesometer, was about 2,800 feet. We had been told we should find water at Padang-batu, but we looked about for it in vain, as we were exceedingly thirsty. At last we turned to the pitcher-plants, but the water contained in the pitchers (about half a pint in each) was full of insects and otherwise uninviting. On tasting it, however, we found it very palatable, though rather warm, and we all quenched our thirst from these natural jugs. Farther on we came to forest again, but of a more dwarfed and

stunted character than below; and alternately passing along
ridges and descending into valleys, we reached a peak separated
from the true summit of the mountain by a considerable chasm.
Here our porters gave in, and declared they could carry their
loads no further; and certainly the ascent to the highest peak
was very precipitous. But on the spot where we were there
was no water, whereas it was well known that there was a
spring close to the summit, so we determined to go on without
them, and carry with us only what was absolutely necessary.
We accordingly took a blanket each, and divided our food and
other articles among us, and went on with only the old Malay
and his son.

After descending into the saddle between the two peaks we
found the ascent very laborious, the slope being so steep as
often to necessitate hand-climbing. Besides a bushy vegetation
the ground was covered knee-deep with mosses on a foundation
of decaying leaves and rugged rock, and it was a hard hour's
climb to the small ledge just below the summit, where an overhanging rock forms a convenient shelter, and a little basin
collects the trickling water. Here we put down our loads, and
in a few minutes more stood on the summit of Mount Ophir,
4,000 feet above the sea. The top is a small rocky platform
covered with rhododendrons and other shrubs. The afternoon
was clear, and the view fine in its way—ranges of hill and
valley everywhere covered with interminable forest, with glistening rivers winding among them. In a distant view a forest
country is very monotonous, and no mountain I have ever
ascended in the tropics presents a panorama equal to that from
Snowdon, while the views in Switzerland are immeasurably
superior. When boiling our coffee I took observations with a
good boiling-point thermometer, as well as with the sympiesometer, and we then enjoyed our evening meal and the noble
prospect that lay before us. The night was calm and very
mild, and having made a bed of twigs and branches over which
we laid our blankets, we passed a very comfortable night. Our
porters had followed us after a rest, bringing only their rice to
cook, and luckily we did not require the baggage they left
behind them. In the morning I caught a few butterflies and
beetles, and my friend got a few land-shells; and we then
descended, bringing with us some specimens of the ferns and
pitcher-plants of Padang-batu.

The place where we had first encamped at the foot of the
mountain being very gloomy, we chose another in a kind of
swamp near a stream overgrown with Zingiberaceous plants, in
which a clearing was easily made. Here our men built two
little huts without sides, that would just shelter us from the
rain; and we lived in them for a week, shooting and insect-
hunting, and roaming about the forests at the foot of the
mountain. This was the country of the great Argus pheasant,
and we continually heard its cry. On asking the old Malay to

try and shoot one for me, he told me that although he had been for twenty years shooting birds in these forests he had never yet shot one, and had never even seen one except after it had been caught. The bird is so exceedingly shy and wary, and runs along the ground in the densest parts of the forest so quickly, that it is impossible to get near it; and its sober colours and rich eye-like spots, which are so ornamental when seen in a museum, must harmonize well with the dead leaves among which it dwells, and render it very inconspicuous. All the specimens sold in Malacca are caught in snares, and my informant, though he had shot none, had snared plenty.

The tiger and rhinoceros are still found here, and a few years ago elephants abounded, but they have lately all disappeared. We found some heaps of dung, which seemed to be that of elephants, and some tracks of the rhinoceros, but saw none of the animals. We, however, kept a fire up all night in case any of these creatures should visit us, and two of our men declared that they did one day see a rhinoceros. When our rice was finished, and our boxes full of specimens, we returned to Ayer-panas, and a few days afterwards went on to Malacca, and thence to Singapore. Mount Ophir has quite a reputation for fever, and all our friends were astonished at our recklessness in staying so long at its foot; but we none of us suffered in the least, and I shall ever look back with pleasure to my trip, as being my first introduction to mountain scenery in the Eastern tropics.

The meagreness and brevity of the sketch I have here given of my visit to Singapore and the Malay Peninsula is due to my having trusted chiefly to some private letters and a note-book, which were lost; and to a paper on Malacca and Mount Ophir which was sent to the Royal Geographical Society, but which was neither read nor printed, owing to press of matter at the end of a session, and the MSS. of which cannot now be found. I the less regret this, however, as so many works have been written on these parts; and I always intended to pass lightly over my travels in the western and better known portions of the Archipelago, in order to devote more space to the remoter districts, about which hardly anything has been written in the English language.

CHAPTER IV.

BORNEO—THE ORANG-UTAN.

I ARRIVED at Saráwak on November 1st, 1854, and left it on January 25th, 1856. In the interval I resided at many different localities, and saw a good deal of the Dyak tribes as well as of the Bornean Malays. I was hospitably entertained by Sir

James Brooke, and lived in his house whenever I was at the town of Saráwak in the intervals of my journeys. But so many books have been written about this part of Borneo since I was there, that I shall avoid going into details of what I saw and heard and thought of Saráwak and its ruler, confining myself chiefly to my experiences as a naturalist in search of shells, insects, birds, and the Orang-utan, and to an account of a journey through a part of the interior seldom visited by Europeans.

The first four months of my visit were spent in various parts of the Saráwak River, from Santubong at its mouth up to the picturesque limestone mountains and Chinese gold-fields of Bow and Bedé. This part of the country has been so frequently described that I shall pass it over, especially as, owing to its being the height of the wet season, my collections were comparatively poor and insignificant.

In March, 1855, I determined to go to the coal-works which were being opened near the Simunjon River, a small branch of the Sádong, a river east of Saráwak and between it and the Batang-Lupar. The Simunjon enters the Sádong River about twenty miles up. It is very narrow and very winding, and much overshadowed by the lofty forest, which sometimes almost meets over it. The whole country between it and the sea is a perfectly level forest-covered swamp, out of which rise a few isolated hills, at the foot of one of which the works are situated. From the landing-place to the hill a Dyak road had been formed, which consisted solely of tree-trunks laid end to end. Along these the bare-footed natives walk and carry heavy burdens with the greatest ease, but to a booted European it is very slippery work, and when one's attention is constantly attracted by the various objects of interest around, a few tumbles into the bog are almost inevitable. During my first walk along this road I saw few insects or birds, but noticed some very handsome orchids in flower, of the genus Cœlogyne, a group which I afterwards found to be very abundant, and characteristic of the district. On the slope of the hill near its foot a patch of forest had been cleared away, and several rude houses erected, in which were residing Mr. Coulson, the engineer, and a number of Chinese workmen. I was at first kindly accommodated in Mr. Coulson's house, but finding the spot very suitable for me, and offering great facilities for collecting, I had a small house of two rooms and a verandah built for myself. Here I remained nearly nine months, and made an immense collection of insects, to which class of animals I devoted my chief attention, owing to the circumstances being especially favourable.

In the tropics a large proportion of the insects of all orders, and especially of the large and favourite group of beetles, are more or less dependent on vegetation, and particularly on timber, bark, and leaves in various stages of decay. In the untouched virgin forest, the insects which frequent such situations are scattered over an immense extent of country, at spots where trees have fallen through decay and old age, or have succumbed to the fury of the tempest ; and twenty square miles of country may not contain so many fallen and decayed trees as are to be found in any small clearing. The quantity and the variety of beetles and of many other insects that can be collected at a given time in any tropical locality, will depend, first upon the immediate vicinity of a great extent of virgin forest, and secondly upon the quantity of trees that for some months past have been, and which are still being cut down, and left to dry and decay upon the ground. Now, during my whole twelve years' collecting in the western and eastern tropics, I never enjoyed such advantages in this respect as at the Simunjon coal-works. For several months from twenty to fifty Chinamen and Dyaks were employed almost exclusively in clearing a large space in the forest, and in making a wide opening for a railroad to the Sádong River, two miles distant. Besides this, sawpits were established at various points in the jungle, and large trees were felled to be cut up into beams and planks. For hundreds of miles in every direction a magnificent forest extended over plain and mountain, rock and morass, and I arrived at the spot just as the rains began to diminish and the daily sunshine to increase ; a time which I have always found the most favourable season for collecting. The number of openings and sunny places and of pathways, were also an attraction to wasps and butterflies ; and by paying a cent each for all insects that were brought me, I obtained from the Dyaks and the Chinamen many fine locusts and Phasmidæ, as well as numbers of handsome beetles.

When I arrived at the mines, on the 14th of March, I had collected in the four preceding months, 320 different kinds of beetles. In less than a fortnight I had doubled this number, an average of about twenty-four new species every day. On one day I collected seventy-six different kinds, of which thirty-four were new to me. By the end of April I had more than a thousand species, and they then went on increasing at a slower rate ; so that I obtained altogether in Borneo about two thousand distinct kinds, of which all but about a hundred were collected at this place, and on scarcely more than a square mile of ground. The most numerous and most interesting groups of beetles were the Longicorns and Rhynchophora, both pre-eminently wood-feeders. The former, characterized by their graceful forms and long antennæ, were especially numerous, amounting to nearly three hundred species, nine-tenths of which were entirely new, and many of them remarkable for their large size, strange forms, and beautiful colouring. The latter correspond to our weevils and allied groups, and in the tropics are exceedingly numerous and varied, often swarming upon dead timber, so that I sometimes obtained fifty or sixty different kinds in a day. My Bornean collections of this group exceeded five hundred species.

My collection of butterflies was not large ; but I obtained

Megacriodes Saundersii.
Cyriopalpus Wallacei.

Diurus furcellatus.
Ectatorhinus Wallacei.

Neocerambyx æneas.
Cladognathus tarandus.

REMARKABLE BEETLES FOUND AT SIMUNJON, BORNEO.

some rare and very handsome insects, the most remarkable being the Ornithoptera Brookeana, one of the most elegant species known. This beautiful creature has very long and pointed wings, almost resembling a sphinx moth in shape. It is deep velvety black, with a curved band of spots of a brilliant metallic-green colour extending across the wings from tip to tip, each spot being shaped exactly like a small triangular feather, and having very much the effect of a row of the wing coverts of the Mexican trogon laid upon black velvet. The only other marks are a broad neck-collar of vivid crimson, and a few delicate white touches on the outer margins of the hind wings. This species, which was then quite new and which I named after Sir James Brooke, was very rare. It was seen occasionally flying swiftly in the clearings, and now and then settling for an instant at puddles and muddy places, so that I only succeeded in capturing two or three specimens. In some other parts of the country I was assured it was abundant, and a good many specimens have been sent to England; but as yet all have been males, and we are quite unable to conjecture what the female may be like, owing to the extreme isolation of the species, and its want of close affinity to any other known insect.[1]

One of the most curious and interesting reptiles which I met with in Borneo was a large tree-frog, which was brought me by one of the Chinese workmen. He assured me that he had seen it come down, in a slanting direction, from a high tree, as if it flew. On examining it, I found the toes very long and fully webbed to their very extremity, so that when expanded they offered a surface much larger than that of the body. The fore legs were also bordered by a membrane, and the body was capable of considerable inflation. The back and limbs were of a very deep shining green colour, the under surface and the inner toes yellow, while the webs were black, rayed with yellow. The body was about four inches long, while the webs of each hind foot, when fully expanded, covered a surface of four square inches, and the webs of all the feet together about twelve square inches. As the extremities of the toes have dilated discs for adhesion, showing the creature to be a true tree-frog, it is difficult to imagine that this immense membrane of the toes can be for the purpose of swimming only, and the account of the Chinaman, that it flew down from the tree, becomes more credible. This is, I believe, the first instance known of a "flying frog," and it is very interesting to Darwinians as showing, that the variability of the toes which have been already modified for purposes of swimming and adhesive climbing, have been taken advantage of to enable an allied species to pass through the air like the flying lizard. It would appear to be a new species of the genus

---

[1] Females have since been captured in some plenty. They resemble the male, but have more white and less brilliant colours.

Rhacophorus, which consists of several frogs of a much smaller
size than this, and having the webs of the toes less developed.

During my stay in Borneo I had no hunter to shoot for me
regularly, and, being myself fully occupied with insects, I did
not succeed in obtaining a very good collection of the birds or
Mammalia, many of which, however, are well known, being
identical with species found in Malacca. Among the Mammalia
were five squirrels, two tiger-cats, the Gymnurus Rafflesii, which

FLYING FROG.

looks like a cross between a pig and a polecat, and the Cynogale
Bennetti—a rare, otter-like animal, with very broad muzzle
clothed with long bristles.

One of my chief objects in coming to stay at Simunjon was to
see the Orang-utan (or great man-like ape of Borneo) in his native
haunts, to study his habits, and obtain good specimens of the
different varieties and species of both sexes, and of the adult and
young animals. In all these objects I succeeded beyond my expectations, and will now give some account of my experience in

hunting the Orang-utan, or " Mias," as it is called by the natives ;
and as this name is short, and easily pronounced, I shall
generally use it in preference to Simia satyrus, or Orang-utan.

Just a week after my arrival at the mines, I first saw a Mias.
I was out collecting insects, not more than a quarter of a mile
from the house, when I heard a rustling in a tree near, and,
looking up, saw a large red-haired animal moving slowly along,
hanging from the branches by its arms. It passed on from tree to
tree till it was lost in the jungle, which was so swampy that I
could not follow it. This mode of progression was, however, very
unusual, and is more characteristic of the Hylobates than of the
Orang. I suppose there was some individual peculiarity in this
animal, or the nature of the trees just in this place rendered it
the most easy mode of progression.

About a fortnight afterwards I heard that one was feeding in
a tree in the swamp just below the house, and, taking my gun, was
fortunate enough to find it in the same place. As soon as I
approached, it tried to conceal itself among the foliage ; but I
got a shot at it, and the second barrel caused it to fall down
almost dead, the two balls having entered the body. This was
a male, about half-grown, being scarcely three feet high. On
April 26th, I was out shooting with two Dyaks, when we found
another about the same size. It fell at the first shot, but did not
seem much hurt, and immediately climbed up the nearest tree,
when I fired, and it again fell, with a broken arm and a wound
in the body. The two Dyaks now ran up to it, and each seized
hold of a hand, telling me to cut a pole, and they would secure
it. But although one arm was broken, and it was only a half-
grown animal, it was too strong for these young savages, drawing
them up towards its mouth notwithstanding all their efforts, so
that they were again obliged to leave go, or they would have
been seriously bitten. It now began climbing up the tree again ;
and, to avoid trouble, I shot it through the heart.

On May 2nd, I again found one on a very high tree, when I
had only a small 80-bore gun with me. However, I fired at it,
and on seeing me, it began howling in a strange voice like a
cough, and seemed in a great rage, breaking off branches with its
hands and throwing them down, and then soon made off over
the tree-tops. I did not care to follow it, as it was swampy, and
in parts dangerous, and I might easily have lost myself in the
eagerness of pursuit.

On the 12th of May I found another, which behaved in a very
similar manner, howling and hooting with rage, and throwing
down branches. I shot at it five times, and it remained dead
on the top of the tree, supported in a fork in such a manner
that it would evidently not fall. I therefore returned home,
and luckily found some Dyaks, who came back with me, and
climbed up the tree for the animal. This was the first full-
grown specimen I had obtained ; but it was a female, and not
nearly so large or remarkable as the full-grown males. It was,

however, 3ft. 6in. high, and its arms stretched out to a width of 6ft. 6in. I preserved the skin of this specimen in a cask of arrack, and prepared a perfect skeleton, which was afterwards purchased for the Derby Museum.

Only four days afterwards some Dyaks saw another Mias near the same place, and came to tell me. We found it to be a rather large one, very high up on a tall tree. At the second shot it fell rolling over, but almost immediately got up again and began to climb. At a third shot it fell dead. This was also a full-grown female, and while preparing to carry it home, we

FEMALE ORANG-UTAN.   (*From a photograph.*)

found a young one face downwards in the bog. This little creature was only about a foot long, and had evidently been hanging to its mother when she first fell. Luckily it did not appear to have been wounded, and after we had cleaned the mud out of its mouth it began to cry out, and seemed quite strong and active. While carrying it home it got its hands in my beard, and grasped so tightly that I had great difficulty in getting free, for the fingers are habitually bent inwards at the last joint so as to form complete hooks. At this time it had not a single tooth, but a few days afterwards it cut its two lower front teeth. Unfortunately, I had no milk to give it, as neither Malays, Chinese, nor Dyaks ever use the article, and I in vain inquired for any female animal that could suckle my little infant. I was therefore obliged to give it rice-water from a bottle with a quill in the cork, which after a few trials it learned to suck very well. This was very meagre diet, and the little creature did not thrive well on it, although I added sugar and cocoa-nut milk occasionally, to make it more nourishing. When I put my finger in its mouth it sucked with great vigour, drawing in its cheeks with all its might in the vain effort to extract some milk, and only after persevering a long time would it give up in disgust, and set up a scream very like that of a baby in similar circumstances.

When handled or nursed, it was very quiet and contented, but when laid down by itself would invariably cry; and for the first few nights was very restless and noisy. I fitted up a little box for a cradle, with a soft mat for it to lie upon, which was changed and washed every day; and I soon found it necessary to wash the little Mias as well. After I had done so a few times, it came to like the operation, and as soon as it was dirty would begin crying, and not leave off till I took it out and carried it to the spout, when it immediately became quiet, although it would wince a little at the first rush of the cold water and make ridiculously wry faces while the stream was running over its head. It enjoyed the wiping and rubbing dry amazingly, and when I brushed its hair seemed to be perfectly happy, lying quite still with its arms and legs stretched out while I thoroughly brushed the long hair of its back and arms. For the first few days it clung desperately with all four hands to whatever it could lay hold of, and I had to be careful to keep my beard out of its way, as its fingers clutched hold of hair more tenaciously than anything else, and it was impossible to free myself without assistance. When restless, it would struggle about with its hands up in the air trying to find something to take hold of, and, when it had got a bit of stick or rag in two or three of its hands, seemed quite happy. For want of something else, it would often seize its own feet, and after a time it would constantly cross its arms and grasp with each hand the long hair that grew just below the opposite shoulder. The great tenacity of its grasp soon diminished, and I was obliged to invent some means to give it exercise and strengthen its limbs. For this purpose I made a short ladder of three or four rounds, on which I put it to hang for a quarter of an hour at a time. At first it seemed much pleased, but it could not get all four hands in a comfortable position, and, after changing about several times, would leave hold of one hand after the other, and drop on to the floor. Sometimes, when hanging only by two hands, it would loose one, and cross it to the opposite shoulder, grasping its own hair; and, as this seemed much more agreeable than the stick, it would then loose the other and tumble down, when it would cross both and lie on its back quite contentedly, never seeming to be hurt by its numerous tumbles. Finding it so fond of hair, I endeavoured to make an artificial mother, by wrapping up a piece of buffalo-skin into a bundle, and suspending it about a foot from the floor. At first this seemed to suit it admirably, as it could sprawl its legs about and always find some hair, which it grasped with the greatest tenacity. I was now in hopes that I had made the little orphan quite happy ; and so it seemed for some time, till it began to remember its lost parent, and try to suck. It would pull itself up close to the skin, and try about everywhere for a likely place ; but, as it only succeeded in getting mouthfuls of hair and wool, it would be greatly disgusted, and scream violently, and after two or three attempts, let go altogether. One day it got some wool into its throat, and I thought it would have choked, but after much gasping it recovered, and I was obliged to take the imitation mother to pieces again, and give up this last attempt to exercise the little creature.

After the first week I found I could feed it better with a spoon, and give it a little more varied and more solid food. Well-soaked biscuit mixed with a little egg and sugar, and sometimes sweet potatoes, were readily eaten ; and it was a never-failing amusement to observe the curious changes of countenance by which it would express its approval or dislike of what was given to it. The poor little thing would lick its lips, draw in its cheeks, and turn up its eyes with an expression of the most supreme satisfaction when it had a mouthful particularly to its taste. On the other hand, when its food was not sufficiently sweet or palatable, it would turn the mouthful about with its tongue for a moment as if trying to extract what flavour there was, and then push it all out between its lips. If the same food was continued, it would set up a scream and kick about violently, exactly like a baby in a passion.

After I had had the little Mias about three weeks, I fortunately obtained a young hare-lip monkey (Macacus cynomolgus), which, though small, was very active, and could feed itself. I placed it in the same box with the Mias, and they immediately became excellent friends, neither exhibiting the least fear of the other. The little monkey would sit upon the other's stomach, or even on its face, without the least regard to its feelings. While I was feeding the Mias, the monkey would sit by, picking up all that was spilt, and occasionally putting out its hands to intercept the spoon ; and as soon as I had finished would pick off what was left sticking to the Mias's lips, and then pull open its mouth and see if any still remained inside ; afterwards lying down on the poor creature's stomach as on a comfortable cushion. The little helpless Mias would submit to all these insults with the most exemplary patience, only too glad to have something warm near it, which it could clasp affectionately in its arms. It sometimes, however, had its revenge ; for when the monkey wanted to go away, the Mias would hold on as long as it could by the loose skin of its back or head, or by its tail, and it was only after many vigorous jumps that the monkey could make his escape.

It was curious to observe the different actions of these two animals, which could not have differed much in age. The Mias, like a very young baby, lying on its back quite helpless, rolling lazily from side to side, stretching out all four hands into the air, wishing to grasp something, but hardly able to guide its fingers to any definite object; and when dissatisfied, opening wide its almost toothless mouth, and expressing its wants by a most infantine scream. The little monkey, on the other hand, in constant motion ; running and jumping about wherever it pleased, examining everything around it, seizing hold of the smallest objects with the greatest precision, balancing itself on the edge of the box, or running up a post, and helping itself to anything eatable that came in its way. There could hardly be a greater contrast, and the baby Mias looked more baby-like by the comparison.

When I had had it about a month, it began to exhibit some signs of learning to run alone. When laid upon the floor it would push itself along by its legs, or roll itself over, and thus make an unwieldy progression. When lying in the box it would lift itself up to the edge into almost an erect position, and once or twice succeeded in tumbling out. When left dirty, or hungry, or otherwise neglected, it would scream violently till attended to, varied by a kind of coughing or pumping noise, very similar to that which is made by the adult animal. If no one was in the house, or its cries were not attended to, it would be quiet after a little while, but the moment it heard a footstep would begin again harder than ever.

After five weeks it cut its two upper front teeth, but in all this time it had not grown the least bit, remaining both in size and weight the same as when I first procured it. This was no doubt owing to the want of milk or other equally nourishing food. Rice-water, rice, and biscuits were but a poor substitute, and the expressed milk of the cocoa-nut which I sometimes gave it did not quite agree with its stomach. To this I imputed an attack of diarrhœa from which the poor little creature suffered greatly, but a small dose of castor-oil operated well, and cured it. A week or two afterwards it was again taken ill, and this time more seriously. The symptoms were exactly those of intermittent fever, accompanied by watery swellings on the feet and head. It lost all appetite for its food, and, after lingering for a week a most pitiable object, died, after being in my possession nearly three months. I much regretted the loss of my little pet, which I had at one time looked forward to bringing up to years of maturity, and taking home to England. For several months it had afforded me daily amusement by its curious ways and the inimitably ludicrous expression of its little countenance. Its weight was three pounds nine ounces, its height fourteen inches, and the spread of its arms twenty-three inches. I preserved its skin and skeleton, and in doing so found that when it fell from the tree it must have broken an arm and a leg, which had, however, united so rapidly that I had only noticed the hard swellings on the limbs where the irregular junction of the bones had taken place.

Exactly a week after I had caught this interesting little animal I succeeded in shooting a full-grown male Orang-utan. I had just come home from an entomologizing excursion when Charles[1] rushed in out of breath with running and excitement, and exclaimed, interrupted by gasps, "Get the gun, sir,—be quick,—such a large Mias!" "Where is it?" I asked, taking hold of my gun as I spoke, which happened luckily to have one barrel loaded with ball. "Close by, sir—on the path to the mines—he can't get away." Two Dyaks chanced to be in the house at the time, so I called them to accompany me, and started off, telling Charley to bring all the ammunition after me as soon as possible. The path from our clearing to the mines led along the side of the hill a little way up its slope, and parallel with it at the foot a wide opening had been made for a road, in which several Chinamen were working, so that the animal could not escape into the swampy forests below without descending to cross the road or ascending to get round the clearings. We walked cautiously along, not making the least noise, and listening attentively for any sound which might betray the presence of the Mias, stopping at intervals to gaze upwards. Charley soon joined us at the place where he had seen the creature, and having taken the ammunition and put a bullet in the other barrel we dispersed a little, feeling sure that it must be somewhere near, as it had probably descended the hill, and would not be likely to return again. After a short time I heard a very slight rustling sound overhead, but on gazing up could see nothing. I moved about in every direction to get a full view into every part of the tree under which I had been standing, when I again heard the same noise but louder, and saw the leaves shaking as if caused by the motion of some heavy animal which moved off to an adjoining tree. I immediately shouted for all of them to come up and try and get a view, so as to allow me to have a shot. This was not an easy matter, as the Mias had a knack of selecting places with dense foliage beneath. Very soon, however, one of the Dyaks called me and pointed upwards, and on looking I saw a great red hairy body and a huge black face gazing down from a great height, as if wanting to know what was making such a disturbance below. I instantly fired, and he made off at once, so that I could not then tell whether I had hit him.

He now moved very rapidly and very noiselessly for so large an animal, so I told the Dyaks to follow and keep him in sight while I loaded. The jungle was here full of large angular fragments of rock from the mountain above, and thick with hanging

[1] Charles Allen, an English lad of sixteen, accompanied me as an assistant.

and twisted creepers. Running, climbing, and creeping among these, we came up with the creature on the top of a high tree near the road, where the Chinamen had discovered him, and were shouting their astonishment with open mouth : "Ya Ya, Tuan ; Orang-utan, Tuan." Seeing that he could not pass here without descending, he turned up again towards the hill, and I got two shots, and following quickly had two more by the time he had again reached the path ; but he was always more or less concealed by foliage, and protected by the large branch on which he was walking. Once while loading I had a splendid view of him, moving along a large limb of a tree in a semi-erect posture, and showing him to be an animal of the largest size. At the path he got on to one of the loftiest trees in the forest, and we could see one leg hanging down useless, having been broken by a ball. He now fixed himself in a fork, where he was hidden by thick foliage, and seemed disinclined to move. I was afraid he would remain and die in this position, and as it was nearly evening I could not have got the tree cut down that day. I therefore fired again, and he then moved off, and going up the hill was obliged to get on to some lower trees, on the branches of one of which he fixed himself in such a position that he could not fall, and lay all in a heap as if dead, or dying.

I now wanted the Dyaks to go up and cut off the branch he was resting on, but they were afraid, saying he was not dead, and would come and attack them. We then shook the adjoining tree, pulled the hanging creepers, and did all we could to disturb him, but without effect, so I thought it best to send for two Chinamen with axes to cut down the tree. While the messenger was gone, however, one of the Dyaks took courage and climbed towards him, but the Mias did not wait for him to get near, moving off to another tree, where he got on to a dense mass of branches and creepers which almost completely hid him from our view. The tree was luckily a small one, so when the axes came we soon had it cut through ; but it was so held up by jungle ropes and climbers to adjoining trees that it only fell into a sloping position. The Mias did not move, and I began to fear that after all we should not get him, as it was near evening, and half a dozen more trees would have to be cut down before the one he was on would fall. As a last resource we all began pulling at the creepers, which shook the tree very much, and, after a few minutes, when we had almost given up all hopes, down he came with a crash and a thud like the fall of a giant. And he was a giant, his head and body being full as large as a man's. He was of the kind called by the Dyaks "Mias Chappan," or "Mias Pappan," which has the skin of the face broadened out to a ridge or fold at each side. His outstretched arms measured seven feet three inches across, and his height, measuring fairly from the top of the head to the heel, was four feet two inches. The body just below the arms was three feet two inches round, and was quite as long as a man's, the legs being exceedingly short in proportion. On examination we found he had been dreadfully wounded. Both legs were broken. One hip-joint and the root of the spine completely shattered, and two bullets were found flattened in his neck and jaws! Yet he was still alive when he fell. The two Chinamen carried him home tied to a pole, and I was occupied with Charley the whole of the next day, preparing the skin and boiling the bones to make a perfect skeleton, which are now preserved in the Museum at Derby.

About ten days after this, on June 4th, some Dyaks came to tell us that the day before a Mias had nearly killed one of their companions. A few miles down the river there is a Dyak house, and the inhabitants saw a large Orang feeding on the young shoots of a palm by the river-side. On being alarmed he retreated towards the jungle which was close by, and a number of the men, armed with spears and choppers, ran out to intercept him. The man who was in front tried to run his spear through the animal's body, but the Mias seized it in his hands, and in an instant got hold of the man's arm, which he seized in his mouth, making his teeth meet in the flesh above the elbow, which he tore and lacerated in a dreadful manner. Had not the others been close behind, the man would have been more seriously injured, if not killed, as he was quite powerless; but they soon destroyed the creature with their spears and choppers. The man remained ill for a long time, and never fully recovered the use of his arm.

They told me the dead Mias was still lying where it had been killed, so I offered them a reward to bring it up to our landing-place immediately, which they promised to do. They did not come, however, till the next day, and then decomposition had commenced, and great patches of the hair came off, so that it was useless to skin it. This I regretted much, as it was a very fine full-grown male. I cut off the head and took it home to clean, while I got my men to make a close fence about five feet high round the rest of the body, which would soon be devoured by maggots, small lizards, and ants, leaving me the skeleton. There was a great gash in his face, which had cut deep into the bone, but the skull was a very fine one, and the teeth remarkably large and perfect.

On June 18th I had another great success, and obtained a fine adult male. A Chinaman told me he had seen him feeding by the side of the path to the river, and I found him at the same place as the first individual I had shot. He was feeding on an oval green fruit having a fine red arillus, like the mace which surrounds the nutmeg, and which alone he seemed to eat, biting off the thick outer rind and dropping it in a continual shower. I had found the same fruit in the stomach of some others which I had killed. Two shots caused this animal to loose his hold, but he hung for a considerable time by one hand, and then fell flat on his face and was half buried in the swamp. For several minutes he lay groaning and panting, while we stood close

round, expecting every breath to be his last. Suddenly, however, by a violent effort he raised himself up, causing us all to
step back a yard or two, when, standing nearly erect, he caught
hold of a small tree, and began to ascend it. Another shot
through the back caused him to fall down dead. A flattened
bullet was found in his tongue, having entered the lower part of
the abdomen and completely traversed the body, fracturing the
first cervical vertebra. Yet it was after this fearful wound that
he had risen, and begun climbing with considerable facility.
This also was a full-grown male of almost exactly the same
dimensions as the other two I had measured.

On June 21st I shot another adult female, which was eating
fruit in a low tree, and was the only one which I ever killed by
a single ball.

On June 24th I was called by a Chinaman to shoot a Mias,
which, he said, was on a tree close by his house, at the coal-
mines. Arriving at the place, we had some difficulty in finding
the animal, as he had gone off into the jungle, which was very
rocky and difficult to traverse. At last we found him up a very
high tree, and could see that he was a male of the largest size.
As soon as I had fired, he moved higher up the tree, and while
he was doing so I fired again ; and we then saw that one arm
was broken. He had now reached the very highest part of an
immense tree, and immediately began breaking off boughs all
around, and laying them across and across to make a nest. It
was very interesting to see how well he had chosen his place,
and how rapidly he stretched out his unwounded arm in every
direction, breaking off good-sized boughs with the greatest ease,
and laying them back across each other, so that in a few minutes
he had formed a compact mass of foliage, which entirely concealed him from our sight. He was evidently going to pass the
night here, and would probably get away early the next morning, if not wounded too severely. I therefore fired again several
times, in hopes of making him leave his nest ; but, though I felt
sure I had hit him, as at each shot he moved a little, he would
not go away. At length he raised himself up, so that half his
body was visible, and then gradually sank down, his head alone
remaining on the edge of the nest. I now felt sure he was dead,
and tried to persuade the Chinaman and his companion to cut
down the tree ; but it was a very large one, and they had been
at work all day, and nothing would induce them to attempt it.
The next morning, at daybreak, I came to the place, and found
that the Mias was evidently dead, as his head was visible in
exactly the same position as before. I now offered four Chinamen a day's wages each to cut the tree down at once, as a few
hours of sunshine would cause decomposition on the surface of
the skin ; but, after looking at it and trying it, they determined
that it was very big and very hard, and would not attempt it.
Had I doubled my offer, they would probably have accepted it,
as it would not have been more than two or three hours' work ;

and had I been on a short visit only I would have done so; but as I was a resident, and intended remaining several months longer, it would not have answered to begin paying too exorbitantly, or I should have got nothing done in future at a lower rate.

For some weeks after, a cloud of flies could be seen all day, hovering over the body of the dead Mias; but in about a month all was quiet, and the body was evidently drying up under the influence of a vertical sun alternating with tropical rains. Two or three months later two Malays, on the offer of a dollar, climbed the tree, and let down the dried remains. The skin was almost entire, enclosing the skeleton, and inside were millions of the pupa-cases of flies and other insects, with thousands of two or three species of small necrophagous beetles. The skull had been much shattered by balls, but the skeleton was perfect, except one small wrist-bone, which had probably dropped out and been carried away by a lizard.

Three days after I had shot this one and lost it, Charles found three small Orangs feeding together. We had a long chase after them, and had a good opportunity of seeing how they make their way from tree to tree, by always choosing those limbs whose branches are intermingled with those of some other tree, and then grasping several of the small twigs together before they venture to swing themselves across. Yet they do this so quickly and certainly, that they make way among the trees at the rate of full five or six miles an hour, as we had continually to run to keep up with them. One of these we shot and killed, but it remained high up in the fork of a tree; and, as young animals are of comparatively little interest, I did not have the tree cut down to get it.

At this time I had the misfortune to slip among some fallen trees, and hurt my ankle, and, not being careful enough at first, it became a severe inflamed ulcer, which would not heal, and kept me a prisoner in the house the whole of July and part of August. When I could get out again, I determined to take a trip up a branch of the Simunjon River to Semábang, where there was said to be a large Dyak house, a mountain with abundance of fruit, and plenty of Orangs and fine birds. As the river was very narrow, and I was obliged to go in a very small boat with little luggage, I only took with me a Chinese boy as a servant. I carried a cask of medicated arrack to put Mias skins in, and stores and ammunition for a fortnight. After a few miles, the stream became very narrow and winding, and the whole country on each side was flooded. On the banks were abundance of monkeys—the common Macacus cynomolgus, a black Semnopithecus, and the extraordinary long-nosed monkey (Nasalis larvatus), which is as large as a three-year-old child, has a very long tail, and a fleshy nose, longer than that of the biggest-nosed man. The further we went on the narrower and more winding the stream became; fallen trees sometimes blocked up our passage, and sometimes tangled branches and creepers met completely across it, and had to be cut away before we could get on. It took us two days to reach Semábang, and we hardly saw a bit of dry land all the way. In the latter part of the journey I could touch the bushes on each side for miles ; and we were often delayed by the screw-pines (Pandanus), which grew abundantly in the water, falling across the stream. In other places dense rafts of floating grass completely filled up the channel, making our journey a constant succession of difficulties.

Near the landing-place we found a fine house, 250 feet long, raised high above the ground on posts, with a wide verandah and still wider platform of bamboo in front of it. Almost all the people, however, were away on some excursion after edible birds'-nests or bees'-wax, and there only remained in the house two or three old men and women with a lot of children. The mountain or hill was close by, covered with a complete forest of fruit-trees, among which the Durian and Mangusteen were very abundant ; but the fruit was not yet quite ripe, except a little here and there. I spent a week at this place, going out every day in various directions about the mountain, accompanied by a Malay, who had stayed with me while the other boatmen returned. For three days we found no Orangs, but shot a deer and several monkeys. On the fourth day, however, we found a Mias feeding on a very lofty Durian tree, and succeeded in killing it, after eight shots. Unfortunately it remained in the tree, hanging by its hands, and we were obliged to leave it and return home, as it was several miles off. As I felt pretty sure it would fall during the night, I returned to the place early the next morning, and found it on the ground beneath the tree. To my astonishment and pleasure, it appeared to be a different kind from any I had yet seen, for although a full-grown male by its fully developed teeth and very large canines, it had no sign of the lateral protuberance on the face, and was about one-tenth smaller in all its dimensions than the other adult males. The upper incisors, however, appeared to be broader than in the larger species, a character distinguishing the Simia morio of Professor Owen, which he had described from the cranium of a female specimen. As it was too far to carry the animal home, I set to work and skinned the body on the spot, leaving the head, hands, and feet attached, to be finished at home. This specimen is now in the British Museum.

At the end of a week, finding no more Orangs, I returned home ; and, taking in a few fresh stores, and this time accompanied by Charles, went up another branch of the river, very similar in character, to a place called Menyille, where there were several small Dyak houses and one large one. Here the landing-place was a bridge of rickety poles, over a considerable distance of water ; and I thought it safer to leave my cask of arrack securely placed in the fork of a tree. To prevent the natives from drinking it, I let several of them see me put in a number of snakes and lizards; but I rather think this did not prevent them from tasting it. We were accommodated here in the verandah of the large house, in which were several great baskets of dried human heads, the trophies of past generations of head-hunters. Here also there was a little mountain covered with fruit-trees, and there were some magnificent Durian trees close by the house, the fruit of which was ripe; and as the Dyaks looked upon me as a benefactor in killing the Mias which destroys a great deal of their fruit, they let us eat as much as we liked, and we revelled in this emperor of fruits in its greatest perfection.

The very day after my arrival in this place, I was so fortunate as to shoot another adult male of the small Orang, the Mias-kassir of the Dyaks. It fell when dead, but caught in a fork of the tree and remained fixed. As I was very anxious to get it, I tried to persuade two young Dyaks who were with me to cut down the tree, which was tall, perfectly straight and smooth-barked, and without a branch for fifty or sixty feet. To my surprise, they said they would prefer climbing up it, but it would be a good deal of trouble, and, after a little talking to-gether, they said they would try. They first went to a clump of bamboo that stood near, and cut down one of the largest stems. From this they chopped off a short piece, and splitting it, made a couple of stout pegs, about a foot long, and sharp at one end. Then cutting a thick piece of wood for a mallet, they drove one of the pegs into the tree and hung their weight upon it. It held, and this seemed to satisfy them, for they imme-diately began making a quantity of pegs of the same kind, while I looked on with great interest, wondering how they could possibly ascend such a lofty tree by merely driving pegs in it, the failure of any one of which at a good height would certainly cause their death. When about two dozen pegs were made, one of them began cutting some very long and slender bamboo from another clump, and also prepared some cord from the bark of a small tree. They now drove in a peg very firmly at about three feet from the ground, and bringing one of the long bamboos, stood it upright close to the tree, and bound it firmly to the two first pegs, by means of the bark cord, and small notches near the head of each peg. One of the Dyaks now stood on the first peg and drove in a third, about level with his face, to which he tied the bamboo in the same way, and then mounted another step, standing on one foot, and holding by the bamboo at the peg immediately above him, while he drove in the next one. In this manner he ascended about twenty feet, when the upright bamboo becoming thin, another was handed up by his companion, and this was joined on by tying both bamboos to three or four of the pegs. When this was also nearly ended, a third was added, and shortly after, the lowest branches of the tree were reached, along which the young Dyak scrambled, and soon sent

the Mias tumbling headlong down. I was exceedingly struck
by the ingenuity of this mode of climbing, and the admirable
manner in which the peculiar properties of the bamboo were
made available. The ladder itself was perfectly safe, since if
any one peg were loose or faulty, and gave way, the strain would
be thrown on several others above and below it. I now understood the use of the line of bamboo pegs sticking in trees, which
I had often seen, and wondered for what purpose they could
have been put there. This animal was almost identical in size
and appearance with the one I had obtained at Semábang, and
was the only other male specimen of the Simia morio which I
obtained. It is now in the Derby Museum.

I afterwards shot two adult females and two young ones of
different ages, all of which I preserved. One of the females,
with several young ones, was feeding on a Durian tree with unripe fruit ; and as soon as she saw us she began breaking off
branches and the great spiny fruits with every appearance of
rage, causing such a shower of missiles as effectually kept us
from approaching too near the tree. This habit of throwing
down branches when irritated has been doubted, but I have, as
here narrated, observed it myself on at least three separate
occasions. It was however always the female Mias who behaved
in this way, and it may be that the male, trusting more to his
great strength and his powerful canine teeth, is not afraid of
any other animal, and does not want to drive them away, while
the parental instinct of the female leads her to adopt this mode
of defending herself and her young ones.

In preparing the skins and skeletons of these animals, I was
much troubled by the Dyak dogs, which, being always kept in
a state of semi starvation, are ravenous for animal food. I had
a great iron pan, in which I boiled the bones to make skeletons,
and at night I covered this over with boards, and put heavy
stones upon it ; but the dogs managed to remove these and carried
away the greater part of one of my specimens. On another
occasion they gnawed away a good deal of the upper leather of
my strong boots, and even ate a piece of my mosquito-curtain,
where some lamp-oil had been spilt over it some weeks before.

On our return down the stream, we had the fortune to fall in
with a very old male Mias, feeding on some low trees growing in
the water. The country was flooded for a long distance, but so
full of trees and stumps that the laden boat could not be got in
among them, and if it could have been we should only have
frightened the Mias away. I therefore got into the water, which
was nearly up to my waist, and waded on till I was near enough
for a shot. The difficulty then was to load my gun again, for I
was so deep in the water that I could not hold the gun sloping
enough to pour the powder in. I therefore had to search for a
shallow place, and after several shots under these trying circumstances, I was delighted to see the monstrous animal roll over
into the water. I now towed him after me to the stream, but the

Malays objected to have the animal put into the boat, and he was so heavy that I could not do it without their help. I looked about for a place to skin him, but not a bit of dry ground was to be seen, till at last I found a clump of two or three old trees and stumps, between which a few feet of soil had collected just above the water, and which was just large enough for us to drag the animal upon it. I first measured him, and found him to be by far the largest I had yet seen, for, though the standing height was the same as the others (4 feet 2 inches), yet the outstretched arms were 7 feet 9 inches, which was six inches more than the previous one, and the immense broad face was 13½ inches wide, whereas the widest I had hitherto seen was only 11½ inches. The girth of the body was 3 feet 7½ inches. I am inclined to believe, therefore, that the length and strength of the arms, and the width of the face, continues increasing to a very great age, while the standing height, from the sole of the foot to the crown of the head, rarely if ever exceeds 4 feet 2 inches.

As this was the last Mias I shot, and the last time I saw an adult living animal, I will give a sketch of its general habits, and any other facts connected with it. The Orang-utan is known to inhabit Sumatra and Borneo, and there is every reason to believe that it is confined to these two great islands, in the former of which, however, it seems to be much more rare. In Borneo it has a wide range, inhabiting many districts on the south-west, south-east, north-east, and north-west coasts, but appears to be chiefly confined to the low and swampy forests. It seems at first sight very inexplicable that the Mias should be quite unknown in the Saráwak valley, while it is abundant in Sambas, on the west, and Sádong, on the east. But when we know the habits and mode of life of the animal, we see a sufficient reason for this apparent anomaly in the physical features of the Saráwak district. In the Sádong, where I observed it, the Mias is only found where the country is low, level, and swampy, and at the same time covered with a lofty virgin forest. From these swamps rise many isolated mountains, on some of which the Dyaks have settled, and covered with plantations of fruit trees. These are a great attraction to the Mias, which comes to feed on the unripe fruits, but always retires to the swamp at night. Where the country becomes slightly elevated, and the soil dry, the Mias is no longer to be found. For example, in all the lower part of the Sádong valley it abounds, but as soon as we ascend above the limits of the tides where the country, though still flat, is high enough to be dry, it disappears. Now the Saráwak valley has this peculiarity—the lower portion though swampy is not covered with continuous lofty forest, but is principally occupied by the Nipa palm; and near the town of Saráwak where the country becomes dry, it is greatly undulated in many parts, and covered with small patches of virgin forest, and much second-growth jungle on ground which has once been cultivated by the Malays or Dyaks.

Now it seems to me probable, that a wide extent of unbroken and equally lofty virgin forest is necessary to the comfortable existence of these animals. Such forests form their open country, where they can roam in every direction with as much facility as the Indian on the prairie, or the Arab on the desert ; passing from tree-top to tree-top without ever being obliged to descend upon the earth. The elevated and the drier districts are more frequented by man, more cut up by clearings and low second-growth jungle not adapted to its peculiar mode of progression, and where it would therefore be more exposed to danger, and more frequently obliged to descend upon the earth. There is probably also a greater variety of fruit in the Mias district, the small mountains which rise like islands out of it serving as a sort of gardens or plantations, where the trees of the uplands are to be found in the very midst of the swampy plains.

It is a singular and very interesting sight to watch a Mias making his way leisurely through the forest. He walks deliberately along some of the larger branches, in the semi-erect attitude which the great length of his arms and the shortness of his legs cause him naturally to assume ; and the disproportion between these limbs is increased by his walking on his knuckles, not on the palm of the hand, as we should do. He seems always to choose those branches which intermingle with an adjoining tree, on approaching which he stretches out his long arms, and, seizing the opposing boughs, grasps them together with both hands, seems to try their strength, and then deliberately swings himself across to the next branch, on which he walks along as before. He never jumps or springs, or even appears to hurry himself, and yet manages to get along almost as quickly as a person can run through the forest beneath. The long and powerful arms are of the greatest use to the animal, enabling it to climb easily up the loftiest trees, to seize fruits and young leaves from slender boughs which will not bear its weight, and to gather leaves and branches with which to form its nest. I have already described how it forms a nest when wounded, but it uses a similar one to sleep on almost every night. This is placed low down, however, on a small tree not more than from twenty to fifty feet from the ground, probably because it is warmer and less exposed to wind than higher up. Each Mias is said to make a fresh one for himself every night ; but I should think that is hardly probable, or their remains would be much more abundant ; for though I saw several about the coal-mines, there must have been many Orangs about every day, and in a year their deserted nests would become very numerous. The Dyaks say that, when it is very wet, the Mias covers himself over with leaves of pandanus, or large ferns, which has perhaps led to the story of his making a hut in the trees.

The Orang does not leave his bed till the sun has well risen and has dried up the dew upon the leaves. He feeds all through the middle of the day, but seldom returns to the same tree two days running. They do not seem much alarmed at man, as they often stared down upon me for several minutes, and then only moved away slowly to an adjacent tree. After seeing one, I have often had to go half a mile or more to fetch my gun, and in nearly every case have found it on the same tree, or within a hundred yards, when I returned. I never saw two full-grown animals together, but both males and females are sometimes accompanied by half-grown young ones, while, at other times, three or four young ones were seen in company. Their food consists almost exclusively of fruit, with occasionally leaves, buds, and young shoots. They seem to prefer unripe fruits, some of which were very sour, others intensely bitter, particularly the large red, fleshy arillus of one which seemed an especial favourite. In other cases they eat only the small seed of a large fruit, and they almost always waste and destroy more than they eat, so that there is a continual rain of rejected portions below the tree they are feeding on. The Durian is an especial favourite, and quantities of this delicious fruit are destroyed wherever it grows surrounded by forest, but they will not cross clearings to get at them. It seems wonderful how the animal can tear open this fruit, the outer covering of which is so thick and tough, and closely covered with strong conical spines. It probably bites off a few of these first, and then, making a small hole, tears open the fruit with its powerful fingers.

The Mias rarely descends to the ground, except when pressed by hunger it seeks for succulent shoots by the river side ; or, in very dry weather, has to search after water, of which it generally finds sufficient in the hollows of leaves. Once only I saw two half-grown Orangs on the ground in a dry hollow at the foot of the Simunjon hill. They were playing together, standing erect, and grasping each other by the arms. It may be safely stated, however, that the Orang never walks erect, unless when using its hands to support itself by branches overhead or when attacked. Representations of its walking with a stick are entirely imaginary.

The Dyaks all declare that the Mias is never attacked by any animal in the forest, with two rare exceptions ; and the accounts I received of these are so curious that I give them nearly in the words of my informants, old Dyak chiefs, who had lived all their lives in the places where the animal is most abundant. The first of whom I inquired said : " No animal is strong enough to hurt the Mias, and the only creature he ever fights with is the crocodile. When there is no fruit in the jungle, he goes to seek food on the banks of the river, where there are plenty of young shoots that he likes, and fruits that grow close to the water. Then the crocodile sometimes tries to seize him, but the Mias gets upon him, and beats him with his hands and feet, and tears him and kills him." He added that he had once seen such a fight, and that he believes that the Mias is always the victor.

My next informant was the Orang Kaya, or chief of the Balow Dyaks, on the Simunjon River. He said: "The Mias has no enemies; no animals dare attack it but the crocodile and the python. He always kills the crocodile by main strength, standing upon it, pulling open its jaws, and ripping up its throat. If a python attacks a Mias, he seizes it with his hands, and then bites it, and soon kills it. The Mias is very strong; there is no animal in the jungle so strong as he."

It is very remarkable that an animal so large, so peculiar, and of such a high type of form as the Orang-utan, should be confined to so limited a district—to two islands, and those almost the last inhabited by the higher Mammalia; for, eastward of Borneo and Java, the Quadrumania, Ruminants, Carnivora, and many other groups of Mammalia, diminish rapidly, and soon entirely disappear. When we consider, further, that almost all other animals have in earlier ages been represented by allied yet distinct forms—that, in the latter part of the tertiary period, Europe was inhabited by bears, deer, wolves, and cats; Australia by kangaroos and other Marsupials; South America by gigantic sloths and ant-eaters; all different from any now existing, though intimately allied to them—we have every reason to believe that the Orang-utan, the Chimpanzee, and the Gorilla have also had their forerunners. With what interest must every naturalist look forward to the time when the caves and tertiary deposits of the tropics may be thoroughly examined, and the past history and earliest appearance of the great man-like apes be at length made known.

I will now say a few words as to the supposed existence of a Bornean Orang as large as the Gorilla. I have myself examined the bodies of seventeen freshly-killed Orangs, all of which were carefully measured; and of seven of them I preserved the skeleton. I also obtained two skeletons killed by other persons. Of this extensive series, sixteen were fully adult, nine being males, and seven females. The adult males of the large Orangs only varied from 4 feet 1 inch to 4 feet 2 inches in height, measured fairly to the heel, so as to give the height of the animal if it stood perfectly erect; the extent of the outstretched arms, from 7 feet 2 inches to 7 feet 8 inches; and the width of the face, from 10 inches to 13½ inches. The dimensions given by other naturalists closely agree with mine. The largest Orang measured by Temminck was 4 feet high. Of twenty-five specimens collected by Schlegel and Müller, the largest old male was 4 feet 1 inch; and the largest skeleton in the Calcutta Museum was, according to Mr. Blyth, 4 feet 1½ inch. My specimens were all from the north-west coast of Borneo; those of the Dutch from the west and south coasts; and no specimen has yet reached Europe exceeding these dimensions, although the total number of skins and skeletons must amount to over a hundred.

Strange to say, however, several persons declare that they

have measured Orangs of a much larger size. Temminck, in his Monograph of the Orang, says, that he has just received news of the capture of a specimen 5 feet 3 inches high. Unfortunately, it never seems to have reached Holland, for nothing has since been heard of any such animal. Mr. St. John, in his *Life in the Forests of the Far East*, vol. ii. p. 237, tells us of an Orang shot by a friend of his, which was 5 feet 2 inches from the heel to the top of the head, the arm 17 inches in girth, and the wrist 12 inches! The head alone was brought to Saráwak, and Mr. St. John tells us that he assisted to measure this, and that it was 15 inches broad by 14 long. Unfortunately, even this skull appears not to have been preserved, for no specimen corresponding to these dimensions has yet reached England.

In a letter from Sir James Brooke, dated October, 1857, in which he acknowledges the receipt of my Papers on the Orang, published in the *Annals and Magazine of Natural History*, he sends me the measurements of a specimen killed by his nephew, which I will give exactly as I received it : "September 3rd, 1867, killed female Orang-utan. Height, from head to heel, 4 feet 6 inches. Stretch from fingers to fingers across body, 6 feet 1 inch. Breadth of face, including callosities, 11 inches." Now, in these dimensions, there is palpably one error ; for in every Orang yet measured by any naturalist, an expanse of arms of 6 feet 1 inch corresponds to a height of about 3 feet 6 inches, while the largest specimens of 4 feet to 4 feet 2 inches high, always have the ex-tended arms as much as 7 feet 3 inches to 7 feet 8 inches. It is, in fact, one of the characters of the genus to have the arms so long that an animal standing nearly erect can rest its fingers on the ground. A height of 4 feet 6 inches would therefore require a stretch of arms of at least 8 feet! If it were only 6 feet to that height, as given in the dimensions quoted, the animal would not be an Orang at all, but a new genus of apes, differing materially in habits and mode of progression. But Mr. Johnson, who shot this animal, and who knows Orangs well, evidently considered it to be one ; and we have therefore to judge whether it is more probable that he made a mistake of *two feet* in the stretch of the arms, or of *one foot* in the height. The latter error is certainly the easiest to make, and it will bring his animal into agreement, as to proportions and size, with all those which exist in Europe. How easy it is to be deceived in the height of these animals is well shown in the case of the Sumatran Orang, the skin of which was described by Dr. Clarke Abel. The captain and crew who killed this animal declared, that when alive he exceeded the tallest man, and looked so gigantic that they thought he was 7 feet high ; but that, when he was killed and lay upon the ground, they found he was only about 6 feet. Now it will hardly be credited that the skin of this identical animal exists in the Calcutta Museum, and Mr. Blyth, the late curator, states " that it is by no means one of the largest size ; " which means that it is about 4 feet high !

Having these undoubted examples of error in the dimensions of Orangs, it is not too much to conclude that Mr. St. John's friend made a similar error of measurement, or rather, perhaps, of memory ; for we are not told that the dimensions were noted down *at the time they were made.* The only figures given by Mr. St. John on his own authority are that "the head was 15 inches broad by 14 inches long." As my largest male was 13½ broad across the face, measured as soon as the animal was killed, I can quite understand that when the head arrived at Saráwak from the Batang Lupar, after two if not three days' voyage, it was so swollen by decomposition as to measure an inch more than when it was fresh. On the whole, therefore, I think it will be allowed, that up to this time we have not the least reliable evidence of the existence of Orangs in Borneo more than 4 feet 2 inches high.

CHAPTER V.

BORNEO—JOURNEY IN THE INTERIOR.

(NOVEMBER 1855 TO JANUARY 1856.)

As the wet season was approaching I determined to return to Saráwak, sending all my collections with Charles Allen round by sea, while I myself proposed to go up to the sources of the Sádong River, and descend by the Saráwak valley. As the route was somewhat difficult, I took the smallest quantity of baggage, and only one servant, a Malay lad named Bujon, who knew the language of the Sádong Dyaks, with whom he had traded. We left the mines on the 27th of November, and the next day reached the Malay village of Gúdong, where I stayed a short time to buy fruit and eggs, and called upon the Datu Bandar, or Malay governor of the place. He lived in a large and well-built house, very dirty outside and in, and was very inquisitive about my business, and particularly about the coal mines. These puzzle the natives exceedingly, as they cannot understand the extensive and costly preparations for working coal, and cannot believe it is to be used only as fuel when wood is so abundant and so easily obtained. It was evident that Europeans seldom came here, for numbers of women skeltered away as I walked through the village ; and one girl about ten or twelve years old, who had just brought a bamboo full of water from the river, threw it down with a cry of horror and alarm the moment she caught sight of me, turned round and jumped into the stream. She swam beautifully, and kept looking back as if expecting I would follow her, screaming violently all the time ; while a number of men and boys were laughing at her ignorant terror.

At Jahi, the next village, the stream became so swift in con-

sequence of a flood, that my heavy boat could make no way,
and I was obliged to send it back and go on in a very small
open one. So far the river had been very monotonous, the
banks being cultivated as rice-fields, and little thatched huts
alone breaking the unpicturesque line of muddy bank crowned
with tall grasses, and backed by the top of the forest behind
the cultivated ground. A few hours beyond Jahi we passed the
limits of cultivation, and had the beautiful virgin forest coming
down to the water's edge, with its palms and creepers, its noble

PORTRAIT OF DYAK YOUTH.

trees, its ferns, and
epiphytes. The banks
of the river were, how-
ever, still generally
flooded, and we had
some difficulty in find-
ing a dry spot to sleep
on. Early in the morn-
ing we reached Empug-
nan, a small Malay vil-
lage situated at the foot
of an isolated mountain
which had been visible
from the mouth of the
Simunjon River. Be-
yond here the tides are
not felt and we now
entered upon a district
of elevated forest, with
a   finer   vegetation.
Large trees stretch out
their arms across the
stream, and the steep,
earthy banks are cloth-
ed with ferns and Zin-
giberaceous plants.

Early in the after-
noon we arrived at
Tabókan, the first vil-
lage of the Hill Dyaks.
On an open space near
the river about twenty boys were playing at a game something
like what we call "prisoner's base"; their ornaments of beads
and brass wire and their gay-coloured kerchiefs and waist-cloths
showing to much advantage, and forming a very pleasing sight.
On being called by Bujon, they immediately left their game to
carry my things up to the "head-house,"—a circular building
attached to most Dyak villages, and serving as a lodging for
strangers, the place for trade, the sleeping-room of the unmarried youths, and the general council-chamber. It is elevated
on lofty posts, has a large fireplace in the middle and windows

in the roof all round, and forms a very pleasant and comfortable abode. In the evening it was crowded with young men and boys, who came to look at me. They were mostly fine young fellows, and I could not help admiring the simplicity and elegance of their costume. Their only dress is the long "chawat," or waist-cloth, which hangs down before and behind. It is generally of blue cotton, ending in three broad bands of red, blue, and white. Those who can afford it wear a handkerchief on the head, which is either red, with a narrow border of gold lace, or of three colours, like the "chawat." The large flat moon-shaped brass earrings, the heavy necklace of white or black beads, rows of brass rings on the arms and legs, and armlets of white shell, all serve to relieve and set off the pure reddish brown skin and jet-black hair. Add to this the little pouch containing materials for betel-chewing and a long slender knife, both invariably worn at the side, and you have the everyday dress of the young Dyak gentleman.

The "Orang Kaya," or rich man, as the chief of the tribe is called, now came in with several of the older men ; and the "bitchára" or talk commenced, about getting a boat and men to take me on the next morning. As I could not understand a word of their language, which is very different from Malay, I took no part in the proceedings, but was represented by my boy Bujon, who translated to me most of what was said. A Chinese trader was in the house, and he, too, wanted men the next day ; but on his hinting this to the Orang Kaya, he was sternly told that a white man's business was now being discussed, and he must wait another day before his could be thought about.

After the "bitchára" was over and the old chiefs gone, I asked the young men to play or dance, or amuse themselves in their accustomed way ; and after some little hesitation they agreed to do so. They first had a trial of strength, two boys sitting opposite each other, foot being placed against foot, and a stout stick grasped by both their hands. Each then tried to throw himself back, so as to raise his adversary up from the ground, either by main strength or by a sudden effort. Then one of the men would try his strength against two or three of the boys ; and afterwards they each grasped their own ankle with a hand, and while one stood as firm as he could, the other swung himself round on one leg, so as to strike the other's free leg, and try to overthrow him. When these games had been played all round with varying success, we had a novel kind of concert. Some placed a leg across the knee, and struck the fingers sharply on the ankle, others beat their arms against their sides like a cock when he is going to crow, thus making a great variety of clapping sounds, while another with his hand under his armpit produced a deep trumpet note ; and, as they all kept time very well, the effect was by no means unpleasing. This seemed quite a favourite amusement with them, and they kept it up with much spirit.

The next morning we started in a boat about thirty feet long, and only twenty-eight inches wide. The stream here suddenly changes its character. Hitherto, though swift, it had been deep and smooth, and confined by steep banks. Now it rushed and rippled over a pebbly, sandy, or rocky bed, occasionally forming miniature cascades and rapids, and throwing up on one side or the other broad banks of finely coloured pebbles. No paddling could make way here, but the Dyaks with bamboo poles propelled us along with great dexterity and swiftness, never losing their balance in such a narrow and unsteady vessel, though standing up and exerting all their force. It was a brilliant day, and the cheerful exertions of the men, the rushing of the sparkling waters, with the bright and varied foliage which from either bank stretched over our heads, produced an exhilarating sensation which recalled my canoe voyages on the grander waters of South America.

Early in the afternoon we reached the village of Borotói, and, though it would have been easy to reach the next one before night, I was obliged to stay, as my men wanted to return and others could not possibly go on with me without the preliminary talking. Besides, a white man was too great a rarity to be allowed to escape them, and their wives would never have forgiven them if, when they returned from the fields, they found that such a curiosity had not been kept for them to see. On entering the house to which I was invited, a crowd of sixty or seventy men, women, and children gathered round me, and I sat for half an hour like some strange animal submitted for the first time to the gaze of an inquiring public. Brass rings were here in the greatest profusion, many of the women having their arms completely covered with them, as well as their legs from the ankle to the knee. Round the waist they wear a dozen or more coils of fine rattan stained red, to which the petticoat is attached. Below this are generally a number of coils of brass wire, a girdle of small silver coins, and sometimes a broad belt of brass ring armour. On their heads they wear a conical hat without a crown, formed of variously coloured beads, kept in shape by rings of rattan, and forming a fantastic but not unpicturesque head-dress.

Walking out to a small hill near the village, cultivated as a rice-field, I had a fine view of the country, which was becoming quite hilly, and towards the south, mountainous. I took bearings and sketches of all that was visible, an operation which caused much astonishment to the Dyaks who accompanied me, and produced a request to exhibit the compass when I returned. I was then surrounded by a larger crowd than before, and when I took my evening meal in the midst of a circle of about a hundred spectators anxiously observing every movement and criticising every mouthful, my thoughts involuntarily recurred to the lions at feeding time. Like those noble animals, I too was used to it, and it did not affect my appetite. The children here were more shy than at Tabokan, and I could not persuade them to play. I therefore turned showman myself, and exhibited the shadow of a dog's head eating, which pleased them so much that all the village in succession came out to see it. The "rabbit on the wall" does not do in Borneo, as there is no animal it resembles. The boys had tops shaped something like whipping-tops, but spun with a string.

The next morning we proceeded as before, but the river had become so rapid and shallow, and the boats were all so small, that though I had nothing with me but a change of clothes, a gun, and a few cooking utensils, two were required to take me on. The rock which appeared here and there on the river-bank was an indurated clay-slate, sometimes crystalline, and thrown up almost vertically. Right and left of us rose isolated limestone mountains, their white precipices glistening in the sun and contrasting beautifully with the luxuriant vegetation that elsewhere clothed them. The river bed was a mass of pebbles, mostly pure white quartz, but with abundance of jaspar and agate, presenting a beautifully variegated appearance. It was only ten in the morning when we arrived at Budw, and, though there were plenty of people about, I could not induce them to allow me to go on to the next village. The Orang Kaya said that if I insisted on having men of course he would get them; but when I took him at his word and said I must have them, there came a fresh remonstrance; and the idea of my going on that day seemed so painful that I was obliged to submit. I therefore walked out over the rice-fields, which are here very extensive, covering a number of the little hills and valleys into which the whole country seems broken up, and obtained a fine view of hills and mountains in every direction.

In the evening the Orang Kaya came in full dress (a spangled velvet jacket, but no trousers), and invited me over to his house, where he gave me a seat of honour under a canopy of white calico and coloured handkerchiefs. The great verandah was crowded with people, and large plates of rice with cooked and fresh eggs were placed on the ground as presents for me. A very old man then dressed himself in bright-coloured cloths and many ornaments, and sitting at the door, murmured a long prayer or invocation, sprinkling rice from a basin he held in his hand, while several large gongs were loudly beaten and a salute of muskets fired off. A large jar of rice wine, very sour but with an agreeable flavour, was then handed round, and I asked to see some of their dances. These were, like most savage performances, very dull and ungraceful affairs; the men dressing themselves absurdly like women, and the girls making themselves as stiff and ridiculous as possible. All the time six or eight large Chinese gongs were being beaten by the vigorous arms of as many young men, producing such a deafening discord that I was glad to escape to the round house, where I slept very comfortable with half a dozen smoke-dried human skulls suspended over my head.

The river was now so shallow that boats could hardly get along. I therefore preferred walking to the next village, expecting to see something of the country, but was much disappointed, as the path lay almost entirely through dense bamboo thickets. The Dyaks get two crops off the ground in succession; one of rice and the other of sugar-cane, maize, and vegetables. The ground then lies fallow eight or ten years, and becomes covered with bamboos and shrubs, which often completely arch over the path and shut out everything from the view. Three hours' walking brought us to the village of Senánkan, where I was again obliged to remain the whole day, which I agreed to do on the promise of the Orang Kaya that his men should next day take me through two other villages across to Sénna, at the head of the Saráwak River. I amused myself as I best could till evening, by walking about the high ground near, to get views of the country and bearings of the chief mountains. There was then another public audience, with gifts of rice and eggs, and drinking of rice wine. These Dyaks cultivate a great extent of ground, and supply a good deal of rice to Saráwak. They are rich in gongs, brass trays, wire, silver coins, and other articles in which a Dyak's wealth consists; and their women and children are highly ornamented with bead necklaces, shells, and brass wire.

In the morning I waited some time, but the men that were to accompany me did not make their appearance. On sending to the Orang Kaya I found that both he and another head-man had gone out for the day, and on inquiring the reason was told that they could not persuade any of their men to go with me because the journey was a long and fatiguing one. As I was determined to get on, I told the few men that remained that the chiefs had behaved very badly, and that I should acquaint the Rajah with their conduct, and I wanted to start immediately. Every man present made some excuse, but others were sent for, and by dint of threats and promises, and the exertion of all Bujon's eloquence, we succeeded in getting off after two hours' delay.

For the first few miles our path lay over a country cleared for rice-fields, consisting entirely of small but deep and sharply-cut ridges and valleys, without a yard of level ground. After crossing the Kayan River, a main branch of the Sádong, we got on to the lower slopes of the Seboran Mountain, and the path lay along a sharp and moderately steep ridge, affording an excellent view of the country. Its features were exactly those of the Himalayas in miniature, as they are described by Dr. Hooker and other travellers; and looked like a natural model of some parts of those vast mountains on a scale of about a tenth, thousands of feet being here represented by hundreds. I now discovered the source of the beautiful pebbles which had so pleased me in the river-bed. The slaty rocks had ceased, and these mountains seemed to consist of a sand-stone conglomerate, which was in some places a mere mass of pebbles cemented together. I might have known that such small streams could not produce such vast quantities of well-rounded pebbles of the very hardest materials. They had evidently been formed in past ages, by the action of some continental stream or seabeach, before the great island of Borneo had risen from the ocean. The existence of such a system of hills and valleys reproducing in miniature all the features of a great mountain region, has an important bearing on the modern theory, that the form of the ground is mainly due to atmospheric rather than to subterranean action. When we have a number of branching valleys and ravines running in many different directions within a square mile, it seems hardly possible to impute their formation, or even their origination, to rents and fissures produced by earthquakes. On the other hand, the nature of the rock, so easily decomposed and removed by water, and the known action of the abundant tropical rains, are in this case, at least, quite sufficient causes for the production of such valleys. But the resemblance between their forms and outlines, their mode of divergence, and the slopes and ridges that divide them, and those of the grand mountain scenery of the Himalayas, is so remarkable, that we are forcibly led to the conclusion that the forces at work in the two cases have been the same, differing only in the time they have been in action, and the nature of the material they have had to work upon.

About noon we reached the village of Menyerry, beautifully situated on a spur of the mountain about 600 feet above the valley, and affording a delightful view of the mountains of this part of Borneo. I here got a sight of Penrissen Mountain at the head of the Saráwak River, and one of the highest in the district, rising to about 6,000 feet above the sea. To the south the Rowan, and further off the Untowan Mountains in the Dutch territory, appeared equally lofty. Descending from Menyerry we again crossed the Kayan, which bends round the spur, and ascended to the pass which divides the Sádong and Saráwak valleys, and which is about 2,000 feet high. The descent from this point was very fine. A stream, deep in a rocky gorge, rushed on each side of us, to one of which we gradually descended, passing over many lateral gulleys and along the faces of some precipices by means of native bamboo bridges. Some of these were several hundred feet long and fifty or sixty high, a single smooth bamboo four inches in diameter forming the only pathway, while a slender handrail of the same material was often so shaky that it could only be used as a guide rather than a support.

Late in the afternoon we reached Sodos, situated on a spur between two streams, but so surrounded by fruit trees that little could be seen of the country. The house was spacious, clean, and comfortable, and the people very obliging. Many of the women and children had never seen a white man before, and were very sceptical as to my being the same colour all over, as my face. They begged me to show them my arms and body, and they were so kind and good-tempered that I felt bound to give them some satisfaction, so I turned up my trousers and let them see the colour of my leg, which they examined with great interest.

In the morning early we continued our descent along a fine valley, with mountains rising 2,000 or 3,000 feet in every direction. The little river rapidly increased in size till we reached Sénna, when it had become a fine pebbly stream navigable for small canoes. Here again the upheaved slaty rock appeared, with the same dip and direction as in the Sádong River. On inquiring for a boat to take me down the stream, I was told that the Sénna Dyaks, although living on the river-banks, never made or used boats. They were mountaineers who had only come down into the valley about twenty years before, and had not yet got into new habits. They are of the same tribe as the people of Menyerry and Sodos. They make good paths and bridges, and cultivate much mountain land, and thus give a more pleasing and civilized aspect to the country than where the people move about only in boats, and confine their cultivation to the banks of the streams.

After some trouble I hired a boat from a Malay trader, and found three Dyaks who had been several times with Malays to Saráwak, and thought they could manage it very well. They turned out very awkward, constantly running aground, striking against rocks, and losing their balance so as almost to upset themselves and the boat ; offering a striking contrast to the skill of the Sea Dyaks. At length we came to a really dangerous rapid where boats were often swamped, and my men were afraid to pass it. Some Malays with a boat-load of rice here overtook us, and after safely passing down kindly sent back one of their men to assist me. As it was, my Dyaks lost their balance in the critical part of the passage, and had they been alone would certainly have upset the boat. The river now became exceedingly picturesque, the ground on each side being partially cleared for rice-fields, affording a good view of the country. Numerous little granaries were built high up in trees overhanging the river, and having a bamboo bridge sloping up to them from the bank ; and here and there bamboo suspension bridges crossed the stream, where overhanging trees favoured their construction.

I slept that night in the village of the Sebungow Dyaks, and the next day reached Saráwak, passing through a most beautiful country, where limestone mountains with their fantastic forms and white precipices shot up on every side, draped and festooned with a luxuriant vegetation. The banks of the Saráwak River are everywhere covered with fruit trees, which supply the Dyaks with a great deal of their food. The Mangosteen, Lansat, Rambutan, Jack, Jambou, and Blimbing, are all abundant ; but most abundant and most esteemed is the Durian, a fruit about which very little is known in England, but which both by natives and Europeans in the Malay Archipelago is reckoned superior to all

others. The old traveller Linschott, writing in 1599, says:—
"It is of such an excellent taste that it surpasses in flavour all
the other fruits of the world, according to those who have tasted
it." And Doctor Paludanus adds:—"This fruit is of a hot and
humid nature. To those not used to it, it seems at first to smell
like rotten onions, but immediately they have tasted it they
prefer it to all other food. The natives give it honourable titles,
exalt it, and make verses on it." When brought into a house the
smell is often so offensive that some persons can never bear to
taste it. This was my own case when I first tried it in Malacca,
but in Borneo I found a ripe fruit on the ground, and, eating it
out of doors, I at once became a confirmed Durian eater.

The Durian grows on a large and lofty forest tree, somewhat
resembling an elm in its general character, but with a more
smooth and scaly bark. The fruit is round or slightly oval,
about the size of a large cocoanut, of a green colour, and covered
all over with short stout spines, the bases of which touch each
other, and are consequently somewhat hexagonal, while the
points are very strong and sharp. It is so completely armed,
that if the stalk is broken off it is a difficult matter to lift one
from the ground. The outer rind is so thick and tough, that
from whatever height it may fall it is never broken. From the
base to the apex five very faint lines may be traced, over which
the spines arch a little ; these are the sutures of the carpels, and
show where the fruit may be divided with a heavy knife and a
strong hand. The five cells are satiny white within, and are
each filled with an oval mass of cream-coloured pulp, imbedded
in which are two or three seeds about the size of chestnuts.
This pulp is the eatable part, and its consistence and flavour are
indescribable. A rich butter-like custard highly flavoured with
almonds gives the best general idea of it, but intermingled with
it come wafts of flavour that call to mind cream-cheese, onion-
sauce, brown sherry, and other incongruities. Then there is a
rich glutinous smoothness in the pulp which nothing else
possesses, but which adds to its delicacy. It is neither acid, nor
sweet, nor juicy, yet one feels the want of none of these qualities,
for it is perfect as it is. It produces no nausea or other bad
effect, and the more you eat of it the less you feel inclined to
stop. In fact to eat Durians is a new sensation, worth a voyage
to the East to experience.

When the fruit is ripe it falls of itself, and the only way to eat
Durians in perfection is to get them as they fall ; and the smell
is then less overpowering. When unripe, it makes a very good
vegetable if cooked, and it is also eaten by the Dyaks raw. In
a good fruit season large quantities are preserved salted, in jars
and bamboos, and kept the year round, when it acquires a most
disgusting odour to Europeans, but the Dyaks appreciate it
highly as a relish with their rice. There are in the forest two
varieties of wild Durians with much smaller fruits, one of them
orange-coloured inside ; and these are probably the origin of

the large and fine Durians, which are never found wild. It would not, perhaps, be correct to say that the Durian is the best of all fruits, because it cannot supply the place of the subacid juicy kinds, such as the orange, grape, mango, and mangosteen, whose refreshing and cooling qualities are so wholesome and grateful ; but as producing a food of the most exquisite flavour it is unsurpassed. If I had to fix on two only, as representing the perfection of the two classes, I should certainly choose the Durian and the Orange as the king and queen of fruits.

The Durian is, however, sometimes dangerous. When the fruit begins to ripen it falls daily and almost hourly, and accidents not unfrequently happen to persons walking or working under the trees. When a Durian strikes a man in its fall, it produces a dreadful wound, the strong spines tearing open the flesh, while the blow itself is very heavy ; but from this very circumstance death rarely ensues, the copious effusion of blood preventing the inflammation which might otherwise take place. A Dyak chief informed me that he had been struck down by a Durian falling on his head, which he thought would certainly have caused his death, yet he recovered in a very short time.

Poets and moralists, judging from our English trees and fruits, have thought that small fruits always grew on lofty trees, so that their fall should be harmless to man, while the large ones trailed on the ground. Two of the largest and heaviest fruits known, however, the Brazil-nut fruit (Bertholletia) and Durian, grow on lofty forest trees, from which they fall as soon as they are ripe, and often wound or kill the native inhabitants. From this we may learn two things : first, not to draw general conclusions from a very partial view of nature ; and secondly, that trees and fruits, no less than the varied productions of the animal kingdom, do not appear to be organized with exclusive reference to the use and convenience of man.

During my many journeys in Borneo, and especially during my various residences among the Dyaks, I first came to appreciate the admirable qualities of the Bamboo. In those parts of South America which I had previously visited, these gigantic grasses were comparatively scarce, and where found but little used, their place being taken as to one class of uses by the great variety of Palms, and as to another by calabashes and gourds. Almost all tropical countries produce Bamboos, and wherever they are found in abundance the natives apply them to a variety of uses. Their strength, lightness, smoothness, straightness, roundness, and hollowness, the facility and regularity with which they can be split, their many different sizes, the varying length of their joints, the ease with which they can be cut and with which holes can be made through them, their hardness outside, their freedom from any pronounced taste or smell, their great abundance, and the rapidity of their growth and increase, are all qualities which render them useful for a hundred different purposes, to serve which other materials would require much more labour and preparation. The Bamboo is one of the most wonderful and most beautiful productions of the tropics, and one of nature's most valuable gifts to uncivilized man.

The Dyak houses are all raised on posts, and are often two or three hundred feet long and forty or fifty wide. The floor is always formed of strips split from large Bamboos, so that each may be nearly flat and about three inches wide, and these are firmly tied down with rattan to the joists beneath. When well made, this is a delightful floor to walk upon barefooted, the rounded surfaces of the Bamboo being very smooth and agreeable to the feet, while at the same time affording a firm hold. But, what is more important, they form with a mat over them an excellent bed, the elasticity of the Bamboo and its ·rounded surface being far superior to a more rigid and a flatter floor. Here we at once find a use for Bamboo which cannot be supplied so well by another material without a vast amount of labour, palms and other substitutes requiring much cutting and smoothing, and not being equally good when finished. When, however, a flat, close floor is required, excellent boards are made by splitting open large Bamboos on one side only, and flattening them out so as to form slabs eighteen inches wide and six feet long, with which some Dyaks floor their houses. These with constant rubbing of the feet and the smoke of years become dark and polished, like walnut or old oak, so that their real material can hardly be recognized. What labour is here saved to a savage whose only tools are an axe and a knife, and who, if he wants boards, must hew them out of the solid trunk of a tree, and must give days and weeks of labour to obtain a surface as smooth and beautiful as the Bamboo thus treated affords him. Again, if a temporary house is wanted, either by the native in his plantation or by the traveller in the forest, nothing is so convenient as the Bamboo, with which a house can be constructed with a quarter of the labour and time than if other materials are used.

As I have already mentioned, the Hill Dyaks in the interior of Saráwak make paths for long distances from village to village and to their cultivated grounds, in the course of which they have to cross many gullies and ravines, and even rivers ; or sometimes, to avoid a long circuit, to carry the path along the face of a precipice. In all these cases the bridges they construct are of Bamboos, and so admirably adapted is the material for this purpose, that it seems doubtful whether they ever would have attempted such works if they had not possessed it. The Dyak bridge is simple but well designed. It consists merely of stout Bamboos crossing each other at the roadway like the letter X, and rising a few feet above it. At the crossing they are firmly bound together, and to a large Bamboo which lays upon them and forms the only pathway, with a slender and often very shaky one to serve as a handrail. When a river is to be crossed an over-

hanging tree is chosen, from which the bridge is partly suspended
and partly supported by diagonal struts from the banks, so as to
avoid placing posts in the stream itself, which would be liable to
be carried away by floods.   In carrying a path along the face of
a precipice, trees and roots are made use of for suspension;
struts arise from suitable notches or crevices in the rocks, and
if these are not sufficient, immense Bamboos fifty or sixty feet
long are fixed on the banks or on the branch of a tree below.
These bridges are traversed daily by men and women carrying
heavy loads, so that any insecurity is soon discovered, and, as the
materials are close at hand, immediately repaired.   When a path
goes over very steep ground, and becomes slippery in very wet

DYAK CROSSING A BAMBOO BRIDGE.

or very dry weather, the Bamboo is used in another way.   Pieces
are cut about a yard long, and opposite notches being made at
each end, holes are formed through which pegs are driven, and
firm and convenient steps are thus formed with the greatest ease
and celerity.   It is true that much of this will decay in one or
two seasons, but it can be so quickly replaced as to make it more
economical than using a harder and more durable wood.

One of the most striking uses to which Bamboo is applied by
the Dyaks, is to assist them in climbing lofty trees, by driving
in pegs in the way I have already described at page 42.   This
method is constantly used in order to obtain wax, which is one
of the most valuable products of the country.   The honey-bee
of Borneo very generally hangs its combs under the branches of

the Tappan, a tree which towers above all others in the forest, and whose smooth cylindrical trunk often rises a hundred feet without a branch. The Dyaks climb these lofty trees at night, building up their Bamboo ladder as they go, and bringing down gigantic honeycombs. These furnish them with a delicious feast of honey and young bees, besides the wax, which they sell to traders, and with the proceeds buy the much-coveted brass wire, earrings, and gold-edged handkerchiefs with which they love to decorate themselves. In ascending Durian and other fruit trees which branch at from thirty to fifty feet from the ground, I have seen them use the Bamboo pegs only, without the upright Bamboo which renders them so much more secure.

The outer rind of the Bamboo, split and shaved thin, is the strongest material for baskets; hen-coops, bird-cages, and conical fish-traps are very quickly made from a single joint, by splitting off the skin in narrow strips left attached to one end, while rings of the same material or of rattan are twisted in at regular distances. Water is brought to the houses by little aqueducts formed of large Bamboos split in half and supported on crossed sticks of various heights so as to give it a regular fall. Thin, long-jointed Bamboos form the Dyaks' only water-vessels, and a dozen of them stand in the corner of every house. They are clean, light, and easily carried, and are in many ways superior to earthen vessels for the same purpose. They also make excellent cooking utensils; vegetables and rice can be boiled in them to perfection, and they are often used when travelling. Salted fruit or fish, sugar, vinegar, and honey are preserved in them instead of in jars or bottles. In a small Bamboo case, prettily carved and ornamented, the Dyak carries his sirih and lime for betel chewing, and his little long-bladed knife has a Bamboo sheath. His favourite pipe is a huge hubble-bubble, which he will construct in a few minutes by inserting a small piece of Bamboo for a bowl obliquely into a large cylinder about six inches from the bottom containing water, through which the smoke passes to a long, slender Bamboo tube. There are many other small matters for which Bamboo is daily used, but enough has now been mentioned to show its value. In other parts of the Archipelago I have myself seen it applied to many new uses, and it is probable that my limited means of observation did not make me acquainted with one-half the ways in which it is serviceable to the Dyaks of Saráwak.

While upon the subject of plants I may here mention a few of the more striking vegetable productions of Borneo. The wonderful Pitcher-plants, forming the genus Nepenthes of botanists, here reach their greatest development. Every mountain-top abounds with them, running along the ground, or climbing over shrubs and stunted trees; their elegant pitchers hanging in every direction. Some of these are long and slender, resembling in form the beautiful Philippine lace-sponge

(Euplectella), which has now become so common ; others are broad and short. Their colours are green, variously tinted and mottled with red or purple. The finest yet known were obtained on the summit of Kini-balou, in North-west Borneo. One of the broad sort, Nepenthes rajah, will hold two quarts of water in its pitcher. Another, Nepenthes Edwardsiania, has a narrow pitcher twenty inches long ; while the plant itself grows to a length of twenty feet.

Ferns are abundant, but are not so varied as on the volcanic mountains of Java ; and Tree-ferns are neither so plentiful nor so large as in that island They grow, however, quite down to the level of the sea, and are generally slender and graceful plants from eight to fifteen feet high. Without devoting much time to the search I collected fifty species of Ferns in Borneo, and I have no doubt a good botanist would have obtained twice the number. The interesting group of Orchids is very abundant, but, as is generally the case, nine-tenths of the species have small and inconspicuous flowers. Among the exceptions are the fine Cœlogynes, whose large clusters of yellow flowers orna-ment the gloomiest forests, and that most extraordinary plant, Vanda Lowii, which last is particularly abundant near some hot springs at the foot of the Peninjauh Mountain. It grows on the lower branches of trees, and its strange pendant flower-spikes often hang down so as almost to reach the ground. These are generally six or eight feet long, bearing large and handsome flowers three inches across, and varying in colour from orange to red, with deep purple-red spots. I measured one spike, which reached the extraordinary length of nine feet eight inches, and bore thirty-six flowers, spirally arranged upon a slender thread-like stalk. Specimens grown in our English hot-houses have produced flower-spikes of equal length, and with a much larger number of blossoms.

Flowers were scarce, as is usual in equatorial forests, and it was only at rare intervals that I met with anything striking. A few fine climbers were sometimes seen, especially a handsome crimson and yellow Æschynanthus, and a fine leguminous plant with clusters of large Cassia-like flowers of a rich purple colour. Once I found a number of small Anonaceous trees of the genus Polyalthea, producing a most striking effect in the gloomy forest shades. They were about thirty feet high, and their slender trunks were covered with large star-like crimson flowers, which clustered over them like garlands, and resembled some artificial decoration more than a natural product.

The forests abound with gigantic trees with cylindrical, but-tressed, or furrowed stems, while occasionally the traveller comes upon a wonderful fig-tree, whose trunk is itself a forest of stems and aërial roots. Still more rarely are found trees which appear to have begun growing in mid-air, and from the same point send out wide-spreading branches above and a com-plicated pyramid of roots descending for seventy or eighty feet to the ground below, and so spreading on every side, that one can stand in the very centre with the trunk of the tree immediately overhead. Trees of this character are found all over the Archipelago, and the preceding illustration (taken from one which I often visited in the Aru Islands) will convey some idea of their general character. I believe that they originate as parasites, from seeds carried by birds and dropped in the fork of some lofty tree. Hence descend aërial roots, clasping and ultimately destroying the supporting tree, which is in time entirely replaced by the humble plant which was at first dependent upon it. Thus we have an actual struggle for life in the vegetable kingdom, not less fatal to the vanquished than the struggles among animals which we can so much more easily observe and understand. The advantage of quicker access to light and warmth and air, which is gained in one way by climbing plants, is here obtained by a forest tree, which has the means of starting in life at an elevation which others can only attain after many years of growth, and then only when the fall of some other tree has made room for them. Thus it is that in the warm and moist and equable climate of the tropics, each available station is seized upon, and becomes the means of developing new forms of life especially adapted to occupy it.

On reaching Saráwak early in December I found there would not be an opportunity of returning to Singapore till the latter end of January. I therefore accepted Sir James Brooke's invitation to spend a

week with him and Mr. St. John at his cottage on Peninjauh.
This is a very steep pyramidal mountain of crystalline basaltic

POLYALTHEA.   STRANGE FOREST TREE.   TREE-FERN.

rock, about a thousand feet high, and covered with luxuriant
forest. There are three Dyak villages upon it, and on a little
platform near the summit is the rude wooden lodge where the

English Rajah was accustomed to go for relaxation and cool fresh air. It is only twenty miles up the river, but the road up the mountain is a succession of ladders on the face of precipices, bamboo bridges over gullies and chasms, and slippery paths over rocks and tree-trunks and huge boulders as big as houses. A cool spring under an overhanging rock just below the cottage furnished us with refreshing baths and delicious drinking water, and the Dyaks brought us daily heaped-up baskets of Mangusteens and Lansats, two of the most delicious of the subacid tropical fruits. We returned to Saráwak for Christmas (the second I had spent with Sir James Brooke), when all the Europeans both in the town and from the out-stations enjoyed the hospitality of the Rajah, who possessed in a pre-eminent degree the art of making every one around him comfortable and happy.

A few days afterwards I returned to the mountain with Charles and a Malay boy named Ali, and stayed there three weeks for the purpose of making a collection of land-shells, butterflies and moths, ferns and orchids. On the hill itself ferns were tolerably plentiful, and I made a collection of about forty species. But what occupied me most was the great abundance of moths which on certain occasions I was able to capture. As during the whole of my eight years' wanderings in the East I never found another spot where these insects were at all plentiful, it will be interesting to state the exact conditions under which I here obtained them.

On one side of the cottage there was a verandah, looking down the whole side of the mountain and to its summit on the right, all densely clothed with forest. The boarded sides of the cottage were whitewashed, and the roof of the verandah was low, and also boarded and whitewashed. As soon as it got dark I placed my lamp on a table against the wall, and with pins, insect-forceps, net, and collecting-boxes by my side, sat down with a book. Sometimes during the whole evening only one solitary moth would visit me, while on other nights they would pour in, in a continual stream, keeping me hard at work catching and pinning till past midnight. They came literally by thousands. These good nights were very few. During the four weeks that I spent altogether on the hill I only had four really good nights, and these were always rainy, and the best of them soaking wet. But wet nights were not always good, for a rainy moonlight night produced next to nothing. All the chief tribes of moths were represented, and the beauty and variety of the species was very great. On good nights I was able to capture from a hundred to two hundred and fifty moths, and these comprised on each occasion from half to two-thirds that number of distinct species. Some of them would settle on the wall, some on the table, while many would fly up to the roof and give me a chase all over the verandah before I could secure them. In order to show the curious connexion between the state of the weather and the degree in which moths were attracted to light, I add a list of my captures each night of my stay on the hill.

Date.	No. of Moths.	Remarks.
1855.		
Dec. 13th	1	Fine ; starlight.
,, 14th	75	Drizzly and fog.
,, 15th	41	Showery ; cloudy.
,, 16th	158	(120 species.) Steady rain.
,, 17th	82	Wet ; rather moonlight.
,, 18th	9	Fine ; moonlight.
,, 19th	2	Fine ; clear moonlight.
,, 31st	200	(130 species.) Dark and windy; heavy rain.
1856.		
Jan. 1st	185	Very wet.
,, 2nd	68	Cloudy and showers.
,, 3rd	50	Cloudy.
,, 4th	12	Fine.
,, 5th	10	Fine.
,, 6th	8	Very fine.
,, 7th	8	Very fine.
,, 8th	10	Fine.
,, 9th	36	Showery.
,, 10th	30	Showery.
,, 11th	260	Heavy rain all night, and dark.
,, 12th	56	Showery.
,, 13th	44	Showery ; some moonlight.
,, 14th	4	Fine ; moonlight.
,, 15th	24	Rain ; moonlight.
,, 16th	6	Showers ; moonlight.
,, 17th	6	Showers ; moonlight.
,, 18th	1	Showers ; moonlight.
Total . .	1,386	

It thus appears that on twenty-six nights I collected 1,386 moths, but that more than 800 of them were collected on four very wet and dark nights. My success here led me to hope that, by similar arrangements, I might in every island be able to obtain abundance of these insects; but, strange to say, during the six succeeding years I was never once able to make any collections at all approaching those at Saráwak. The reason of this I can pretty well understand to be owing to the absence of some one or other essential condition that were here all combined. Sometimes the dry season was the hindrance; more frequently residence in a town or village not close to virgin forest, and surrounded by other houses whose lights were a counter-attraction ; still more frequently residence in a dark palm-thatched house, with a lofty roof, in whose recesses every moth was lost the instant it entered. This last was the greatest drawback, and the real reason why I never again was able to make a collection of moths ; for I never afterwards lived in a solitary jungle-house with a low-boarded and whitewashed verandah, so constructed as to prevent insects at once escaping into the upper part of the house, quite out of reach. After my long experience, my numerous failures, and my one success, I feel sure that if any party of naturalists ever make a yacht-voyage to explore the Malayan Archipelago, or any other tropical region, making entomology one of their chief pursuits, it would well repay them to carry a small framed verandah, or a verandah-shaped tent of white canvas, to set up in every favourable situation, as a means of making a collection of noc-turnal Lepidoptera, and also of obtaining rare specimens of Coleoptera and other insects. I make the suggestion here, be-cause no one would suspect the enormous difference in results that such an apparatus would produce ; and because I consider it one of the curiosities of a collector's experience to have found out that some such apparatus is required.

When I returned to Singapore I took with me the Malay lad named Ali, who subsequently accompanied me all over the Archipelago. Charles Allen preferred staying at the Mission-house, and afterwards obtained employment in Sarawak and in Singapore, till he again joined me four years later at Amboyna in the Moluccas.

CHAPTER VI.

BORNEO—THE DYAKS.

THE manners and customs of the aborigines of Borneo have been described in great detail, and with much fuller information than I possess, in the writings of Sir James Brooke, Messrs. Low, St. John, Johnson Brooke, and many others. I do not propose to go over the ground again, but shall confine myself to a sketch, from personal observation, of the general character of the Dyaks, and of such physical, moral, and social characteristics as have been less frequently noticed.

The Dyak is closely allied to the Malay, and more remotely to the Siamese, Chinese, and other Mongol races. All these are characterized by a reddish-brown or yellowish-brown skin of various shades, by jet-black straight hair, by the scanty or de-ficient beard, by the rather small and broad nose, and high cheekbones ; but none of the Malayan races have the oblique eyes which are characteristic of the more typical Mongols. The average stature of the Dyaks is rather more than that of the Malays, while it is considerably under that of most Europeans.

Their forms are well proportioned, their feet and hands small, and they rarely or never attain the bulk of body so often seen in Malays and Chinese.

I am inclined to rank the Dyaks above the Malays in mental capacity, while in moral character they are undoubtedly superior to them. They are simple and honest, and become the prey of the Malay and Chinese traders, who cheat and plunder them continually. They are more lively, more talkative, less secretive, and less suspicious than the Malay, and are therefore pleasanter companions. The Malay boys have little inclination for active sports and games, which form quite a feature in the life of the Dyak youths, who, besides outdoor games of skill and strength, possess a variety of indoor amusements. One wet day, in a Dyak house, when a number of boys and young men were about me, I thought to amuse them with something new, and showed them how to make "cat's cradle" with a piece of string. Greatly to my surprise, they knew all about it, and more than I did ; for, after I and Charles had gone through all the changes we could make, one of the boys took it off my hand, and made several new figures which quite puzzled me. They then showed me a number of other tricks with pieces of string, which seemed a favourite amusement with them.

Even these apparently trifling matters may assist us to form a truer estimate of the Dyaks' character and social condition. We learn thereby, that these people have passed beyond that first stage of savage life in which the struggle for existence absorbs the whole faculties, and in which every thought and idea is connected with war or hunting, or the provision for their immediate necessities. These amusements indicate a capability of civilization, an aptitude to enjoy other than mere sensual pleasures, which might be taken advantage of to elevate their whole intellectual and social life.

The moral character of the Dyaks is undoubtedly high—a statement which will seem strange to those who have heard of them only as head-hunters and pirates. The Hill Dyaks of whom I am speaking, however, have never been pirates, since they never go near the sea ; and head-hunting is a custom originating in the petty wars of village with village, and tribe with tribe, which no more implies a bad moral character than did the custom of the slave-trade a hundred years ago imply want of general morality in all who participated in it. Against this one stain on their character (which in the case of the Sarawak Dyaks no longer exists) we have to set many good points. They are truthful and honest to a remarkable degree. From this cause it is very often impossible to get from them any definite information, or even an opinion. They say, "If I were to tell you what I don't know, I might tell a lie ;" and whenever they voluntarily relate any matter of fact, you may be sure they are speaking the truth. In a Dyak village the fruit trees have each their owner, and it has often happened to me, on asking an inhabitant to gather me some fruit, to be answered, " I can't do that, for the owner of the tree is not here ; " never seeming to contemplate the possibility of acting otherwise. Neither will they take the smallest thing belonging to an European. When living at Simunjon, they continually came to my house, and would pick up scraps of torn newspaper or crooked pins that I had thrown away, and ask as a great favour whether they might have them. Crimes of violence (other than head-hunting) are almost unknown ; for in twelve years, under Sir James Brooke's rule, there had been only one case of murder in a Dyak tribe, and that one was committed by a stranger who had been adopted into the tribe. In several other matters of morality they rank above most uncivilized, and even above many civilized nations. They are temperate in food and drink, and the gross sensuality of the Chinese and Malays is unknown among them. They have the usual fault of all people in a half-savage state—apathy and dilatoriness ; but, however annoying this may be to Europeans who come in contact with them, it cannot be considered a very grave offence, or be held to outweigh their many excellent qualities.

During my residence among the Hill Dyaks, I was much struck by the apparent absence of those causes which are generally supposed to check the increase of population, although there were plain indications of stationary or but slowly increasing numbers. The conditions most favourable to a rapid increase of population are, an abundance of food, a healthy climate, and early marriages. Here these conditions all exist. The people produce far more food than they consume, and exchange the surplus for gongs and brass cannon, ancient jars, and gold and silver ornaments, which constitute their wealth. On the whole, they appear very free from disease, marriages take place early (but not too early), and old bachelors and old maids are alike unknown. Why, then, we must inquire, has not a greater population been produced ? Why are the Dyak villages so small and so widely scattered, while nine-tenths of the country is still covered with forest ?

Of all the checks to population among savage nations mentioned by Malthus—starvation, disease, war, infanticide, immorality, and infertility of the women—the last is that which he seems to think least important, and of doubtful efficacy ; and yet it is the only one that seems to me capable of accounting for the state of the population among the Sarawak Dyaks. The population of Great Britain increases so as to double itself in about fifty years. To do this it is evident that each married couple must average three children who live to be married at the age of about twenty-five. Add to these those who die in infancy, those who never marry, or those who marry late in life and have no offspring, the number of children born to each marriage must average four or five ; and we know that families of seven or eight are very common, and of ten and twelve by no means rare. But from inquiries at almost every Dyak tribe I visited, I ascertained that the women rarely had more than three or four children, and an old chief assured me that he had never known a woman have more than seven. In a village consisting of a hundred and fifty families, only one consisted of six children living, and only six of five children, the majority appearing to be two, three, or four. Comparing this with the known proportions in European countries, it is evident that the number of children to each marriage can hardly average more than three or four; and as even in civilized countries half the population die before the age of twenty-five, we should have only two left to replace their parents; and so long as this state of things continued, the population must remain stationary. Of course this is a mere illustration; but the facts I have stated seem to indicate that something of the kind really takes place; and if so, there is no difficulty in understanding the smallness and almost stationary population of the Dyak tribes.

We have next to inquire what is the cause of the small number of births and of living children in a family. Climate and race may have something to do with this, but a more real and efficient cause seems to me to be the hard labour of the women, and the heavy weights they constantly carry. A Dyak woman generally spends the whole day in the field, and carries home every night a heavy load of vegetables and firewood, often for several miles, over rough and hilly paths; and not unfrequently has to climb up a rocky mountain by ladders, and over slippery stepping-stones, to an elevation of a thousand feet. Besides this, she has an hour's work every evening to pound the rice with a heavy wooden stamper, which violently strains every part of the body. She begins this kind of labour when nine or ten years old, and it never ceases but with the extreme decrepitude of age. Surely we need not wonder at the limited number of her progeny, but rather be surprised at the successful efforts of nature to prevent the extermination of the race.

One of the surest and most beneficial effects of advancing civilization, will be the amelioration of the condition of these women. The precept and example of higher races will make the Dyak ashamed of his comparatively idle life, while his weaker partner labours like a beast of burthen. As his wants become increased and his taste refined, the women will have more household duties to attend to, and will then cease to labour in the field—a change which has already to a great extent taken place in the allied Malay, Javanese, and Bugis tribes. Population will then certainly increase more rapidly, improved systems of agriculture and some division of labour will become necessary in order to provide the means of existence, and a more complicated social state will take the place of the simple conditions of society which now obtain among them. But, with the sharper struggle for existence that will then occur, will the happiness of the people as a whole be increased or diminished? Will not evil passions be aroused by the spirit of competition, and crimes and vices, now unknown or dormant, be called into active existence? These are problems that time alone can solve; but it is to be hoped that education and a high-class European example may obviate much of the evil that too often arises in analogous cases, and that we may at length be able to point to one instance of an uncivilized people who have not become demoralized and finally exterminated, by contact with European civilization.

A few words in conclusion, about the government of Saráwak. Sir James Brooke found the Dyaks oppressed and ground down by the most cruel tyranny. They were cheated by the Malay traders, and robbed by the Malay chiefs. Their wives and children were often captured and sold into slavery, and hostile tribes purchased permission from their cruel rulers to plunder, enslave, and murder them. Anything like justice or redress for these injuries was utterly unattainable. From the time Sir James obtained possession of the country, all this was stopped. Equal justice was awarded to Malay, Chinaman, and Dyak. The remorseless pirates from the rivers farther east were punished, and finally shut up within their own territories, and the Dyak, for the first time, could sleep in peace. His wife and children were now safe from slavery; his house was no longer burnt over his head; his crops and his fruits were now his own, to sell or consume as he pleased. And the unknown stranger who had done all this for them, and asked for nothing in return, what could he be? How was it possible for them to realize his motives? Was it not natural that they should refuse to believe he was a man? for of pure benevolence combined with great power, they had had no experience among men. They naturally concluded that he was a superior being, come down upon earth to confer blessings on the afflicted. In many villages where he had not been seen, I was asked strange questions about him. Was he not as old as the mountains? Could he not bring the dead to life? And they firmly believe that he can give them good harvests, and make their fruit-trees bear an abundant crop.

In forming a proper estimate of Sir James Brooke's government, it must ever be remembered that he held Saráwak solely by the goodwill of the native inhabitants. He had to deal with two races, one of whom, the Mahometan Malays, looked upon the other race, the Dyaks, as savages and slaves, only fit to be robbed and plundered. He has effectually protected the Dyaks, and has invariably treated them as, in his sight, equal to the Malays; and yet he has secured the affection and goodwill of both. Notwithstanding the religious prejudices of Mahometans, he has induced them to modify many of their worst laws and customs, and to assimilate their criminal code to that of the civilized world. That his government still continues, after twenty-seven years—notwithstanding his frequent absences from ill-health, notwithstanding conspiracies of Malay chiefs, and insurrections of Chinese gold-diggers, all of which have been overcome by the support of the native population, and notwithstanding financial, political, and domestic troubles—is due, I believe, solely to the many admirable qualities which Sir James Brooke possessed, and especially to his having convinced the native population, by every action of his life, that he ruled them, not for his own advantage, but for their good.

Since these lines were written, his noble spirit has passed away. But though, by those who knew him not, he may be sneered at as an enthusiastic adventurer, or abused as a hard-hearted despot, the universal testimony of every one who came in contact with him in his adopted country, whether European, Malay, or Dyak, will be, that Rajah Brooke was a great, a wise, and a good ruler—a true and faithful friend—a man to be admired for his talents, respected for his honesty and courage, and loved for his genuine hospitality, his kindness of disposition, and his tenderness of heart.[1]

CHAPTER VII.

JAVA.

I SPENT three months and a half in Java, from July 18th to October 31st, 1861, and shall briefly describe my own movements, and my observations on the people and the natural history of the country. To all those who wish to understand how the Dutch now govern Java, and how it is that they are enabled to derive a large annual revenue from it, while the population increases, and the inhabitants are contented, I recommend the study of Mr. Money's excellent and interesting work, *How to Manage a Colony*. The main facts and conclusions of that work I most heartily concur in, and I believe that the Dutch system is the very best that can be adopted, when a European nation conquers or otherwise acquires possession of a country inhabited by an industrious but semi-barbarous people. In my account of Northern Celebes, I shall show how successfully the same system has been applied to a people in a very different state of civilization from the Javanese ; and in the meanwhile will state in the fewest words possible what that system is.

The mode of government now adopted in Java is to retain the whole series of native rulers, from the village chief up to princes, who, under the name of Regents, are the heads of districts about the size of a small English county. With each Regent is

---

[1] The present Rajah, Charles Johnson Brooke, nephew of Sir James, seems to have continued the government in the spirit of its founder. Its territories have been extended by friendly arrangement with the Sultan of Bruni so as to include the larger part of the north-west district of Borneo, and peace and prosperity have everywhere been maintained. Fifty years of government of alien and antagonistic races, with their own consent, and with the continued support of the native chiefs, is a success of which the friends and countrymen of Sir James Brooke may well be proud.

placed a Dutch Resident, or Assistant Resident, who is considered to be his "elder brother," and whose "orders" take the form of "recommendations," which are however implicitly obeyed. Along with each Assistant Resident is a Controller, a kind of inspector of all the lower native rulers, who periodically visits every village in the district, examines the proceedings of the native courts, hears complaints against the head-men or other native chiefs, and superintends the Government plantations. This brings us to the "culture system," which is the source of all the wealth the Dutch derive from Java, and is the subject of much abuse in this country because it is the reverse of "free trade." To understand its uses and beneficial effects, it is necessary first to sketch the common results of free European trade with uncivilized peoples.

Natives of tropical climates have few wants, and, when these are supplied, are disinclined to work for superfluities without some strong incitement. With such a people the introduction of any new or systematic cultivation is almost impossible, except by the despotic orders of chiefs whom they have been accustomed to obey, as children obey their parents. The free competition of European traders, however, introduces two powerful inducements to exertion. Spirits or opium is a temptation too strong for most savages to resist, and to obtain these he will sell whatever he has, and will work to get more. Another temptation he cannot resist, is goods on credit. The trader offers him gay cloths, knives, gongs, guns, and gunpowder, to be paid for by some crop perhaps not yet planted, or some product yet in the forest. He has not sufficient forethought to take only a moderate quantity, and not enough energy to work early and late in order to get out of debt ; and the consequence is that he accumulates debt upon debt, and often remains for years, or for life, a debtor and almost a slave. This is a state of things which occurs very largely in every part of the world in which men of a superior race freely trade with men of a lower race. It extends trade no doubt for a time, but it demoralizes the native, checks true civilization, and does not lead to any permanent increase in the wealth of the country ; so that the European government of such a country must be carried on at a loss.

The system introduced by the Dutch was to induce the people, through their chiefs, to give a portion of their time to the cultivation of coffee, sugar, and other valuable products. A fixed rate of wages—low indeed, but about equal to that of all places where European competition has not artificially raised it —was paid to the labourers engaged in clearing the ground and forming the plantations under Government superintendence The produce is sold to the Government at a low fixed price. Out of the net profits a percentage goes to the chiefs, and the remainder is divided among the workmen. This surplus in good years is something considerable. On the whole, the people are well fed and decently clothed ; and have acquired habits of steady industry and the art of scientific cultivation, which must be of service to them in the future. It must be remembered, that the Government expended capital for years before any return was obtained ; and if they now derive a large revenue, it is in a way which is far less burthensome, and far more beneficial to the people, than any tax that could be levied.

But although the system may be a good one, and as well adapted to the development of arts and industry in a half-civilized people, as it is to the material advantage of the governing country, it is not pretended that in practice it is perfectly carried out. The oppressive and servile relations between chiefs and people, which have continued for perhaps a thousand years, cannot be at once abolished ; and some evil must result from those relations, till the spread of education and the gradual infusion of European blood causes it naturally and insensibly to disappear. It is said that the Residents, desirous of showing a large increase in the products of their districts, have sometimes pressed the people to such continued labour on the plantations that their rice crops have been materially diminished, and famine has been the result. If this has happened, it is certainly not a common thing, and is to be set down to the abuse of the system, by the want of judgment or want of humanity in the Resident.

A tale has lately been written in Holland, and translated into English, entitled *Max Havelaar ; or, The Coffee Auctions of the Dutch Trading Company*, and with our usual one-sidedness in all relating to the Dutch Colonial System, this work has been excessively praised, both for its own merits, and for its supposed crushing exposure of the iniquities of the Dutch government of Java. Greatly to my surprise, I found it a very tedious and long-winded story, full of rambling digressions ; and whose only point is to show that the Dutch Residents and Assistant Residents wink at the extortions of the native princes ; and that in some districts the natives have to do work without payment, and have their goods taken away from them without compensation. Every statement of this kind is thickly interspersed with italics and capital letters ; but as the names are all fictitious, and neither dates, figures, nor details are ever given, it is impossible to verify or answer them. Even if not exaggerated, the facts stated are not nearly so bad as those of the oppression by free-trade indigo-planters, and torturing by native tax-gatherers under British rule in India, with which the readers of English newspapers were familiar a few years ago. Such oppression, however, is not fairly to be imputed in either case to the particular form of government, but is rather due to the infirmity of human nature, and to the impossibility of at once destroying all trace of ages of despotism on the one side, and of slavish obedience to their chiefs on the other.

It must be remembered, that the complete establishment of the Dutch power in Java is much more recent than that of our rule in India, and that there have been several changes of government, and in the mode of raising revenue. The inhabitants have been so recently under the rule of their native princes, that it is not easy at once to destroy the excessive reverence they feel for their old masters, or to diminish the oppressive exactions which the latter have always been accustomed to make. There is, however, one grand test of the prosperity, and even of the happiness, of a community, which we can apply here—the rate of increase of the population.

It is universally admitted, that when a country increases rapidly in population, the people cannot be very greatly oppressed or very badly governed. The present system of raising a revenue by the cultivation of coffee and sugar, sold to Government at a fixed price, began in 1832. Just before this, in 1826, the population by census was 5,500,000, while at the beginning of the century it was estimated at 3,500,000. In 1850, when the cultivation system had been in operation eighteen years, the population by census was over 9,500,000, or an increase of seventy-three per cent. in twenty-four years. At the last census, in 1865, it amounted to 14,168,416, an increase of very nearly fifty per cent. in fifteen years—a rate which would double the population in about twenty-six years. As Java (with Madura) contains about 38,500 geographical square miles, this will give an average of 368 persons to the square mile, just double that of the populous and fertile Bengal Presidency as given in Thornton's *Gazetteer of India*, and fully one-third more than that of Great Britain and Ireland at the last census. If, as I believe, this vast population is on the whole contented and happy, the Dutch Government should consider well before abruptly changing a system which has led to such great results.[1]

Taking it as a whole, and surveying it from every point of view, Java is probably the very finest and most interesting tropical island in the world. It is not first in size, but it is more than 600 miles long, and from sixty to 120 miles wide, and in area is nearly equal to England; and it is undoubtedly the most fertile, the most productive, and the most populous island within the tropics. Its whole surface is magnificently varied with mountain and forest scenery. It possesses thirty-eight volcanic mountains, several of which rise to ten or twelve thousand feet high. Some of these are in constant activity, and one or other of them displays almost every phenomenon produced by the action of subterranean fires, except regular lava streams, which never occur in Java. The abundant moisture and tropical heat of the climate causes these mountains to be clothed with luxuriant vegetation, often to their very summits, while forests and plantations cover their lower slopes. The animal productions, especially the birds and insects, are

[1] In 1879 the population had still further increased to over nineteen millions, and in 1894 to twenty-five millions.

beautiful and varied, and present many peculiar forms found nowhere else upon the globe. The soil throughout the island is exceedingly fertile, and all the productions of the tropics, together with many of the temperate zones, can be easily culti- vated. Java too possesses a civilization, a history and anti- quities of its own, of great interest. The Brahminical religion flourished in it from an epoch of unknown antiquity till about the year 1478, when that of Mahomet superseded it. The former religion was accompanied by a civilization which has not been equalled by the conquerors ; for, scattered through the country, especially in the eastern part of it, are found buried in lofty forests, temples, tombs, and statues of great beauty and grandeur ; and the remains of extensive cities, where the tiger, the rhinoceros, and the wild bull now roam undisturbed. A modern civilization of another type is now spreading over the land. Good roads run through the country from end to end ; European and native rulers work harmoniously together; and life and property are as well secured as in the best governed states of Europe. I believe, therefore, that Java may fairly claim to be the finest tropical island in the world, and equally interesting to the tourist seeking after new and beautiful scenes ; to the naturalist who desires to examine the variety and beauty of tropical nature ; or to the moralist and the politician who want to solve the problem of how man may be best governed under new and varied conditions.

The Dutch mail steamer brought me from Ternate to Soura- baya, the chief town and port in the eastern part of Java, and after a fortnight spent in packing up and sending off my last collections, I started on a short journey into the interior. Travelling in Java is very luxurious but very expensive, the only way being to hire or borrow a carriage, and then pay half- a-crown a mile for post-horses, which are changed at regular posts every six miles, and will carry you at the rate of ten miles an hour from one end of the island to the other. Bullock carts or coolies are required to carry all extra baggage. As this kind of travelling would not suit my means, I determined on making only a short journey to the district at the foot of Mount Arjuna, where I was told there were extensive forests, and where I hoped to be able to make some good collections. The country for many miles behind Sourabaya is perfectly flat and everywhere culti- vated, being a delta or alluvial plain watered by many branch- ing streams. Immediately around the town the evident signs of wealth and of an industrious population were very pleasing ; but as we went on, the constant succession of open fields skirted by rows of bamboos, with here and there the white buildings and tall chimney of a sugar-mill, became monotonous. The roads run in straight lines for several miles at a stretch, and are bordered by rows of dusty tamarind-trees. At each mile there are little guard-houses, where a policeman is stationed ; and there is a wooden gong, which by means of concerted signals may be made to convey information over the country with great rapidity. About every six or seven miles is the post-house, where the horses are changed as quickly as were those of the mail in the old coaching days in England.

I stopped at Modjo-kerto, a small town about forty miles south of Sourabaya, and the nearest point on the high road to the district I wished to visit. I had a letter of introduction to Mr. Ball, an Englishman long resident in Java and married to a Dutch lady, and he kindly invited me to stay with him till I could fix on a place to suit me. A Dutch Assistant Resident as well as a Regent or native Javanese prince lived here. The town was neat, and had a nice open grassy space like a village green, on which stood a magnificent fig-tree (allied to the Banyan of India, but more lofty), under whose shade a kind of market is continually held, and where the inhabitants meet together to lounge and chat. The day after my arrival, Mr. Ball drove me over to the village of Modjo-agong, where he was building a house and premises for the tobacco trade, which is carried on here by a system of native cultivation and advance purchase, somewhat similar to the indigo trade in British India. On our way we stayed to look at a fragment of the ruins of the ancient city of Modjo-pahit, consisting of two lofty brick masses, apparently the sides of a gateway. The extreme perfection and beauty of the brickwork astonished me. The bricks are exceedingly fine and hard, with sharp angles and true surfaces. They are laid with great exactness, without visible mortar or cement, yet somehow fastened together so that the joints are hardly perceptible, and sometimes the two surfaces coalesce in a most incomprehensible manner. Such admirable brickwork I have never seen before or since. There was no sculpture here, but abundance of bold projections and finely-worked mouldings. Traces of buildings exist for many miles in every direction, and almost every road and pathway shows a foundation of brickwork beneath it—the paved roads of the old city. In the house of the Waidono or district chief at Modjo-agong, I saw a beautiful figure carved in high relief out of a block of lava, and which had been found buried in the ground near the village. On my expressing a wish to obtain some such specimen, Mr. B. asked the chief for it, and much to my surprise he immediately gave it me. It represented the Hindoo goddess Durga, called in Java, Lora Jonggrang (the exalted virgin). She has eight arms, and stands on the back of a kneeling bull. Her lower right hand holds the tail of the bull, while the corresponding left hand grasps the hair of a captive, Dewth Mahikusor, the personification of vice, who has attempted to slay her bull. He has a cord round his waist, and crouches at her feet in an attitude of supplication. The other hands of the goddess hold, on her right side, a double hook or small anchor, a broad straight sword, and a noose of thick cord ; on her left, a girdle or armlet of large beads or shells, an unstrung bow, and a standard or war flag. This deity was a special favourite among the old Javanese, and her image is often found in the ruined temples which abound in the eastern part of the island.

The specimen I had obtained was a small one, about two feet high, weighing perhaps a hundredweight; and the next day we had it conveyed to Modjo-kerto to await my return to Sourabaya. Having decided to stay some time at Wonosalem, on the lower

ANCIENT BAS-RELIEF.

slopes of the Arjuna Mountain, where I was informed I should find forest and plenty of game, I had first to obtain a recommendation from the Assistant Resident to the Regent, and then an order from the Regent to the Waidono; and when after a week's delay I arrived with my baggage and men at Modjo-agong, I found them all in the midst of a five days' feast, to celebrate the circumcision of the Waidono's younger brother and cousin, and had a small room in an outhouse given me to stay in.

The courtyard and the great open reception-shed were full of natives coming and going and making preparations for a feast which was to take place at midnight, to which I was invited, but preferred going to bed. A native band, or Gamelang, was playing almost all the evening, and I had a good opportunity of seeing the instruments and musicians. The former are chiefly gongs of various sizes, arranged in sets of from eight to twelve, on low wooden frames. Each set is played by one performer with one or two drumsticks. There are also some very large gongs, played singly or in pairs, and taking the place of our drums and kettledrums. Other instruments are formed by broad metallic bars, supported on strings stretched across frames; and others again of strips of bamboo similarly placed and producing the highest notes. Besides these there were a flute and a curious two-stringed violin, requiring in all twenty-four performers. There was a conductor, who led off and regulated the time, and each performer took his part, coming in occasionally with a few bars so as to form a harmonious combination. The pieces played were long and complicated, and some of the players were mere boys, who took their parts with great precision. The general effect was very pleasing, but, owing to the similarity of most of the instruments, more like a gigantic musical box than one of our bands; and in order to enjoy it thoroughly it is necessary to watch the large number of performers who are engaged in it. The next morning, while I was waiting for the men and horses who were to take me and my baggage to my destination, the two lads, who were about fourteen years old, were brought out, clothed in a sarong from the waist downwards, and having the whole body covered with a yellow powder, and profusely decked with white blossoms in wreaths, necklaces, and armlets, looking at first sight very like savage brides. They were conducted by two priests to a bench placed in front of the house in the open air, and the ceremony of circumcision was then performed before the assembled crowd.

The road to Wonosalem led through a magnificent forest, in the depths of which we passed a fine ruin of what appeared to have been a royal tomb or mausoleum. It is formed entirely of stone, and elaborately carved. Near the base is a course of boldly projecting blocks, sculptured in high relief, with a series of scenes which are probably incidents in the life of the defunct. These are all beautifully executed, some of the figures of animals in particular being easily recognizable and very accurate. The general design, as far as the ruined state of the upper part will permit of its being seen, is very good, effect being given by an immense number and variety of projecting or retreating courses of squared stones in place of mouldings. The size of this structure is about thirty feet square by twenty high, and as the traveller comes suddenly upon it on a small elevation by the roadside, overshadowed by gigantic trees, overrun with plants and creepers, and closely backed by the gloomy forest, he is struck by the solemnity and picturesque beauty of the scene, and is led to ponder on the strange law of progress, which looks so like retrogression, and which in so many distant parts of the world has exterminated or driven out a highly artistic and constructive race, to make room for one which, as far as we can judge, is very far its inferior.

Few Englishmen are aware of the number and beauty of the architectural remains in Java. They have never been popularly illustrated or described, and it will therefore take most persons by surprise to learn that they far surpass those of Central America, perhaps even those of India. To give some idea of these ruins, and perchance to excite wealthy amateurs to explore them thoroughly and obtain by photography an accurate record of their beautiful sculptures before it is too late, I will enumerate the most important, as briefly described in Sir Stamford Raffles' *History of Java.*

BRAMBANAM.—Near the centre of Java, between the native capitals of Djoko-kerta and Surakerta, is the village of Brambanam, near which are abundance of ruins, the most important being the temples of Loro-Jongran and Chandi Sewa. At Loro-Jongran there were twenty separate buildings, six large and fourteen small temples. They are now a mass of ruins, but the largest temples are supposed to have been ninety feet high. They were all constructed of solid stone, everywhere decorated with carvings and bas-reliefs, and adorned with numbers of statues, many of which still remain entire. At Chandi Sewa, or the "Thousand Temples," are many fine colossal figures. Captain Baker, who surveyed these ruins, said he had never in his life seen "such stupendous and finished specimens of human labour, and of the science and taste of ages long since forgot, crowded together in so small a compass as in this spot." They cover a space of nearly six hundred feet square, and consist of an outer row of eighty-four small temples, a second row of seventy-six, a third of sixty-four, a fourth of forty-four, and the fifth forming an inner parallelogram of twenty-eight; in all two hundred and ninety-six small temples, disposed in five regular parallelograms. In the centre is a large cruciform temple surrounded by lofty flights of steps richly ornamented with sculpture, and containing many apartments. The tropical vegetation has ruined most of the smaller temples, but some remain tolerably perfect, from which the effect of the whole may be imagined.

About half a mile off is another temple called Chandi Kali Bening, seventy-two feet square and sixty feet high, in very fine preservation, and covered with sculptures of Hindoo mythology surpassing any that exist in India. Other ruins of palaces, halls and temples, with abundance of sculptured deities, are found in the same neighbourhood.

BOROBODO.—About eighty miles westward, in the province of Kedu, is the great temple of Borobodo. It is built upon a small hill, and consists of a central dome and seven ranges of terraced walls covering the slope of the hill and forming open galleries each below the other, and communicating by steps and gateways. The central dome is fifty feet in diameter ; around it is a triple circle of seventy-two towers, and the whole building is six hundred and twenty feet square, and about one hundred feet high. In the terrace walls are niches containing cross-legged figures larger than life to the number of about four hundred, and both sides of all the terrace walls are covered with bas-reliefs crowded with figures, and carved in hard stone ; and which must therefore occupy an extent of nearly three miles in length ! The amount of human labour and skill expended on the Great Pyramids of Egypt sinks into insignificance when compared with that required to complete this sculptured hill-temple in the interior of Java.

GUNONG PRAU.—About forty miles south-west of Samarang, on a mountain called Gunong Prau, an extensive plateau is covered with ruins. To reach these temples four flights of stone steps were made up the mountain from opposite directions, each flight consisting of more than a thousand steps. Traces of nearly four hundred temples have been found here, and many (perhaps all) were decorated with rich and delicate sculptures. The whole country between this and Brambanam, a distance of sixty miles, abounds with ruins ; so that fine sculptured images may be seen lying in the ditches, or built into the walls of enclosures.

In the eastern part of Java, at Kediri and in Malang, there are equally abundant traces of antiquity, but the buildings themselves have been mostly destroyed. Sculptured figures, however, abound ; and the ruins of forts, palaces, baths, aqueducts and temples, can be everywhere traced. It is altogether contrary to the plan of this book to describe what I have not myself seen ; but, having been led to mention them, I felt bound to do something to call attention to these marvellous works of art. One is overwhelmed by the contemplation of these innumerable sculptures, worked with delicacy and artistic feeling in a hard, intractable, trachytic rock, and all found in one tropical island. What could have been the state of society, what the amount of population, what the means of subsistence which rendered such gigantic works possible, will, perhaps, ever remain a mystery ; and it is a wonderful example of the power of religious ideas in social life, that in the very country where, five hundred years ago, these grand works were being yearly executed, the inhabitants now only build rude houses of bamboo and thatch, and look upon these relics of their forefathers with ignorant amazement, as the undoubted productions of giants or of demons. It is much to be regretted that the Dutch Government do not take vigorous steps for the preservation of these ruins from the destroying agency of tropical vegetation ; and for the collection of the fine sculptures which are everywhere scattered over the land.

Wonosalem is situated about a thousand feet above the sea, but unfortunately it is at a distance from the forest, and is surrounded by coffee-plantations, thickets of bamboo, and coarse grasses. It was too far to walk back daily to the forest, and in other directions I could find no collecting ground for insects. The place was, however, famous for peacocks, and my boy soon shot several of these magnificent birds, whose flesh we found to be tender, white, and delicate, and similar to that of a turkey. The Java peacock is a different species from that of India, the neck being covered with scale-like green feathers, and the crest of a different form ; but the eyed train is equally large and equally beautiful. It is a singular fact in geographical distribution that the peacock should not be found in Sumatra or Borneo, while the superb Argus, Fire-backed, and Ocellated pheasants of those islands are equally unknown in Java. Exactly parallel is the fact that in Ceylon and Southern India, where the peacock abounds, there are none of the splendid Lophophori and other gorgeous pheasants which inhabit Northern India. It would seem as if the peacock can admit of no rivals in its domain. Were these birds rare in their native country, and unknown alive in Europe, they would assuredly be considered as the true princes of the feathered tribes, and altogether unrivalled for stateliness and beauty. As it is, I suppose scarcely any one if asked to fix upon the most beautiful bird in the world would name the peacock, any more than the Papuan savage or the Bugis trader would fix upon the bird of paradise for the same honour.

Three days after my arrival at Wonosalem, my friend Mr. Ball came to pay me a visit. He told me that two evenings before, a boy had been killed and eaten by a tiger close to Modjo-agong. He was riding on a cart drawn by bullocks, and was coming home about dusk on the main road ; and when not half a mile from the village a tiger sprang upon him, carried him off into the jungle close by, and devoured him. Next morning his remains were discovered, consisting only of a few mangled bones. The Waidono had got together about seven hundred men, and was in chase of the animal, which, I afterwards heard, they found and killed. They only use spears when in pursuit of a tiger in this way. They surround a large tract of country, and draw gradually together till the animal is enclosed in a compact ring of armed men. When he sees there is no escape he generally makes a spring, and is received on a dozen spears, and almost instantly stabbed to death. The skin of an animal thus killed is, of course, worthless, and in this case the skull, which I had begged Mr. Ball to secure for me, was hacked to pieces to divide the teeth, which are worn as charms.

After a week at Wonosalem, I returned to the foot of the mountain, to a village named Djapannan, which was surrounded by several patches of forest, and seemed altogether pretty well suited to my pursuits. The chief of the village had prepared two small bamboo rooms on one side of his own courtyard to accommodate me, and seemed inclined to assist me as much as he could. The weather was exceedingly hot and dry, no rain having fallen for several months, and there was, in consequence, a great scarcity of insects, and especially of beetles. I therefore devoted myself chiefly to obtaining a good set of the birds, and succeeded in making a tolerable collection. All the peacocks we had hitherto shot had had short or imperfect tails, but I now obtained two magnificent specimens more than seven feet long, one of which I preserved entire, while I kept the train only attached to the tail of two or three others. When this bird is seen feeding on the ground, it appears wonderful how it can rise into the air with such a long and cumbersome train of feathers. It does so, however, with great ease, by running quickly for a short distance, and then rising obliquely ; and will fly over trees of a considerable height. I also obtained here a specimen of the rare green jungle-fowl (Gallus furcatus), whose back and neck are beautifully scaled with bronzy feathers, and whose smooth-edged oval comb is of a violet purple colour, changing to green at the base. It is also remarkable in possessing a single large wattle beneath its throat, brightly coloured in three patches of red, yellow, and blue. The common jungle-cock (Gallus bankiva) was also obtained here. It is almost exactly like a common gamecock, but the voice is different, being much shorter and more abrupt ; whence its native name is Bekéko. Six different kinds of woodpeckers and four kingfishers were found here, the fine hornbill, Buceros lunatus, more than four feet long, and the pretty little lorikeet, Loriculus pusillus, scarcely more than as many inches.

One morning, as I was preparing and arranging my speci-mens, I was told there was to be a trial ; and presently four or five men came in and squatted down on a mat under the audience-shed in the court. The chief then came in with his clerk, and sat down opposite them. Each spoke in turn, telling his own tale, and then I found out that those who first entered were the prisoner, accuser, policeman, and witness, and that the prisoner was indicated solely by having a loose piece of cord twined round his wrists, but not tied. It was a case of robbery, and after the evidence was given, and a few questions had been asked by the chief, the accused said a few words, and then sen-tence was pronounced, which was a fine. The parties then got up and walked away together, seeming quite friendly ; and throughout there was nothing in the manner of any one present indicating passion or ill-feeling—a very good illustration of the Malayan type of character.

In a month's collecting at Wonosalem and Djapannan I accu-mulated ninety-eight species of birds, but a most miserable lot of insects. I then determined to leave East Java and try the more moist and luxuriant districts at the western extremity of the island. I returned to Sourabaya by water, in a roomy boat which brought myself, servants, and baggage at one-fifth the expense it had cost me to come to Modjo-kerto. The river has been rendered navigable by being carefully banked up, but with the usual effect of rendering the adjacent country liable occasionally to severe floods. An immense traffic passes down this river; and at a lock we passed through, a mile of laden boats were waiting two or three deep, which pass through in their turn six at a time.

PORTRAIT OF JAVANESE CHIEF.

A few days afterwards I went by steamer to Batavia, where I stayed about a week at the chief hotel, while I made arrangements for a trip into the interior. The business part of the city is near the harbour, but the hotels and all the residences of the officials and European merchants are in a suburb two miles off, laid out in wide streets and squares so as to cover a great extent of ground. This is very inconvenient for visitors, as the only public conveyances are handsome two-horse carriages, whose lowest charge is five guilders (8s. 4d.) for half a day, so that an hour's business in the morning and a visit in the evening costs 16s. 8d. a day for carriage hire alone.

Batavia agrees very well with Mr. Money's graphic account of it, except that his "clear canals" were all muddy, and his "smooth gravel drives" up to the houses were one and all formed of coarse pebbles, very painful to walk upon, and hardly explained by the fact that in Batavia everybody drives, as it can hardly be supposed that people never walk in their gardens. The Hôtel des Indes was very comfortable, each visitor having a sitting-room and bedroom opening on a verandah, where he can take his morning coffee and afternoon tea. In the centre of the quadrangle is a building containing a number of marble baths always ready for use ; and there is an excellent *table d'hôte* breakfast at ten, and dinner at six, for all which there is a moderate charge per day.

I went by coach to Buitenzorg, forty miles inland and about a thousand feet above the sea, celebrated for its delicious climate and its Botanical Gardens. With the latter I was somewhat disappointed. The walks were all of loose pebbles, making any lengthened wanderings about them very tiring and painful under a tropical sun. The gardens are no doubt wonderfully rich in tropical and especially in Malayan plants, but there is a great absence of skilful laying-out ; there are not enough men to keep the place thoroughly in order, and the plants themselves are seldom to be compared for luxuriance and beauty to the same species grown in our hothouses. This can easily be explained. The plants can rarely be placed in natural or very favourable conditions. The climate is either too hot or too cool, too moist or too dry, for a large proportion of them, and they seldom get the exact quantity of shade or the right quality of soil to suit them. In our stoves these varied conditions can be supplied to each individual plant far better than in a large garden, where the fact that the plants are most of them growing in or near their native country is supposed to preclude the necessity of giving them much individual attention. Still, however, there is much to admire here. There are avenues of stately palms, and clumps of bamboos of perhaps fifty different kinds ; and an endless variety of tropical shrubs and trees with strange and beautiful foliage. As a change from the excessive heats of Batavia, Buitenzorg is a delightful abode. It is just elevated enough to have deliciously cool evenings and nights, but not so much as to require any change of clothing ; and to a person long resident in the hotter climate of the plains, the air is always fresh and pleasant, and admits of walking at almost any hour of the day. The vicinity is most picturesque and luxuriant, and the great volcano of Gunung-Salak, with its truncated and jagged summit, forms a characteristic background to many of the landscapes. A great mud eruption took place in 1699, since which date the mountain has been entirely inactive.

On leaving Buitenzorg, I had coolies to carry my baggage and a horse for myself, both to be changed every six or seven miles. The road rose gradually, and after the first stage the hills closed in a little on each side, forming a broad valley ; and the temperature was so cool and agreeable, and the country so interesting, that I preferred walking. Native villages imbedded in fruit trees, and pretty villas inhabited by planters or retired Dutch officials, gave this district a very pleasing and civilized aspect ; but what most attracted my attention was the system of terrace-cultivation, which is here universally adopted, and which is, I should think, hardly equalled in the world. The slopes of the main valley, and of its branches, were everywhere cut in terraces up to a considerable height, and when they wound round the recesses of the hills produced all the effect of magnificent amphitheatres. Hundreds of square miles of country are thus terraced, and convey a striking idea of the industry of the people and the antiquity of their civilization. These terraces are extended year by year as the population increases, by the inhabitants of each village working in concert under the direction of their chiefs ; and it is perhaps by this system of village culture alone, that such extensive terracing and irrigation has been rendered possible. It was probably introduced by the Brahmins from India, since in those Malay countries where there is no trace of a previous occupation by a civilized people, the terrace system is unknown. I first saw this mode of cultivation in Bali and Lombock, and, as I shall have to describe it in some detail there (see Chapter X.), I need say no more about it in this place, except that, owing to the finer outlines and greater luxuriance of the country in West Java, it produces there the most striking and picturesque effect. The lower slopes of the mountains in Java possess such a delightful climate and luxuriant soil ; living is so cheap and life and property are so secure, that a considerable number of Europeans who have been engaged in Government service, settle permanently in the country instead of returning to Europe. They are scattered everywhere throughout the more accessible parts of the island, and tend greatly to the gradual improvement of the native population, and to the continued peace and prosperity of the whole country.

Twenty miles beyond Buitenzorg the post road passes over the Megamendong Mountain, at an elevation of about 4,500 feet. The country is finely mountainous, and there is much virgin forest still left upon the hills, together with some of the oldest coffee-plantations in Java, where the plants have attained almost the dimensions of forest trees. About 500 feet below the summit level of the pass there is a road-keeper's hut, half of which I hired for a fortnight, as the country looked promising for making collections. I almost immediately found that the productions of West Java were remarkably different from those of the eastern part of the island ; and that all the more remarkable and characteristic Javanese birds and insects were to be found here. On the very first day, my hunters obtained for me the elegant yellow and green trogon (Harpactes Reinwardti), the gorgeous little minivet flycatcher (Pericrocotus miniatus), which looks like a flame of fire as it flutters among the bushes, and the rare and curious black and crimson oriole (Analcipus sanguinolentus), all of them species which are found only in Java, and even seem to be confined to its western portion. In a week I obtained no less than twenty-four species of birds, which I had not found in the east of the island, and in a fortnight this number increased to forty species, almost all of which

are peculiar to the Javanese fauna. Large and handsome
butterflies were also tolerably abundant. In dark ravines, and
occasionally on the roadside, I captured the superb Papilio
arjuna, whose wings seem powdered with grains of golden green,
condensed into bands and moon-shaped spots ; while the ele-
gantly-formed Papilio cöon was sometimes to be found fluttering
slowly along the shady pathways (see figure at page 99). One
day a boy brought me a butterfly between his fingers, perfectly
unhurt. He had caught it as it was sitting with wings erect

CALLIPER BUTTERFLY.

sucking up the liquid from a muddy spot by the roadside.
Many of the finest tropical butterflies have this habit, and they
are generally so intent upon their meal that they can be easily
approached and captured. It proved to be the rare and curious
Charaxes kadenii, remarkable for having on each hind wing
two curved tails like a pair of callipers. It was the only specimen I ever saw, and is still the only representative of its kind
in English collections.

In the east of Java I had suffered from the intense heat and
drought of the dry season, which had been very inimical to

insect life. Here I had got into the other extreme of damp, wet, and cloudy weather, which was equally unfavourable. During the month which I spent in the interior of West Java, I never had a really hot fine day throughout. It rained almost every afternoon, or dense mists came down from the mountains, which equally stopped collecting, and rendered it most difficult to dry my specimens, so that I really had no chance of getting a fair sample of Javanese entomology.

By far the most interesting incident in my visit to Java was a trip to the summit of the Pangerango and Gedeh mountains ; the former an extinct volcanic cone about 10,000 feet high, the latter an active crater on a lower portion of the same mountain range. Tchipanas, about four miles over the Megamendong Pass, is at the foot of the mountain. A small country house for the Governor-General and a branch of the Botanic Gardens are situated here, the keeper of which accommodated me with a bed for a night. There are many beautiful trees and shrubs planted here, and large quantities of European vegetables are grown for the Governor-General's table. By the side of a little torrent that bordered the garden, quantities of orchids were cultivated, attached to the trunks of trees, or suspended from the branches, forming an interesting open-air orchid-house. As I intended to stay two or three nights on the mountain I engaged two coolies to carry my baggage, and with my two hunters we started early the next morning. The first mile was over open country, which brought us to the forest that covers the whole mountain from a height of about 5,000 feet. The next mile or two was a tolerably steep ascent through a grand virgin forest, the trees being of great size, and the undergrowth consisting of fine herbaceous plants, tree-ferns, and shrubby vegetation. I was struck by the immense number of ferns that grew by the side of the road. Their variety seemed endless, and I was continually stopping to admire some new and interesting forms. I could now well understand what I had been told by the gardener, that 300 species had been found on this one mountain. A little before noon we reached the small plateau of Tjiburong at the foot of the steeper part of the mountain, where there is a plank-house for the accommodation of travellers. Close by is a picturesque waterfall and a curious cavern, which I had not time to explore. Continuing our ascent the road became narrow, rugged and steep, winding zigzag up the cone, which is covered with ir- regular masses of rock, and overgrown with a dense luxuriant but less lofty vegetation. We passed a torrent of water which is not much lower than the boiling point, and has a most singular appearance as it foams over its rugged bed, sending up clouds of steam, and often concealed by the overhanging herbage of ferns and lycopodia, which here thrive with more luxuriance than elsewhere.

At about 7,500 feet we came to another hut of open bamboos, at a place called Kandang Badak, or "Rhinoceros-field," which we were going to make our temporary abode. Here was a small clearing, with abundance of tree-ferns and some young plantations of Cinchona. As there was now a thick mist and drizzling rain, I did not attempt to go on to the summit that evening, but made two visits to it during my stay, as well as one to the active crater of Gedeh. This is a vast semicircular chasm, bounded by black perpendicular walls of rock, and surrounded by miles of rugged scoria-covered slopes. The crater itself is not very deep. It exhibits patches of sulphur and variously-coloured volcanic products, and emits from several vents continual streams of smoke and vapour. The extinct cone of Pangerango was to me more interesting. The summit is an irregular undulating plain with a low bordering ridge, and one deep lateral chasm. Unfortunately there was perpetual mist and rain either above or below us all the time I was on the mountain ; so that I never once saw the plain below, or had a glimpse of the magnificent view which in fine weather is to be obtained from its summit. Notwithstanding this drawback I enjoyed the excursion exceedingly, for it was the first time I had been high enough on a mountain near the Equator to watch the change from a tropical to a temperate flora. I will now briefly sketch these changes as I observed them in Java.

On ascending the mountain, we first met with temperate forms of herbaceous plants, so low as 3,000 feet, where strawberries and violets begin to grow, but the former are tasteless and the latter have very small and pale flowers. Weedy Compositæ also begin to give a European aspect to the wayside herbage. It is between 2,000 and 5,000 feet that the forests and ravines exhibit the utmost development of tropical luxuriance and beauty. The abundance of noble Tree-ferns, sometimes fifty feet high, contributes greatly to the general effect, since of all the forms of tropical vegetation they are certainly the most striking and beautiful. Some of the deep ravines which have been cleared of large timber are full of them from top to bottom ; and where the road crosses one of these valleys, the view of their feathery crowns, in varied positions above and below the eye, offers a spectacle of picturesque beauty never to be forgotten. The splendid foliage of the broad-leaved Musaceæ and Zingiberaceæ, with their curious and brilliant flowers, and the elegant and varied forms of plants allied to Begonia and Melastoma, continually attract the attention in this region. Filling up the spaces between the trees and larger plants, on every trunk and stump and branch, are hosts of Orchids, Ferns and Lycopods, which wave and hang and intertwine in ever-varying complexity. At about 5,000 feet I first saw horsetails (Equisetum), very like our own species. At 6,000 feet, Raspberries abound, and thence to the summit of the mountain there are three species of eatable Rubus. At 7,000 feet Cypresses appear, and the forest trees become reduced in size, and more covered with mosses and lichens. From this point upward these rapidly increase, so that the blocks of rock and scoria that form the mountain slope are completely hidden in a mossy vegetation. At about 8,000 feet European forms of plants become abundant. Several species of Honeysuckle, St. John's-wort, and Guelder-rose abound, and at about 9,000 feet we first meet with the rare and beautiful Royal Cowslip (Primula imperialis), which is said to be found nowhere else in the world but on this solitary mountain summit. It has a tall, stout stem, sometimes more than three feet high, the root leaves are eighteen inches long, and it bears several whorls of cowslip-like flowers, instead of a terminal cluster only. The forest trees, gnarled and dwarfed to the dimensions of bushes, reach up to the very rim of the old crater, but do not extend over the hollow on its summit. Here we find a good deal of open ground, with thickets of shrubby Artemisias and Gnaphaliums, like our southernwood and cudweed, but six or eight feet high; while Buttercups, Violets, Whortle-berries, Sowthistles, Chickweed, white and yellow Cruciferæ, Plantain, and annual grasses everywhere abound. Where there are bushes and shrubs the St. John's-wort and Honeysuckle grow abundantly, while the Imperial Cowslip only exhibits its elegant blossoms under the damp shade of the thickets.

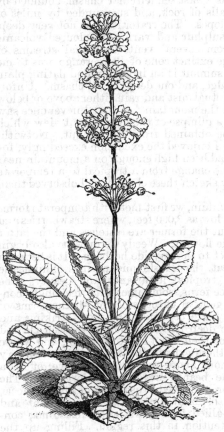

PRIMULA IMPERIALIS.

Mr. Motley, who visited the mountain in the dry season, and paid much attention to botany, gives the following list of genera characteristic of distant and more temperate regions :—Two species of Violet, three of Ranunculus, three of Impatiens, eight or ten of Rubus, and species of Primula, Hypericum, Swertia, Convallaria (Lily of the Valley), Vaccinium (Cranberry), Rhododendron, Gnaphalium, Polygonum, Digitalis, (Foxglove), Lonicera (Honeysuckle), Plantago (Ribgrass), Artemisia (Wormwood), Lobelia, Oxalis (Wood-sorrel), Quercus (Oak), and Taxus (Yew). A few of the smaller plants (Plantago major and lanceolata, Sonchus oleraceus, and Artemisia vulgaris) are identical with European species.

The fact of a vegetation so closely allied to that of Europe occurring on isolated mountain peaks, in an island south of the Equator, while all the lowlands for thousands of miles around are occupied by a flora of a totally different character is very extraordinary, and has only recently received an intelligible explanation. The Peak of Teneriffe, which rises to a greater height and is much nearer to Europe, contains no such Alpine flora ; neither do the mountains of Bourbon and Mauritius. The case of the volcanic peaks of Java is therefore somewhat exceptional, but there are several analogous, if not exactly parallel cases, that will enable us better to understand in what way the phenomena may possibly have been brought about. The higher peaks of the Alps, and even of the Pyrenees, contain a number of plants absolutely identical with those of Lapland, but nowhere found in the intervening plains. On the summit of the White Mountains, in the United States, every plant is identical with species growing in Labrador. In these cases all ordinary means of transport fail. Most of the plants have heavy seeds, which could not possibly be carried such immense distances by the wind ; and the agency of birds in so effectually stocking these Alpine heights is equally out of the question. The difficulty was so great, that some naturalists were driven to believe that these species were all separately created twice over on these distant peaks. The determination of a recent glacial epoch, however, soon offered a much more satisfactory solution, and one that is now universally accepted by men of science. At this period, when the mountains of Wales were full of glaciers, and the mountainous parts of Central Europe, and much of America north of the great lakes, were covered with snow and ice, and had a climate resembling that of Labrador and Greenland at the present day, an Arctic flora covered all these regions. As this epoch of cold passed away, and the snowy mantle of the country, with the glaciers that descended from every mountain summit, receded up their slopes and towards the north pole, the plants receded also, always clinging as now to the margins of the perpetual snow line. Thus it is that the same species are now found on the summits of the mountains of temperate Europe and America, and in the barren north-polar regions.

But there is another set of facts, which help us on another step towards the case of the Javenese mountain flora. On the higher slopes of the Himalaya, on the tops of the mountains of Central India and of Abyssinia, a number of plants occur which, though not identical with those of European mountains, belong to the same genera, and are said by botanists to represent them ; and most of these could not exist in the warm intervening plains. Mr. Darwin believed that this class of facts can be explained in the same way ; for, during the greatest severity of the glacial epoch, temperate forms of plants will have extended to the confines of the tropics, and on its departure, will have retreated up these southern mountains, as well as northward to the plains and hills of Europe. But in this case, the time elapsed, and the great change of conditions, have allowed many of these plants to become so modified that we now consider them to be distinct species. A variety of other facts of a similar nature, have led him to believe that the depression of temperature was at one time sufficient to allow a few north-temperate plants to cross the Equator (by the most elevated routes) and to reach the Antarctic regions, where they are now found. The evidence on which this belief rests, will be found in the latter part of Chapter II. of the *Origin of Species ;* and, accepting it for the present as an hypothesis, it enables us to account for the presence of a flora of European type on the volcanoes of Java.

It will, however, naturally be objected that there is a wide expanse of sea between Java and the continent, which would have effectually prevented the immigration of temperate forms of plants during the glacial epoch. This would undoubtedly be a fatal objection, were there not abundant evidence to show that Java has been formerly connected with Asia, and that the union must have occurred at about the epoch required. The most striking proof of such a junction is, that the great Mammalia of Java, the rhinoceros, the tiger, and the Banteng or wild ox, occur also in Siam and Burmah, and these would certainly not have been introduced by man. The Javanese peacock and several other birds are also common to these two countries ; but, in the majority of cases, the species are distinct, though closely allied, indicating that a considerable time (required for such modification) has elapsed since the separation, while it has not been so long as to cause an entire change. Now this exactly corresponds with the time we should require since the temperate forms of plants entered Java. These are almost all now distinct species ; but the changed conditions under which they are now forced to exist, and the probability of some of them having since died out on the continent of India, sufficiently accounts for the Javanese species being different.[1]

In my more special pursuits, I had very little success upon the mountain, owing, perhaps, to the excessively unpropitious weather and the shortness of my stay. At from 7,000 to 8,000 feet elevation, I obtained one of the most lovely of the small fruit pigeons (Ptilonopus roseicollis), whose entire head and

---

[1] I have now arrived at another explanation of these and analogous facts, and one which seems to me more complete and less improbable. (See my *Island Life*, chap. xxiii., and *Darwinism*, pp. 362-373.)

neck are of an exquisite rosy pink colour, contrasting finely
with its otherwise green plumage ; and on the very summit,
feeding on the ground among the strawberries that have been
planted there, I obtained a dull-coloured thrush, with the form
and habits of a starling (Turdus fumidus). Insects were almost
entirely absent, owing no doubt to the extreme dampness, and
I did not get a single butterfly the whole trip ; yet I feel sure
that, during the dry season, a week's residence on this mountain
would well repay the collector in every department of natural
history.

After my return to Toego, I endeavoured to find another
locality to collect in, and removed to a coffee-plantation some
miles to the north, and tried in succession higher and lower
stations on the mountain ; but I never succeeded in obtaining
insects in any abundance, and birds were far less plentiful than
on the Megamendong Mountain. The weather now became
more rainy than ever, and as the wet season seemed to have set
in in earnest, I returned to Batavia, packed up and sent off my
collections, and left by steamer on November 1st for Banca and
Sumatra.

CHAPTER VIII.

SUMATRA.

(NOVEMBER 1861 TO JANUARY 1862.)

THE mail steamer from Batavia to Singapore took me to
Muntok (or as on English maps, "Minto"), the chief town and
port of Banca. Here I stayed a day or two, till I could obtain
a boat to take me across the straits, and up the river to Palembang. A few walks into the country showed me that it was
very hilly, and full of granitic and laterite rocks, with a dry
and stunted forest vegetation ; and I could find very few insects.
A good-sized open sailing-boat took me across to the mouth of
the Palembang River, where at a fishing village, a rowing-boat
was hired to take me up to Palembang, a distance of nearly a
hundred miles by water. Except when the wind was strong and
favourable we could only proceed with the tide, and the banks
of the river were generally flooded Nipa-swamps, so that the
hours we were obliged to lie at anchor passed very heavily.
Reaching Palembang on the 8th of November, I was lodged by
the Doctor, to whom I had brought a letter of introduction, and
endeavoured to ascertain where I could find a good locality for
collecting. Every one assured me that I should have to go a
very long way further to find any dry forest, for at this season
the whole country for many miles inland was flooded. I there-

fore had to stay a week at Palembang before I could determine on my future movements.

The city is a large one, extending for three or four miles along a fine curve of the river, which is as wide as the Thames at Greenwich. The stream is, however, much narrowed by the houses which project into it upon piles, and within these, again, there is a row of houses built upon great bamboo rafts, which are moored by rattan cables to the shore or to piles, and rise and fall with the tide. The whole river-front on both sides is chiefly formed of such houses, and they are mostly shops open to the water, and only raised a foot above it, so that by taking a small boat it is easy to go to market and purchase anything that is to be had in Palembang. The natives are true Malays, never building a house on dry land if they can find water to set it in, and never going anywhere on foot if they can reach the place in a boat. A considerable portion of the population are Chinese and Arabs, who carry on all the trade ; while the only Europeans are the civil and military officials of the Dutch Government. The town is situated at the head of the delta of the river, and between it and the sea there is very little ground elevated above high-water mark ; while for many miles further inland, the banks of the main stream and its numerous tributaries are swampy, and in the wet season flooded for a considerable distance. Palembang is built on a patch of elevated ground, a few miles in extent, on the north bank of the river. At a spot about three miles from the town this rises into a little hill, the top of which is held sacred by the natives, and is shaded by some fine trees, inhabited by a colony of squirrels, which have become half tame. On holding out a few crumbs of bread or any fruit, they come running down the trunk, take the morsel out of your fingers, and dart away instantly. Their tails are carried erect, and the hair, which is ringed with grey, yellow, and brown, radiates uniformly around them, and looks exceedingly pretty. They have somewhat of the motions of mice, coming on with little starts, and gazing intently with their large black eyes, before venturing to advance further. The manner in which Malays often obtain the confidence of wild animals is a very pleasing trait in their character, and is due in some degree to the quiet deliberation of their manners, and their love of repose rather than of action. The young are obedient to the wishes of their elders, and seem to feel none of that propensity to mischief which European boys exhibit. How long would tame squirrels continue to inhabit trees in the vicinity of an English village, even if close to the church ? They would soon be pelted and driven away, or snared and confined in a whirling cage. I have never heard of these pretty animals being tamed in this way in England, but I should think it might be easily done in any gentleman's park, and they would certainly be as pleasing and attractive as they would be uncommon.

After many inquiries, I found that a day's journey by water

above Palembang there commenced a military road, which extended up to the mountains and even across to Bencoolen, and I determined to take this route and travel on till I found some tolerable collecting ground. By this means I should secure dry land and a good road, and avoid the rivers, which at this season are very tedious to ascend owing to the powerful currents, and very unproductive to the collector owing to most of the lands in their vicinity being under water. Leaving early in the morning we did not reach Lorok, the village where the road begins, till late at night. I stayed there a few days, but found that almost all the ground in the vicinity not under water was cultivated, and that the only forest was in swamps which were now inaccessible. The only bird new to me which I obtained at Lorok was the fine long-tailed parroquet (Palæcrnis longicauda). The people here assured me that the country was just the same as this for a very long way—more than a week's journey, and they seemed hardly to have any conception of an elevated forest-clad country, so that I began to think it would be useless going on, as the time at my disposal was too short to make it worth my while to spend much more of it in moving about. At length, however, I found a man who knew the country, and was more intelligent ; and he at once told me that if I wanted forest I must go to the district of Rembang, which I found on inquiry was about twenty-five or thirty miles off.

The road is divided into regular stages, of ten or twelve miles each, and, without sending on in advance to have coolies ready, only this distance can be travelled in a day. At each station there are houses for the accommodation of passengers, with cooking-house and stables, and six or eight men always on guard. There is an established system for coolies at fixed rates, the inhabitants of the surrounding villages all taking their turn to be subject to coolie service, as well as that of guards at the station for five days at a time. This arrangement makes travelling very easy, and was a great convenience for me. I had a pleasant walk of ten or twelve miles in the morning, and the rest of the day could stroll about and explore the village and neighbourhood, having a house ready to occupy without any formalities whatever. In three days I reached Moera-dua, the first village in Rembang, and finding the country dry and undulating, with a good sprinkling of forest, I determined to remain a short time and try the neighbourhood. Just opposite the station was a small but deep river, and a good bathing-place ; and beyond the village was a fine patch of forest, through which the road passed, overshadowed by magnificent trees, which partly tempted me to stay ; but after a fortnight I could find no good place for insects, and very few birds different from the common species of Malacca. I therefore moved on another stage to Lobo Raman, where the guard-house is situated quite by itself in the forest, nearly a mile from each of three villages. This was very agreeable to me, as I could move about without having every motion watched by crowds of men, women, and children, and I had also a much greater variety of walks to each of the villages and the plantations around them.

The villages of the Sumatran Malays are somewhat peculiar and very picturesque. A space of some acres is surrounded with a high fence, and over this area the houses are thickly strewn without the least attempt at regularity. Tall cocoa-nut trees grow abundantly between them, and the ground is bare and smooth with the trampling of many feet. The houses are raised about six feet on posts, the best being entirely built of planks, others of bamboo. The former are always more or less ornamented with carving, and have high-pitched roofs and

CHIEF'S HOUSE AND RICE SHED IN A SUMATRAN VILLAGE.

overhanging eaves. The gable ends and all the chief posts and beams are sometimes covered with exceedingly tasteful carved work, and this is still more the case in the district of Menangkabo, further west. The floor is made of split bamboo, and is rather shaky, and there is no sign of anything we should call furniture. There are no benches or chairs or stools, but merely the level floor covered with mats, on which the inmates sit or lie. The aspect of the village itself is very neat, the ground being often swept before the chief houses; but very bad odours abound, owing to there being under every house a stinking mud-hole, formed by all waste liquids and refuse matter, poured down through the floor above. In most other things Malays are tolerably clean—in some scrupulously so; and this peculiar and nasty custom, which is almost universal, arises, I have little doubt, from their having been originally a maritime and water-loving people, who built their houses on posts in the water, and only migrated gradually inland, first up the rivers and streams, and then into the dry interior. Habits which were at once so convenient and so cleanly, and which had been so long practised as to become a portion of the domestic life of the nation, were of course continued when the first settlers built their houses inland ; and without a regular system of drainage, the arrangement of the villages is such, that any other system would be very inconvenient.

In all these Sumatran villages I found considerable difficulty in getting anything to eat. It was not the season for vegetables, and when, after much trouble, I managed to procure some yams of a curious variety, I found them hard and scarcely eatable. Fowls were very scarce ; and fruit was reduced to one of the poorest kinds of banana. The natives (during the wet season at least) live exclusively on rice, as the poorer Irish do on potatoes. A pot of rice cooked very dry and eaten with salt and red peppers, twice a day, forms their entire food during a large part of the year. This is no sign of poverty, but is simply custom ; for their wives and children are loaded with silver armlets from wrist to elbow, and carry dozens of silver coins strung round their necks or suspended from their ears.

As I had moved away from Palembang, I had found the Malay spoken by the common people less and less pure, till at length it became quite unintelligible, although the continual recurrence of many well-known words assured me it was a form of Malay, and enabled me to guess at the main subject of conversation. This district had a very bad reputation a few years ago, and travellers were frequently robbed and murdered. Fights between village and village were also of frequent occurrence, and many lives were lost, owing to disputes about boundaries or intrigues with women. Now, however, since the country has been divided into districts under " Controlleurs," who visit every village in turn to hear complaints and settle disputes, such things are no more heard of. This is one of the numerous examples I have met with of the good effects of the Dutch Government. It exercises a strict surveillance over its most distant possessions, establishes a form of government well adapted to the character of the people, reforms abuses, punishes crimes, and makes itself everywhere respected by the native population.

Lobo Raman is a central point of the east end of Sumatra, being about a hundred and twenty miles from the sea to the east, north, and west. The surface is undulating, with no mountains or even hills, and there is no rock, the soil being generally a red friable clay. Numbers of small streams and rivers intersect the country, and it is pretty equally divided between open clearings and patches of forest, both virgin and second growth, with abundance of fruit trees ; and there is no

lack of paths to get about in any direction. Altogether it is the
very country that would promise most for a naturalist, and I
feel sure that at a more favourable time of year it would prove
exceedingly rich ; but it was now the rainy season, when, in the
very best of localities, insects are always scarce, and there being
no fruit on the trees there was also a scarcity of birds. During
a month's collecting, I added only three or four new species to
my list of birds, although I obtained very fine specimens of
many which were rare and interesting. In butterflies I was

DIFFERENT FEMALES OF PAPILIO MEMNON.

rather more successful, obtaining several fine species quite new
to me, and a considerable number of very rare and beautiful
insects. I will give here some account of two species of butterflies, which, though very common in collections, present us with
peculiarities of the highest interest.

The first is the handsome Papilio memnon, a splendid butterfly
of a deep black colour, dotted over with lines and groups of
scales of a clear ashy blue. Its wings are five inches in expanse
and the hind wings are rounded, with scalloped edges. This

applies to the males ; but the females are very different, and
vary so much that they were once supposed to form several
distinct species. They may be divided into two groups—those
which resemble the male in shape, and those which differ entirely
from him in the outline of the wings. The first vary much in
colour, being often nearly white with dusky yellow and red
markings, but such differences often occur in butterflies. The
second group are much more extraordinary, and would never be
supposed to be the same insect, since the hind wings are lengthened out into large spoon-shaped tails, no rudiment of which is
ever to be perceived in the males or in the ordinary form of
females. These tailed females are never of the dark and blue-
glossed tints which prevail in the male and often occur in the

PAPILIO COON.

females of the same form, but are invariably ornamented with
stripes and patches of white or buff, occupying the larger part
of the surface of the hind wings. This peculiarity of colouring
led me to discover that this extraordinary female closely resembles (when flying) another butterfly of the same genus but
of a different group (Papilio coon) ; and that we have here a
case of mimicry similar to those so well illustrated and explained
by Mr. Bates.[1] That the resemblance is not accidental is
sufficiently proved by the fact, that in the North of India, where
Papilio coon is replaced by an allied form (Papilio Doubledayi)
having red spots in place of yellow, a closely-allied species or
variety of Papilio memnon (P. androgeus), has the tailed female
also red spotted. The use and reason of this resemblance appears
to be, that the butterflies imitated belong to a section of the

[1] Trans. Linn. Soc. vol. xviii. p. 495 ; *Naturalist on the Amazons,* vol. i. p. 290.

genus Papilio which from some cause or other are not attacked by birds, and by so closely resembling these in form and colour the female of Memnon and its ally also escape persecution. Two other species of this same section (Papilio antiphus and Papilio polyphontes) are so closely imitated by two female forms of Papilio theseus (which comes in the same section with Memnon), that they completely deceived the Dutch entomologist De Haan, and he accordingly classed them as the same species !

But the most curious fact connected with these distinct forms is, that they are both the offspring of either form. A single brood of larvæ were bred in Java by a Dutch entomologist, and produced males as well as tailed and tailless females, and there is every reason to believe that this is always the case, and that forms intermediate in character never occur. To illustrate these phenomena, let us suppose a roaming Englishman in some remote island to have two wives—one a black-haired, red-skinned Indian, the other a woolly-headed, sooty-skinned negress; and that instead of the children being mulattoes of brown or dusky tints, mingling the characteristics of each parent in varying degrees, all the boys should be as fair-skinned and blue-eyed as their father, while the girls should altogether resemble their mothers. This would be thought strange enough, but the case of these butterflies is yet more extraordinary, for each mother is capable not only of producing male offspring like the father, and female like herself, but also other females like her fellow wife, and altogether differing from herself !

The other species to which I have to direct attention is the Kallima paralekta, a butterfly of the same family group as our Purple Emperor, and of about the same size or larger. Its upper surface is of a rich purple, variously tinged with ash colour, and across the fore wings there is a broad bar of deep orange, so that when on the wing it is very conspicuous. This species was not uncommon in dry woods and thickets, and I often endeavoured to capture it without success, for after flying a short distance it would enter a bush among dry or dead leaves, and however carefully I crept up to the spot I could never discover it till it would suddenly start out again and then disappear in a similar place. At length I was fortunate enough to see the exact spot where the butterfly settled, and though I lost sight of it for some time, I at length discovered that it was close before my eyes, but that in its position of repose it so closely resembled a dead leaf attached to a twig as almost certainly to deceive the eye even when gazing full upon it. I captured several specimens on the wing, and was able fully to understand the way in which this wonderful resemblance is produced.

The end of the upper wings terminates in a fine point, just as the leaves of many tropical shrubs and trees are pointed, while the lower wings are somewhat more obtuse, and are lengthened out into a short thick tail. Between these two points there runs a dark curved line exactly representing the midrib of a leaf, and

from this radiate on each side a few oblique marks which well
imitate the lateral veins.   These marks are more clearly seen on
the outer portion of the base of the wings, and on the inner side

LEAF BUTTERFLY IN FLIGHT AND REPOSE.

towards the middle and apex, and they are produced by striæ
and markings which are very common in allied species, but
which are here modified and strengthened so as to imitate more
exactly the venation of a leaf.   The tint of the under surface

varies much, but it is always some ashy brown or reddish colour, which matches with those of dead leaves. The habit of the species is always to rest on a twig and among dead or dry leaves, and in this position with the wings closely pressed together, their outline is exactly that of a moderately-sized leaf, slightly curved or shrivelled. The tail of the hind wings forms a perfect stalk, and touches the stick while the insect is supported by the middle pair of legs, which are not noticed among the twigs and fibres that surround it. The head and antennæ are drawn back between the wings so as to be quite concealed, and there is a little notch hollowed out at the very base of the wings, which allows the head to be retracted sufficiently. All these varied details combine to produce a disguise that is so complete and marvellous as to astonish every one who observes it ; and the habits of the insects are such as to utilize all these peculiarities, and render them available in such a manner as to remove all doubt of the purpose of this singular case of mimicry, which is undoubtedly a protection to the insect. Its strong and swift flight is sufficient to save it from its enemies when on the wing, but if it were equally conspicuous when at rest it could not long escape extinction, owing to the attacks of the insectivorous birds and reptiles that abound in the tropical forests. A very closely allied species, Kallima inachis, inhabits India, where it is very common, and specimens are sent in every collection from the Himalayas. On examining a number of these, it will be seen that no two are alike, but all the variations correspond to those of dead leaves. Every tint of yellow, ash, brown, and red is found here, and in many specimens there occur patches and spots formed of small black dots, so closely resembling the way in which minute fungi grow on leaves that it is almost impossible at first not to believe that fungi have grown on the butterflies themselves !

If such an extraordinary adaptation as this stood alone, it would be very difficult to offer any explanation of it ; but although it is perhaps the most perfect case of protective imitation known, there are hundreds of similar resemblances in nature, and from these it is possible to deduce a general theory of the manner in which they have been slowly brought about. The principle of variation and that of "natural selection," or survival of the fittest, as elaborated by Mr. Darwin in his celebrated *Origin of Species*, offers the foundation for such a theory ; and I have myself endeavoured to apply it to all the chief cases of imitation in an article published in the *Westminster Review* for 1867, entitled "Mimicry, and other Protective Resemblances among Animals," to which any reader is referred who wishes to know more about this subject.[1]

In Sumatra, monkeys are very abundant, and at Lobo Raman they used to frequent the trees which overhang the guard-house,

[1] This article forms the third chapter of my *Natural Selection and Tropical Nature*.

and give me a fine opportunity of observing their gambols. Two species of Semnopithecus were most plentiful—monkeys of a slender form, with very long tails. Not being much shot at they are rather bold, and remain quite unconcerned when natives alone are present; but when I came out to look at them, they would stare for a minute or two and then make off. They take tremendous leaps from the branches of one tree to those of another a little lower, and it is very amusing when one strong leader takes a bold jump, to see the others following with more or less trepidation; and it often happens that one or two of the last seem quite unable to make up their minds to leap till the rest are disappearing, when, as if in desperation at being left alone, they throw themselves frantically into the air, and often go crashing through the slender branches and fall to the ground.

A very curious ape, the Siamang, was also rather abundant, but it is much less bold than the monkeys, keeping to the virgin forests and avoiding villages. This species is allied to the little long-armed apes of the genus Hylobates, but is considerably larger, and differs from them by having the two first fingers of the feet united together, nearly to the end, whence its Latin name, Siamanga syndactyla. It moves much more slowly than the active Hylobates, keeping lower down in trees, and not indulging in such tremendous leaps; but it is still very active, and by means of its immense long arms, five feet six inches across in an adult about three feet high, can swing itself along among the trees at a great rate. I purchased a small one, which had been caught by the natives and tied up so tightly as to hurt it. It was rather savage at first, and tried to bite; but when we had released it and given it two poles under the verandah to hang upon, securing it by a short cord running along the pole with a ring so that it could move easily, it became more contented, and would swing itself about with great rapidity. It ate almost any kind of fruit and rice, and I was in hopes to have brought it to England but it died just before I started. It took a dislike to me at first which I tried to get over by feeding it constantly myself. One day, however, it bit me so sharply while giving it food, that I lost patience and gave it rather a severe beating, which I regretted afterwards, as from that time it disliked me more than ever. It would allow my Malay boys to play with it, and for hours together would swing by its arms from pole to pole and on to the rafters of the verandah, with so much ease and rapidity that it was a constant source of amusement to us. When I returned to Singapore it attracted great attention, as no one had seen a Siamang alive before, although it is not uncommon in some parts of the Malay peninsula.

As the Orang-utan is known to inhabit Sumatra, and was in fact first discovered there, I made many inquiries about it; but none of the natives had ever heard of such an animal, nor could I find any of the Dutch officials who knew anything about it. We may conclude, therefore, that it does not inhabit the great forest plains in the east of Sumatra where one would naturally expect to find it, but is probably confined to a limited region in the north-west—a part of the island entirely in the hands of native rulers. The other great Mammalia of Sumatra, the elephant and the rhinoceros, are more widely distributed ; but the former is much more scarce than it was a few years ago, and seems to retire rapidly before the spread of cultivation. About Lobo Raman tusks and bones are occasionally found in the forest, but the living animal is now never seen. The rhinoceros (Rhinoceros sumatranus) still abounds, and I continually saw its tracts and its dung, and once disturbed one feeding, which went crashing away through the jungle, only permitting me a momentary glimpse of it through the dense underwood. I obtained a tolerably perfect cranium, and a number of teeth, which were picked up by the natives.

Another curious animal, which I had met with in Singapore and in Borneo, but which was more abundant here, is the Galeopithecus, or flying lemur. This creature has a broad membrane extending all round its body to the extremities of the toes, and to the point of the rather long tail. This enables it to pass obliquely through the air from one tree to another. It is sluggish in its motions, at least by day, going up a tree by short runs of a few feet, and then stopping a moment as if the action was difficult. It rests during the day clinging to the trunks of trees, where its olive or brown fur, mottled with irregular whitish spots and blotches, resembles closely the colour of mottled bark, and no doubt helps to protect it. Once, in a bright twilight, I saw one of these animals run up a trunk in a rather open place, and then glide obliquely through the air to another tree, on which it alighted near its base, and immediately began to ascend. I paced the distance from the one tree to the other, and found it to be seventy yards ; and the amount of descent I estimated at not more than thirty-five or forty feet, or less than one in five. This I think proves that the animal must have some power of guiding itself through the air, otherwise in so long a distance it would have little chance of alighting exactly upon the trunk. Like the Cuscus of the Moluccas, the Galeopithecus feeds chiefly on leaves, and possesses a very voluminous stomach and long convoluted intestines. The brain is very small, and the animal possesses such remarkable tenacity of life, that it is exceedingly difficult to kill it by any ordinary means. The tail is prehensile, and is probably made use of as an additional support while feeding. It is said to have only a single young one at a time, and my own observation confirms this statement, for I once shot a female, with a very small blind and naked little creature clinging closely to its breast, which was quite bare and much wrinkled, reminding me of the young of Marsupials, to which it seemed to form a transition. On the back, and extending over the limbs and membrane, the fur of these animals is short but exquisitely soft, resembling in its texture that of the Chinchilla.

I returned to Palembang by water, and while staying a day

FEMALE HORNBILL, AND YOUNG BIRD.

at a village while a boat was being made watertight, I had the good fortune to obtain a male, female, and young bird of one of the large hornbills. I had sent my hunters to shoot, and while I was at breakfast they returned, bringing me a fine large male,

of the Buceros bicornis, which one of them assured me he had
shot while feeding the female, which was shut up in a hole in a
tree. I had often read of this curious habit, and immediately
returned to the place, accompanied by several of the natives.
After crossing a stream and a bog, we found a large tree leaning over some water, and on its lower side, at a height of about
twenty feet, appeared a small hole, and what looked like a
quantity of mud, which I was assured had been used in stopping
up the large hole. After a while we heard the harsh cry of a
bird inside, and could see the white extremity of its beak put
out. I offered a rupee to any one who would go up and get out
the bird, with the egg or young one; but they all declared it
was too difficult, and they were afraid to try. I therefore very
reluctantly came away. In about an hour afterwards, much to
my surprise, a tremendous loud hoarse screaming was heard
and the bird was brought me, together with a young one which
had been found in the hole. This was a most curious object, as
large as a pigeon, but without a particle of plumage on any
part of it. It was exceedingly plump and soft, and with a semi-
transparent skin, so that it looked more like a bag of jelly,
with head and feet stuck on, than like a real bird.

The extraordinary habit of the male, in plastering up the
female with her egg, and feeding her during the whole time of
incubation, and till the young one is fledged, is common to
several of the large hornbills, and is one of those strange facts
in natural history which are "stranger than fiction."

CHAPTER IX.

NATURAL HISTORY OF THE INDO-MALAY ISLANDS.

In the first chapter of this work I have stated generally the
reasons which lead us to conclude that the large islands in the
western portion of the Archipelago—Java, Sumatra, and
Borneo—as well as the Malay peninsula and the Philippine
islands, have been recently separated from the continent of
Asia. I now propose to give a sketch of the Natural History
of these, which I term the Indo-Malay islands, and to show how
far it supports this view, and how much information it is able
to give us of the antiquity and origin of the separate islands.

The flora of the Archipelago is at present so imperfectly
known, and I have myself paid so little attention to it, that I
cannot draw from it many facts of importance. The Malayan
type of vegetation is however a very important one; and Dr.
Hooker informs us, in his *Flora Indica*, that it spreads over
all the moister and more equable parts of India, and that many
plants found in Ceylon, the Himalayas, the Nilghiri, and Khasia
mountains are identical with those of Java and the Malay

peninsula. Among the more characteristic forms of this flora are the rattans—climbing palms of the genus Calamus, and a great variety of tall as well as stemless palms. Orchids, Araceæ, Zingiberaceæ, and ferns are especially abundant, and the genus Grammatophyllum—a gigantic epiphytal orchid, whose clusters of leaves and flower-stems are ten or twelve feet long—is peculiar to it. Here, too, is the domain of the wonderful pitcher plants (Nepenthaceæ), which are only represented else-

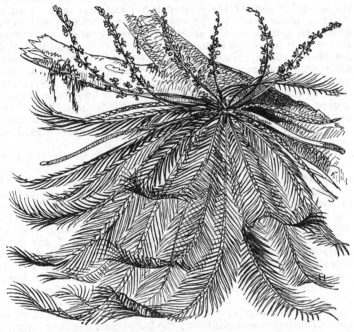

GRAMMATOPHYLLUM, A GIGANTIC ORCHID.

where by solitary species in Ceylon, Madagascar, the Seychelles, Celebes, and the Moluccas. Those celebrated fruits, the Mangosteen and the Durian, are natives of this region, and will hardly grow out of the Archipelago. The mountain plants of Java have already been alluded to as showing a former connexion with the continent of Asia; and a still more extraordinary and more ancient connexion with Australia has been indicated by Mr. Low's collections from the summit of Kinibalou, the loftiest mountain in Borneo.

Plants have much greater facilities for passing across arms of the sea than animals. The lighter seeds are easily carried by the winds, and many of them are specially adapted to be so

carried. Others can float a long time unhurt in the water, and
are drifted by winds and currents to distant shores. Pigeons,
and other fruit-eating birds, are also the means of distributing
plants, since the seeds readily germinate after passing through
their bodies. It thus happens that plants which grow on shores
and lowlands have a wide distribution, and it requires an
extensive knowledge of the species of each island to determine
the relations of their floras with any approach to accuracy. At
present we have no such complete knowledge of the botany of
the several islands of the Archipelago ; and it is only by such
striking phenomena as the occurrence of northern and even
European genera on the summits of the Javanese mountains
that we can prove the former connexion of that island with the
Asiatic continent. With land animals, however, the case is very
different. Their means of passing a wide expanse of sea are far
more restricted. Their distribution has been more accurately
studied, and we possess a much more complete knowledge of such
groups as mammals and birds in most of the islands, than we do
of the plants. It is these two classes which will supply us with
most of our facts as to the geographical distribution of organized
beings in this region.

The number of Mammalia known to inhabit the Indo-Malay
region is very considerable, ~~exceeding 170~~ species. With the
exception of the bats, none of these have any regular means of
passing arms of the sea many miles in extent, and a consideration
of their distribution must therefore greatly assist us in determining whether these islands have ever been connected with each
other or with the continent since the epoch of existing species.

probably 250

The Quadrumana or monkey tribe form one of the most
characteristic features of this region. Twenty-four distinct
species are known to inhabit it, and these are distributed with
tolerable uniformity over the islands, nine being found in Java,
ten in the Malay peninsula, eleven in Sumatra, and thirteen in
Borneo. The great man-like Orang-utans are found only in
Sumatra and Borneo ; the curious Siamang (next to them in
size) in Sumatra and Malacca ; the long-nosed monkey only in
Borneo ; while every island has representatives of the Gibbons
or long-armed apes, and of monkeys. The lemur-like animals,
Nycticebus, Tarsius, and Galeopithecus, are found in all the
islands.

Seven species found on the Malay peninsula extend also into
Sumatra, four into Borneo, and three into Java ; while two range
into Siam and Burmah, and one into North India. With the
exception of the Orang-utan, the Siamang, the Tarsius spectrum,
and the Galeopithecus, all the Malayan genera of Quadrumana
are represented in India by closely allied species, although,
owing to the limited range of most of these animals, so few are
absolutely identical.

Of Carnivora, thirty-three species are known from the Indo-
Malay region, of which about eight are found also in Burmah

and India. Among these are the tiger, leopard, a tiger-cat, civet, and otter ; while out of the twenty genera of Malayan Carnivora, thirteen are represented in India by more or less closely allied species. As an example, the curious Malayan glutton (Helictis orientalis) is represented in Northern India by a closely allied species, Helictis nipalensis.

The hoofed animals are twenty-two in number, of which about seven extend into Burmah and India. All the deer are of peculiar species, except two, which range from Malacca into India. Of the cattle, one Indian species reaches Malacca, while the Bos sondaicus of Java and Borneo is also found in Siam and Burmah. A goat-like animal is found in Sumatra which has its representative in India ; while the two-horned rhinoceros of Sumatra and the single-horned species of Java, long supposed to be peculiar to these islands, are now both ascertained to exist in Burmah, Pegu, and Moulmein. The elephant of Sumatra, Borneo, and Malacca is now considered to be identical with that of Ceylon and India.

In all other groups of Mammalia the same general phenomena recur. A few species are identical with those of India. A much larger number are closely allied or representative forms ; while there are always a small number of peculiar genera, consisting of animals unlike those found in any other part of the world.

There are about fifty bats, of which less than one-fourth are Indian species ; thirty-four Rodents (squirrels, rats, &c.), of which six or eight only are Indian ; and ten Insectivora, with one exception peculiar to the Malay region. The squirrels are very abundant and characteristic, only two species out of twenty-five extending into Siam and Burmah. The Tupaias are curious insect-eaters, which closely resemble squirrels, and are almost confined to the Malay islands, as are the small feather-tailed Ptilocerus lowii of Borneo, and the curious long-snouted and naked-tailed Gymnurus rafflesii.

As the Malay peninsula is a part of the continent of Asia, the question of the former union of the islands to the mainland will be best elucidated by studying the species which are found in the former district, and also in some of the islands. Now, if we entirely leave out of consideration the bats, which have the power of flight, there are still forty-eight species of mammals common to the Malay peninsula and the three large islands. Among these are seven Quadrumana (apes, monkeys, and lemurs), animals which pass their whole existence in forests, which never swim, and which would be quite unable to traverse a single mile of sea ; nineteen Carnivora, some of which no doubt might cross by swimming, but we cannot suppose so large a number to have passed in this way across a strait which, except at one point, is from thirty to fifty miles wide ; and five hoofed animals, including the Tapir, two species of rhinoceros, and an elephant. Besides these there are thirteen Rodents and four Insectivora, including a shrew-mouse and six squirrels, whose unaided passage over twenty miles of sea is even more inconceivable than that of the larger animals.

But when we come to the cases of the same species inhabiting two of the more widely separated islands, the difficulty is much increased. Borneo is distant nearly 150 miles from Biliton, which is about fifty miles from Banca, and this fifteen from Sumatra, yet there are no less than thirty-six species of mammals common to Borneo and Sumatra. Java again is more than 250 miles from Borneo, yet these two islands have twenty-two species in common, including monkeys, lemurs, wild oxen, squirrels, and shrews. These facts seem to render it absolutely certain that there has been at some former period a connexion between all these islands and the mainland, and the fact that most of the animals common to two or more of them show little or no variation, but are often absolutely identical, indicates that the separation must have been recent in a geological sense; that is, not earlier than the Newer Pliocene epoch, at which time land animals began to assimilate closely with those now existing.

Even the bats furnish an additional argument, if one were needed, to show that the islands could not have been peopled from each other and from the continent without some former connexion. For if such had been the mode of stocking them with animals, it is quite certain that creatures which can fly long distances would be the first to spread from island to island, and thus produce an almost perfect uniformity of species over the whole region. But no such uniformity exists, and the bats of each island are almost, if not quite, as distinct as the other mammals. For example, sixteen species are known in Borneo, and of these ten are found in Java and five in Sumatra, a proportion about the same as that of the Rodents, which have no direct means of migration. We learn from this fact, that the seas which separate the islands from each other are wide enough to prevent the passage even of flying animals, and that we must look to the same causes as having led to the present distribution of both groups. The only sufficient cause we can imagine is the former connexion of all the islands with the continent, and such a change is in perfect harmony with what we know of the earth's past history, and is rendered probable by the remarkable fact that a rise of only three hundred feet would convert the wide seas that separate them into an immense winding valley or plain about three hundred miles wide and twelve hundred long.

It may, perhaps, be thought that birds which possess the power of flight in so pre-eminent a degree, would not be limited in their range by arms of the sea, and would thus afford few indications of the former union or separation of the islands they inhabit. This, however, is not the case. A very large number of birds appear to be as strictly limited by watery barriers as are quadrupeds; and as they have been so much more attentively collected, we have more complete materials to work upon, and are enabled to deduce from them still more definite and satisfactory results. Some groups, however, such as the aquatic birds, the waders, and the birds of prey, are great wanderers ; other groups are little known except to ornithologists. I shall therefore refer chiefly to a few of the best known and most remarkable families of birds, as a sample of the conclusions furnished by the entire class.

The birds of the Indo-Malay region have a close resemblance to those of India ; for though a very large proportion of the species are quite distinct, there are only about fifteen peculiar genera, and not a single family group confined to the former district. If, however, we compare the islands with the Burmese, Siamese, and Malayan countries, we shall find still less difference, and shall be convinced that all are closely united by the bond of a former union. In such well-known families as the woodpeckers, parrots, trogons, barbets, kingfishers, pigeons, and pheasants, we find some identical species spreading over all India, and as far as Java and Borneo, while a very large proportion are common to Sumatra and the Malay peninsula.

The force of these facts can only be appreciated when we come to treat of the islands of the Austro-Malay region, and show how similar barriers have entirely prevented the passage of birds from one island to another, so that out of at least three hundred and fifty land birds inhabiting Java and Borneo, not more than ten have passed eastward into Celebes. Yet the straits of Macassar are not nearly so wide as the Java sea, and at least a hundred species are common to Borneo and Java.

I will now give two examples to show how a knowledge of the distribution of animals may reveal unsuspected facts in the past history of the earth. At the eastern extremity of Sumatra, and separated from it by a strait about fifteen miles wide, is the small rocky island of Banca, celebrated for its tin mines. One of the Dutch residents there sent some collections of birds and animals to Leyden, and among them were found several species distinct from those of the adjacent coast of Sumatra. One of these was a squirrel (Sciurus bangkanus), closely allied to three other species inhabiting respectively the Malay peninsula, Sumatra, and Borneo, but quite as distinct from them all as they are from each other. There were also two new ground thrushes of the genus Pitta, closely allied to, but quite distinct from, two other species inhabiting both Sumatra and Borneo, and which did not perceptibly differ in these large and widely separated islands. This is just as if the Isle of Man possessed a peculiar species of thrush and blackbird, distinct from the birds which are common to England and Ireland.

These curious facts would indicate that Banca may have existed as a distinct island even longer than Sumatra and Borneo, and there are some geological and geographical facts which render this not so improbable as it would at first seem to be.

Although on the map Banca appears so close to Sumatra, this does not arise from its having been recently separated from it; for the adjacent district of Palembang is new land, being a great alluvial swamp formed by torrents from the mountains a hundred miles distant. Banca, on the other hand, agrees with Malacca, Singapore, and the intervening island of Lingen, in being formed of granite and laterite; and these have all most likely once formed an extension of the Malay peninsula. As the rivers of Borneo and Sumatra have been for ages filling up the intervening sea, we may be sure that its depth has recently been greater, and it is very probable that those large islands were never directly connected with each other except through the Malay peninsula. At that period the same species of squirrel and Pitta may have inhabited all these countries; but when the subterranean disturbances occurred which led to the elevation of the volcanoes of Sumatra, the small island of Banca may have been separated first, and its productions being thus isolated might be gradually modified before the separation of the larger islands had been completed. As the southern part of Sumatra extended eastward and formed the narrow straits of Banca, many birds and insects and some Mammalia would cross from one to the other, and thus produce a general similarity of productions, while a few of the older inhabitants remained, to reveal by their distinct forms their different origin. Unless we suppose some such changes in physical geography to have occurred, the presence of peculiar species of birds and mammals in such an island as Banca is a hopeless puzzle; and I think I have shown that the changes required are by no means so improbable as a mere glance at the map would lead us to suppose.

For our next example let us take the great islands of Sumatra and Java. These approach so closely together, and the chain of volcanoes that runs through them gives such an air of unity to the two, that the idea of their having been recently dissevered is immediately suggested. The natives of Java, however, go further than this; for they actually have a tradition of the catastrophe which broke them asunder, and fix its date at not much more than a thousand years ago. It becomes interesting, therefore, to see what support is given to this view by the comparison of their animal productions.

The Mammalia have not been collected with sufficient completeness in both islands to make a general comparison of much value, and so many species have been obtained only as live specimens in captivity, that their locality has often been erroneously given—the island in which they were obtained being substituted for that from which they originally came. Taking into consideration only those whose distribution is more accurately known, we learn that Sumatra is, in a zoological sense, more nearly related to Borneo than it is to Java. The great man-like apes, the elephant, the tapir, and the Malay bear, are all common to the two former countries, while they are absent from the latter. Of the three long-tailed monkeys (Semnopithecus) inhabiting Sumatra, one extends into Borneo, but the two species of Java are both peculiar to it. So also the great Malay deer (Rusa equina), and the small Tragulus kanchil, are common to Sumatra and Borneo, but do not extend into Java, where they are replaced by Tragulus javanicus. The tiger, it is true, is found in Sumatra and Java, but not in Borneo. But as this animal is known to swim well, it may have found its way across the Straits of Sunda, or it may have inhabited Java before it was separated from the main land, and from some unknown cause have ceased to exist in Borneo.

In Ornithology there is a little uncertainty owing to the birds of Java and Sumatra being much better known than those of Borneo ; but the ancient separation of Java as an island is well exhibited by the large number of its species which are not found in any of the other islands. It possesses no less than seven pigeons peculiar to itself, while Sumatra has only one. Of its two parrots one extends into Borneo, but neither into Sumatra. Of the fifteen species of woodpeckers inhabiting Sumatra only four reach Java, while eight of them are found in Borneo and twelve in the Malay peninsula. The two Trogons found in Java are peculiar to it, while of those inhabiting Sumatra at least two extend to Malacca and one to Borneo. There are a very large number of birds, such as the great Argus pheasant, the fire-backed and ocellated pheasants, the crested partridge (Rollulus coronatus), the small Malacca parrot (Psittinus incertus), the great helmeted hornbill (Buceroturus galeatus), the pheasant ground-cuckoo (Carpococcyx radiatus), the rose-crested bee-eater (Nyctiornis amicta), the great gaper (Corydon sumatranus), and the green-crested gaper (Calyptomena viridis), and many others, which are common to Malacca, Sumatra, and Borneo, but are entirely absent from Java. On the other hand we have the peacock, the green jungle cock, two blue ground thrushes (Arrenga cyanea and Myophonus flavirostris), the fine pink-headed dove (Ptilonopus porphyreus), three broad-tailed ground pigeons (Macropygia), and many other interesting birds, which are found nowhere in the Archipelago out of Java.

Insects furnish us with similar facts wherever sufficient data are to be had, but owing to the abundant collections that have been made in Java, an unfair preponderance may be given to that island. This does not, however, seem to be the case with the true Papilionidæ or swallow-tailed butterflies, whose large size and gorgeous colouring has led to their being collected more frequently than other insects. Twenty-seven species are known from Java, twenty-nine from Borneo, and only twenty-one from Sumatra. Four are entirely confined to Java, while only two are peculiar to Borneo and one to Sumatra. The isolation of Java will, however, be best shown by grouping the islands in pairs, and indicating the number of species common to each pair. Thus :—

Borneo . . . 29 species ⎫
Sumatra . . 21 do. ⎬ 20 species common to both islands.

Borneo . . . 29 do. ⎫
Java . . . . 27 do. ⎬ 20   do.   do.

Sumatra . . 21 do. ⎫
Java . . . . 27 do. ⎬ 11   do.   do.

Making some allowance for our imperfect knowledge of the Sumatran species, we see that Java is more isolated from the two larger islands than they are from each other, thus entirely confirming the results given by the distribution of birds and Mammalia, and rendering it almost certain that the last-named island was the first to be completely separated from the Asiatic continent, and that the native tradition of its having been recently separated from Sumatra is entirely without foundation.

We are now enabled to trace out with some probability the course of events. Beginning at the time when the whole of the Java sea, the Gulf of Siam, and the Straits of Malacca were dry land, forming with Borneo, Sumatra, and Java, a vast southern prolongation of the Asiatic continent, the first movement would be the sinking down of the Java sea, and the Straits of Sunda, consequent on the activity of the Javanese volcanoes along the southern extremity of the land, and leading to the complete separation of that island. As the volcanic belt of Java and Sumatra increased in activity, more and more of the land was submerged, till first Borneo, and afterwards Sumatra, became entirely severed. Since the epoch of the first disturbance, several distinct elevations and depressions may have taken place, and the islands may have been more than once joined with each other or with the mainland, and again separated. Successive waves of immigration may thus have modified their animal productions, and led to those anomalies in distribution which are so difficult to account for by any single operation of elevation or submergence. The form of Borneo, consisting of radiating mountain chains with intervening broad alluvial valleys, suggests the idea that it has once been much more submerged than it is at present (when it would have somewhat resembled Celebes or Gilolo in outline), and has been increased to its present dimensions by the filling up of its gulfs with sedimentary matter, assisted by gradual elevation of the land. Sumatra has also been evidently much increased in size by the formation of alluvial plains along its north-eastern coasts.

There is one peculiarity in the productions of Java that is very puzzling—the occurrence of several species or groups characteristic of the Siamese countries or of India, but which do not occur in Borneo or Sumatra. Among Mammals the

Rhinoceros javanicus is the most striking example, for a *distinct* species is found in Borneo and Sumatra, while the Javanese species occurs in Burmah and even in Bengal. Among birds, the small ground dove, Geopelia striata, and the curious bronze-coloured magpie, Crypsirhina varians, are common to Java and Siam; while there are in Java species of Pteruthius, Arrenga, Myiophonus, Zoothera, Sturnopastor, and Estrelda, the nearest allies of which are found in various parts of India, while nothing like them is known to inhabit Borneo or Sumatra.

Such a curious phenomenon as this can only be understood by supposing that, subsequent to the separation of Java, Borneo became almost entirely submerged, and on its re-elevation was for a time connected with the Malay peninsula and Sumatra, but not with Java or Siam. Any geologist who knows how strata have been contorted and tilted up, and how elevations and depressions must often have occurred alternately, not once or twice only, but scores and even hundreds of times, will have no difficulty in admitting that such changes as have been here indicated are not in themselves improbable. The existence of extensive coal-beds in Borneo and Sumatra, of such recent origin that the leaves which abound in their shales are scarcely distinguishable from those of the forests which now cover the country, proves that such changes of level actually did take place; and it is a matter of much interest, both to the geologist and to the philosophic naturalist, to be able to form some conception of the order of those changes, and to understand how they may have resulted in the actual distribution of animal life in these countries;—a distribution which often presents phenomena so strange and contradictory, that without taking such changes into consideration we are unable even to imagine how they could have been brought about.

CHAPTER X.

BALI AND LOMBOCK.

(JUNE, JULY, 1856.)

THE islands of Bali and Lombock, situated at the east end of Java, are particularly interesting. They are the only islands of the whole Archipelago in which the Hindoo religion still maintains itself—and they form the extreme points of the two great zoological divisions of the Eastern hemisphere; for although so similar in external appearance and in all physical features, they differ greatly in their natural productions. It was after having spent two years in Borneo, Malacca and Singapore, that I made a somewhat involuntary visit to these islands on my way to Macassar. Had I been able to obtain a passage direct to that place from Singapore, I should probably never have gone near them, and should have missed some of the most important discoveries of my whole expedition to the East.

It was on the 13th of June, 1856, after a twenty days' passage from Singapore in the *Kembang Djepoon* (Rose of Japan), a schooner belonging to a Chinese merchant, manned by a Javanese crew, and commanded by an English captain, that we cast anchor in the dangerous roadstead of Bileling on the north side of the island of Bali. Going on shore with the captain and the Chinese supercargo, I was at once introduced to a novel and interesting scene. We went first to the house of the Chinese Bandar, or chief merchant, where we found a number of natives, well dressed, and all conspicuously armed with krisses, displaying their large handles of ivory or gold, or beautifully grained and polished wood.

The Chinamen had given up their national costume and adopted the Malay dress, and could then hardly be distinguished from the natives of the island—an indication of the close affinity of the Malayan and Mongolian races. Under the thick shade of some mango-trees close by the house, several women-merchants were selling cotton goods ; for here the women trade and work for the benefit of their husbands, a custom which Mahometan Malays never adopt. Fruit, tea, cakes, and sweetmeats were brought us ; many questions were asked about our business and the state of trade in Singapore, and we then took a walk to look at the village. It was a very dull and dreary place ; a collection of narrow lanes bounded by high mud walls, enclosing bamboo houses, into some of which we entered and were very kindly received.

During the two days that we remained here, I walked out into the surrounding country to catch insects, shoot birds, and spy out the nakedness or fertility of the land. I was both astonished and delighted ; for as my visit to Java was some years later, I had never beheld so beautiful and well-cultivated a district out of Europe. A slightly undulating plain extends from the sea-coast about ten or twelve miles inland, where it is bounded by a fine range of wooded and cultivated hills. Houses and villages, marked out by dense clumps of cocoa-nut palms, tamarind and other fruit trees, are dotted about in every direction ; while between them extend luxuriant rice-grounds, watered by an elaborate system of irrigation that would be the pride of the best cultivated parts of Europe. The whole surface of the country is divided into irregular patches, following the undulations of the ground, from many acres to a few perches in extent, each of which is itself perfectly level, but stands a few inches or several feet above or below those adjacent to it. Every one of these patches can be flooded or drained at will, by means of a system of ditches and small channels, into which are diverted the whole of the streams that descend from the mountains. Every patch now bore crops in various stages of growth, some almost ready for cutting, and all in the most flourishing condition and of the most exquisite green tints.

The sides of the lanes and bridle roads were often edged with prickly Cacti and a leafless Euphorbia, but the country being so highly cultivated there was not much room for indigenous vegetation, except upon the sea-beach. We saw plenty of the fine race of domestic cattle descended from the Bos sondaicus of Java, driven by half-naked boys, or tethered in pasture-grounds. They are large and handsome animals, of a light brown colour, with white legs, and a conspicuous oval patch behind of the same colour. Wild cattle of the same race are said to be still found in the mountains. In so well-cultivated a country it was not to be expected that I could do much in natural history, and my ignorance of how important a locality this was for the elucidation of the geographical distribution of animals, caused me to neglect obtaining some specimens which I never met with again. One of these was a weaver bird with a bright yellow head, which built its bottle-shaped nests by dozens on some trees near the beach. It was the Ploceus hypoxanthus, a native of Java ; and here at the extreme limits of its range westerly. I shot and preserved specimens of a wagtail-thrush, an oriole, and some starlings, all species found in Java, and some of them peculiar to that island. I also obtained some beautiful butterflies, richly marked with black and orange on a white ground, and which were the most abundant insects in the country lanes. Among these was a new species, which I have named Pieris tamar.

Leaving Bileling, a pleasant sail of two days brought us to Ampanam in the island of Lombock, where I proposed to remain till I could obtain a passage to Macassar. We enjoyed superb views of the twin volcanoes of Bali and Lombock, each about eight thousand feet high, which form magnificent objects at sun-rise and sunset, when they rise out of the mists and clouds that surround their bases, glowing with the rich and changing tints of these the most charming moments in a tropical day.

The bay or roadstead of Ampanam is extensive, and being at this season sheltered from the prevalent south-easterly winds, was as smooth as a lake. The beach of black volcanic sand is very steep, and there is at all times a heavy surf upon it, which during spring-tides increases to such an extent that it is often impossible for boats to land, and many serious accidents have occurred. Where we lay anchored, about a quarter of a mile from the shore, not the slightest swell was perceptible, but on approaching nearer undulations began, which rapidly increased, so as to form rollers which toppled over on to the beach at regular intervals with a noise like thunder. Sometimes this surf increases suddenly during perfect calms, to as great a force and fury as when a gale of wind is blowing, beating to pieces all boats that may not have been hauled sufficiently high upon the beach, and carrying away incautious natives. This violent surf is probably in some way dependent on the swell of the great southern ocean, and the violent currents that flow through the Straits of Lombock. These are so uncertain that vessels preparing to anchor in the bay are sometimes suddenly swept away into the straits, and are not able to get back again for a fortnight! What seamen call the " ripples " are also very violent in the straits, the sea appearing to boil and foam and dance like the rapids below a cataract ; vessels are swept about helpless, and small ones are occasionally swamped in the finest weather and under the brightest skies.

I felt considerably relieved when all my boxes and myself had passed in safety through the devouring surf, which the natives look upon with some pride, saying, that "their sea is always hungry, and eats up everything it can catch." I was kindly received by Mr. Carter, an Englishman, who is one of the Bandars or licensed traders of the port, who offered me hospitality and every assistance during my stay. His house, store-houses and offices were in a yard surrounded by a tall bamboo fence, and were entirely constructed of bamboo with a thatch of grass, the only available building materials. Even these were now very scarce, owing to the great consumption in rebuilding the place since the great fire some months before, which in an hour or two had destroyed every building in the town.

The next day I went to see Mr. S., another merchant to whom I had brought letters of introduction, and who lived about seven miles off. Mr. Carter kindly lent me a horse, and I was accompanied by a young Dutch gentleman residing at Ampanam, who offered to be my guide. We first passed through the towns and suburbs along a straight road bordered by mud walls and a fine avenue of lofty trees ; then through rice-fields irrigated in the same manner as I had seen them at Bileling, and afterwards over sandy pastures near the sea, and occasionally along the beach itself. Mr. S. received us kindly, and offered me a residence at his house should I think the neighbourhood favourable for my pursuits. After an early breakfast we went out to explore, taking guns and insect-net. We reached some low hills which seemed to offer the most favourable ground, passing over swamps, sandy flats overgrown with coarse sedges, and through pastures and cultivated grounds, finding however very little in the way of either birds or insects. On our way we passed one or two human skeletons, enclosed within a small bamboo fence, with the clothes, pillow, mat, and betel-box of the unfortunate individual,—who had been either murdered or executed. Returning to the house, we found a Balinese chief and his followers on a visit. Those of higher rank sat on chairs, the others squatted on the floor. The chief very coolly asked for beer and brandy, and helped himself and his followers, apparently more out of curiosity than anything else as regards the beer, for it seemed very distasteful to them, while they drank the brandy in tumblers with much relish.

Returning to Ampanam, I devoted myself for some days to shooting the birds of the neighbourhood. The fine fig-trees of

the avenues, where a market was held, were tenanted by superb
orioles (Oriolus broderpii) of a rich orange colour, and peculiar
to this island and the adjacent ones of Sumbawa and Flores.
All round the town were abundance of the curious Tropidorhynchus timoriensis, allied to the Friar bird of Australia.
They are here called "Quaich-quaich," from their strange loud
voice, which seems to repeat these words in various and not
unmelodious intonations.

Every day boys were to be seen walking along the roads and
by the hedges and ditches, catching dragon flies with bird-lime.
They carry a slender stick, with a few twigs at the end well
anointed, so that the least touch captures the insect, whose
wings are pulled off before it is consigned to a small basket.
The dragon-flies are so abundant at the time of the rice-flowering
that thousands are soon caught in this way. The bodies are
fried in oil with onions and preserved shrimps, or sometimes
alone and are considered a great delicacy. In Borneo, Celebes,
and many other islands, the larvæ of bees and wasps are eaten,
either alive as pulled out of the cells, or fried like the dragon-
flies. In the Moluccas the grubs of the palm-beetle (Calandra)
are regularly brought to market in bamboos, and sold for food ;
and many of the great horned Lamellicorn beetles are slightly
roasted on the embers and eaten whenever met with. The
superabundance of insect life is therefore turned to some account
by these islanders.

Finding that birds were not very numerous, and hearing much
of Labuan Tring at the southern extremity of the bay, where
there was said to be much uncultivated country and plenty of
birds as well as deer and wild pigs, I determined to go there with
my two servants, Ali, the Malay lad from Borneo, and Manuel, a
Portuguese of Malacca accustomed to bird-skinning. I hired a
native boat with outriggers, to take us with our small quantity
of luggage, and a day's rowing and tracking along the shore
brought us to the place.

I had a note of introduction to an Amboynese Malay, and
obtained the use of part of his house to live and work in. His
name was "Inchi Daud" (Mr. David), and he was very civil ;
but his accommodations were limited, and he could only give me
part of his reception-room. This was the front part of a bamboo
house (reached by a ladder of about six rounds very wide apart),
and having a beautiful view over the bay. However, I soon
made what arrangements were possible and then set to work.
The country around was pretty and novel to me, consisting of
abrupt volcanic hills enclosing flat valleys or open plains. The
hills were covered with a dense scrubby bush of bamboos and
prickly trees and shrubs, the plains were adorned with hundreds
of noble palm-trees, and in many places with a luxuriant shrubby
vegetation. Birds were plentiful and very interesting, and I
now saw for the first time many Australian forms that are quite
absent from the islands westward. Small white cockatoos were

abundant, and their loud screams, conspicuous white colour, and pretty yellow crests, rendered them a very important feature in the landscape. This is the most westerly point on the globe where any of the family are to be found. Some small honey-suckers of the genus Ptilotis, and the strange mound-maker (Megapodius gouldii), are also here first met with on the traveller's journey eastward. The last-mentioned bird requires a fuller notice.

one sp. also in Bali —

The Megapodidæ are a small family of birds found only in Australia and the surrounding islands, but extending as far as the Philippines and North-west Borneo. They are allied to the gallinaceous birds, but differ from these and from all others in never sitting upon their eggs, which they bury in sand, earth, or rubbish, and leave to be hatched by the heat of the sun or of fermentation. They are all characterized by very large feet and long curved claws, and most of the species of Megapodius rake and scratch together all kinds of rubbish, dead leaves, sticks, stones, earth, rotten wood, &c., till they form a large mound, often six feet high and twelve feet across, in the middle of which they bury their eggs. The natives can tell by the condition of these mounds whether they contain eggs or not ; and they rob them whenever they can, as the brick-red eggs (as large as those of a swan) are considered a great delicacy. A number of birds are said to join in making these mounds and lay their eggs together, so that sometimes forty or fifty may be found. The mounds are to be met with here and there in dense thickets, and are great puzzles to strangers, who cannot understand who can possibly have heaped together cartloads of rubbish in such out-of-the-way places ; and when they inquire of the natives they are but little wiser, for it almost always appears to them the wildest romance to be told that it is all done by birds. The species found in Lombock is about the size of a small hen, and entirely of dark olive and brown tints. It is a miscellaneous feeder, devouring fallen fruits, earth-worms, snails, and centipedes, but the flesh is white and well-flavoured when properly cooked.

The large green pigeons were still better eating, and were much more plentiful. These fine birds, exceeding our largest tame pigeons in size, abounded on the palm trees, which now bore huge bunches of fruits—mere hard globular nuts, about an inch in diameter, and covered with a dry green skin and a very small portion of pulp. Looking at the pigeon's bill and head, it would seem impossible that it could swallow such large masses, or that it could obtain any nourishment from them ; yet I often shot these birds with several palm-fruits in the crop, which generally burst when they fell to the ground. I obtained here eight species of Kingfishers, among which was a very beautiful new one, named by Mr. Gould, Halcyon fulgidus. It was found always in thickets, away from water, and seemed to feed on snails and insects picked up from the ground after the manner

of the great Laughing Jackass of Australia. The beautiful
little violet and orange species (Ceyx rufidorsa) is found in
similar situations, and darts rapidly along like a flame of fire.
Here also I first met with the pretty Australian Bee-eater
(Merops ornatus). This elegant little bird sits on twigs in open
places, gazing eagerly around, and darting off at intervals to
seize some insect which it sees flying near ; returning afterwards
to the same twig to swallow it. Its long, sharp, curved bill, the
two long narrow feathers in its tail, its beautiful green plumage
varied with rich brown and black and vivid blue on the throat,
render it one of the most graceful and interesting objects a
naturalist can see for the first time.

Of all the birds of Lombock, however, I sought most after
the beautiful ground thrushes (Pitta concinna), and always
thought myself lucky if I obtained one. They were found only
in the dry plains densely covered with thickets, and carpeted at
this season with dead leaves. They were so shy that it was
very difficult to get a shot at them, and it was only after a good
deal of practice that I discovered how to do it. The habit of
these birds is to hop about on the ground, picking up insects,
and on the least alarm to run into the densest thicket or take a
flight close along the ground. At intervals they utter a peculiar
cry of two notes which when once heard is easily recognized, and
they can also be heard hopping along among the dry leaves. My
practice was, therefore, to walk cautiously along the narrow
pathways with which the country abounded, and on detecting
any sign of a Pitta's vicinity to stand motionless and give a
gentle whistle occasionally, imitating the notes as near as
possible. After half an hour's waiting I was often rewarded
by seeing the pretty bird hopping along in the thicket. Then I
would perhaps lose sight of it again, till, having my gun raised
and ready for a shot, a second glimpse would enable me to secure
my prize, and admire its soft puffy plumage and lovely colours.
The upper part is rich soft green, the head jet black with a
stripe of blue and brown over each eye ; at the base of the tail
and on the shoulders are bands of bright silvery blue, and the
under side is delicate buff with a stripe of rich crimson, bordered
with black on the belly. Beautiful grass-green doves, little
crimson and black flower-peckers, large black cuckoos, metallic
king-crows, golden orioles, and the fine jungle-cocks—the origin
of all our domestic breeds of poultry—were among the birds that
chiefly attracted my attention during our stay at Labuan Tring.

The most characteristic feature of the jungle was its thorniness.
The shrubs were thorny ; the creepers were thorny ; the bamboos
even were thorny. Everything grew zigzag and jagged, and in
an inextricable tangle, so that to get through the bush with gun
or net or even spectacles was generally not to be done, and
insect-catching in such localities was out of the question. It
was in such places that the Pittas often lurked, and when shot
it became a matter of some difficulty to secure the bird, and

seldom without a heavy payment of pricks and scratches and torn clothes could the prize be won. The dry volcanic soil and arid climate seem favourable to the production of such stunted and thorny vegetation, for the natives assured me that this was nothing to the thorns and prickles of Sumbawa, whose surface still bears the covering of volcanic ashes thrown out forty years ago by the terrible eruption of Tomboro. Among the shrubs and trees that are not prickly the Apocynaceæ were most abundant, their bilobed fruits of varied form and colour, and often of most tempting appearance, hanging everywhere by the waysides as if to invite to destruction the weary traveller who may be unaware of their poisonous properties. One in particular with a smooth shining skin of a golden orange colour, rivals in appearance the golden apples of the Hesperides, and has great attractions for many birds, from the white cockatoos to the little yellow Zosterops, who feast on the crimson seeds which are displayed when the fruit bursts open. The great palm called "Gubbong" by the natives, a species of Corypha, is the most striking feature of the plains, where it grows by thousands and appears in three different states—in leaf, in flower and fruit, or dead. It has a lofty cylindrical stem about a hundred feet high and two to three feet in diameter ; the leaves are large and fan-shaped, and fall off when the tree flowers, which it does only once in its life in a huge terminal spike, on which are produced masses of a smooth round fruit of a green colour and about an inch in diameter. When these ripen and fall the tree dies, and remains standing a year or two before it falls. Trees in leaf only are far more numerous than those in flower and fruit, while dead trees are scattered here and there among them. The trees in fruit are the resort of the great green fruit pigeons, which have been already mentioned. Troops of monkeys (Macacus cynomolgus) may often be seen occupying a tree, showering down the fruit in great profusion, chattering when disturbed, and making an enormous rustling as they scamper off among the dead palm leaves ; while the pigeons have a loud booming voice more like the roar of a wild beast than the note of a bird.

My collecting operations here were carried on under more than usual difficulties. One small room had to serve for eating, sleeping and working, for storehouse and dissecting-room ; in it were no shelves, cupboards, chairs or tables ; ants swarmed in every part of it, and dogs, cats and fowls entered it at pleasure. Besides this it was the parlour and reception-room of my host, and I was obliged to consult his convenience and that of the numerous guests who visited us. My principal piece of furniture was a box, which served me as a dining-table, a seat while skinning birds, and as the receptacle of the birds when skinned and dried. To keep them free from ants we borrowed, with some difficulty, an old bench, the four legs of which being placed in cocoa-nut shells filled with water kept us tolerably free from these pests. The box and the bench were however literally the only places where anything could be put away, and they were generally well occupied by two insect boxes and about a hundred birds' skins in process of drying. It may therefore be easily conceived that when anything bulky or out of the common way was collected, the question "Where is it to be put?" was rather a difficult one to answer. All animal substances moreover require some time to dry thoroughly, emit a very disagreeable odour while doing so, and are particularly attractive to ants, flies, dogs, rats, cats, and other vermin, calling for especial cautions and constant supervision, which under the circumstances above described were impossible.

My readers may now partially understand why a travelling naturalist of limited means, like myself, does so much less than is expected or than he would himself wish to do. It would be interesting to preserve skeletons of many birds and animals, reptiles and fishes in spirits, skins of the larger animals, remarkable fruits and woods, and the most curious articles of manufacture and commerce; but it will be seen that under the circumstances I have just described it would have been impossible to add these to the collections which were my own more especial favourites. When travelling by boat the difficulties are as great or greater, and they are not diminished when the journey is by land. It was absolutely necessary therefore to limit my collections to certain groups to which I could devote constant personal attention, and thus secure from destruction or decay what had been often obtained by much labour and pains.

While Manuel sat skinning his birds of an afternoon, generally surrounded by a little crowd of Malays and Sassaks (as the indigenes of Lombock are termed), he often held forth to them with the air of a teacher, and was listened to with profound attention. He was very fond of discoursing on the "special providences" of which he believed he was daily the subject. "Allah has been merciful to-day" he would say—for although a Christian he adopted the Mahometan mode of speech—" and has given us some very fine birds; we can do nothing without Him." Then one of the Malays would reply, "To be sure, birds are like mankind; they have their appointed time to die; when that time comes nothing can save them, and if it has not come you cannot kill them." A murmur of assent follows this sentiment, and cries of "Butul! Butul!" (Right, right.) Then Manuel would tell a long story of one of his unsuccessful hunts; —how he saw some fine bird and followed it a long way, and then missed it, and again found it, and shot two or three times at it, but could never hit it. "Ah!" says an old Malay, "its time was not come, and so it was impossible for you to kill it." A doctrine this which is very consoling to the bad marksman, and which quite accounts for the facts, but which is yet somehow not altogether satisfactory.

It is universally believed in Lombock that some men have the power to turn themselves into crocodiles, which they do for the sake of devouring their enemies, and many strange tales are told of such transformations. I was, therefore, rather surprised one evening to hear the following curious fact stated, and as it was not contradicted by any of the persons present, I am inclined to accept it provisionally, as a contribution to the Natural History of the island. A Bornean Malay who had been for many years resident here, said to Manuel, "One thing is strange in this country—the scarcity of ghosts." "How so?" asked Manuel. "Why, you know," said the Malay, "that in our countries to the westward, if a man dies or is killed, we dare not pass near the place at night, for all sorts of noises are heard which show that ghosts are about. But here there are numbers of men killed, and their bodies lie unburied in the fields and by the roadside, and yet you can walk by them at night and never hear or see anything at all, which is not the case in our country, as you know very well." "Certainly I do," said Manuel ; and so it was settled that ghosts were very scarce, if not altogether unknown in Lombock. I would observe, however, that as the evidence is purely negative we should be wanting in scientific caution if we accepted this fact as sufficiently well established.

One evening I heard Manuel, Ali, and a Malay man whispering earnestly together outside the door, and could distinguish various allusions to "krisses," throat-cutting, heads, &c., &c. At length Manuel came in, looking very solemn and frightened, and said to me in English, "Sir—must take care ;—no safe here ;—want cut throat." On further inquiry, I found that the Malay had been telling them that the Rajah had just sent down an order to the village, that they were to get a certain number of heads for an offering in the temples to secure a good crop of rice. Two or three other Malays and Bugis, as well as the Amboyna man in whose house we lived, confirmed this account, and declared that it was a regular thing every year, and that it was necessary to keep a good watch and never go out alone. I laughed at the whole thing, and tried to persuade them that it was a mere tale, but to no effect. They were all firmly persuaded that their lives were in danger. Manuel would not go out shooting alone, and I was obliged to accompany him every morning, but I soon gave him the slip in the jungle. Ali was afraid to go and look for firewood without a companion, and would not even fetch water from the well a few yards behind the house unless armed with an enormous spear. I was quite sure all the time that no such order had been sent or received, and that we were in perfect safety. This was well shown shortly afterwards, when an American sailor ran away from his ship on the east side of the island, and made his way on foot and unarmed across to Ampanam, having met with the greatest hospitality on the whole route. Nowhere would the smallest payment be taken for the food and lodging which were willingly furnished him. On pointing out this fact to Manuel, he replied, "He one bad man—run away from his ship—no one can believe word he say ;" and so I was obliged to leave him in the uncomfortable persuasion that he might any day have his throat cut.

A circumstance occurred here which appeared to throw some light on the cause of the tremendous surf at Ampanam. One evening I heard a strange rumbling noise, and at the same time the house shook slightly. Thinking it might be thunder, I asked, "What is that?" "It is an earthquake," answered Inchi Daud, my host ; and he then told me that slight shocks were occasionally felt there, but he had never known them severe. This happened on the day of the last quarter of the moon, and consequently when the tides were low and the surf usually at its weakest. On inquiry afterwards at Ampanam, I found that no earthquake had been noticed, but that on one night there had been a very heavy surf, which shook the house, and the next day there was a very high tide, the water having flooded Mr. Carter's premises, higher than he had ever known it before. These unusual tides occur every now and then, and are not thought much of ; but by careful inquiry I ascertained that the surf had occurred on the very night I had felt the earthquake at Labuan Tring, nearly twenty miles off. This would seem to indicate, that although the ordinary heavy surf may be due to the swell of the great Southern Ocean confined in a narrow channel, combined with a peculiar form of bottom near the shore, yet the sudden heavy surfs and high tides that occur occasionally in perfectly calm weather, may be due to slight upheavals of the ocean-bed in this eminently volcanic region.

CHAPTER XI.

LOMBOCK : MANNERS AND CUSTOMS OF THE PEOPLE.

HAVING made a very fine and interesting collection of the birds of Labuan Tring, I took leave of my kind host, Inchi Daud, and returned to Ampanam to await an opportunity to reach Macassar. As no vessel had arrived bound for that port, I determined to make an excursion into the interior of the island, accompanied by Mr. Ross, an Englishman born in the Keeling Islands, and now employed by the Dutch Government to settle the affairs of a missionary who had unfortunately become bankrupt here. Mr. Carter kindly lent me a horse, and Mr. Ross took his native groom.

Our route for some distance lay along a perfectly level country, bearing ample crops of rice. The road was straight and generally bordered with lofty trees forming a fine avenue. It was at first sandy, afterwards grassy, with occasional streams

and mud-holes. At a distance of about four miles we reached
Mataram, the capital of the island and the residence of the
Rajah. It is a large village with wide streets bordered by a
magnificent avenue of trees, and low houses concealed behind
mud walls. Within this royal city no native of the lower orders
is allowed to ride, and our attendant, a Javanese, was obliged
to dismount and lead his horse while we rode slowly through.
The abodes of the Rajah and of the High Priest are distinguished
by pillars of red brick constructed with much taste ; but the
palace itself seemed to differ but little from the ordinary houses
of the country. Beyond Mataram and close to it is Karangassam,
the ancient residence of the native or Sassak Rajahs before the
conquest of the island by the Balinese.

Soon after passing Mataram the country began gradually to
rise in gentle undulations, swelling occasionally into low hills
towards the two mountainous tracts in the northern and southern
parts of the island. It was now that I first obtained an adequate
idea of one of the most wonderful systems of cultivation in the
world, equalling all that is related of Chinese industry, and as
far as I know surpassing in the labour that has been bestowed
upon it any tract of equal extent in the most civilized countries
of Europe. I rode through this strange garden utterly amazed,
and hardly able to realize the fact, that in this remote and little
known island, from which all Europeans except a few traders
at the port are jealously excluded, many hundreds of square
miles of irregularly undulating country have been so skilfully
terraced and levelled, and so permeated by artificial channels,
that every portion of it can be irrigated and dried at pleasure.
According as the slope of the ground is more or less rapid, each
terraced plot consists in some places of many acres, in others of
a few square yards. We saw them in every state of cultivation ;
some in stubble, some being ploughed, some with rice-crops in
various stages of growth. Here were luxuriant patches of
tobacco ; there, cucumbers, sweet potatoes, yams, beans or
Indian-corn, varied the scene. In some places the ditches were
dry, in others little streams crossed our road and were distributed over lands about to be sown or planted. The banks
which bordered every terrace rose regularly in horizontal lines
above each other ; sometimes rounding an abrupt knoll and
looking like a fortification, or sweeping round some deep hollow
and forming on a gigantic scale the seats of an amphitheatre.
Every brook and rivulet had been diverted from its bed, and
instead of flowing along the lowest ground were to be found
crossing our road half-way up an ascent, yet bordered by ancient
trees and moss-grown stones so as to have all the appearance of
a natural channel, and bearing testimony to the remote period
at which the work had been done. As we advanced further into
the country, the scene was diversified by abrupt rocky hills, by
steep ravines, and by clumps of bamboos and palm-trees near
houses or villages ; while in the distance the fine range of

mountains of which Lombock peak, eight thousand feet high, is the culminating point, formed a fit background to a view scarcely to be surpassed either in human interest or picturesque beauty.

Along the first part of our road we passed hundreds of women carrying rice, fruit, and vegetables to market ; and further on an almost uninterrupted line of horses laden with rice in bags or in the ear, on their way to the port of Ampanam. At every few miles along the road, seated under shady trees or slight sheds, were sellers of sugar-cane, palm-wine, cooked rice, salted eggs, and fried plantains, with a few other native delicacies. At these stalls a hearty meal may be made for a penny, but we contented ourselves with drinking some sweet palm-wine, a most delicious beverage in the heat of the day. After having travelled about twenty miles we reached a higher and drier region, where, water being scarce, cultivation was confined to the little flats bordering the streams. Here the country was as beautiful as before, but of a different character ; consisting of undulating downs of short turf interspersed with fine clumps of trees and bushes, sometimes the woodland, sometimes the open ground predominating. We only passed through one small patch of true forest, where we were shaded by lofty trees and saw around us a dark and dense vegetation, highly agreeable after the heat and glare of the open country.

At length, about an hour after noon, we reached our destination—the village of Coupang, situated nearly in the centre of the island—and entered the outer court of a house belonging to one of the chiefs with whom my friend Mr. Ross had a slight acquaintance. Here we were requested to seat ourselves under an open shed with a raised floor of bamboo, a place used to receive visitors and hold audiences. Turning our horses to graze on the luxuriant grass of the courtyard, we waited till the great man's Malay interpreter appeared, who inquired our business and informed us that the Pumbuckle (chief) was at the Rajah's house, but would soon be back. As we had not yet breakfasted, we begged he would get us something to eat, which he promised to do as soon as possible. It was however about two hours before anything appeared, when a small tray was brought containing two saucers of rice, four small fried fish, and a few vegetables. Having made as good a breakfast as we could, we strolled about the village, and returning, amused ourselves by conversation, with a number of men and boys who gathered round us ; and by exchanging glances and smiles with a number of women and girls who peeped at us through half-open doors and other crevices. Two little boys named Mousa and Isa (Moses and Jesus) were great friends with us, and an impudent little rascal called Kachang (a bean) made us all laugh by his mimicry and antics.

At length about four o'clock the Pumbuckle made his appearance, and we informed him of our desire to stay with him a few days, to shoot birds and see the country. At this he seemed somewhat disturbed, and asked if we had brought a letter from the Anak Agong (Son of Heaven), which is the title of the Rajah of Lombock. This we had not done, thinking it quite unnecessary; and he then abruptly told us that he must go and speak to his Rajah, to see if we could stay. Hours passed away, night came and he did not return. I began to think we were suspected of some evil designs, for the Pumbuckle was evidently afraid of getting himself into trouble. He is a Sassak prince, and, though a supporter of the present Rajah, is related to some of the heads of a conspiracy which was quelled a few years since.

About five o'clock a pack-horse bearing my guns and clothes arrived, with my men Ali and Manuel, who had come on foot. The sun set, and it soon became dark, and we got rather hungry as we sat wearily under the shed and no one came. Still hour after hour we waited, till about nine o'clock, the Pumbuckle, the Rajah, some priests, and a number of their followers arrived and took their seats around us. We shook hands, and for some minutes there was a dead silence. Then the Rajah asked what we wanted; to which Mr. Ross replied by endeavouring to make them understand who we were, and why we had come, and that we had no sinister intentions whatever; and that we had not brought a letter from the "Anak Agong," merely because we had thought it quite unnecessary. A long conversation in the Bali language then took place, and questions were asked about my guns, and what powder I had, and whether I used shot or bullets; also what the birds were for, and how I preserved them, and what was done with them in England. Each of my answers and explanations was followed by a low and serious conversation which we could not understand, but the purport of which we could guess. They were evidently quite puzzled, and did not believe a word we had told them. They then inquired if we were really English, and not Dutch; and although we strongly asserted our nationality, they did not seem to believe us.

After about an hour, however, they brought us some supper (which was the same as the breakfast, but without the fish), and after it some very weak coffee and pumpkins boiled with sugar. Having discussed this, a second conference took place; questions were again asked, and the answers again commented on. Between whiles lighter topics were discussed. My spectacles (concave glasses) were tried in succession by three or four old men, who could not make out why they could not see through them, and the fact no doubt was another item of suspicion against me. My beard, too, was the subject of some admiration, and many questions were asked about personal peculiarities which it is not the custom to allude to in European society. At length, about one in the morning, the whole party rose to depart, and, after conversing some time at the gate, all went away. We now begged the interpreter, who with a few boys and men remained about us, to show us a place to sleep in, at which he seemed very much surprised, saying he thought we were very well accommodated where we were. It was quite chilly, and we were very thinly clad and had brought no blankets, but all we could get after another hour's talk was a native mat and pillow, and a few old curtains to hang round three sides of the open shed and protect us a little from the cold breeze. We passed the rest of the night very uncomfortably, and determined to return in the morning and not submit any longer to such shabby treatment.

We rose at daybreak, but it was near an hour before the interpreter made his appearance. We then asked to have some coffee and to see the Pumbuckle, as we wanted a horse for Ali, who was lame, and wished to bid him adieu. The man looked puzzled at such unheard-of demands and vanished into the inner court, locking the door behind him and leaving us again to our meditations. An hour passed and no one came, so I ordered the horses to be saddled and the pack-horse to be loaded, and prepared to start. Just then the interpreter came up on horseback, and looked aghast at our preparations. "Where is the Pumbuckle?" we asked. "Gone to the Rajah's," said he. "We are going," said I. "Oh! pray don't," said he; "wait a little; they are having a consultation, and some priests are coming to see you, and a chief is going off to Mataram to ask the permission of the Anak Agong for you to stay." This settled the matter. More talk, more delay, and another eight or ten hours' consultation were not to be endured; so we started at once, the poor interpreter almost weeping at our obstinacy and hurry, and assuring us—"the Pumbuckle would be very sorry, and the Rajah would be very sorry, and if we would but wait all would be right." I gave Ali my horse, and started on foot, but he afterwards mounted behind Mr. Ross's groom, and we got home very well, though rather hot and tired.

At Mataram we called at the house of Gusti Gadioca, one of the princes of Lombock, who was a friend of Mr. Carter's, and who had promised to show me the guns made by native workmen. Two guns were exhibited, one six the other seven feet long, and of a proportionably large bore. The barrels were twisted and well finished, though not so finely worked as ours. The stock was well made, and extended to the end of the barrel. Silver and gold ornament was inlaid over most of the surface, but the locks were taken from English muskets. The Gusti assured me, however, that the Rajah had a man who made locks and also rifled barrels. The workshop where these guns were made and the tool used were next shown us, and were very remarkable. An open shed with a couple of small mud forges were the chief objects visible. The bellows consisted of two bamboo cylinders, with pitsons worked by hand. They move very easily, having a loose stuffing of feathers thickly set round the piston so as to act as a valve, and produce a regular blast. Both cylinders communicate with the same nozzle, one piston rising while the other falls. An oblong piece of iron on the ground was

the anvil, and a small vice was fixed on the projecting root of a
tree outside. These, with a few files and hammers, were literally
the only tools with which an old man makes these fine guns,
finishing them himself from the rough iron and wood.

I was anxious to know how they bored these long barrels,
which seemed perfectly true and are said to shoot admirably;
and, on asking the Gusti, received the enigmatical answer:

GUN-BORING.

"We use a basket full of stones." Being utterly unable to
imagine what he could mean, I asked if I could see how they did
it, and one of the dozen little boys around us was sent to fetch
the .basket. He soon returned with this most extraordinary
boring-machine, the mode of using which the Gusti then explained to me. It was simply a strong bamboo basket, through

the bottom of which was stuck upright a pole about three feet long, kept in its place by a few sticks tied across the top with rattans. The bottom of the pole has an iron ring, and a hole in which four-cornered borers of hardened iron can be fitted. The barrel to be bored is buried upright in the ground, the borer is inserted into it, the top of the stick or vertical shaft is held by a cross-piece of bamboo with a hole in it, and the basket is filled with stones to get the required weight. Two boys turn the bamboo round. The barrels are made in pieces of about eighteen inches long, which are first bored small, and then welded together upon a straight iron rod. The whole barrel is then worked with borers of gradually increasing size, and in three days the boring is finished. The whole matter was explained in such a straightforward manner that I have no doubt the process described to me was that actually used ; although, when examining one of the handsome, well-finished, and serviceable guns, it was very hard to realize the fact, that they had been made from first to last with tools hardly sufficient for an English blacksmith to make a horse-shoe.

The day after we returned from our excursion, the Rajah came to Ampanam to a feast given by Gusti Gadioca, who resides there ; and soon after his arrival we went to have an audience. We found him in a large courtyard sitting on a mat under a shady tree ; and all his followers, to the number of three or four hundred, squatting on the ground in a large circle round him. He wore a sarong or Malay petticoat and a green jacket. He was a man about thirty-five years of age, and of a pleasing countenance, with some appearance of intellect combined with indecision. We bowed, and took our seats on the ground near some chiefs we were acquainted with, for while the Rajah sits no one can stand or sit higher. He first inquired who I was, and what I was doing in Lombock, and then requested to see some of my birds. I accordingly sent for one of my boxes of bird-skins and one of insects, which he examined carefully, and seemed much surprised that they could be so well preserved. We then had a little conversation about Europe and the Russian war, in which all natives take an interest. Having heard much of a country-seat of the Rajah's called Gunong Sari, I took the opportunity to ask permission to visit it and shoot a few birds there, which he immediately granted. I then thanked him and we took our leave.

An hour after, his son came to visit Mr. Carter accompanied by about a hundred followers, who all sat on the ground while he came into the open shed where Manuel was skinning birds. After some time he went into the house, had a bed arranged to sleep a little, then drank some wine, and after an hour or two had dinner brought him from the Gusti's house, which he ate with eight of the principal priests and princes. He pronounced a blessing over the rice and commenced eating first, after which the rest fell to. They rolled up balls of rice in their hands, dipped them in the gravy and swallowed them rapidly, with little pieces of meat and fowl cooked in a variety of ways. A boy fanned the young Rajah while eating. He was a youth of about fifteen, and had already three wives. All wore the kris, or Malay crooked dagger, on the beauty and value of which they greatly pride themselves. A companion of the Rajah's had one with a golden handle, in which were set twenty-eight diamonds and several other jewels. He said it had cost him 700*l*. The sheaths are of ornamental wood and ivory, often covered on one side with gold. The blades are beautifully veined with white metal worked into the iron, and they are kept very carefully. Every man without exception carries a kris, stuck behind into the large waist-cloth which all wear, and it is generally the most valuable piece of property he possesses.

A few days afterwards our long-talked-of excursion to Gunong Sari took place. Our party was increased by the captain and supercargo of a Hamburg ship loading with rice for China. We were mounted on a very miscellaneous lot of Lombock ponies, which we had some difficulty in supplying with the necessary saddles, &c. ; and most of us had to patch up our girths, bridles, or stirrup-leathers, as best we could. We passed through Mataram, where we were joined by our friend Gusti Gadioca, mounted on a handsome black horse, and riding as all the natives do, without saddle or stirrups, using only a handsome saddle-cloth and very ornamental bridle. About three miles further, along pleasant byways, brought us to the place. We entered through a rather handsome brick gateway supported by hideous Hindoo deities in stone. Within was an enclosure with two square fish-ponds and some fine trees ; then another gate-way through which we entered into a park. On the right was a brick house, built somewhat in the Hindoo style, and placed on a high terrace or platform ; on the left a large fish-pond, supplied by a little rivulet which entered it out of the mouth of a gigantic crocodile well executed in brick and stone. The edges of the pond were bricked, and in the centre rose a fantastic and picturesque pavilion ornamented with grotesque statues. The pond was well stocked with fine fish, which come every morning to be fed at the sound of a wooden gong which is hung near for the purpose. On striking it a number of fish immediately came out of the masses of weed with which the pond abounds, and followed us along the margin expecting food. At the same time some deer came out of an adjacent wood, which, from being seldom shot at and regularly fed, are almost tame. The jungle and woods which surrounded the park appearing to abound in birds, I went to shoot a few, and was rewarded by getting several specimens of the fine new kingfisher, Halcyon fulgidus, and the curious and handsome ground thrush, Zoothera andromeda. The former belies its name by not frequenting water or feeding on fish. It lives constantly in low damp thickets picking up ground insects, centipedes, and small mollusca. Altogether I was much pleased with my visit to this place, and it gave me a higher opinion than I had before entertained of the taste of these people, although the style of the buildings and of the sculpture is very much inferior to those of the magnificent ruins in Java. I must now say a few words about the character, manners, and customs of these interesting people.

The aborigines of Lombock are termed Sassaks. They are a Malay race hardly differing in appearance from the people of Malacca or Borneo. They are Mahometans and form the bulk of the population. The ruling classes, on the other hand, are natives of the adjacent island of Bali, and are of the Brahminical religion. The government is an absolute monarchy, but it seems to be conducted with more wisdom and moderation than is usual in Malay countries. The father of the present Rajah conquered the island, and the people seem now quite reconciled to their new rulers, who do not interfere with their religion, and probably do not tax them any heavier than did the native chiefs they have supplanted. The laws now in force in Lombock are very severe. Theft is punished by death. Mr. Carter informed me that a man once stole a metal coffee-pot from his house. He was caught, the pot restored, and the man brought to Mr. Carter to punish as he thought fit. All the natives recommended Mr. Carter to have him "krissed" on the spot ; "for if you don't," said they, "he will rob you again." Mr. Carter, however, let him off, with a warning, that if he ever came inside his premises again he would certainly be shot. A few months afterwards the same man stole a horse from Mr. Carter. The horse was recovered, but the thief was not caught. It is an established rule, that any one found in a house after dark, unless with the owner's knowledge, may be stabbed, his body thrown out into the street or upon the beach, and no questions will be asked.

The men are exceedingly jealous and very strict with their wives. A married woman may not accept a cigar or a sirih leaf from a stranger under pain of death. I was informed that some years ago one of the English traders had a Balinese woman of good family living with him—the connexion being considered quite honourable by the natives. During some festival this girl offended against the law by accepting a flower or some such trifle from another man. This was reported to the Rajah (to some of whose wives the girl was related), and he immediately sent to the Englishman's house, ordering him to give the woman up as she must be "krissed." In vain he begged and prayed, and offered to pay any fine the Rajah might impose, and finally refused to give her up unless he was forced to do so. This the Rajah did not wish to resort to, as he no doubt thought he was acting as much for the Englishman's honour as for his own ; so he appeared to let the matter drop. But some time afterwards, he sent one of his followers to the house, who beckoned the girl to the door, and then saying, "The Rajah sends you this," stabbed her to the heart. More serious infidelity is punished still more cruelly, the woman and her paramour being tied back to back and thrown into the sea, where some large crocodiles are always on the watch to devour the bodies. One such execution took place while I was at Ampanam, but I took a long walk into the country to be out of the way till it was all over, thus missing the opportunity of having a horrible narrative to enliven my somewhat tedious story.

One morning, as we were sitting at breakfast, Mr. Carter's servant informed us that there was an "Amok" in the village —in other words, that a man was "running a muck." Orders were immediately given to shut and fasten the gates of our enclosure; but hearing nothing for some time, we went out, and found there had been a false alarm, owing to a slave having run away, declaring he would "amok," because his master wanted to sell him. A short time before, a man had been killed at a gaming-table, because, having lost half a dollar more than he possessed, he was going to "amok." Another had killed or wounded seventeen people before he could be destroyed. In their wars a whole regiment of these people will sometimes agree to "amok," and then rush on with such energetic desperation as to be very formidable to men not so excited as themselves. Among the ancients these would have been looked upon as heroes or demigods who sacrificed themselves for their country. Here it is simply said—they made "amok."

Macassar is the most celebrated place in the East for "running a muck." There are said to be one or two a month on the average, and five, ten, or twenty persons are sometimes killed or wounded at one of them. It is the national and therefore the honourable mode of committing suicide among the natives of Celebes, and is the fashionable way of escaping from their difficulties. A Roman fell upon his sword, a Japanese rips up his stomach, and an Englishman blows out his brains with a pistol. The Bugis mode has many advantages to one suicidally inclined. A man thinks himself wronged by society—he is in debt and cannot pay—he is taken for a slave or has gambled away his wife or child into slavery—he sees no way of recovering what he has lost, and becomes desperate. He will not put up with such cruel wrongs, but will be revenged on mankind and die like a hero. He grasps his kris-handle, and the next moment draws out the weapon and stabs a man to the heart. He runs on, with the bloody kris in his hand, stabbing at every one he meets. "Amok! Amok!" then resounds through the streets. Spears, krisses, knives and guns are brought out against him. He rushes madly forward, kills all he can—men, women, and children— and dies overwhelmed by numbers amid all the excitement of a battle. And what that excitement is those who have been in one best know, but all who have ever given way to violent passions, or even indulged in violent and exciting exercises, may form a very good idea. It is a delirious intoxication, a temporary

madness that absorbs every thought and every energy. And can
we wonder at the kris-bearing, untaught, brooding Malay preferring such a death, looked upon as almost honourable, to the
cold-blooded details of suicide, if he wishes to escape from over-
whelming troubles, or the merciless clutches of the hangman
and the disgrace of a public execution, when he has taken the
law into his own hands, and too hastily revenged himself upon
his enemy? In either case he chooses rather to "amok."

The great staples of the trade of Lombock as well as of Bali are
rice and coffee; the former grown on the plains, the latter on
the hills. The rice is exported very largely to other islands of
the Archipelago, to Singapore, and even to China, and there are
generally one or more vessels loading in the port. It is brought
into Ampanam on pack-horses, and almost every day a string of
these would come into Mr. Carter's yard. The only money the
natives will take for their rice is Chinese copper cash, twelve
hundred of which go to a dollar. Every morning two large sacks
of this money had to be counted out into convenient sums for payment. From Bali quantities of dried beef and ox-tails are exported, and from Lombock a good many ducks and ponies. The
ducks are a peculiar breed, which have very long flat bodies, and
walk erect almost like penguins. They are generally of a pale
reddish ash colour, and are kept in large flocks. They are very
cheap and are largely consumed by the crews of the rice ships,
by whom they are called Baly-soldiers, but are more generally
known elsewhere as penguin-ducks.

My Portuguese bird-stuffer Fernandez now insisted on breaking his agreement and returning to Singapore; partly from
home-sickness, but more I believe from the idea that his life was
not worth many months purchase among such bloodthirsty and
uncivilized peoples. It was a considerable loss to me, as I had
paid him full three times the usual wages for three months in
advance, half of which was occupied in the voyage and the rest
in a place where I could have done without him, owing to there
being so few insects that I could devote my own time to shooting and skinning. A few days after Fernandez had left, a small
schooner came in bound for Macassar, to which place I took a
passage. As a fitting conclusion to my sketch of these interesting islands, I will narrate an anecdote which I heard of the
present Rajah; and which, whether altogether true or not, well
illustrates native character, and will serve as a means of introducing some details of the manners and customs of the country
to which I have not yet alluded.

CHAPTER XII.

LOMBOCK : HOW THE RAJAH TOOK THE CENSUS.

THE Rajah of Lombock was a very wise man, and he showed his wisdom greatly in the way he took the census. For my readers must know that the chief revenues of the Rajah were derived from a head-tax of rice, a small measure being paid annually by every man, woman, and child in the island. There was no doubt that every one paid this tax, for it was a very light one, and the land was fertile and the people well off; but it had to pass through many hands before it reached the Government storehouses. When the harvest was over the villagers brought their rice to the Kapala kampong, or head of the village ; and no doubt he sometimes had compassion on the poor or sick and passed over their short measure, and sometimes was obliged to grant a favour to those who had complaints against him ; and then he must keep up his own dignity by having his granaries better filled than his neighbours, and so the rice that he took to the "Waidono" that was over his district was generally a good deal less than it should have been. And all the "Waidonos" had of course to take care of themselves, for they were all in debt, and it was so easy to take a little of the Government rice, and there would still be plenty for the Rajah. And the "Gustis" or princes who received the rice from the Waidonos helped themselves likewise, and so when the harvest was all over and the rice tribute was all brought in, the quantity was found to be less each year than the one before. Sickness in one district, and fevers in another, and failure of the crops in a third, were of course alleged as the cause of this falling off ; but when the Rajah went to hunt at the foot of the great mountain, or went to visit a "Gusti" on the other side of the island, he always saw the villages full of people, all looking well-fed and happy. And he noticed that the krisses of his chiefs and officers were getting handsomer and handsomer ; and the handles that were of yellow wood were changed for ivory, and those of ivory were changed for gold, and diamonds and emeralds sparkled on many of them ; and he knew very well which way the tribute-rice went. But as he could not prove it he kept silence, and resolved in his own heart some day to have a census taken, so that he might know the number of his people, and not be cheated out of more rice than was just and reasonable.

But the difficulty was how to get this census. He could not go himself into every village and every house, and count all the people ; and if he ordered it to be done by the regular officers they would quickly understand what it was for, and the census would be sure to agree exactly with the quantity of rice he got last year. It was evident therefore that to answer his purpose no one must suspect why the census was taken; and to make sure of this, no one must know that there was any census taken at all. This was a very hard problem; and the Rajah thought and thought, as hard as a Malay Rajah can be expected to think, but could not solve it; and so he was very unhappy, and did nothing but smoke and chew betel with his favourite wife, and eat scarcely anything; and even when he went to the cock-fight did not seem to care whether his best birds won or lost. For several days he remained in this sad state, and all the court were afraid some evil eye had bewitched the Rajah; and an unfortunate Irish captain who had come in for a cargo of rice and who squinted dreadfully, was very nearly being krissed, but being first brought to the royal presence was graciously ordered to go on board and remain there while his ship stayed in the port.

One morning however, after about a week's continuance of this unaccountable melancholy, a welcome change took place, for the Rajah sent to call together all the chiefs and priests and princes who were then in Mataram, his capital city; and when they were all assembled in anxious expectation, he thus addressed them:

"For many days my heart has been very sick and I knew not why, but now the trouble is cleared away, for I have had a dream. Last night the spirit of the 'Gunong Agong'—the great fire mountain—appeared to me, and told me that I must go up to the top of the mountain. All of you may come with me to near the top, but then I must go up alone, and the great spirit will again appear to me and will tell me what is of great importance to me and to you and to all the people of the island. Now go all of you and make this known through the island, and let every village furnish men to make clear a road for us to go through the forest and up the great mountain."

So the news was spread over the whole island that the Rajah must go to meet the great spirit on the top of the mountain; and every village sent forth its men, and they cleared away the jungle and made bridges over the mountain streams and smoothed the rough places for the Rajah's passage. And when they came to the steep and craggy rocks of the mountain, they sought out the best paths, sometimes along the bed of a torrent, sometimes along narrow ledges of the black rocks; in one place cutting down a tall tree so as to bridge across a chasm, in another constructing ladders to mount the smooth face of a precipice. The chiefs who superintended the work fixed upon the length of each day's journey beforehand according to the nature of the road, and chose pleasant places by the banks of clear streams and in the neighbourhood of shady trees, where they built sheds and huts of bamboo well thatched with the leaves of palm-trees, in which the Rajah and his attendants might eat and sleep at the close of each day.

And when all was ready, the princes and priests and chief men came again to the Rajah, to tell him what had been done and to ask him when he would go up the mountain. And he fixed a day, and ordered every man of rank and authority to accompany him, to do honour to the great spirit who had bid him undertake the journey, and to show how willingly they obeyed his commands. And then there was much preparation throughout the whole island. The best cattle were killed and the meat salted and sun-dried ; and abundance of red peppers and sweet potatoes were gathered ; and the tall pinang-trees were climbed for the spicy betel nut, the sirih-leaf was tied up in bundles, and every man filled his tobacco pouch and lime box to the brim, so that he might not want any of the materials for chewing the refreshing betel during the journey. And the stores of provisions were sent on a day in advance. And on the day before that appointed for starting, all the chiefs both great and small came to Mataram, the abode of the king, with their horses and their servants, and the bearers of their sirih boxes, and their sleeping-mats, and their provisions. And they encamped under the tall Waringin-trees that border all the roads about Mataram, and with blazing fires frighted away the ghouls and evil spirits that nightly haunt the gloomy avenues.

In the morning a great procession was formed to conduct the Rajah to the mountain. And the royal princes and relations of the Rajah mounted their black horses, whose tails swept the ground ; they used no saddle or stirrups, but sat upon a cloth of gay colours ; the bits were of silver and the bridles of many coloured cords. The less important people were on small strong horses of various colours, well suited to a mountain journey ; and all (even the Rajah) were bare-legged to above the knee, wearing only the gay coloured cotton waist-cloth, a silk or cotton jacket, and a large handkerchief tastefully folded round the head. Every one was attended by one or two servants bearing his sirih and betel boxes, who were also mounted on ponies ; and great numbers more had gone on in advance or waited to bring up the rear. The men in authority were numbered by hundreds and their followers by thousands, and all the island wondered what great thing would come of it.

For the first two days they went along good roads and through many villages which were swept clean, and had bright cloths hung out at the windows ; and all the people, when the Rajah came, squatted down upon the ground in respect, and every man riding got off his horse and squatted down also, and many joined the procession at every village. At the place where they stopped for the night, the people had placed stakes along each side of the roads in front of the houses. These were split crosswise at the top, and in the cleft were fastened little clay lamps, and between them were stuck the green leaves of palm-trees, which, dripping with the evening dew, gleamed prettily with the many twinkling lights. And few went to sleep that night till the morning hours, for every house held a knot of eager talkers, and much betel-nut was consumed, and endless were the conjectures what would come of it.

On the second day they left the last village behind them and entered the wild country that surrounds the great mountain, and rested in the huts that had been prepared for them on the banks of a stream of cold and sparkling water. And the Rajah's hunters, armed with long and heavy guns, went in search of deer and wild bulls in the surrounding woods, and brought home the meat of both in the early morning, and sent it on in advance to prepare the mid-day meal. On the third day they advanced as far as horses could go, and encamped at the foot of high rocks, among which narrow pathways only could be found to reach the mountain-top. And on the fourth morning when the Rajah set out, he was accompanied only by a small party of priests and princes with their immediate attendants ; and they toiled wearily up the rugged way, and sometimes were carried by their servants, till they passed up above the great trees, and then among the thorny bushes, and above them again on to the black and burnt rock of the highest part of the mountain.

And when they were near the summit the Rajah ordered them all to halt, while he alone went to meet the great spirit on the very peak of the mountain. So he went on with two boys only who carried his sirih and betel, and soon reached the top of the mountain among great rocks, on the edge of the great gulf whence issue forth continually smoke and vapour. And the Rajah asked for sirih, and told the boys to sit down under a rock and look down the mountain, and not to move till he returned to them. And as they were tired, and the sun was warm and pleasant, and the rock sheltered them from the cold wind, the boys fell asleep. And the Rajah went a little way on under another rock ; and he was tired, and the sun was warm and pleasant, and he too fell asleep.

And those who were waiting for the Rajah thought him a long time on the top of the mountain, and thought the great spirit must have much to say, or might perhaps want to keep him on the mountain always, or perhaps he had missed his way in coming down again. And they were debating whether they should go and search for him, when they saw him coming down with the two boys. And when he met them he looked very grave, but said nothing ; and then all descended together, and the procession returned as it had come ; and the Rajah went to his palace and the chiefs to their villages, and the people to their houses, to tell their wives and children all that had happened, and to wonder yet again what would come of it.

And three days afterwards the Rajah summoned the priests and the princes and the chief men of Mataram, to hear what the great spirit had told him on the top of the mountain. And when they were all assembled, and the betel and sirih had been handed round, he told them what had happened. On the top of the mountain he had fallen into a trance, and the great spirit had appeared to him with a face like burnished gold, and had said—"O Rajah ! much plague and sickness and fevers are coming upon all the earth, upon men and upon horses and upon cattle ; but as you and your people have obeyed me and have come up to my great mountain, I will teach you how you and all the people of Lombock may escape this plague." And all waited anxiously, to hear how they were to be saved from so fearful a calamity. And after a short silence the Rajah spoke again and told them,—that the great spirit had commanded that twelve sacred krisses should be made, and that to make them every village and every district must send a bundle of needles—a needle for every head in the village. And when any grievous disease appeared in any village, one of the sacred krisses should be sent there ; and if every house in that village had sent the right number of needles, the disease would immediately cease ; but if the number of needles sent had not been exact, the kris would have no virtue.

So the princes and chiefs sent to all their villages and communicated the wonderful news ; and all made haste to collect the needles with the greatest accuracy, for they feared that if but one were wanting the whole village would suffer. So one by one the head men of the villages brought in their bundles of needles ; those who were near Mataram came first, and those who were far off came last ; and the Rajah received them with his own hands, and put them away carefully in an inner chamber, in a camphor-wood chest whose hinges and clasps were of silver ; and on every bundle was marked the name of the village and the district from whence it came, so that it might be known that all had heard and obeyed the commands of the great spirit.

And when it was quite certain that every village had sent in its bundle, the Rajah divided the needles into twelve equal parts, and ordered the best steel-worker in Mataram to bring his forge and his bellows and his hammers to the palace, and to make the twelve krisses under the Rajah's eye, and in the sight of all men who chose to see it. And when they were finished, they were wrapped up in new silk and put away carefully until they might be wanted.

Now the journey to the mountain was in the time of the east wind when no rain falls in Lombock. And soon after the krisses were made it was the time of the rice harvest, and the chiefs of districts and of villages brought in their tax to the Rajah according to the number of heads in their villages. And to those that wanted but little of the full amount, the Rajah said nothing ; but when those came who brought only half or a fourth part of what was strictly due, he said to them mildly,

"The needles which you sent from your village were many more than came from such-a-one's village, yet your tribute is less than his ; go back and see who it is that has not paid the tax." And the next year the produce of the tax increased greatly, for they feared that the Rajah might justly kill those who a second time kept back the right tribute. And so the Rajah became very rich, and increased the number of his soldiers, and gave golden jewels. to his wives, and bought fine black horses from the white-skinned Hollanders, and made great feasts when his children were born or were married ; and none of the Rajahs or Sultans among the Malays were so great or so powerful as the Rajah of Lombock.

And the twelve sacred krisses had great virtue. And when any sickness appeared in a village one of them was sent for ; and sometimes the sickness went away, and then the sacred kris was taken back again with great honour, and the head men of the village came to tell the Rajah of its miraculous power, and to thank him. And sometimes the sickness would not go away ; and then everybody was convinced that there had been a mistake in the number of needles sent from that village, and therefore the sacred kris had no effect, and had to be taken back again by the head men with heavy hearts, but still with all honour,—for was not the fault their own ?

CHAPTER XIII.

TIMOR.

(COUPANG, 1857–1859. DELLI, 1861.)

THE island of Timor is about three hundred miles long and sixty wide, and seems to form the termination of the great range of volcanic islands which begins with Sumatra more than two thousand miles to the west. It differs however very remarkably from all the other islands of the chain in not possessing any active volcanoes, with the one exception of Timor Peak near the centre of the island, which was formerly active, but was blown up during an eruption in 1638 and has since been quiescent. In no other part of Timor do there appear to be any recent igneous rocks, so that it can hardly be classed as a volcanic island. Indeed its position is just outside of the great volcanic belt, which extends from Flores through Ombay and Wetter to Banda.

I first visited Timor in 1857, staying a day at Coupang, the chief Dutch town at the west end of the island ; and again in May 1859, when I stayed a fortnight in the same neighbour-hood. In the spring of 1861 I spent four months at Delli, the capital of the Portuguese possessions in the eastern part of the island.

The whole neighbourhood of Coupang appears to have been elevated at a recent epoch, consisting of a rugged surface of coral rock, which rises in a vertical wall between the beach and the town, whose low white red-tiled houses give it an appearance very similar to other Dutch settlements in the East. The vegetation is everywhere scanty and scrubby. Plants of the families Apocynaceæ and Euphorbiaceæ abound ; but there is nothing that can be called a forest, and the whole country has a parched and desolate appearance, contrasting strongly with the lofty forest trees and perennial verdure of the Moluccas or of Singapore. The most conspicuous feature of the vegetation was the abundance of fine fan-leaved palms (Borassus flabelliformis), from the leaves of which are constructed the strong and durable water-buckets in general use, and which are much superior to those formed from any other species of palm. From the same tree, palm-wine and sugar are made, and the common thatch for houses formed of the leaves lasts six or seven years without removal. Close to the town I noticed the foundation of a ruined house below high-water mark, indicating recent subsidence. Earthquakes are not severe here, and are so infrequent and harmless that the chief houses are built of stone.

The inhabitants of Coupang consist of Malays, Chinese, and Dutch, besides the natives ; so that there are many strange and complicated mixtures among the population. There is one resident English merchant, and whalers as well as Australian ships often come here for stores and water. The native Timorese preponderate, and a very little examination serves to show that they have nothing in common with Malays, but are much more closely allied to the true Papuans of the Aru Islands and New Guinea. They are tall, have pronounced features, large, somewhat aquiline noses, and frizzly hair, and are generally of a dusky brown colour. The way in which the women talk to each other and to the men, their loud voices and laughter, and general character of self-assertion, would enable an experienced observer to decide, even without seeing them, that they were not Malays.

Mr. Arndt, a German and the Government doctor, invited me to stay at his house while in Coupang, and I gladly accepted his offer, as I only intended making a short visit. We at first began speaking French, but he got on so badly that we soon passed insensibly into Malay ; and we afterwards held long discussions on literary, scientific, and philosophical questions, in that semi-barbarous language, whose deficiencies we made up by the free use of French or Latin words.

After a few walks in the neighbourhood of the town, I found such a poverty of insects and birds that I determined to go for a few days to the island of Semao at the western extremity of Timor, where I heard that there was forest country with birds not found at Coupang. With some difficulty I obtained a large dug-out boat with out-riggers, to take me over, a distance of about twenty miles. I found the country pretty well wooded, but covered with shrubs and thorny bushes rather than forest trees, and everywhere excessively parched and dried up by the long-continued dry season. I stayed at the village of Oeassa, remarkable for its soap springs. One of these is in the middle of the village bubbling out from a little cone of mud to which the ground rises all round like a volcano in miniature. The water has a soapy feel and produces a strong lather when any greasy substance is washed in it. It contains alkali and iodine, in such quantities as to destroy all vegetation for some distance round. Close by the village is one of the finest springs I have ever seen, contained in several rocky basins communicating by narrow channels. These have been neatly walled where required and partly levelled, and form fine natural baths. The water is well tasted and clear as crystal, and the basins are surrounded by a grove of lofty many-stemmed banyan-trees, which keep them always cool and shady, and add greatly to the picturesque beauty of the scene.

The village consists of curious little houses very different from any I have seen elsewhere. They are of an oval figure, and the walls are made of sticks about four feet high placed close together. From this rises a high conical roof thatched with grass. The only opening is a door about three feet high. The people are like the Timorese with frizzly or wavy hair and of a coppery brown colour. The better class appear to have a mixture of some superior race which has much improved their features. I saw in Coupang some chiefs from the island of Savu further west, who presented characters very distinct from either the Malay or Papuan races. They most resembled Hindoos, having fine well-formed features and straight thin noses with clear brown complexions. As the Brahminical religion once spread over all Java, and even now exists in Bali and Lombock, it is not at all improbable that some natives of India should have reached this island, either by accident or to escape persecution, and formed a permanent settlement there.

I stayed at Oeassa four days, when, not finding any insects and very few new birds, I returned to Coupang to await the next mail steamer. On the way I had a narrow escape of being swamped. The deep coffin-like boat was filled up with my baggage, and with vegetables, cocoa-nuts and other fruit for Coupang market, and when we had got some way across into a rather rough sea, we found that a quantity of water was coming in which we had no means of baling out. This caused us to sink deeper in the water, and then we shipped seas over our sides, and the rowers who had before declared it was nothing now became alarmed, and turned the boat round to get back to the coast of Semao, which was not far off. By clearing away some of the baggage a little of the water could be baled out, but hardly so fast as it came in, and when we neared the coast we found nothing but vertical walls of rock against which the sea was violently beating. We coasted along some distance till we found a little cove into which we ran the boat, hauled it on shore, and emptying it found a large hole in the bottom, which had been temporarily stopped up with a plug of cocoa-nut husk which had come out. Had we been a quarter of a mile further off before we discovered the leak, we should certainly have been obliged to throw most of our baggage overboard, and might easily have lost our lives. After we had put all straight and secure we again started, and when we were half-way across got into such a strong current and high cross sea that we were very nearly being swamped a second time, which made me vow never to trust myself again in such small and miserable vessels.

The mail steamer did not arrive for a week, and I occupied myself in getting as many of the birds as I could, and found some which were very interesting. Among these were five species of pigeons, of as many distinct genera, and most of them peculiar to the island ; two parrots—the fine red-winged broadtail (Platycercus vulneratus) allied to an Australian species, and a green species of the genus Geoffroyus. The Tropidorhynchus timorensis was as ubiquitous and as noisy as I had found it at Lombock ; and the Sphæcothera viridis, a curious green oriole, with bare red orbits, was a great acquisition. There were several pretty finches, warblers, and flycatchers, and among them I obtained the elegant blue and red Cyornis hyacinthina ; but I cannot recognise among my collections the species mentioned by Dampier, who seems to have been much struck by the number of small song-birds in Timor. He says : "One sort of these pretty little birds my men called the ringing bird, because it had six notes, and always repeated all his notes twice, one after the other, beginning high and shrill and ending low. The bird was about the bigness of a lark, having a small sharp black bill and blue wings, the head and breast were of a pale red, and there was a blue streak about its neck." In Semao monkeys are abundant. They are the common hare-lipped monkey (Macacus cynomolgus), which is found all over the western islands of the Archipelago, and may have been introduced by natives, who often carry it about captive. There are also some deer, but it is not quite certain whether they are of the same species as are found in Java.

I arrived at Delli, the capital of the Portuguese possessions in Timor, on January 12, 1861, and was kindly received by Captain Hart, an Englishman and an old resident, who trades in the produce of the country and cultivates coffee on an estate at the foot of the hills. With him I was introduced to Mr. Geach, a mining-engineer who had been for two years endeavouring to discover copper in sufficient quantity to be worth working.

Delli is a most miserable place compared with even the poorest of the Dutch towns. The houses are all of mud and thatch ; the

fort is only a mud inclosure ; and the custom-house and church
are built of the same mean materials, with no attempt at
decoration or even neatness. The whole aspect of the place is
that of a poor native town, and there is no sign of cultivation or
civilization round about it. His Excellency the Governor's
house is the only one that makes any pretensions to appearance,
and that is merely a low white-washed cottage or bungalow.
Yet there is one thing in which civilization exhibits itself.
Officials in black and white European costume, and officers in
gorgeous uniforms, abound in a degree quite disproportionate
to the size or appearance of the place.

The town being surrounded for some distance by swamps
and mud-flats is very unhealthy, and a single night often gives
a fever to new-comers which not unfrequently proves fatal. To
avoid this malaria, Captain Hart always slept at his plantation,
on a slight elevation about two miles from the town, where Mr.
Geach also had a small house, which he kindly invited me to
share. We rode there in the evening ; and in the course of two
days my baggage was brought up, and I was able to look about
me and see if I could do any collecting.

For the first few weeks I was very unwell and could not go
far from the house. The country was covered with low spiny
shrubs and acacias, except in a little valley where a stream came
down from the hills, where some fine trees and bushes shaded
the water and formed a very pleasant place to ramble up. There
were plenty of birds about, and of a tolerable variety of species ;
but very few of them were gaily coloured. Indeed, with one or
two exceptions, the birds of this tropical island were hardly so
ornamental as those of Great Britain. Beetles were so scarce
that a collector might fairly say there were none, as the few
obscure or uninteresting species would not repay him for the
search. The only insects at all remarkable or interesting were
the butterflies, which, though comparatively few in species, were
sufficiently abundant, and comprised a large proportion of new
or rare sorts. The banks of the stream formed my best collecting-
ground, and I daily wandered up and down its shady bed, which
about a mile up became rocky and precipitous. Here I obtained
the rare and beautiful swallow-tail butterflies, Papilio ænomaus
and P. liris ; the males of which are quite unlike each other, and
belong in fact to distinct sections of the genus, while the females
are so much alike that they are undistinguishable on the wing,
and to an uneducated eye equally so in the cabinet. Several
other beautiful butterflies rewarded my search in this place ;
among which I may especially mention the Cethosia leschenaultii,
whose wings of the deepest purple are bordered with buff in such
a manner as to resemble at first sight our own Camberwell beauty,
although it belongs to a different genus. The most abundant
butterflies were the whites and yellows (Pieridæ), several of
which I had already found at Lombock and at Coupang, while
others were new to me.

Early in February we made arrangements to stay for a week at a village called Baliba, situated about four miles off on the mountains, at an elevation of 2,000 feet. We took our baggage and a supply of all necessaries on pack-horses ; and though the distance by the route we took was not more than six or seven miles, we were half a day getting there. The roads were mere tracks, sometimes up steep rocky stairs, sometimes in narrow gullies worn by the horses' feet, and where it was necessary to tuck up our legs on our horses' necks to avoid having them crushed. At some of these places the baggage had to be unloaded, at others it was knocked off. Sometimes the ascent or descent was so steep that it was easier to walk than to cling to our ponies' backs ; and thus we went up and down, over bare hills whose surface was covered with small pebbles and scattered over with Eucalypti, reminding me of what I had read of parts of the interior of Australia rather than of the Malay Archipelago.

The village consisted of three houses only, with low walls, raised a few feet on posts, and very high roofs thatched with grass hanging down to within two or three feet of the ground. A house which was unfinished and partly open at the back was given for our use, and in it we rigged up a table, some benches, and a screen, while an inner enclosed portion served us for a sleeping apartment. We had a splendid view down upon Delli and the sea beyond. The country round was undulating and open, except in the hollows, where there were some patches of forest, which Mr. Geach, who had been all over the eastern part of Timor, assured me was the most luxuriant he had yet seen in the island. I was in hopes of finding some insects here, but was much disappointed, owing perhaps to the dampness of the climate ; for it was not till the sun was pretty high that the mists cleared away, and by noon we were generally clouded up again, so that there was seldom more than an hour or two of fitful sunshine. We searched in every direction for birds and other game, but they were very scarce. On our way I had shot the fine white-headed pigeon, Ptilonopus cinctus, and the pretty little lorikeet, Trichoglossus euteles. I got a few more of these at the blossoms of the Eucalypti, and also the allied species Trichoglossus iris, and a few other small but interesting birds. The common jungle-cock of India (Gallus bankiva) was found here, and furnished us with some excellent meals ; but we could get no deer. Potatoes are grown higher up the mountains in abundance, and are very good. We had a sheep killed every other day, and ate our mutton with much appetite in the cool climate which rendered a fire always agreeable.

Although one-half the European residents in Delli are continually ill from fever, and the Portuguese have occupied the place for three centuries, no one has yet built a house on these fine hills, which, if a tolerable road were made, would be only an hour's ride from the town ; and almost equally good situations might be found on a lower level at half an hour's distance. The fact that potatoes and wheat of excellent quality are grown in abundance at from 3,000 to 3,500 feet elevation, shows what the climate and soil are capable of if properly cultivated. From one to two thousand feet high, coffee would thrive ; and there are hundreds of square miles of country, over which all the varied products which require climates between those of coffee and wheat would flourish ; but no attempt has yet been made to form a single mile of road, or a single acre of plantation !

There must be something very unusual in the climate of Timor to permit of wheat being grown at so moderate an elevation. The grain is of excellent quality, the bread made from it being equal to any I have ever tasted ; and it is universally acknowledged to be unsurpassed by any made from imported European or American flour. The fact that the natives have (quite of their own accord) taken to cultivating such foreign articles as wheat and potatoes, which they bring in small quantities on the backs of ponies by the most horrible mountain tracks, and sell very cheaply at the sea-side, sufficiently indicates what might be done, if good roads were made, and if the people were taught, encouraged, and protected. Sheep also do well on the mountains ; and a breed of hardy ponies in much repute all over the Archipelago, runs half wild ; so that it appears as if this island, so barren-looking and devoid of the usual features of tropical vegetation, were yet especially adapted to supply a variety of products essential to Europeans, which the other islands will not produce, and which they accordingly import from the other side of the globe.

On the 24th of February my friend Mr. Geach left Timor, having finally reported that no minerals worth working were to be found. The Portuguese were very much annoyed, having made up their minds that copper is abundant, and still believing it to be so. It appears that from time immemorial pure native copper has been found at a place on the coast about thirty miles east of Delli. The natives say they find it in the bed of a ravine, and many years ago a captain of a vessel is said to have got some hundreds-weight of it. Now, however, it is evidently very scarce, as during the two years Mr. Geach resided in the country, none was found. I was shown one piece several pounds' weight, having much the appearance of one of the larger Australian nuggets, but of pure copper instead of gold. The natives and the Portuguese have very naturally imagined, that where these fragments come from there must be more ; and they have a report or tradition, that a mountain at the head of the ravine is almost pure copper, and of course of immense value.

After much difficulty a company was at length formed to work the copper mountain, a Portuguese merchant of Singapore supplying most of the capital. So confident were they of the existence of the copper, that they thought it would be waste of

time and money to have any exploration made first; and
accordingly sent to England for a mining-engineer, who was to
bring out all necessary tools, machinery, laboratory utensils,
a number of mechanics, and stores of all kinds for two years,
in order to commence work on a copper-mine which he was
told was already discovered. On reaching Singapore a ship
was freighted to take the men and stores to Timor, where
they at length arrived after much delay, a long voyage, and
very great expense.

A day was then fixed to "open the mines." Captain Hart
accompanied Mr. Geach as interpreter. The Governor, the
Commandante, the Judge, and all the chief people of the place,
went in state to the mountain, with Mr. Geach's assistant and
some of the workmen. As they went up the valley Mr. Geach
examined the rocks, but saw no signs of copper. They went on
and on, but still nothing except a few mere traces of very poor
ore. At length they stood on the copper mountain itself. The
Governor stopped, the officials formed a circle, and he then
addressed them, saying,—that at length the day had arrived they
had all been so long expecting, when the treasures of the soil
of Timor would be brought to light,—and much more in very
grandiloquent Portuguese ; and concluded by turning to Mr.
Geach, and requesting him to point out the best spot for them
to begin work at once, and uncover the mass of virgin copper.
As the ravines and precipices among which they had passed, and
which had been carefully examined, revealed very clearly the
nature and mineral constitution of the country, Mr. Geach
simply told them that there was not a trace of copper there,
and that it was perfectly useless to begin work. The audience
were thunderstruck ! The Governor could not believe his ears.
At length, when Mr. Geach had repeated his statement, the
Governor told him severely that he was mistaken ; that they all
knew there *was* copper there in abundance, and all they wanted
him to tell them, as a mining-engineer, was *how best to get at it;*
and that at all events he was to begin work some where. This Mr.
Geach refused to do, trying to explain, that the ravines had cut
far deeper into the hill than he could do in years, and that he
would not throw away money or time on any such useless
attempt. After this speech had been interpreted to him, the
Governor saw it was no use, and without saying a word
turned his horse and rode away, leaving my friends alone on the
mountain. They all believed there was some conspiracy—that
the Englishman *would not* find the copper, and that they had
been cruelly betrayed.

Mr. Geach then wrote to the Singapore merchant who was his
employer, and it was arranged that he should send the mechanics
home again, and himself explore the country for minerals. At
first the Government threw obstacles in his way and entirely
prevented his moving ; but at length he was allowed to travel
about, and for more than a year he and his assistant explored

the eastern part of Timor, crossing it in several places from sea to sea, and ascending every important valley, without finding any minerals that would pay the expense of working. Copper ore exists in several places, but always too poor in quality. The best would pay well if situated in England ; but in the interior of an utterly barren country, with roads to make, and all skilled labour and materials to import, it would have been a losing concern. Gold also occurs, but very sparingly and of poor quality. A fine spring of pure petroleum was discovered far in the interior, where it can never be available till the country is civilized. The whole affair was a dreadful disappointment to the Portuguese Government, who had considered it such a certain thing that they had contracted for the Dutch mail steamers to stop at Delli ; and several vessels from Australia were induced to come with miscellaneous cargoes, for which they expected to find a ready sale among the population at the newly-opened mines. The lumps of native copper are still, however, a mystery. Mr. Geach has examined the country in every direction without being able to trace their origin ; so that it seems probable that they result from the *débris* of old copper-bearing strata, and are not really more abundant than gold nuggets are in Australia or California. A high reward was offered to any native who should find a piece and show the exact spot where he obtained it, but without effect.

The mountaineers of Timor are a people of Papuan type, having rather slender forms, bushy frizzled hair, and the skin of a dusky brown colour. They have the long nose with overhanging apex which is so characteristic of the Papuan, and so absolutely unknown among races of Malayan origin. On the coast there has been much admixture of some of the Malay races, and perhaps of Hindoo, as well as of Portuguese. The general stature there is lower, the hair wavy instead of frizzled, and the features less prominent. The houses are built on the ground, while the mountaineers raise theirs on posts three or four feet high. The common dress is a long cloth, twisted round the waist and hanging to the knee, as shown in the illustration (page 150), copied from a photograph. Both men carry the national umbrella, made of an entire fan-shaped palm leaf, carefully stitched at the fold of each leaflet to prevent splitting. This is opened out, and held sloping over the head and back during a shower. The small water-bucket is made from an entire unopened leaf of the same palm, and the covered bamboo probably contains honey for sale. A curious wallet is generally carried, consisting of a square of strongly woven cloth, the four corners of which are connected by cords, and often much ornamented with beads and tassels. Leaning against the house behind the figure on the right are bamboos, used instead of water jars.

A prevalent custom is the " pomali," exactly equivalent to the " taboo " of the Pacific islanders, and equally respected. It is used on the commonest occasions, and a few palm leaves stuck outside a garden as a sign of the "pomali" will preserve its produce from thieves as effectually as the threatening notice of man-traps, spring guns, or a savage dog, would do with us. The dead are placed on a stage, raised six or eight feet above

TIMOR MEN. (*From a photograph.*)

the ground, sometimes open and sometimes covered with a roof. Here the body remains till the relatives can afford to make a feast, when it is buried. The Timorese are generally great thieves, but are not bloodthirsty. They fight continually among themselves and take every opportunity of kidnapping unprotected people of other tribes for slaves ; but Europeans may pass anywhere through the country in safety. Except a few half-breeds in the town, there are no native Christians in the island of Timor. The people retain their independence in a great measure, and both dislike and despise their would-be rulers, whether Portuguese or Dutch.

The Portuguese government in Timor is a most miserable one. Nobody seems to care the least about the improvement of the country, and at this time, after three hundred years of occupation, there has not been a mile of road made beyond the town, and there is not a solitary European resident anywhere in the interior. All the Government officials oppress and rob the natives as much as they can, and yet there is no care taken to render the town defensible should the Timorese attempt to attack it. So ignorant are the military officers, that having received a small mortar and some shells, no one could be found who knew how to use them ; and during an insurrection of the natives (while I was at Delli) the officer who expected to be sent against the insurgents was instantly taken ill ! and they were allowed to get possession of an important pass within three miles of the town, where they could defend themselves against ten times the force. The result was that no provisions were brought down from the hills ; a famine was imminent, and the Governor had to send off to beg for supplies from the Dutch Governor of Amboyna.

In its present state Timor is more trouble than profit to its Dutch and Portuguese rulers, and it will continue to be so unless a different system is pursued. A few good roads into the elevated districts of the interior ; a conciliatory policy and strict justice towards the natives, and the introduction of a good system of cultivation as in Java and Northern Celebes, might yet make Timor a productive and valuable island. Rice grows well on the marshy flats which often fringe the coast, and maize thrives in all the lowlands, and is the common food of the natives as it was when Dampier visited the island in 1699. The small quantity of coffee now grown is of very superior quality, and it might be increased to any extent. Sheep thrive, and would always be valuable as fresh food for whalers and to supply the adjacent islands with mutton, if not for their wool ; although it is probable that on the mountains this product might soon be obtained by judicious breeding. Horses thrive amazingly ; and enough wheat might be grown to supply the whole Archipelago if there were sufficient induce- ments to the natives to extend its cultivation, and good roads by which it could be cheaply transported to the coast. Under such a system the natives would soon perceive that European government was advantageous to them. They would begin to save money, and property being rendered secure they would rapidly acquire new wants and new tastes, and become large consumers of European goods. This would be a far surer source of profit to their rulers than imposts and extortion, and would be at the same time more likely to produce peace and obedience, than the mock-military rule which has hitherto proved most ineffective. To inaugurate such a system would however require an immediate outlay of capital, which neither Dutch nor Portuguese seem inclined to make,—and a number of honest and energetic officials, which the latter nation at least seems unable to produce ; so that it is much to be feared that Timor will for many years to come remain in its present state of chronic insurrection and mis-government.[1]

Morality at Delli is at as low an ebb as in the far interior of Brazil, and crimes are connived at which would entail infamy and criminal prosecution in Europe. While I was there it was generally asserted and believed in the place, that two officers had poisoned the husbands of women with whom they were carrying on intrigues, and with whom they immediately co-habited on the death of their rivals. Yet no one ever thought for a moment of showing disapprobation of the crime, or even of considering it a crime at all, the husbands in question being low half-castes, who of course ought to make way for the pleasures of their superiors.

Judging from what I saw myself and by the descriptions of Mr. Geach, the indigenous vegetation of Timor is poor and monotonous. The lower ranges of the hills are everywhere covered with scrubby Eucalypti, which only occasionally grow into lofty forest trees. Mingled with these in smaller quantities are acacias and the fragrant sandal-wood, while the higher mountains, which rise to about six or seven thousand feet, are either covered with coarse grass or are altogether barren. In the lower grounds are a variety of weedy bushes, and open waste places are covered everywhere with a nettle-like wild mint. Here is found the beautiful crown lily, Gloriosa superba, winding among the bushes, and displaying its magnificent blossoms in great profusion. A wild vine also occurs, bearing great irregular bunches of hairy grapes of a coarse but very luscious flavour. In some of the valleys where the vegetation is richer, thorny shrubs and climbers are so abundant as to make the thickets quite impenetrable.[2]

The soil seems very poor, consisting chiefly of decomposing clayey shales ; and the bare earth and rock is almost everywhere visible. The drought of the hot season is so severe that most of the streams dry up in the plains before they reach the sea ; everything becomes burnt up, and the leaves of the larger

---

[1] When Mr. H. O. Forbes visited Delli in 1883, a slight improvement had taken place under a more energetic governor.

[2] Mr. H. O. Forbes collected plants assiduously for six months in the eastern portion of Timor, and obtained about 255 species of flowering plants, a very small number for a tropical island. The total number of species known from the whole island is considerably less than a thousand, although it was visited by the celebrated Robert Brown in 1803, and later by numerous continental botanists and collectors. (*See* Forbes' *Naturalist's Wanderings in the Eastern Archipelago*, pp. 497-523.)

trees fall as completely as in our winter. On the mountains from two to four thousand feet elevation there is a much moister atmosphere, so that potatoes and other European products can be grown all the year round. Besides ponies, almost the only exports of Timor are sandal-wood and bees'-wax. The sandal-wood (Santalum sp.) is the produce of a small tree, which grows sparingly in the mountains of Timor and many of the other islands in the far East. The wood is of a fine yellow colour, and possesses a well-known delightful fragrance which is wonderfully permanent. It is brought down to Delli in small logs, and is chiefly exported to China, where it is largely used to burn in the temples, and in the houses of the wealthy.

The bees'-wax is a still more important and valuable product, formed by the wild bees (Apis dorsata), which build huge honeycombs, suspended in the open air from the under-side of the lofty branches of the highest trees. These are of a semicircular form, and often three or four feet in diameter. I once saw the natives take a bees' nest, and a very interesting sight it was. In the valley where I used to collect insects, I one day saw three or four Timorese men and boys under a high tree, and, looking up, saw on a very lofty horizontal branch three large bees' combs. The tree was straight and smooth-barked and without a branch, till at seventy or eighty feet from the ground it gave out the limb which the bees had chosen for their home. As the men were evidently looking after the bees, I waited to watch their operations. One of them first produced a long piece of wood apparently the stem of a small tree or creeper, which he had brought with him, and began splitting it through in several directions, which showed that it was very tough and stringy. He then wrapped it in palm-leaves, which were secured by twisting a slender creeper round them. He then fastened his cloth tightly round his loins, and producing another cloth wrapped it round his head, neck, and body, and tied it firmly round his neck, leaving his face, arms, and legs completely bare. Slung to his girdle he carried a long thin coil of cord ; and while he had been making these preparations one of his companions had cut a strong creeper or bush-rope eight or ten yards long, to one end of which the wood-torch was fastened, and lighted at the bottom, emitting a steady stream of smoke. Just above the torch a chopping-knife was fastened by a short cord.

The bee-hunter now took hold of the bush-rope just above the torch and passed the other end round the trunk of the tree, holding one end in each hand. Jerking it up the tree a little above his head he set his foot against the trunk, and leaning back began walking up it. It was wonderful to see the skill with which he took advantage of the slightest irregularities of the bark or obliquity of the stem to aid his ascent, jerking the stiff creeper a few feet higher when he had found a firm hold for his bare foot. It almost made me giddy to look at him as he rapidly got up—thirty, forty, fifty feet above the ground ; and I kept wondering how he could possibly mount the next few feet of straight smooth trunk. Still, however, he kept on with as much coolness and apparent certainty as if he were going up a ladder, till he got within ten or fifteen feet of the bees. Then he stopped a moment, and took care to swing the torch (which hung just at his feet) a little towards these dangerous insects, so as to send up the stream of smoke between him and them. Still going on, in a minute more he brought himself under the limb, and, in a manner quite unintelligible to me, seeing that both hands were occupied in supporting himself by the creeper, managed to get upon it.

By this time the bees began to be alarmed, and formed a dense buzzing swarm just over him, but he brought the torch up closer to him, and coolly brushed away those that settled on his arms or legs. Then stretching himself along the limb, he crept towards the nearest comb and swung the torch just under it. The moment the smoke touched it, its colour changed in a most curious manner from black to white, the myriads of bees that had covered it flying off and forming a dense cloud above and around. The man then lay at full length along the limb, and brushed off the remaining bees with his hand, and then drawing his knife cut off the comb at one slice close to the tree, and attaching the thin cord to it, let it down to his companions below. He was all this time enveloped in a crowd of angry bees, and how he bore their stings so coolly, and went on with his work at that giddy height so deliberately, was more than I could understand. The bees were evidently not stupefied by the smoke or driven away far by it, and it was impossible that the small stream from the torch could protect his whole body when at work. There were three other combs on the same tree, and all were successively taken, and furnished the whole party with a luscious feast of honey and young bees, as well as a valuable lot of wax.

After two of the combs had been let down, the bees became rather numerous below, flying about wildly and stinging viciously. Several got about me, and I was soon stung, and had to run away, beating them off with my net and capturing them for specimens. Several of them followed me for at least half a mile, getting into my hair and persecuting me most pertinaciously, so that I was more astonished than ever at the immunity of the natives. I am inclined to think that slow and deliberate motion, and no attempt at escape, are perhaps the best safeguards. A bee settling on a passive native probably behaves as it would on a tree or other inanimate substance, which it does not attempt to sting. Still they must often suffer, but they are used to the pain and learn to bear it impassively, as without doing so no man could be a bee-hunter.

CHAPTER XIV.

THE NATURAL HISTORY OF THE TIMOR GROUP.

IF we look at a map of the Archipelago, nothing seems more unlikely than that the closely connected chain of islands from Java to Timor should differ materially in their natural productions. There are, it is true, certain differences of climate and of physical geography, but these do not correspond with the division the naturalist is obliged to make. Between the two ends of the chain there is a great contrast of climate, the west being exceedingly moist and having only a short and irregular dry season, the east being as dry and parched up, and having but a short wet season. This change, however, occurs about the middle of Java, the eastern portion of that island having as strongly marked seasons as Lombock and Timor. There is also a difference in physical geography; but this occurs at the eastern termination of the chain, where the volcanoes which are the marked feature of Java, Bali, Lombock, Sumbawa, and Flores, turn northwards through Gunong Api to Banda, leaving Timor with only one volcanic peak near its centre; while the main portion of the island consists of old sedimentary rocks. Neither of these physical differences corresponds with the remarkable change in natural productions which occurs at the Straits of Lombock, separating the island of that name from Bali; and which is at once so large in amount and of so fundamental a character, as to form an important feature in the zoological geography of our globe.

The Dutch naturalist Zollinger, who resided a long time in the island of Bali, informs us that its productions completely assimilate with those of Java, and that he is not aware of a single animal found in it which does not inhabit the larger island. During the few days which I stayed on the north coast of Bali on my way to Lombock, I saw several birds highly characteristic of Javan ornithology. Among these were the yellow-headed weaver (Ploceus hypoxanthus), the black grasshopper thrush (Copsychus amoenus), the rosy barbet (Megalæma rosea), the Malay oriole (Oriolus horsfieldi), the Java ground starling (Sturnopastor jalla), and the Javanese three-toed woodpecker (Chrysonotus tiga). On crossing over to Lombock, separated from Bali by a strait less than twenty miles wide, I naturally expected to meet with some of these birds again; but during a stay there of three months I never saw one of them, but found a totally different set of species, most of which were utterly unknown not only in Java, but also in Borneo, Sumatra, and Malacca. For example, among the commonest birds in

Lombock were white cockatoos and three species of Melipha-
gidæ or honey-suckers, belonging to family groups which are
entirely absent from the western or Indo-Malayan region of
the Archipelago. On passing to Flores and Timor the distinctness from the Javanese productions increases, and we find that
these islands form a natural group, whose birds are related to
those of Java and Australia, but are quite distinct from either.
Besides my own collections in Lombock and Timor, my assistant
Mr. Allen made a good collection in Flores ; and these, with a
few species obtained by the Dutch naturalists, enable us to
form a very good idea of the natural history of this group of
islands, and to derive therefrom some very interesting results.

The number of birds known from these islands up to this
date, is,—63 from Lombock, 86 from Flores, and 118 from Timor ;
and from the whole group 188 species.[1] With the exception of
two or three species which appear to have been derived from
the Moluccas, all these birds can be traced, either directly or by
close allies, to Java on the one side or to Australia on the other ;
although no less than 82 of them are found nowhere out of this
small group of islands. There is not, however, a single genus
peculiar to the group, or even one which is largely represented
in it by peculiar species ; and this is a fact which indicates that
the fauna is strictly derivative, and that its origin does not go
back beyond one of the most recent geological epochs. Of
course there are a large number of species (such as most of the
waders, many of the raptorial birds, some of the kingfishers,
swallows, and a few others), which range so widely over a large
part of the Archipelago, that it is impossible to trace them as
having come from any one part rather than from another.
There are fifty-seven such species in my list, and besides these
there are thirty-five more which, though peculiar to the Timor
group, are yet allied to wide-ranging forms. Deducting these
ninety-two species, we have nearly a hundred birds left whose
relations with those of other countries we will now consider.

If we first take those species which, as far as we yet know,
are absolutely confined to each island, we find, in—

Lombock 4, belonging to 2 genera, of which 1 is Australian, 1 Indian.
Flores . 12   ,,   7   ,,   5 are   ,,   2 ,,
Timor . 42   ,,   20   ,,   16   ,,   4 ,,

The actual number of peculiar species in each island I do not
suppose to be at all accurately determined, since the rapidly
increasing numbers evidently depend upon the more extensive
collections made in Timor than in Flores, and in Flores than in
Lombock ; but what we can depend more upon, and what is of
more especial interest, is the greatly increased proportion of
Australian forms and decreased proportion of Indian forms, as

[1] Four or five new species have been since added from the island of Sumbawa. (See
Guillemard's *Cruise of the Marchesa*, Vol. II., p. 864.)

we go from west to east. We shall show this in a yet more striking manner by counting the number of species identical with those of Java and Australia respectively in each island, thus :

	In Lombock.	In Flores.	In Timor.
Javan birds	33	23	11
Australian birds	4	5	10

Here we see plainly the course of the migration which has been going on for hundreds or thousands of years, and is still going on at the present day. Birds entering from Java are most numerous in the island nearest Java ; each strait of the sea to be crossed to reach another island offers an obstacle, and thus a smaller number get over to the next island.[1] It will be observed that the number of birds that appear to have entered from Australia is much less than those which have come from Java ; and we may at first sight suppose that this is due to the wide sea that separates Australia from Timor. But this would be a hasty and, as we shall soon see, an unwarranted supposition. Besides these birds identical with species inhabiting Java and Australia, there are a considerable number of others very closely allied to species peculiar to those countries, and we must take these also into account before we form any conclusion on the matter. It will be as well to combine these with the former table thus :

	In Lombock.	In Flores.	In Timor.
Javan birds	33	23	11
Closely allied to Javan birds	1	5	6
Total	34	28	17
Australian birds	4	5	10
Closely allied to Australian birds	3	9	26
Total	7	14	36

We now see that the total number of birds which seem to have been derived from Java and Australia is very nearly equal, but there is this remarkable difference between the two series : that whereas the larger proportion by far of the Java set are identical with those still inhabiting that country, an almost equally large proportion of the Australian set are distinct, though often very closely allied species. It is to be observed also, that these representative or allied species diminish in number as they recede from Australia, while they increase in number as they recede from Java. There are two reasons for this, one being that the islands decrease rapidly in

[1] The names of all the birds inhabiting these islands are to be found in the "Proceedings of the Zoological Society of London" for the year 1863.

size from Timor to Lombock, and can therefore support a decreasing number of species ; the other and the more important
is, that the distance of Australia from Timor cuts off the supply
of fresh immigrants, and has thus allowed variation to have
full play ; while the vicinity of Lombock to Bali and Java has
allowed a continual influx of fresh individuals which, by
crossing with the earlier immigrants, has checked variation.

To simplify our view of the derivative origin of the birds of
these islands let us treat them as a whole, and thus perhaps
render more intelligible their respective relations to Java and
Australia.

The Timor group of islands contains :—

Javan birds 36	Australian birds 13
Closely allied species . . . 11	Closely allied species . . . 35
Derived from Java . . . 47	Derived from Australia . . 48

We have here a wonderful agreement in the number of birds
belonging to Australian and Javanese groups, but they are
divided in exactly a reverse manner, three-fourths of the Javan
birds being identical species and one-fourth representatives,
while only one-fourth of the Australian forms are identical and
three-fourths representatives. This is the most important fact
which we can elicit from a study of the birds of these islands,
since it gives us a very complete clue to much of their past
history.[1]

Change of species is a slow process. On that we are all
agreed, though we may differ about how it has taken place.
The fact that the Australian species in these islands have
mostly changed, while the Javan species have almost all remained unchanged, would therefore indicate that the district
was first peopled from Australia. But, for this to have been
the case, the physical conditions must have been very different
from what they are now. Nearly three hundred miles of open
sea now separate Australia from Timor, which island is connected with Java by a chain of broken land divided by straits
which are nowhere more than about twenty miles wide.
Evidently there are now great facilities for the natural productions of Java to spread over and occupy the whole of these
islands, while those of Australia would find very great difficulty in getting across. To account for the present state of
things, we should naturally suppose that Australia was once
much more closely connected with Timor than it is at present ;
and that this was the case is rendered highly probable by the
fact of a submarine bank extending along all the north and

[1] The new species of birds discovered in the group since this was written are so few,
and so equally distributed between the two regions, that they will not affect the
conclusions here arrived at.

west coast of Australia, and at one place approaching within
twenty miles of the coast of Timor. This indicates a recent
subsidence of North Australia, which probably once extended
as far as the edge of this bank, between which and Timor there
is an unfathomed depth of ocean.

I do not think that Timor was ever actually connected with
Australia, because such a large number of very abundant and
characteristic groups of Australian birds are quite absent, and
not a single Australian mammal has entered Timor ; which
would certainly not have been the case had the lands been
actually united. Such groups as the bower birds (Ptilono-
rhynchus), the black and red cockatoos (Calyptorhynchus), the
blue wrens (Malurus), the crowshrikes (Cracticus), the Australian
shrikes (Falcunculus and Colluricincla), and many others, which
abound all over Australia, would certainly have spread into
Timor if it had been united to that country, or even if for any
long time it had approached nearer to it than twenty miles.
Neither do any of the most characteristic groups of Australian
insects occur in Timor ; so that everything combines to indicate
that a strait of the sea has always separated it from Australia,
but that at one period this strait was reduced to a width of
about twenty miles.

But at the time when this narrowing of the sea took place in
one direction, there must have been a greater separation at the
other end of the chain, or we should find more equality in the
numbers of identical and representative species derived from
each extremity. It is true that the widening of the strait at the
Australian end by subsidence, would, by putting a stop to immigration and inter-crossing of individuals from the mother
country, have allowed full scope to the causes which have led to
the modification of the species ; while the continued stream of
immigrants from Java, would by continual intercrossing, check
such modification. This view will not, however, explain all the
facts ; for the character of the fauna of the Timorese group is
indicated as well by the forms which are absent from it as by
those which it contains, and is by this kind of evidence shown
to be much more Australian than Indian. No less than twenty-
nine genera, all more or less abundant in Java, and most of
which range over a wide area, are altogether absent ; while of
the equally diffused Australian genera only about fourteen are
wanting. This would clearly indicate that there has been, till
recently, a wide separation from Java ; and the fact that the
islands of Bali and Lombock are small, and are almost wholly
volcanic, and contain a smaller number of modified forms than
the other islands, would point them out as of comparatively
recent origin. A wide arm of the sea probably occupied their
place at the time when Timor was in the closest proximity to
Australia ; and as the subterranean fires were slowly piling up
the now fertile islands of Bali and Lombock, the northern shores
of Australia would be sinking beneath the ocean. Some such

changes as have been here indicated, enable us to understand
how it happens, that though the birds of this group are on the
whole almost as much Indian as Australian, yet the species
which are peculiar to the group are mostly Australian in
character ; and also why such a large number of common Indian
forms which extend through Java and Bali, should not have
transmitted a single representative to the islands further east.

The Mammalia of Timor as well as those of the other islands
of the group are exceedingly scanty, with the exception of bats.
These last are tolerably abundant, and no doubt many more
remain to be discovered. Out of fifteen species known from
Timor, nine are found also in Java, or the islands west of it ;
three are Moluccan species, most of which are also found in
Australia, and the rest are peculiar to Timor.

The land mammals are only six in number, as follows : 1. The
common monkey, Macacus cynomolgus, which is found in all the
Indo-Malayan islands, and has spread from Java through Bali
and Lombock to Timor. This species is very frequent on the
banks of rivers, and may have been conveyed from island to
island on trees carried down by floods. 2. Paradoxurus fas-
ciatus ; a civet cat, very common over a large part of the
Archipelago. 3. Cervus timoriensis ; a deer, closely allied to the
Javan and Moluccan species, if distinct. 4. A wild pig, Sus
timoriensis ; perhaps the same as some of the Moluccan species.
5. A shrew mouse, Sorex tenuis ; supposed to be peculiar to
Timor. 6. An Eastern opossum, Cuscus orientalis ; found also
in the Moluccas, if not a distinct species.

The fact that not one of these species is Australian, or nearly
allied to any Australian form, is strongly corroborative of the
opinion that Timor has never formed a part of that country ; as
in that case some kangaroo or other marsupial animal would
almost certainly be found there. It is no doubt very difficult
to account for the presence of some of the few mammals that
do exist in Timor, especially the deer. We must consider,
however, that during thousands, and perhaps hundreds of
thousands of years, these islands and the seas between them
have been subjected to volcanic action. The land has been
raised and has sunk again ; the straits have been narrowed or
widened ; many of the islands may have been joined and dissevered again ; violent floods have again and again devastated
the mountains and plains, carrying out to sea hundreds of forest
trees, as has often happened during volcanic eruptions in Java ;
and it does not seem improbable that once in a thousand, or ten
thousand years, there should have occurred such a favourable
combination of circumstances as would lead to the migration of
two or three land animals from one island to another. This is
all that we need ask to account for the very scanty and fragmentary group of Mammalia which now inhabit the large island
of Timor. The deer may very probably have been introduced
by man, for the Malays often keep tame fawns ; and it may not

require a thousand, or even five hundred years, to establish new characters in an animal removed to a country so different in climate and vegetation as is Timor from the Moluccas. I have not mentioned horses, which are often thought to be wild in Timor, because there are no grounds whatever for such a belief. The Timor ponies have every one an owner, and are quite as much domesticated animals as the cattle on a South American hacienda.

I have dwelt at some length on the origin of the Timorese fauna, because it appears to me a most interesting and instructive problem. It is very seldom that we can trace the animals of a district so clearly as we can in this case, to two definite sources; and still more rarely that they furnish such decisive evidence, of the time, and the manner, and the proportions of their introduction. We have here a group of Oceanic Islands in miniature—islands which have never formed part of the adjacent lands, although so closely approaching them ; and their productions have the characteristics of true Oceanic Islands slightly modified. These characteristics are, the absence of all Mammalia except bats, and the occurrence of peculiar species of birds, insects, and land shells, which, though found nowhere else, are plainly related to those of the nearest land. Thus, we have an entire absence of Australian mammals, and the presence of only a few stragglers from the west, which can be accounted for in the manner already indicated. Bats are tolerably abundant. Birds have many peculiar species, with a decided relationship to those of the two nearest masses of land. The insects have similar relations with the birds. As an example, four species of the Papilionidæ are peculiar to Timor, three others are also found in Java, and one in Australia. Of the four peculiar species two are decided modifications of Javanese forms, while the others seem allied to those of the Moluccas and Celebes. The very few land shells known are all, curiously enough, allied to or identical with Moluccan or Celebes forms. The Pieridæ (white and yellow butterflies) which wander more, and from frequenting open grounds are more liable to be blown out to sea, seem about equally related to those of Java, Australia, and the Moluccas.

It has been objected to Mr. Darwin's theory,—of Oceanic Islands having never been connected with the mainland,—that this would imply that their animal population was a matter of chance ; it has been termed the "*flotsam* and *jetsam* theory," and it has been maintained that nature does not work by the "chapter of accidents." But in the case which I have here described, we have the most positive evidence that such *has* been the mode of peopling the islands. Their productions *are* of that miscellaneous character which we should expect from such an origin ; and to suppose that they have been portions of Australia or of Java will introduce perfectly gratuitous difficulties, and render it quite impossible to explain those curious relations which the best known group of animals (the birds) have been shown to exhibit. On the other hand, the depth of the surrounding seas, the form of the submerged banks, and the volcanic character of most of the islands, all point to an independent origin.

Before concluding, I must make one remark to avoid misapprehension. When I say that Timor has never formed part of Australia, I refer only to recent geological epochs. In Secondary or even Eocene or Miocene times, Timor and Australia may have been connected ; but if so, all record of such a union has been lost by subsequent submergence ; and in accounting for the present land-inhabitants of any country we have only to consider these changes which have occurred since its last elevation above the waters. Since such last elevation, I feel confident that Timor has not formed part of Australia.

CHAPTER XV.

CELEBES.

(MACASSAR. SEPTEMBER TO NOVEMBER, 1856.)

I LEFT Lombock on the 30th of August, and reached Macassar in three days. It was with great satisfaction that I stepped on a shore which I had been vainly trying to reach since February, and where I expected to meet with so much that was new and interesting.

The coast of this part of Celebes is low and flat, lined with trees and villages so as to conceal the interior, except at occasional openings which show a wide extent of bare and marshy rice-fields. A few hills, of no great height, were visible in the background ; but owing to the perpetual haze over the land at this time of the year, I could nowhere discern the high central range of the peninsula, or the celebrated peak of Bontyne at its southern extremity. In the roadstead of Macassar there was a fine 42-gun frigate, the guardship of the place, as well as a small war steamer and three or four little cutters used for cruising after the pirates which infest these seas. There were also a few square-rigged trading-vessels, and twenty or thirty native praus of various sizes. I brought letters of introduction to a Dutch gentleman, Mr. Mesman, and also to a Danish shopkeeper, who could both speak English, and who promised to assist me in finding a place to stay at, suitable for my pursuits. In the meantime, I went to a kind of club-house, in default of any hotel in the place.

Macassar was the first Dutch town I had visited, and I found it prettier and cleaner than any I had yet seen in the East. The Dutch have some admirable local regulations. All European houses must be kept well whitewashed, and every person must, at four in the afternoon, water the road in front of his house. The streets are kept clear of refuse, and covered drains carry away all impurities into large open sewers, into which the tide is admitted at high-water and allowed to flow out when it has ebbed, carrying all the sewage with it into the sea. The town consists chiefly of one long narrow street, along the sea-side, devoted to business, and principally occupied by the Dutch and Chinese merchants' offices and warehouses, and the native shops or bazaars. This extends northwards for more than a mile, gradually merging into native houses, often of a most miserable description, but made to have a neat appearance by being all built up exactly to the straight line of the street, and being generally backed by fruit trees. This street is usually thronged with a native population of Bugis and Macassar men, who wear cotton trousers about twelve inches long, covering only from the hip to half-way down the thigh, and the universal Malay sarong, of gay checked colours, worn round the waist or across the shoulders in a variety of ways. Parallel to this street run two short ones, which form the old Dutch town, and are enclosed by gates. These consist of private houses, and at their southern end is the fort, the church, and a road at right angles to the beach, containing the houses of the Governor and of the principal officials. Beyond the fort again, along the beach, is another long street of native huts and many country houses of the tradesmen and merchants. All around extend the flat rice-fields, now bare and dry and forbidding, covered with dusty stubble and weeds. A few months back these were a mass of verdure, and their barren appearance at this season offered a striking contrast to the perpetual crops on the same kind of country in Lombock and Bali, where the seasons are exactly similar, but where an elaborate system of irrigation produces the effect of a perpetual spring.

The day after my arrival I paid a visit of ceremony to the Governor, accompanied by my friend the Danish merchant, who spoke excellent English. His Excellency was very polite, and offered me every facility for travelling about the country and prosecuting my researches in natural history. We conversed in French, which all Dutch officials speak very well.

Finding it very inconvenient and expensive to stay in the town, I removed at the end of a week to a little bamboo house, kindly offered me by Mr. Mesman. It was situated about two miles away, on a small coffee-plantation and farm, and about a mile beyond Mr. M.'s own country house. It consisted of two rooms raised about seven feet above the ground, the lower part being partly open (and serving excellently to skin birds in) and partly used as a granary for rice. There was a kitchen and other outhouses, and several cottages near were occupied by men in Mr. M.'s employ.

After being settled a few days in my new house, I found that no collections could be made without going much further into the country. The rice-fields for some miles round resembled

English stubbles late in autumn, and were almost as unproductive of bird or insect life. There were several native villages scattered about, so embosomed in fruit trees that at a distance they looked like clumps or patches of forest. These were my only collecting places, but they produced a very limited number of species, and were soon exhausted. Before I could move to any more promising district it was necessary to obtain permission from the Rajah of Goa, whose territories approach to within two miles of the town of Macassar. I therefore presented myself at the Governor's office and requested a letter to the Rajah, to claim his protection, and permission to travel in his territories whenever I might wish to do so. This was immediately granted, and a special messenger was sent with me to carry the letter.

My friend Mr. Mesman kindly lent me a horse, and accompanied me on my visit to the Rajah, with whom he was great friends. We found his Majesty seated out of doors, watching the erection of a new house. He was naked from the waist up, wearing only the usual short trousers and sarong. Two chairs were brought out for us, but all the chiefs and other natives were seated on the ground. The messenger, squatting down at the Rajah's feet, produced the letter, which was sewn up in a covering of yellow silk. It was handed to one of the chief officers, who ripped it open and returned it to the Rajah, who read it, and then showed it to Mr. M., who both speaks and reads the Macassar language fluently, and who explained fully what I required. Permission was immediately granted me to go where I liked in the territories of Goa, but the Rajah desired, that should I wish to stay any time at a place I would first give him notice, in order that he might send some one to see that no injury was done me. Some wine was then brought us, and afterwards some detestable coffee and wretched sweetmeats, for it is a fact that I have never tasted good coffee where people grow it themselves.

Although this was the height of the dry season, and there was a fine wind all day, it was by no means a healthy time of year. My boy Ali had hardly been a day on shore when he was attacked by fever, which put me to great inconvenience, as at the house where I was staying nothing could be obtained but at meal-times. After having cured Ali, and with much difficulty got another servant to cook for me, I was no sooner settled at my country abode than the latter was attacked with the same disease ; and, having a wife in the town, left me. Hardly was he gone than I fell ill myself, with strong intermittent fever every other day. In about a week I got over it, by a liberal use of quinine, when scarcely was I on my legs than Ali became worse than ever. His fever attacked him daily, but early in the morning he was pretty well, and then managed to cook me enough for the day. In a week I cured him and also succeeded in getting another boy who could cook and shoot, and had no objection to go into the interior. His name was Baderoon, and as he was unmarried and had been used to a roving life, having been several voyages to North Australia to catch trepang or "bêche de mer," I was in hopes of being able to keep him. I also got hold of a little impudent rascal of twelve or fourteen, who could speak some Malay, to carry my gun or insect-net and make himself generally useful. Ali had by this time become a pretty good bird-skinner, so that I was fairly supplied with servants.

I made many excursions into the country, in search of a good station for collecting birds and insects. Some of the villages a few miles inland are scattered about in woody ground which has once been virgin forest, but of which the constituent trees have been for the most part replaced by fruit trees, and particularly by the large palm, Arenga saccharifera, from which wine and sugar are made, and which also produces a coarse black fibre used for cordage. That necessary of life, the bamboo, has also been abundantly planted. In such places I found a good many birds, among which were the fine cream-coloured pigeon, Carpophaga luctuosa, and the rare blue-headed roller, Coracias temmincki, which has a most discordant voice, and generally goes in pairs, flying from tree to tree, and exhibiting while at rest that all-in-a-heap appearance and jerking motion of the head and tail which are so characteristic of the great Fissirostral group to which it belongs. From this habit alone, the kingfishers, bee-eaters, rollers, trogons, and South American puff-birds, might be grouped together by a person who had observed them in a state of nature, but who had never had an opportunity of examining their form and structure in detail. Thousands of crows, rather smaller than our rook, keep up a constant cawing in these plantations ; the curious wood-swallows (Artami), which closely resemble swallows in their habits and flight but differ much in form and structure, twitter from the tree-tops ; while a lyre-tailed drongo-shrike, with brilliant black plumage and milk-white eyes, continually deceives the naturalist by the variety of its unmelodious notes.

In the more shady parts butterflies were tolerably abundant ; the most common being species of Euplæa and Danais, which frequent gardens and shrubberies, and owing to their weak flight are easily captured. A beautiful pale blue and black butterfly, which flutters along near the ground among the thickets, and settles occasionally upon flowers, was one of the most striking ; and scarcely less so, was one with a rich orange band on a blackish ground : these both belong to the Pieridæ, the group that contains our common white butterflies, although differing so much from them in appearance. Both were quite new to European naturalists.[1] Now and then I extended my walks some miles further, to the only patch of true forest I

---

[1] The former has been named Eronia tritæa ; the latter Tachyris ithome.

could find, accompanied by my two boys with guns and insect-
net. We used to start early, taking our breakfast with us, and
eating it wherever we could find shade and water. At such
times my Macassar boys would put a minute fragment of rice
and meat or fish on a leaf, and lay it on a stone or stump as an
offering to the deity of the spot; for though nominal Mahometans
the Macassar people retain many pagan superstitions, and are
but lax in their religious observances. Pork, it is true, they
hold in abhorrence, but will not refuse wine when offered them,
and consume immense quantities of "sagueir," or palm-wine,
which is about as intoxicating as ordinary beer or cider. When
well made it is a very refreshing drink, and we often took a
draught at some of the little sheds dignified by the name of
bazaars, which are scattered about the country wherever there
is any traffic.

One day Mr. Mesman told me of a larger piece of forest
where he sometimes went to shoot deer, but he assured me it
was much further off, and that there were no birds. However,
I resolved to explore it, and the next morning at five o'clock we
started, carrying our breakfast and some other provisions with
us, and intending to stay the night at a house on the borders of
the wood. To my surprise two hours' hard walking brought us
to this house, where we obtained permission to pass the night.
We then walked on, Ali and Baderoon with a gun each, Baso
carrying our provisions and my insect-box, while I took only my
net and collecting-bottle and determined to devote myself
wholly to the insects. Scarcely had I entered the forest
when I found some beautiful little green and gold speckled
weevils allied to the genus Pachyrhynchus, a group which is
almost confined to the Philippine Islands, and is quite unknown
in Borneo, Java, or Malacca. The road was shady and apparently
much trodden by horses and cattle, and I quickly obtained some
butterflies I had not before met with. Soon a couple of reports
were heard, and coming up to my boys I found they had
shot two specimens of one of the finest of known cuckoos,
Phœnicophaus callirhynchus. This bird derives its name from
its large bill being coloured of a brilliant yellow, red, and black,
in about equal proportions. The tail is exceedingly long, and
of a fine metallic purple, while the plumage of the body is light
coffee brown. It is one of the characteristic birds of the island
of Celebes, to which it is confined.

After sauntering along for a couple of hours we reached a
small river, so deep that horses could only cross it by swimming,
so we had to turn back; but as we were getting hungry, and the
water of the almost stagnant river was too muddy to drink, we
went towards a house a few hundred yards off. In the plantation we saw a small raised hut, which we thought would do
well for us to breakfast in, so I entered, and found inside a
young woman with an infant. She handed me a jug of water,
but looked very much frightened. However, I sat down on the

door-step, and asked for the provisions. In handing them up, Baderoon saw the infant, and started back as if he had seen a serpent. It then immediately struck me that this was a hut in which, as among the Dyaks of Borneo and many other savage tribes, the women are secluded for some time after the birth of their child, and that we did very wrong to enter it ; so we walked off and asked permission to eat our breakfast in the family mansion close at hand, which was of course granted. While I ate, three men, two women, and four children watched every motion, and never took eyes off me till I had finished.

On our way back in the heat of the day I had the good fortune to capture three specimens of a fine Ornithoptera, the largest, the most perfect, and the most beautiful of butterflies. I trembled with excitement as I took the first out of my net and found it to be in perfect condition. The ground colour of this superb insect was a rich shining bronzy black, the lower wings delicately grained with white, and bordered by a row of large spots of the most brilliant and satiny yellow. The body was marked with shaded spots of white, yellow, and fiery orange, while the head and thorax were intense black. On the underside the lower wings were satiny white, with the marginal spots half black and half yellow. I gazed upon my prize with extreme interest, as I at first thought it was quite a new species. It proved however to be a variety of Ornithoptera remus, one of the rarest and most remarkable species of this highly esteemed group. I also obtained several other new and pretty butterflies. When we arrived at our lodging-house, being particularly anxious about my insect treasures, I suspended the box from a bamboo on which I could detect no sign of ants, and then began skinning some of my birds. During my work I often glanced at my precious box to see that no intruders had arrived, till after a longer spell of work than usual I looked again, and saw to my horror that a column of small red ants were descending the string and entering the box. They were already busy at work at the bodies of my treasures, and another half-hour would have seen my whole day's collection destroyed. As it was, I had to take every insect out, clean them thoroughly as well as the box, and then seek for a place of safety for them. As the only effectual one I begged a plate and a basin from my host, filled the former with water, and standing the latter in it placed my box on the top, and then felt secure for the night ; a few inches of clean water or oil being the only barrier these terrible pests are not able to pass.

On returning home to Mamájam (as my house was called) I had a slight return of intermittent fever, which kept me some days indoors. As soon as I was well, I again went to Goa, accompanied by Mr. Mesman, to beg the Rajah's assistance in getting a small house built for me near the forest. We found him at a cock-fight in a shed near his palace, which however he immediately left to receive us, and walked with us up an inclined plane of boards which serves for stairs to his house. This was large, well built, and lofty, with bamboo floor and glass windows. The greater part of it seemed to be one large hall divided by the supporting posts. Near a window sat the Queen squatting on a rough wooden arm-chair, chewing the everlasting sirih and betel-nut, while a brass spittoon by her side and a sirih-box in front were ready to administer to her wants. The Rajah seated himself opposite to her in a similar chair, and a similar spittoon and sirih-box were held by a little boy squatting at his side. Two other chairs were brought for us. Several young women, some the Rajah's daughters, others slaves, were standing about; a few were working at frames making sarongs, but most of them were idle.

And here I might (if I followed the example of most travellers) launch out into a glowing description of the charms of these damsels, the elegant costumes they wore, and the gold and silver ornaments with which they were adorned. The jacket or body of purple gauze would figure well in such a description, allowing the heaving bosom to be seen beneath it, while "sparkling eyes," and "jetty tresses," and "tiny feet" might be thrown in profusely. But, alas! regard for truth will not permit me to expatiate too admiringly on such topics, determined as I am to give as far as I can a true picture of the people and places I visit. The princesses were, it is true, sufficiently good-looking, yet neither their persons nor their garments had that appearance of freshness and cleanliness without which no other charms can be contemplated with pleasure. Everything had a dingy and faded appearance, very disagreeable and unroyal to a European eye. The only thing that excited some degree of admiration was the quiet and dignified manner of the Rajah, and the great respect always paid to him. None can stand erect in his presence, and when he sits on a chair, all present (Europeans of course excepted) squat upon the ground. The highest seat is literally, with these people, the place of honour and the sign of rank. So unbending are the rules in this respect, that when an English carriage which the Rajah of Lombock had sent for arrived, it was found impossible to use it because the driver's seat was the highest, and it had to be kept as a show in its coach-house. On being told the object of my visit, the Rajah at once said that he would order a house to be emptied for me, which would be much better than building one, as that would take a good deal of time. Bad coffee and sweetmeats were given us as before.

Two days afterwards I called on the Rajah, to ask him to send a guide with me to show me the house I was to occupy. He immediately ordered a man to be sent for, gave him instructions, and in a few minutes we were on our way. My conductor could speak no Malay, so we walked on in silence for an hour, when we turned into a pretty good house and I was asked to sit down. The head man of the district lived here, and in about half an hour we started again, and another hour's walk brought us to the village where I was to be lodged. We went to the residence of the village chief, who conversed with my conductor for some time. Getting tired, I asked to be shown the house that was prepared for me, but the only reply I could get was, "Wait a little," and the parties went on talking as before. So I told them I could not wait, as I wanted to see the house and then to go shooting in the forest. This seemed to puzzle them, and at length, in answer to questions, very poorly explained by one or two bystanders who knew a little Malay, it came out that no house was ready, and no one seemed to have the least idea where to get one. As I did not want to trouble the Rajah any more, I thought it best to try to frighten them a little ; so I told them that if they did not immediately find me a house as the Rajah had ordered, I should go back and complain to him, but that if a house was found me I would pay for the use of it. This had the desired effect, and one of the head men of the village asked me to go with him and look for a house. He showed me one or two of the most miserable and ruinous description, which I at once rejected, saying, " I must have a good one, and near to the forest." The next he showed me suited very well, so I told him to see that it was emptied the next day, for that the day after I should come and occupy it.

On the day mentioned, as I was not quite ready to go, I sent my two Macassar boys with brooms to sweep out the house thoroughly. They returned in the evening and told me, that when they got there the house was inhabited, and not a single article removed. However, on hearing they had come to clean and take possession, the occupants made a move, but, with a good deal of grumbling, which made me feel rather uneasy as to how the people generally might take my intrusion into their village. The next morning we took our baggage on three pack-horses, and, after a few break-downs, arrived about noon at our destination.

After getting all my things set straight, and having made a hasty meal, I determined if possible to make friends with the people. I therefore sent for the owner of the house and as many of his acquaintances as liked to come, to have a "bitchara," or talk. When they were all seated, I gave them a little tobacco all round, and having my boy Baderoon for interpreter, tried to explain to them why I came there ; that I was very sorry to turn them out of the house, but that the Rajah had ordered it rather than build a new one, which was what I had asked for, and then placed five silver rupees in the owner's hand as one month's rent. I then assured them that my being there would be a benefit to them, as I should buy their eggs and fowls and fruit ; and if their children would bring me shells and insects, of which I showed them specimens, they also might earn a good many coppers. After all this had been fully explained to them, with a long talk and discussion between every sentence, I could

see that I had made a favourable impression ; and that very
afternoon, as if to test my promise to buy even miserable little
snail-shells, a dozen children came one after another, bringing
me a few specimens each of a small Helix, for which they duly
received " coppers," and went away amazed but rejoicing.

A few days' exploration made me well acquainted with the
surrounding country. I was a long way from the road in the
forest which I had first visited, and for some distance round my
house were old clearings and cottages. I found a few good
butterflies, but beetles were very scarce, and even rotten timber
and newly-felled trees (generally so productive) here produced
scarcely anything. This convinced me that there was not a
sufficient extent of forest in the neighbourhood to make the
place worth staying at long, but it was too late now to think of
going further, as in about a month the wet season would begin ;
so I resolved to stay here and get what was to be had. Unfortunately, after a few days I became ill with a low fever which
produced excessive lassitude and disinclination to all exertion.
In vain I endeavoured to shake it off ; all I could do was to stroll
quietly each day for an hour about the gardens near, and to the
well, where some good insects were occasionally to be found ;
and the rest of the day to wait quietly at home, and receive
what beetles and shells my little corps of collectors brought me
daily. I imputed my illness chiefly to the water, which was
procured from shallow wells, around which there was almost
always a stagnant puddle in which the buffaloes wallowed.
Close to my house was an inclosed mudhole where three buffaloes
were shut up every night, and the effluvia from which freely
entered through the open bamboo floor. My Malay boy Ali
was affected with the same illness, and as he was my chief bird-
skinner I got on but slowly with my collections.

The occupations and mode of life of the villagers differed but
little from those of all other Malay races. The time of the
women was almost wholly occupied in pounding and cleaning
rice for daily use, in bringing home firewood and water, and in
cleaning, dyeing, spinning, and weaving the native cotton into
sarongs. The weaving is done in the simplest kind of frame
stretched on the floor, and is a very slow and tedious process.
To form the checked pattern in common use, each patch of
coloured threads has to be pulled up separately by hand and the
shuttle passed between them ; so that about an inch a day is
the usual progress in stuff a yard and a half wide. The men
cultivate a little sirih (the pungent pepper leaf used for chewing
with betel-nut) and a few vegetables ; and once a year rudely
plough a small patch of ground with their buffaloes and plant
rice, which then requires little attention till harvest time. Now
and then they have to see to the repairs of their houses, and
make mats, baskets, or other domestic utensils, but a large part
of their time is passed in idleness.

Not a single person in the village could speak more than a few words of Malay, and hardly any of the people appeared to have seen a European before. One most disagreeable result of this was, that I excited terror alike in man and beast. Wherever I went, dogs barked, children screamed, women ran away, and men stared as though I were some strange and terrible cannibal monster. Even the pack-horses on the roads and paths would start aside when I appeared and rush into the jungle ; and as to those horrid, ugly brutes, the buffaloes, they could never be approached by me ; not for fear of my own but of others' safety. They would first stick out their necks and stare at me, and then on a nearer view break loose from their halters or tethers, and rush away helter-skelter as if a demon were after them, without any regard for what might be in their way. Whenever I met buffaloes carrying packs along a pathway, or being driven home to the village, I had to turn aside into the jungle and hide myself till they had passed, to avoid a catastrophe which would increase the dislike with which I was already regarded. Every day about noon the buffaloes were brought into the village and were tethered in the shade around the houses ; and then I had to creep about like a thief by back ways, for no one could tell what mischief they might do to children and houses were I to walk among them. If I came suddenly upon a well where women were drawing water or children bathing, a sudden flight was the certain result ; which things occurring day after day, were very unpleasant to a person who does not like to be disliked, and who had never been accustomed to be treated as an ogre.

About the middle of November, finding my health no better, and insects, birds, and shells all very scarce, I determined to return to Mamájam, and pack up my collections before the heavy rains commenced. The wind had already begun to blow from the west, and many signs indicated that the rainy season might set in earlier than usual ; and then everything becomes very damp, and it is almost impossible to dry collections properly. My kind friend, Mr. Mesman, again lent me his pack-horses, and with the assistance of a few men to carry my birds and insects, which I did not like to trust on horses' backs, we got everything home safe. Few can imagine the luxury it was to stretch myself on a sofa, and to take my supper comfortably at table seated in my easy bamboo chair, after having for five weeks taken all my meals uncomfortably on the floor. Such things are trifles in health, but when the body is weakened by disease the habits of a lifetime cannot be so easily set aside.

My house, like all bamboo structures in this country, was a leaning one, the strong westerly winds of the wet season having set all its posts out of the perpendicular to such a degree, as to make me think it might some day possibly go over altogether. It is a remarkable thing that the natives of Celebes have not discovered the use of diagonal struts in strengthening buildings. I doubt if there is a native house in the country two years old and at all exposed to the wind, which stands upright ; and no wonder, as they merely consist of posts and joists all placed upright or horizontal, and fastened rudely together with rattans. They may be seen in every stage of the process of tumbling down, from the first slight inclination, to such a dangerous slope that it becomes a notice to quit to the occupiers.

The mechanical geniuses of the country have only discovered two ways of remedying the evil. One is, after it has commenced, to tie the house to a post in the ground on the windward side by a rattan or bamboo cable. The other is a preventive, but how they ever found it out and did not discover the true way is a mystery. This plan is, to build the house in the usual way, but instead of having all the principal supports of straight posts, to have two or three of them chosen as crooked as possible. I had often noticed these crooked posts in houses, but imputed it to the scarcity of good straight timber, till one day I met some men carrying home a post shaped something like a dog's hind leg, and inquired of my native boy what they were going to do with such a piece of wood. "To make a post for a house," said he. "But why don't they get a straight one, there are plenty here?" said I. "Oh," replied he, "they prefer some like that in a house, because then it won't fall," evidently imputing the effect to some occult property of crooked timber. A little consideration and a diagram will, however, show, that the effect imputed to the crooked post may be really produced by it. A true square changes its figure readily into a rhomboid or oblique figure, but when one or two of the uprights are bent or sloping, and placed so as to oppose each other, the effect of a strut is produced, though in a rude and clumsy manner.

Just before I had left Mamájam the people had sown a considerable quantity of maize, which appears above ground in two or three days, and in favourable seasons ripens in less than two months. Owing to a week's premature rains the ground was all flooded when I returned, and the plants just coming into ear were yellow and dead. Not a grain would be obtained by the whole village, but luckily it is only a luxury, not a necessary of life. The rain was the signal for ploughing to begin, in order to sow rice on all the flat lands between us and the town. The plough used is a rude wooden instrument, with a very short single handle, a tolerably well-shaped coulter, and the point formed of a piece of hard palm-wood fastened in with wedges. One or two buffaloes draw it at a very slow pace. The seed is sown broadcast, and a rude wooden harrow is used to smooth the surface.

By the beginning of December the regular wet season had set in. Westerly winds and driving rains sometimes continued for days together; the fields for miles around were under water, and the ducks and buffaloes enjoyed themselves amazingly. All along the road to Macassar ploughing was daily going on in the mud and water, through which the wooden plough easily makes its way, the ploughman holding the plough-handle with one hand while a long bamboo in the other serves to guide the buffaloes. These animals require an immense deal of driving to get them on at all ; a continual shower of exclamations is kept up at them, and " Oh ! ah ! gee ! ugh !" are to be heard in various keys and in an uninterrupted succession all day long. At night we were favoured with a different kind of concert. The dry ground around my house had become a marsh tenanted by frogs, who kept up a most incredible noise from dusk to dawn. They were somewhat musical too, having a deep vibrating note which at times closely resembles the tuning of two or three bass-viols in an orchestra. In Malacca and Borneo I had heard no such sounds as these, which indicates that the frogs, like most of the animals of Celebes, are of species peculiar to it.

My kind friend and landlord, Mr. Mesman, was a good specimen of the Macassar-born Dutchman. He was about thirty-five

NATIVE WOODEN PLOUGH.

years of age, had a large family, and lived in a spacious house near the town, situated in the midst of a grove of fruit trees, and surrounded by a perfect labyrinth of offices, stables, and native cottages occupied by his numerous servants, slaves, or dependants. He usually rose before the sun, and after a cup of coffee looked after his servants, horses, and dogs, till seven, when a substantial breakfast of rice and meat was ready in a cool verandah. Putting on a clean white linen suit, he then drove to town in his buggy, where he had an office, with two or three Chinese clerks who looked after his affairs. His business was that of a coffee and opium merchant. He had a coffee estate at Bontyne, and a small prau which traded to the Eastern islands near New Guinea, for mother-of-pearl and tortoiseshell. About one he would return home, have coffee and cake or fried plantain, first changing his dress for a coloured cotton shirt and trousers and bare feet, and then take a siesta with a book. About four, after a cup of tea, he would walk round his premises, and generally stroll down to Mamájam, to pay me a visit and look after his farm.

This consisted of a coffee plantation and an orchard of fruit trees, a dozen horses and a score of cattle, with a small village of Timorese slaves and Macassar servants. One family looked after the cattle and supplied the house with milk, bringing me also a large glassful every morning, one of my greatest luxuries. Others had charge of the horses, which were brought in every afternoon and fed with cut grass. Others had to cut grass for their master's horses at Macassar—not a very easy task in the dry season, when all the country looks like baked mud ; or in the rainy season, when miles in every direction are flooded. How they managed it was a mystery to me, but they know grass must be had, and they get it. One lame woman had charge of a flock of ducks. Twice a day she took them out to feed in the marshy places, let them waddle and gobble for an hour or two, and then drove them back and shut them up in a small dark shed to digest their meal, whence they gave forth occasionally a melancholy quack. Every night a watch was set, principally for the sake of the horses, the people of Goa, only two miles off, being notorious thieves, and horses offering the easiest and most valuable spoil. This enabled me to sleep in security, although many people in Macassar thought I was running a great risk, living alone in such a solitary place and with such bad neighbours.

My house was surrounded by a kind of straggling hedge of roses, jessamines, and other flowers, and every morning one of the women gathered a basketful of the blossoms for Mr. Mesman's family. I generally, took a couple for my own breakfast table, and the supply never failed during my stay, and I suppose never does. Almost every Sunday Mr. M. made a shooting excursion with his eldest son, a lad of fifteen, and I generally accompanied him ; for though the Dutch are Protestants, they do not observe Sunday in the rigid manner practised in England and English colonies. The Governor of the place has his public reception every Sunday evening, when card-playing is the regular amusement.

On December 13th I went on board a prau bound for the Aru Islands, a journey which will be described in the latter part of this work.

On my return, after a seven months' absence, I visited another district to the north of Macassar, which will form the subject of the next chapter.

CHAPTER XVI.

CELEBES.

(MACASSAR.   JULY TO NOVEMBER, 1857.)

I REACHED Macassar again on the 11th of July, and established myself in my old quarters at Mamájam, to sort, arrange, clean, and pack up my Aru collections. This occupied me a month ; and having shipped them off for Singapore, had my guns repaired, and received a new one from England, together with a stock of pins, arsenic, and other collecting requisites, I began to feel eager for work again, and had to consider where I should spend my time till the end of the year. I had left Macassar, seven months before, a flooded marsh, being ploughed up for the rice-sowing. The rains had continued for five months, yet now all the rice was cut, and dry and dusty stubbles covered the country just as when I had first arrived there.

After much inquiry I determined to visit the district of Máros, about thirty miles north of Macassar, where Mr. Jacob Mesman, a brother of my friend, resided, who had kindly offered to find me house-room and give me assistance should I feel inclined to visit him. I accordingly obtained a pass from the Resident, and having hired a boat set off one evening for Máros. My boy Ali was so ill with fever that I was obliged to leave him in the hospital, under the care of my friend the German doctor, and I had to make shift with two new servants utterly ignorant of everything. We coasted along during the night, and at daybreak enter the Máros river, and by three in the afternoon reached the village. I immediately visited the Assistant Resident, and applied for ten men to carry my baggage, and a horse for myself. These were promised to be ready that night, so that I could start as soon as I liked in the morning. After having taken a cup of tea I took my leave, and slept in the boat. Some of the men came at night as promised, but others did not arrive till the next morning. It took some time to divide my baggage fairly among them, as they all wanted to shirk the heavy boxes, and would seize hold of some light article and march off with it, till made to come back and wait till the whole had been fairly apportioned. At length about eight o'clock all was arranged, and we started for our walk to Mr. M.'s farm.

The country was at first a uniform plain of burnt-up rice-grounds, but at a few miles distance precipitous hills appeared, backed by the lofty central range of the peninsula. Towards these our path lay, and after having gone six or eight miles the hills began to advance into the plain right and left of us, and the ground became pierced here and there with blocks and pillars of limestone rock, while a few abrupt conical hills and peaks rose like islands. Passing over an elevated tract forming the shoulder of one of the hills, a picturesque scene lay before us. We looked down into a little valley almost entirely surrounded by mountains, rising abruptly in huge precipices, and forming a succession of knolls and peaks and domes of the most varied and fantastic shapes. In the very centre of the valley was a large bamboo house, while scattered around were a dozen cottages of the same material.

I was kindly received by Mr. Jacob Mesman in an airy saloon detached from the house, and entirely built of bamboo and thatched with grass. After breakfast he took me to his foreman's house, about a hundred yards off, half of which was given up to me till I should decide where to have a cottage built for my own use. I soon found that this spot was too much exposed to the wind and dust, which rendered it very difficult to work with papers or insects. It was also dreadfully hot in the afternoon, and after a few days I got a sharp attack of fever, which determined me to move. I accordingly fixed on a place about a mile off, at the foot of a forest-covered hill, where in a few days Mr. M. built for me a nice little house, consisting of a good-sized enclosed verandah or open room, and a small inner sleeping-room, with a little cook-house outside. As soon as it was finished I moved into it, and found the change most agreeable.

The forest which surrounded me was open and free from underwood, consisting of large trees, widely scattered with a great quantity of palm-trees (Arenga saccharifera), from which palm wine and sugar are made. There were also great numbers of a wild Jack-fruit tree (Artocarpus), which bore abundance of large reticulated fruit, serving as an excellent vegetable. The ground was as thickly covered with dry leaves as it is in an English wood in November; the little rocky streams were all dry, and scarcely a drop of water or even a damp place was anywhere to be seen. About fifty yards below my house, at the foot of the hill, was a deep hole in a watercourse where good water was to be had, and where I went daily to bathe, by having buckets of water taken out and pouring it over my body.

My host Mr. M. enjoyed a thoroughly country life, depending almost entirely on his gun and dogs to supply his table. Wild pigs of large size were very plentiful, and he generally got one or two a week, besides deer occasionally, and abundance of jungle-fowl, hornbills, and great fruit pigeons. His buffaloes supplied plenty of milk, from which he made his own butter; he grew his own rice and coffee, and had ducks, fowls, and their eggs in profusion. His palm-trees supplied him all the year round with "sagueir," which takes the place of beer; and the sugar made from them is an excellent sweetmeat. All the fine tropical vegetables and fruits were abundant in their season, and his cigars were made from tobacco of his own raising. He kindly sent me a bamboo of buffalo-milk every morning; it was as thick as cream, and required diluting with water to keep it fluid during the day. It mixes very well with tea and coffee, although it has a slight peculiar flavour, which after a time is

SUGAR PALM. (*Arenga saccharifera.*)

not disagreeable. I also got as much sweet "sagueir" as I liked to drink, and Mr. M. always sent me a piece of each pig he killed, which with fowls, eggs, and the birds we shot ourselves, and buffalo beef about once a fortnight, kept my larder sufficiently well supplied.

Every bit of flat land was cleared and used as rice-fields, and on the lower slopes of many of the hills tobacco and vegetables were grown. Most of the slopes are covered with huge blocks of rock, very fatiguing to scramble over, while a number of the hills are so precipitous as to be quite inaccessible. These circumstances, combined with the excessive drought, were very unfavourable for my pursuits. Birds were scarce, and I got but few new to me. Insects were tolerably plentiful, but unequal. Beetles, usually so numerous and interesting, were exceedingly scarce, some of the families being quite absent and others only represented by very minute species. The Flies and Bees, on the other hand, were abundant, and of these I daily obtained new and interesting species. The rare and beautiful Butterflies of Celebes were the chief object of my search, and I found many species altogether new to me, but they were generally so active and shy as to render their capture a matter of great difficulty. Almost the only good place for them was in the dry beds of the streams in the forest, where, at damp places, muddy pools, or even on the dry rocks, all sorts of insects could be found. In these rocky forests dwell some of the finest butterflies in the world. Three species of Ornithoptera, measuring seven or eight inches across the wings, and beautifully marked with spots or masses of satiny yellow on a black ground, wheel through the thickets with a strong sailing flight. About the damp places are swarms of the beautiful blue-banded Papilios, miletus and telephus, the superb golden green P. macedon, and the rare little swallow-tail Papilio rhesus, of all of which, though very active, I succeeded in capturing fine series of specimens.

I have rarely enjoyed myself more than during my residence here. As I sat taking my coffee at six in the morning, rare birds would often be seen on some tree close by, when I would hastily sally out in my slippers, and perhaps secure a prize I had been seeking after for weeks. The great hornbills of Celebes (Buceros cassidix) would often come with loud-flapping wings, and perch upon a lofty tree just in front of me ; and the black baboon monkeys, Cynopithecus nigrescens, often stared down in astonishment at such an intrusion into their domains ; while at night herds of wild pigs roamed about the house, devouring refuse, and obliging us to put away everything eatable or breakable from our little cooking-house. A few minutes' search on the fallen trees around my house at sunrise and sunset, would often produce me more beetles than I would meet with in a day's collecting, and odd moments could be made valuable which when living in villages or at a distance from the forest are inevitably wasted. Where the sugar-palms were dripping with sap, flies congregated in immense numbers, and it was by spending half an hour at these when I had the time to spare, that I obtained the finest and most remarkable collection of this group of insects that I have ever made.

Then what delightful hours I passed wandering up and down the dry river-courses, full of water-holes and rocks and fallen trees, and overshadowed by magnificent vegetation! I soon got to know every hole and rock and stump, and came up to each with cautious step and bated breath to see what treasures it would produce. At one place I would find a little crowd of the rare butterfly Tachyris zarinda, which would rise up at my approach, and display their vivid orange and cinnabar-red wings, while among them would flutter a few of the fine blue-banded Papilios. Where leafy branches hung over the gully, I might expect to find a grand Ornithoptera at rest and an easy prey. At certain rotten trunks I was sure to get the curious little tiger beetle, Therates flavilabris. In the denser thickets I would capture the small metallic blue butterflies (Amblypodia) sitting on the leaves, as well as some rare and beautiful leaf-beetles of the families Hispidæ and Chrysomelidæ.

I found that the rotten jack-fruits were very attractive to many beetles, and used to split them partly open and lay them about in the forest near my house to rot. A morning's search at these often produced me a score of species,—Staphylinidæ, Nitidulidæ, Onthophagi, and minute Carabidæ being the most abundant. Now and then the "sagueir" makers brought me a fine rosechafer (Sternoplus schaumii) which they found licking up the sweet sap. Almost the only new birds I met with for some time were a handsome ground thrush (Pitta celebensis), and a beautiful violet-crowned dove (Ptilonopus celebensis), both very similar to birds I had recently obtained at Aru, but of distinct species.

About the latter part of September a heavy shower of rain fell, admonishing us that we might soon expect wet weather, much to the advantage of the baked-up country. I therefore determined to pay a visit to the falls of the Máros river, situated at the point where it issues from the mountains—a spot often visited by travellers and considered very beautiful. Mr. M. lent me a horse, and I obtained a guide from a neigh-bouring village; and taking one of my men with me, we started at six in the morning, and after a ride of two hours over the flat rice-fields skirting the mountains which rose in grand pre-cipices on our left, we reached the river about half-way between Máros and the falls, and thence had a good bridle-road to our destination, which we reached in another hour. The hills had closed in round us as we advanced; and when we reached a ruinous shed which had been erected for the accommodation of visitors, we found ourselves in a flat-bottomed valley about a quarter of a mile wide, bounded by precipitous and often over-hanging limestone rocks. So far the ground had been cultivated, but it now became covered with bushes and large scattered trees.

As soon as my scanty baggage had arrived and was duly deposited in the shed, I started off alone for the fall, which was about a quarter of a mile further on. The river is here about twenty yards wide, and issues from a chasm between two vertical walls of limestone, over a rounded mass of basaltic rock about forty feet high, forming two curves separated by a slight ledge. The water spreads beautifully over this surface in a thin sheet of foam, which curls and eddies in a succession of concentric cones till it falls into a fine deep pool below. Close to the very edge of the fall a narrow and very rugged path leads to the river above, and thence continues close under the precipice along the water's edge, or sometimes in the water, for a few hundred yards, after which the rocks recede a little, and leave a wooded bank on one side, along which the path is continued, till in about half a mile a second and smaller fall is reached. Here the river seems to issue from a cavern, the rocks having fallen from above so as to block up the channel and bar further progress. The fall itself can only be reached by a path which ascends behind a huge slice of rock which has partly fallen away from the mountain, leaving a space two or three feet wide, but disclosing a dark chasm descending into the bowels of the mountain, and which, having visited several such, I had no great curiosity to explore.

Crossing the stream a little below the upper fall, the path ascends a steep slope for about five hundred feet, and passing through a gap enters a narrow valley, shut in by walls of rock absolutely perpendicular and of great height. Half a mile further this valley turns abruptly to the right, and becomes a mere rift in the mountain. This extends another half mile, the walls gradually approaching till they are only two feet apart, and the bottom rising steeply to a pass which leads probably into another valley, but which I had no time to explore. Returning to where this rift had begun, the main path turns up to the left in a sort of gully, and reaches a summit over which a fine natural arch of rock passes at a height of about fifty feet. Thence was a steep descent through thick jungle with glimpses of precipices and distant rocky mountains, probably leading into the main river valley again. This was a most tempting region to explore, but there were several reasons why I could go no further. I had no guide, and no permission to enter the Bugis territories, and as the rains might at any time set in, I might be prevented from returning by the flooding of the river. I therefore devoted myself during the short time of my visit to obtaining what knowledge I could of the natural productions of the place.

The narrow chasms produced several fine insects quite new to me, and one new bird, the curious Phlægenas tristigmata, a large ground pigeon with yellow breast and crown, and purple neck. This rugged path is the highway from Máros to the Bugis country beyond the mountains. During the rainy season it is quite impassable, the river filling its bed and rushing between perpendicular cliffs many hundred feet high. Even at the time of my visit it was most precipitous and fatiguing, yet women and children came over it daily, and men carrying heavy loads of palm sugar of very little value. It was along the path between the lower and the upper falls, and about the margin of the upper pool, that I found most insects. The large semi-transparent butterfly, Idea tondana, flew lazily along by dozens, and it was here that I at length obtained an insect which I had hoped but hardly expected to meet with—the magnificent Papilio androcles, one of the largest and rarest known swallow-tailed butterflies. During my four days' stay at the falls I was so fortunate as to obtain six good specimens. As this beautiful creature flies, the long white tails flicker like streamers, and when settled on the beach it carries them raised upwards, as if to preserve them from injury. It is scarce even here, as I did not see more than a dozen specimens in all, and had to follow many of them up and down the river's bank repeatedly before I succeeded in their capture. When the sun shone hottest about noon, the moist beach of the pool below the upper fall presented a beautiful sight, being dotted with groups of gay butterflies,—orange, yellow, white, blue, and green,—which on being disturbed rose into the air by hundreds, forming clouds of variegated colours.

Such gorges, chasms, and precipices as here abound, I have nowhere seen in the Archipelago. A sloping surface is scarcely anywhere to be found, huge walls and rugged masses of rock terminating all the mountains and inclosing the valleys. In many parts there are vertical or even overhanging precipices five or six hundred feet high, yet completely clothed with a tapestry of vegetation. Ferns, Pandanaceæ, shrubs, creepers, and even forest trees, are mingled in an evergreen network, through the interstices of which appears the white limestone rock or the dark holes and chasms with which it abounds. These precipices are enabled to sustain such an amount of vegetation by their peculiar structure. Their surfaces are very irregular, broken into holes and fissures, with ledges overhanging the mouths of gloomy caverns; but from each projecting part have descended stalactites, often forming a wild gothic tracery over the caves and receding hollows, and affording an admirable support to the roots of the shrubs, trees, and creepers, which luxuriate in the warm pure atmosphere and the gentle moisture which constantly exudes from the rocks. In places where the precipice offers smooth surfaces of solid rock, it remains quite bare, or only stained with lichens and dotted with clumps of ferns that grow on the small ledges and in the minutest crevices.

The reader who is familiar with tropical nature only through the medium of books and botanical gardens, will picture to himself in such a spot many other natural beauties. He will think that I have unaccountably forgotten to mention the brilliant flowers, which, in gorgeous masses of crimson, gold, or azure, must spangle these verdant precipices, hang over the cascade, and adorn the margin of the mountain stream. But

what is the reality? In vain did I gaze over these vast walls of
verdure, among the pendant creepers and bushy shrubs, all
around the cascade, on the river's bank, or in the deep caverns
and gloomy fissures,—not one single spot of bright colour could
be seen ; not one single tree or bush or creeper bore a flower
sufficiently conspicuous to form an object in the landscape. In
every direction the eye rested on green foliage and mottled
rock.   There was infinite variety in the colour and aspect of the
foliage, there was grandeur in the rocky masses and in the
exuberant luxuriance of the vegetation, but there was no
brilliancy of colour, none of those bright flowers and gorgeous
masses of blossom, so generally considered to be everywhere
present in the tropics.   I have here given an accurate sketch of
a luxuriant tropical scene as noted down on the spot, and its
general characteristics as regards colour have been so often
repeated, both in South America and over many thousand miles
in the Eastern tropics, that I am driven to conclude that it
represents the general aspect of nature in the equatorial (that
is, the most tropical) parts of the tropical regions. How is it
then, that the descriptions of travellers generally give a very
different idea ? And where, it may be asked, *are* the glorious
flowers that we know do exist in the tropics ? These questions
can be easily answered.   The fine tropical flowering-plants
cultivated in our hot-houses ; have been culled from the most
varied regions, and therefore give a most erroneous idea of
their abundance in any one region. Many of them are very
rare, others extremely local, while a considerable number inhabit
the more arid regions of Africa and India, in which tropical
vegetation does not exhibit itself in its usual luxuriance. Fine
and varied foliage, rather than gay flowers, is more characteristic of those parts where tropical vegetation attains its highest
development, and in such districts each kind of flower seldom
lasts in perfection more than a few weeks, or sometimes a few
days.   In every locality a lengthened residence will show an
abundance of magnificent and gaily-blossomed plants, but they
have to be sought for, and are rarely at any one time or place
so abundant as to form a perceptible feature in the landscape.
But it has been the custom of travellers to describe and group
together all the fine plants they have met with during a long
journey, and thus produce the effect of a gay and flower-painted
landscape.   They have rarely studied and described individual
scenes where vegetation was most luxuriant and beautiful, and
fairly stated what effect was produced in them by flowers. I
have done so frequently, and the result of these examinations
has convinced me, that the bright colours of flowers have a much
greater influence on the general aspect of nature in temperate
than in tropical climates. During twelve years spent amid the
grandest tropical vegetation, I have seen nothing comparable to
the effect produced on our landscapes by gorse, broom, heather,
wild hyacinths, hawthorn, purple orchises, and buttercups.

The geological structure of this part of Celebes is interesting. The limestone mountains, though of great extent, seem to be entirely superficial, resting on a basis of basalt which in some places forms low rounded hills between the more precipitous mountains. In the rocky beds of the streams basalt is almost always found, and it is a step in this rock which forms the cascade already described. From it the limestone precipices rise abruptly ; and in ascending the little stairway along the side of the fall, you step two or three times from the one rock on to the other,—the limestone dry and rough, being worn by the water and rains into sharp ridges and honeycombed holes,—the basalt moist, even, and worn smooth and slippery by the passage of bare-footed pedestrians. The solubility of the limestone by rain-water is well seen in the little blocks and peaks which rise thickly through the soil of the alluvial plains as you approach the mountains. They are all skittle-shaped, larger in the middle than at the base, the greatest diameter occurring at the height to which the country is flooded in the wet season, and thence decreasing regularly to the ground. Many of them overhang considerably, and some of the slenderer pillars appear to stand upon a point. When the rock is less solid it becomes curiously honeycombed by the rains of successive winters, and I noticed some masses reduced to a complete network of stone, through which light could be seen in every direction. From these mountains to the sea extends a perfectly flat alluvial plain, with no indication that water would accumulate at a great depth beneath it, yet the authorities at Macassar have spent much money in boring a well a thousand feet deep in hope of getting a supply of water like that obtained by the Artesian wells in the London and Paris basins. It is not to be wondered at that the attempt was unsuccessful.

Returning to my forest hut, I continued my daily search after birds and insects. The weather however became dreadfully hot and dry, every drop of water disappearing from the pools and rock-holes, and with it the insects which frequented them. Only one group remained unaffected by the intense drought ; the Diptera, or two-winged flies, continued as plentiful as ever, and on these I was almost compelled to concentrate my attention for a week or two, by which means I increased my collection of that Order to about two hundred species. I also continued to obtain a few new birds, among which were two or three kinds of small hawks and falcons, a beautiful brush-tongued paroquet, Trichoglossus ornatus, and a rare black and white crow, Corvus advena.

At length about the middle of October, after several gloomy days, down came a deluge of rain, which continued to fall almost every afternoon, showing that the early part of the wet season had commenced. I hoped now to get a good harvest of insects, and in some respects I was not disappointed. Beetles became much more numerous, and under a thick bed of leaves that had accumulated on some rocks by the side of a forest stream, I found abundance of Carabidæ, a family generally scarce in the tropics. The butterflies however disappeared. Two of my servants were attacked with fever, dysentery, and swelled feet, just at the time that the third had left me, and for some days they both lay groaning in the house. When they got a little better I was attacked myself, and as my stores were nearly finished and everything was getting very damp, I was obliged to prepare for my return to Macassar, especially as the strong westerly winds would render the passage in a small open boat disagreeable if not dangerous.

Since the rains began, numbers of huge millipedes, as thick as one's finger and eight or ten inches long, crawled about everywhere, in the paths, on trees, about the house,—and one morning when I got up I even found one in my bed! They were generally of a dull lead colour or of a deep brick red, and were very nasty-looking things to be coming everywhere in one's way, although quite harmless. Snakes too began to show themselves. I killed two of a very abundant species, big-headed and of a bright green colour, which lie coiled up on leaves and shrubs and can scarcely be seen till one is close upon them. Brown snakes got into my net while beating among dead leaves for insects, and made me rather cautious about inserting my hand till I knew what kind of game I had captured. The fields and meadows which had been parched and sterile now became suddenly covered with fine long grass; the river-bed where I had so many times walked over burning rocks, was now a deep and rapid stream; and numbers of herbaceous plants and shrubs were everywhere springing up and bursting into flower. I found plenty of new insects, and if I had had a good, roomy, water-and-wind-proof house, I should perhaps have stayed during the wet season, as I feel sure many things can then be obtained which are to be found at no other time. With my summer hut, however, this was impossible. During the heavy rains a fine drizzly mist penetrated into every part of it, and I began to have the greatest difficulty in keeping my specimens dry.

Early in November I returned to Macassar, and having packed up my collections, started in the Dutch mail steamer for Amboyna and Ternate. Leaving this part of my journey for the present, I will in the next chapter conclude my account of Celebes by describing the extreme northern part of the island which I visited two years later.

CHAPTER XVII.

CELEBES.

(MENADO.   JUNE TO SEPTEMBER, 1859.)

It was after my residence at Timor-Coupang that I visited the north-eastern extremity of Celebes, touching on my way at Banda, Amboyna, and Ternate. I reached Menado on the 10th of June, 1859, and was very kindly received by Mr. Tower, an Englishman, but a very old resident in Menado, where he carries on a general business. He introduced me to Mr. L. Duivenboden (whose father had been my friend at Ternate), who had much taste for natural history ; and to Mr. Neys, a native of Menado, but who was educated at Calcutta, and to whom Dutch, English, and Malay were equally mother-tongues. All these gentlemen showed me the greatest kindness, accompanied me in my earliest walks about the country, and assisted me by every means in their power. I spent a week in the town very pleasantly, making explorations and inquiries after a good collecting station, which I had much difficulty in finding, owing to the wide cultivation of coffee and cacao, which has led to the clearing away of the forests for many miles round the town, and over extensive districts far into the interior.

The little town of Menado is one of the prettiest in the East. It has the appearance of a large garden containing rows of rustic villas, with broad paths between, forming streets generally at right angles with each other. Good roads branch off in several directions towards the interior, with a succession of pretty cottages, neat gardens, and thriving plantations, interspersed with wildernesses of fruit trees. To the west and south the country is mountainous, with groups of fine volcanic peaks 6,000 or 7,000 feet high, forming grand and picturesque backgrounds to the landscape.

The inhabitants of Minahasa (as this part of Celebes is called) differ much from those of all the rest of the island, and in fact from any other people in the Archipelago. They are of a light-brown or yellow tint, often approaching the fairness of a European ; of a rather short stature, stout and well-made ; of an open and pleasing countenance, more or less disfigured as age increases by projecting cheek-bones ; and with the usual long, straight, jet-black hair of the Malayan races. In some of the inland villages where they may be supposed to be of the purest race, both men and women are remarkably handsome ; while nearer the coasts where the purity of their blood has been destroyed by the intermixture of other races, they approach to the ordinary types of the wild inhabitants of the surrounding countries.

In mental and moral characteristics they are also highly peculiar. They are remarkably quiet and gentle in disposition, submissive to the authority of those they consider their superiors, and easily induced to learn and adopt the habits of civilized people. They are clever mechanics, and seem capable of acquiring a considerable amount of intellectual education.

Up to a very recent period these people were thorough savages, and there are persons now living in Menado who remember a state of things identical with that described by the writers of the sixteenth and seventeenth centuries. The inhabitants of the several villages were distinct tribes, each under its own chief, speaking languages unintelligible to each other, and almost always at war. They built their houses elevated upon lofty posts to defend themselves from the attacks of their enemies. They were head hunters like the Dyaks of Borneo, and were said to be sometimes cannibals. When a chief died, his tomb was adorned with two fresh human heads ; and if those of enemies could not be obtained, slaves were killed for the occasion. Human skulls were the great ornaments of the chiefs' houses. Strips of bark were their only dress. The country was a pathless wilderness, with small cultivated patches of rice and vegetables, or clumps of fruit-trees, diversifying the otherwise unbroken forest. Their religion was that naturally engendered in the undeveloped human mind by the contemplation of grand natural phenomena and the luxuriance of tropical nature. The burning mountain, the torrent and the lake, were the abode of their deities ; and certain trees and birds were supposed to have especial influence over men's actions and destiny. They held wild and exciting festivals to propitiate these deities or demons ; and believed that men could be changed by them into animals, either during life or after death.

Here we have a picture of true savage life ; of small isolated communities at war with all around them, subject to the wants and miseries of such a condition, drawing a precarious existence from the luxuriant soil, and living on from generation to generation, with no desire for physical amelioration, and no prospect of moral advancement.

Such was their condition down to the year 1822, when the coffee-plant was first introduced, and experiments were made as to its cultivation. It was found to succeed admirably at from fifteen hundred up to four thousand feet above the sea. The chiefs of villages were induced to undertake its cultivation. Seed and native instructors were sent from Java ; food was supplied to the labourers engaged in clearing and planting ; a fixed price was established at which all coffee brought to the government collectors was to be paid for, and the village chiefs who now received the title of " Majors " were to receive five per cent. of the produce. After a time, roads were made from the

port of Menado up to the plateau, and smaller paths were cleared
from village to village; missionaries settled in the more populous
districts and opened schools, and Chinese traders penetrated to
the interior and supplied clothing and other luxuries in exchange
for the money which the sale of the coffee had produced. At the
same time, the country was divided into districts, and the system
of "Controlleurs," which had worked so well in Java, was introduced. The "Controlleur" was a European, or a native of
European blood, who was the general superintendent of the
cultivation of the district, the adviser of the chiefs, the protector
of the people, and the means of communication between both
and the European Government. His duties obliged him to visit
every village in succession once a month, and to send in a report
on their condition to the Resident. As disputes between adjacent
villages were now settled by appeal to a superior authority, the
old and inconvenient semi-fortified houses were disused, and
under the direction of the "Controlleurs" most of the houses
were rebuilt on a neat and uniform plan. It was this interesting
district which I was now about to visit.

Having decided on my route, I started at 8 A.M. on the 22nd of
June. Mr. Tower drove me the first three miles in his chaise,
and Mr. Neys accompanied me on horseback three miles further
to the village of Lotta. Here we met the Controlleur of the
district of Tondáno, who was returning home from one of his
monthly tours, and who had agreed to act as my guide and companion on the journey. From Lotta we had an almost continual
ascent for six miles, which brought us on to the plateau of
Tondáno at an elevation of about 2,400 feet. We passed through
three villages whose neatness and beauty quite astonished me.
The main road, along which all the coffee is brought down from
the interior in carts drawn by buffaloes, is always turned aside
at the entrance of a village, so as to pass behind it, and thus
allow the village street itself to be kept neat and clean. This
is bordered by neat hedges often formed entirely of rose-trees,
which are perpetually in blossom. There is a broad central path
and a border of fine turf, which is kept well swept and neatly
cut. The houses are all of wood, raised about six feet on substantial posts neatly painted blue, while the walls are whitewashed. They all have a verandah enclosed with a neat balus-
trade, and are generally surrounded by orange-trees and flowering shrubs. The surrounding scenery is verdant and picturesque.
Coffee plantations of extreme luxuriance, noble palms and tree
ferns, wooded hills and volcanic peaks, everywhere meet the eye.
I had heard much of the beauty of this country, but the reality
far surpassed my expectations.

About one o'clock we reached Tomohón, the chief place of a
district, having a native chief now called the "Major," at whose
house we were to dine. Here was a fresh surprise for me. The
house was large, airy, and very substantially built of hard native
timber, squared and put together in a most workmanlike manner.

It was furnished in European style, with handsome chandelier lamps, and the chairs and tables all well made by native workmen. As soon as we entered, madeira and bitters were offered us. Then two handsome boys, neatly dressed in white and with smoothly brushed jet-black hair, handed us each a basin of water and a clean napkin on a salver. The dinner was excellent. Fowls cooked in various ways, wild pig roasted, stewed and fried, a fricassee of bats, potatoes, rice, and other vegetables, all served on good china, with finger glasses and fine napkins, and abundance of good claret and beer, seemed to me rather curious at the table of a native chief on the mountains of Celebes. Our host was dressed in a suit of black with patent-leather shoes, and really looked comfortable and almost gentlemanly in them. He sat at the head of the table and did the honours well, though he did not talk much. Our conversation was entirely in Malay, as that is the official language here, and in fact the mother-tongue and only language of the Controlleur, who is a native-born half-breed. The Major's father, who was chief before him, wore, I was informed, a strip of bark as his sole costume, and lived in a rude hut raised on lofty poles, and abundantly decorated with human heads. Of course we were expected, and our dinner was prepared in the best style, but I was assured that the chiefs all take a pride in adopting European customs, and in being able to receive their visitors in a handsome manner.

After dinner and coffee, the Controlleur went on to Tondáno, and I strolled about the village waiting for my baggage, which was coming in a bullock-cart and did not arrive till after midnight. Supper was very similar to dinner, and on retiring I found an elegant little room with a comfortable bed, gauze curtains with blue and red hangings, and every convenience. Next morning at sunrise the thermometer in the verandah stood at 69°, which I was told is about the usual lowest temperature at this place, 2,500 feet above the sea. I had a good breakfast of coffee, eggs, and fresh bread and butter, which I took in the spacious verandah, amid the odour of roses, jessamine, and other sweet-scented flowers, which filled the garden in front ; and about eight o'clock left Tomohón with a dozen men carrying my baggage.

Our road lay over a mountain ridge about 4,000 feet above the sea, and then descended about 500 feet to the little village of Rurúkan, the highest in the district of Minahasa, and probably in all Celebes. Here I had determined to stay for some time to see whether this elevation would produce any change in the zoology. The village had only been formed about ten years, and was quite as neat as those I had passed through and much more picturesque. It is placed on a small level spot, from which there is an abrupt wooded descent down to the beautiful lake of Tondáno, with volcanic mountains beyond. On one side is a ravine, and beyond it a fine mountainous and wooded country.

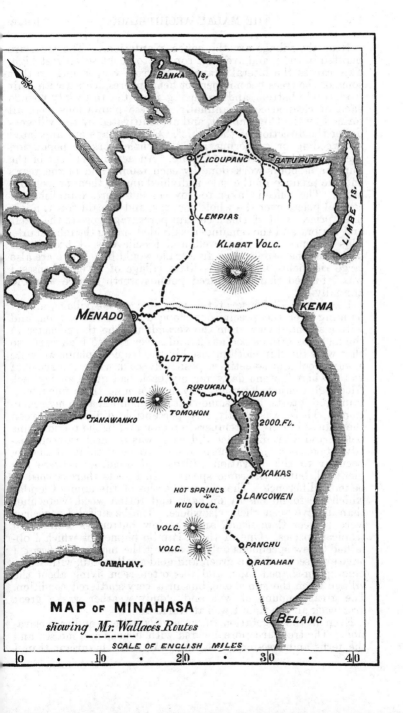

BANKA Is.

LICOUPANG · · · · · BATUPUTIH

LIMBE Is.

LEMPIAS

KLABAT VOLC.

MENADO · · · · · · · · · · · · KEMA

LOTTA

RURUKAN · TONDANO
LOKON VOLC. · · TOMOHON
TANAWANKO · · · 2000.Ft.

KAKAS

HOT SPRINGS
· MUD VOLC. · · LANGOWEN

VOLC.
VOLC. · · · · PANGHU
· · · · RATAHAN
O.AMAHAY.

MAP OF MINAHASA
showing Mr. Wallace's Routes
- - - - - - -
SCALE OF ENGLISH MILES

BELANG

0 · 10 · 20 · 30 · 40

Near the village are the coffee plantations. The trees are planted in rows, and are kept topped to about seven feet high. This causes the lateral branches to grow very strong, so that some of the trees become perfect hemispheres, loaded with fruit from top to bottom, and producing from ten to twenty pounds each of cleaned coffee annually. These plantations were all formed by the Government, and are cultivated by the villagers under the direction of their chief. Certain days are appointed for weeding or gathering, and the whole working population are summoned by sound of gong. An account is kept of the number of hours work done by each family, and at the year's end the produce of the sale is divided among them proportionately. The coffee is taken to Government stores established at central places over the whole country, and is paid for at a low fixed price. Out of this a certain percentage goes to the chiefs and majors, and the remainder is divided among the inhabitants. This system works very well, and I believe is at present far better for the people than free-trade would be. There are also large rice-fields, and in this little village of seventy houses I was informed that a hundred pounds worth of rice was sold annually.

I had a small house at the very end of the village, almost hanging over the precipitous slope down to the stream, and with a splendid view from the verandah. The thermometer in the morning often stood at 62° and never rose so high as 80°, so that with the thin clothing used in the tropical plains we were always cool and sometimes positively cold, while the spout of water where I went daily for my bath had quite an icy feel. Although I enjoyed myself very much among these fine mountains and forests, I was somewhat disappointed as to my collections. There was hardly any perceptible difference between the animal life in this temperate region and in the torrid plains below, and what difference did exist was in most respects disadvantageous to me. There seemed to be nothing absolutely peculiar to this elevation. Birds and quadrupeds were less plentiful, but of the same species. In insects there seemed to be more difference. The curious beetles of the family Cleridæ, which are found chiefly on bark and rotten wood, were finer than I have seen them elsewhere. The beautiful Longicorns were scarcer than usual, and the few butterflies were all of tropical species. One of these, Papilio blumei, of which I obtained a few specimens only, is among the most magnificent I have ever seen. It is a green and gold swallow-tail, with azure-blue spoon-shaped tails, and was often seen flying about the village when the sun shone, but in a very shattered condition. The great amount of wet and cloudy weather was a great drawback all the time I was at Rurúkan.

Even in the vegetation there is very little to indicate elevation. The trees are more covered with lichens and mosses, and the ferns and tree-ferns are finer and more luxuriant than I had been accustomed to see them on the low grounds, both probably attributable to the almost perpetual moisture that here prevails. Abundance of a tasteless raspberry, with blue and yellow Compositæ, have somewhat of a temperate aspect; and minute ferns and Orchideæ, with dwarf Begonias on the rocks, make some approach to a sub-alpine vegetation. The forest however is most luxuriant. Noble palms, Pandani, and tree-ferns are abundant in it, while the forest trees are completely festooned with Orchideæ, Bromeliæ, Araceæ, Lycopodiums, and mosses. The ordinary stemless ferns abound; some with gigantic fronds ten or twelve feet long, others barely an inch high; some with entire and massive leaves, others elegantly waving their finely-cut foliage, and adding endless variety and interest to the forest paths. The cocoa-nut palm still produces fruit abundantly, but is said to be deficient in oil. Oranges thrive better than below, producing abundance of delicious fruit; but the shaddock or pumplemous (Citrus decumana) requires the full force of a tropical sun, for it will not thrive even at Tondáno a thousand feet lower. On the hilly slopes rice is cultivated largely, and ripens well, although the temperature rarely or never rises to 80°, so that one would think it might be grown even in England in fine summers, especially if the young plants were raised under glass.

The mountains have an unusual quantity of earth or vegetable mould spread over them. Even on the steepest slopes there is everywhere a covering of clays and sands, and generally a good thickness of vegetable soil. It is this which perhaps contributes to the uniform luxuriance of the forest, and delays the appearance of that sub-alpine vegetation which depends almost as much on the abundance of rocky and exposed surfaces as on difference of climate. At a much lower elevation on Mount Ophir in Malacca, Dacrydiums and Rhododendrons with abundance of Nepenthes, ferns, and terrestrial orchids suddenly took the place of the lofty forest; but this was plainly due to the occurrence of an extensive slope of bare granitic rock at an elevation of less than 3,000 feet. The quantity of vegetable soil, and also of loose sands and clays, resting on steep slopes, hill-tops and the sides of ravines, is a curious and important phenomenon. It may be due in part to constant slight earthquake shocks, facilitating the disintegration of rock; but would also seem to indicate that the country has been long exposed to gentle atmospheric action, and that its elevation has been exceedingly slow and continuous.

During my stay at Rurúkan my curiosity was satisfied by experiencing a pretty sharp earthquake-shock. On the evening of June 29th, at a quarter after eight, as I was sitting reading, the house began shaking with a very gentle, but rapidly increasing motion. I sat still enjoying the novel sensation for some seconds; but in less than half a minute it became strong enough to shake me in my chair, and to make the house visibly rock about, and creak and crack as if it would fall to pieces. Then began a cry throughout the village of "Tana goyang! tana goyang!" (Earthquake! earthquake!) Everybody rushed out of their houses—women screamed and children cried—and I thought it prudent to go out too. On getting up, I found my head giddy and my steps unsteady, and could hardly walk without falling. The shock continued about a minute, during which time I felt as if I had been turned round and round, and was almost sea-sick. Going into the house again, I found a lamp and a bottle of arrack upset. The tumbler which formed the lamp had been thrown out of the saucer in which it had stood. The shock appeared to be nearly vertical, rapid, vibratory, and jerking. It was sufficient, I have no doubt, to have thrown down brick chimneys and walls and church towers; but as the houses here are all low, and strongly framed of timber, it is impossible for them to be much injured, except by a shock that would utterly destroy a European city. The people told me it was ten years since they had had a stronger shock than this, at which time many houses were thrown down and some people killed.

At intervals of ten minutes to half an hour, slight shocks and tremors were felt, sometimes strong enough to send us all out again. There was a strange mixture of the terrible and the ludicrous in our situation. We might at any moment have a much stronger shock, which would bring down the house over us, or—what I feared more—cause a landslip, and send us down into the deep ravine on the very edge of which the village is built; yet I could not help laughing each time we ran out at a slight shock, and then in a few moments ran in again. The sublime and the ridiculous were here literally but a step apart. On the one hand, the most terrible and destructive of natural phenomena was in action around us—the rocks, the mountains, the solid earth were trembling and convulsed, and we were utterly impotent to guard against the danger that might at any moment overwhelm us. On the other hand was the spectacle of a number of men, women, and children running in and out of their houses, on what each time proved a very unnecessary alarm, as each shock ceased just as it became strong enough to frighten us. It seemed really very much like "playing at earthquakes," and made many of the people join me in a hearty laugh, even while reminding each other that it really might be no laughing matter.

At length the evening got very cold, and I became very sleepy, and determined to turn in, leaving orders to my boys, who slept nearer the door, to wake me in case the house was in danger of falling. But I miscalculated my apathy, for I could not sleep much. The shocks continued at intervals of half an hour or an hour all night, just strong enough to wake me thoroughly each time and keep me on the alert ready to jump up in case of danger. I was therefore very glad when morning

came. Most of the inhabitants had not been to bed at all, and
some had stayed out of doors all night. For the next two days
and nights shocks still continued at short intervals, and several
times a day for a week, showing that there was some very extensive disturbance beneath our portion of the earth's crust. How
vast the forces at work really are can only be properly appreciated
when, after feeling their effects, we look abroad over the wide
expanse of hill and valley, plain and mountain, and thus realize
in a slight degree the immense mass of matter heaved and shaken.
The sensation produced by an earthquake is never to be forgotten. We feel ourselves in the grasp of a power to which the
wildest fury of the winds and waves are as nothing ; yet the
effect is more a thrill of awe than the terror which the more
boisterous war of the elements produces. There is a mystery
and an uncertainty as to the amount of danger we incur, which
gives greater play to the imagination, and to the influences of
hope and fear. These remarks apply only to a moderate earthquake. A severe one is the most destructive and the most
horrible catastrophe to which human beings can be exposed.

A few days after the earthquake I took a walk to Tondáno, a
large village of about 7,000 inhabitants, situated at the lower
end of the lake of the same name. I dined with the Controlleur,
Mr. Bensneider, who had been my guide to Tomohón. He had
a fine large house, in which he often received visitors ; and his
garden was the best for flowers which I had seen in the tropics,
although there was no great variety. It was he who introduced
the rose hedges which give such a charming appearance to the
villages ; and to him is chiefly due the general neatness and
good order that everywhere prevail. I consulted him about a
fresh locality, as I found Rurúkan too much in the clouds,
dreadfully damp and gloomy, and with a general stagnation of
bird and insect life. He recommended me a village some
distance beyond the lake, near which was a large forest, where
he thought I should find plenty of birds. As he was going
himself in a few days I decided to accompany him.

After dinner I asked him for a guide to the celebrated waterfall on the outlet stream of the lake. It is situated about a
mile and a half below the village, where a slight rising ground
closes in the basin, and evidently once formed the shore of the
lake. Here the river enters a gorge, very narrow and tortuous,
along which it rushes furiously for a short distance and then
plunges into a great chasm, forming the head of a large valley.
Just above the fall the channel is not more than ten feet wide,
and here a few planks are thrown across, whence, half hid by
luxuriant vegetation, the mad waters may be seen rushing
beneath, and a few feet farther plunge into the abyss. Both
sight and sound are grand and impressive. It was here that,
four years before my visit, a former Governor of the Moluccas
committed suicide, by leaping into the torrent. This at least is
the general opinion, as he suffered from a painful disease which

was supposed to have made him weary of his life. His body was found next day in the stream below.

Unfortunately, no good view of the fall could now be obtained, owing to the quantity of wood and high grass that lined the margins of the precipices. There are two falls, the lower being the most lofty ; and it is possible, by a long circuit, to descend into the valley and see them from below. Were the best points of view searched for and rendered accessible, these falls would probably be found to be the finest in the Archipelago. The chasm seems to be of great depth, probably 500 or 600 feet. Unfortunately I had no time to explore this valley, as I was anxious to devote every fine day to increasing my hitherto scanty collections.

Just opposite my abode in Rurukan was the school-house. The schoolmaster was a native, educated by the Missionary at Tomohón. School was held every morning for about three hours, and twice a week in the evening there was catechising and preaching. There was also a service on Sunday morning. The children were all taught in Malay, and I often heard them repeating the multiplication-table up to twenty times twenty very glibly. They always wound up with singing, and it was very pleasing to hear many of our old-psalm-tunes in these remote mountains, sung with Malay words. Singing is one of the real blessings which Missionaries introduce among savage nations, whose native chants are almost always monotonous and melancholy.

On catechising evenings the schoolmaster was a great man, preaching and teaching for three hours at a stretch much in the style of an English ranter. This was pretty cold work for his auditors, however warming to himself ; and I am inclined to think that these native teachers, having acquired facility of speaking and an endless supply of religious platitudes to talk about, ride their hobby rather hard, without much consideration for their flock. The Missionaries, however, have much to be proud of in this country. They have assisted the Government in changing a savage into a civilized community in a wonderfully short space of time. Forty years ago the country was a wilderness, the people naked savages, garnishing their rude houses with human heads. Now it is a garden, worthy of its sweet native name of "Minahasa." Good roads and paths traverse it in every direction ; some of the finest coffee plantations in the world surround the villages, interspersed with extensive rice-fields more than sufficient for the support of the population.

The people are now the most industrious, peaceable, and civilized in the whole Archipelago. They are the best clothed, the best housed, the best fed, and the best educated ; and they have made some progress towards a higher social state. I believe there is no example elsewhere of such striking results being produced in so short a time—results which are entirely due to the system of government now adopted by the Dutch in their Eastern possessions. The system is one which may be called a "paternal despotism." Now we Englishmen do not like despotism—we hate the name and the thing, and we would rather see people ignorant, lazy, and vicious, than use any but moral force to make them wise, industrious, and good. And we are right when we are dealing with men of our own race, and of similar ideas and equal capacities with ourselves. Example and precept, the force of public opinion, and the slow, but sure spread of education, will do everything in time; without engendering any of those bitter feelings, or producing any of that servility, hypocrisy, and dependence, which are the sure results of a despotic government. But what should we think of a man who should advocate these principles of perfect freedom in a family or a school? We should say that he was applying a good general principle to a case in which the conditions rendered it inapplicable—the case in which the governed are in an admitted state of mental inferiority to those who govern them, and are unable to decide what is best for their permanent welfare. Children must be subjected to some degree of authority and guidance; and if properly managed they will cheerfully submit to it, because they know their own inferiority, and believe their elders are acting solely for their good. They learn many things the use of which they cannot comprehend, and which they would never learn without some moral and social if not physical pressure. Habits of order, of industry, of cleanliness, of respect and obedience, are inculcated by similar means. Children would never grow up into well-behaved and and well-educated men, if the same absolute freedom of action that is allowed to men were allowed to them. Under the best aspect of education, children are subjected to a mild despotism for the good of themselves and of society; and their confidence in the wisdom and goodness of those who ordain and apply this despotism, neutralizes the bad passions and degrading feelings, which under less favourable conditions are its general results.

Now, there is not merely an analogy,—there is in many respects an identity of relation, between master and pupil or parent and child on the one hand, and an uncivilized race and its civilized rulers on the other. We know (or think we know) that the education and industry, and the common usages of civilized man, are superior to those of savage life; and, as he becomes acquainted with them, the savage himself admits this. He admires the superior acquirements of the civilized man, and it is with pride that he will adopt such usages as do not interfere too much with his sloth, his passions, or his prejudices. But as the wilful child or the idle schoolboy, who was never taught obedience, and never made to do anything which of his own free will he was not inclined to do, would in most cases obtain neither education nor manners; so it is much more unlikely that the savage, with all the confirmed habits of manhood and the traditional prejudices of race, should ever do more than copy a few of the least beneficial customs of civilization, without some stronger stimulus than precept, very imperfectly backed by example.

If we are satisfied that we are right in assuming the government over a savage race, and occupying their country; and if we further consider it our duty to do what we can to improve our rude subjects and raise them up towards our own level, we must not be too much afraid of the cry of "despotism" and "slavery," but must use the authority we possess to induce them to do work, which they may not altogether like, but which we know to be an indispensable step in their moral and physical advancement. The Dutch have shown much good policy in the means by which they have done this. They have in most cases upheld and strengthened the authority of the native chiefs, to whom the people have been accustomed to render a voluntary obedience; and by acting on the intelligence and self-interest of these chiefs, have brought about changes in the manners and customs of the people, which would have excited ill-feeling and perhaps revolt had they been directly enforced by foreigners.

In carrying out such a system, much depends upon the character of the people; and the system which succeeds admirably in one place could only be very partially worked out in another. In Minahasa the natural docility and intelligence of the race have made their progress rapid; and how important this is, is well illustrated by the fact, that in the immediate vicinity of the town of Menado are a tribe called Banteks, of a much less tractable disposition, who have hitherto resisted all efforts of the Dutch Government to induce them to adopt any systematic cultivation. These remain in a ruder condition, but engage themselves willingly as occasional porters and labourers, for which their greater strength and activity well adapt them.

No doubt the system here sketched seems open to serious objection. It is to a certain extent despotic, and interferes with free trade, free labour, and free communication. A native cannot leave his village without a pass, and cannot engage himself to any merchant or captain without a Government permit. The coffee has all to be sold to Government, at less than half the price that the local merchant would give for it, and he consequently cries out loudly against "monoply" and "oppression." He forgets, however, that the coffee plantations were established by the Government at great outlay of capital and skill; that it gives free education to the people, and that the monopoly is in lieu of taxation. He forgets that the product he wants to purchase and make a profit by, is the creation of the Government, without whom the people would still be savages. He knows very well that free trade would, as its first result, lead to the importation of whole cargoes of arrack, which would be carried over the country and exchanged for coffee.

That drunkenness and poverty would spread over the land ; that the public coffee plantations would not be kept up ; that the quality and quantity of the coffee would soon deteriorate ; that traders and merchants would get rich, but that the people would relapse into poverty and barbarism. That such is invariably the result of free trade with any savage tribes who possess a valuable product, native or cultivated, is well known to those who have visited such people ; but we might even anticipate from general principles that evil results would happen. If there is one thing rather than another to which the grand law of continuity or development will apply, it is to human progress. There are certain stages through which society must pass in its onward march from barbarism to civilization. Now one of these stages has always been some form or other of despotism, such as feudalism or servitude, or a despotic paternal government ; and we have every reason to believe that it is not possible for humanity to leap over this transition epoch, and pass at once from pure savagery to free civilization. The Dutch system attempts to supply this missing link, and to bring the people on by gradual steps to that higher civilization, which we (the English) try to force upon them at once. Our system has always failed. We demoralize and we extirpate, but we never really civilize. Whether the Dutch system can permanently succeed is but doubtful, since it may not be possible to compress the work of ten centuries into one ; but at all events it takes nature as a guide, and is therefore more deserving of success, and more likely to succeed, than ours.[1]

There is one point connected with this question which I think the Missionaries might take up with great physical and moral results. In this beautiful and healthy country, and with abundance of food and necessaries, the population does not increase as it ought to do. I can only impute this to one cause. Infant mortality, produced by neglect while the mothers are working in the plantations, and by general ignorance of the conditions of health in infants. Women all work, as they have always been accustomed to do. It is no hardship to them, but I believe is often a pleasure and relaxation. They either take their infants with them, in which case they leave them in some shady spot on the ground, going at intervals to give them nourishment, or they leave them at home in the care of other children too young to work. Under neither of these circumstances can infants be properly attended to, and great mortality is the result, keeping down the increase of population far below the rate which the general prosperity of the country and the universality of marriage would lead us to expect. This is a matter in which the Government is directly interested, since it

---

[1] Dr. Guillemard, who visited Minahasa twenty-five years later, found the country in much the same condition as I describe—drunkenness and crime were almost unknown, and the people were contented and happy. (*See The Cruise of the "Marchesa,"* Vol. II., p. 181.)

is by the increase of the population alone that there can be any large and permanent increase in the produce of coffee. The Missionaries should take up the question, because, by inducing married women to confine themselves to domestic duties, they will decidedly promote a higher civilization, and directly increase the health and happiness of the whole community. The people are so docile, and so willing to adopt the manners and customs of Europeans, that the change might be easily effected, by merely showing them that it was a question of morality and civilization, and an essential step in their progress towards an equality with their white rulers.

After a fortnight's stay at Rurúkan, I left that pretty and interesting village in search of a locality and climate more productive of birds and insects. I passed the evening with the Controlleur of Tondáno, and the next morning at nine left in a small boat for the head of the lake, a distance of about ten miles. The lower end of the lake is bordered by swamps and marshes of considerable extent, but a little further on the hills come down to the water's edge and give it very much the appearance of a great river, the width being about two miles. At the upper end is the village of Kákas, where I dined with the head-man in a good house like those I have already described ; and then went on to Langówan, four miles distant over a level plain. This was the place where I had been recommended to stay, and I accordingly unpacked my baggage and made myself comfortable in the large house devoted to visitors. I obtained a man to shoot for me, and another to accompany me the next day to the forest, where I was in hopes of finding a good collecting ground.

In the morning after breakfast I started off, but found I had four miles to walk over a wearisome straight road through coffee plantations before I could get to the forest, and as soon as I did so it came on to rain heavily, and did not cease till night. This distance to walk every day was too far for any profitable work, especially when the weather was so uncertain. I therefore decided at once that I must go further on, till I found some place close to or in a forest country. In the afternoon my friend Mr. Bensneider arrived, together with the Controlleur of the next district, called Belang, from whom I learnt that six miles further on there was a village called Panghu, which had been recently formed and had a good deal of forest close to it ; and he promised me the use of a small house if I liked to go there.

The next morning I went to see the hot springs and mud volcanoes, for which this place is celebrated. A picturesque path, among plantations and ravines, brought us to a beautiful circular basin about forty feet diameter, bordered by a calcareous ledge, so uniform and truly curved that it looked like a work of art. It was filled with clear water very near the boiling point, and emitting clouds of steam with a strong sulphureous odour.

It overflows at one point and forms a little stream of hot water, which at a hundred yards distance is still too hot to hold the hand in. A little further on, in a piece of rough wood, were two other springs not so regular in outline, but appearing to be much hotter, as they were in a continual state of active ebullition. At intervals of a few minutes a great escape of steam or gas took place, throwing up a column of water three or four feet high.

We then went to the mud-springs, which are about a mile off, and are still more curious. On a sloping tract of ground in a slight hollow is a small lake of liquid mud, in patches of blue, red, or white, and in many places boiling and bubbling most furiously. All around on the indurated clay are small wells and craters full of boiling mud. These seem to be forming continually, a small hole appearing first, which emits jets of steam and boiling mud, which on hardening forms a little cone with a crater in the middle. The ground for some distance is very unsafe, as it is evidently liquid at a small depth, and bends with pressure like thin ice. At one of the smaller marginal jets which I managed to approach, I held my hand to see if it was really as hot as it looked, when a little drop of mud that spurted on to my finger scalded like boiling water. A short distance off there was a flat bare surface of rock, as smooth and hot as an oven floor, which was evidently an old mud-pool dried up and hardened. For hundreds of yards round there were banks of reddish and white clay used for whitewash, and these were so hot close to the surface that the hand could hardly bear to be held in cracks a few inches deep, from which arose a strong sulphureous vapour. I was informed that some years back a French gentleman who visited these springs ventured too near the liquid mud, when the crust gave way and he was engulfed in the horrible cauldron.

This evidence of intense heat so near the surface over a large tract of country was very impressive, and I could hardly divest myself of the notion that some terrible catastrophe might at any moment devastate the country. Yet it is probable that all these apertures are really safety-valves, and that the inequalities of the resistance of various parts of the earth's crust will always prevent such an accumulation of force as would be required to upheave and overwhelm any extensive area. About seven miles west of this is a volcano which was in eruption about thirty years before my visit, presenting a magnificent appearance and covering the surrounding country with showers of ashes. The plains around the lake formed by the intermingling and decomposition of volcanic products are of amazing fertility, and with a little management in the rotation of crops might be kept in continual cultivation. Rice is now grown on them for three or four years in succession, when they are left fallow for the same period, after which rice or maize can be again grown. Good rice produces thirty-fold, and coffee trees continue bearing abundantly for ten or fifteen years, without any manure and with scarcely any cultivation.

I was delayed a day by incessant rain, and then proceeded to Panghu, which I reached just before the daily rain began at 11 A.M. After leaving the summit level of the lake basin, the road is carried along the slope of a fine forest ravine. The descent is a long one, so that I estimated the village to be not more than 1,500 feet above the sea, yet I found the morning temperature often 69°, the same as at Tondáno at least 600 or 700 feet higher. I was pleased with the appearance of the place, which had a good deal of forest and wild country around it, and found prepared for me a little house consisting only of a verandah and a back room. This was only intended for visitors to rest in, or to pass a night, but it suited me very well. I was so unfortunate, however, as to lose both my hunters just at this time. One had been left at Tondáno with fever and diarrhœa, and the other was attacked at Langówan with inflammation of the chest, and as his case looked rather bad I had him sent back to Menado. The people here were all so busy with their rice-harvest, which it was important for them to finish owing to the early rains, that I could get no one to shoot for me.

During the three weeks that I stayed at Panghu, it rained nearly every day, either in the afternoon only, or all day long ; but there were generally a few hours sunshine in the morning, and I took advantage of these to explore the roads and paths, the rocks and ravines, in search of insects. These were not very abundant, yet I saw enough to convince me that the locality was a good one, had I been there at the beginning instead of at the end of the dry season. The natives brought me daily a few insects obtained at the Sagueir palms, including some fine Cetonias and stag-beetles. Two little boys were very expert with the blowpipe, and brought me a good many small birds, which they shot with pellets of clay. Among these was a pretty little flower-pecker of a new species (Prionochilus aureolimbatus), and several of the loveliest honeysuckers I had yet seen. My general collection of birds was, however, almost at a standstill ; for though I at length obtained a man to shoot for me, he was not good for much, and seldom brought me more than one bird a day. The best thing he shot was the large and rare fruit-pigeon peculiar to Northern Celebes (Carpophaga forsteni), which I had long been seeking after.

I was myself very successful in one beautiful group of insects, the tiger-beetles, which seem more abundant and varied here than anywhere else in the Archipelago. I first met with them on a cutting in the road, where a hard clayey bank was partially overgrown with mosses and small ferns. Here I found running about a small olive-green species which never took flight ; and more rarely a fine purplish black wingless insect, which was always found motionless in crevices, and was therefore probably nocturnal. It appeared to me to form a new genus. About

the roads in the forest, I found the large and handsome
Cicindela heros, which I had before obtained sparingly at
Macassar; but it was in the mountain torrent of the ravine
itself that I got my finest things. On dead trunks overhanging
the water, and on the banks and foliage, I obtained three very
pretty species of Cicindela, quite distinct in size, form, and
colour, but having an almost identical pattern of pale spots. I
also found a single specimen of a most curious species with very
long antennæ. But my finest discovery here was the Cicindela
gloriosa, which I found on mossy stones just rising above the
water. After obtaining my first specimen of this elegant insect,
I used to walk up the stream, watching carefully every moss-
covered rock and stone. It was rather shy, and would often
lead me a long chase from stone to stone, becoming invisible
every time it settled on the damp moss, owing to its rich
velvety green colour. On some days I could only catch a few
glimpses of it, on others I got a single specimen, and on a
few occasions two, but never without a more or less active
pursuit. This and several other species I never saw but in this
one ravine.

Among the people here I saw specimens of several types,
which, with the peculiarities of the languages, gives me some
notion of their probable origin. A striking illustration of the
low state of civilization of these people till quite recently, is to
be found in the great diversity of their languages. Villages
three or four miles apart have separate dialects, and each group
of three or four such villages has a distinct language quite
unintelligible to all the rest; so that, till the recent introduction
of Malay by the Missionaries, there must have been a bar to all
free communication. These languages offer many peculiarities.
They contain a Celebes-Malay element and a Papuan element,
along with some radical peculiarities found also in the languages
of the Siau and Sanguir islands further north, and therefore
probably derived from the Philippine Islands. Physical
characters correspond. There are some of the less civilized
tribes which have semi-Papuan features and hair, while in some
villages the true Celebes or Bugis physiognomy prevails. The
plateau of Tondáno is chiefly inhabited by people nearly as
white as the Chinese, and with very pleasing semi-European
features. The people of Siau and Sanguir much resemble
these, and I believe them to be perhaps immigrants from some
of the islands of North Polynesia. The Papuan type will
represent the remnant of the aborigines, while those of the
Bugis character show the extension northward of the superior
Malay races.

As I was wasting valuable time at Panghu owing to the bad
weather and the illness of my hunters, I returned to Menado
after a stay of three weeks. Here I had a little touch of fever,
and what with drying and packing away my collections and
getting fresh servants, it was a fortnight before I was again

ready to start. I now went eastward over an undulating
country skirting the great volcano of Klábat, to a village called
Lempias, situated close to the extensive forest that covers the
lower slopes of that mountain. My baggage was carried from
village to village by relays of men, and as each change involved
some delay, I did not reach my destination (a distance of
eighteen miles) till sunset. I was wet through, and had to wait
for an hour in an uncomfortable state till the first instalment of
my baggage arrived, which luckily contained my clothes, while
the rest did not come in till midnight.

This being the district inhabited by that singular animal the
Babirusa (Hog-deer) I inquired about skulls, and soon obtained
several in tolerable condition, as well as a fine one of the rare
and curious "Sapi-utan" (Anoa depressicornis). Of this animal
I had seen two living specimens at Menado, and was surprised
at their great resemblance to small cattle, or still more to the
Eland of South Africa. Their Malay name signifies "forest ox,"
and they differ from very small high-bred oxen principally by
the low-hanging dewlap, and straight pointed horns which
slope back over the neck. I did not find the forest here so rich in
insects as I had expected, and my hunters got me very few birds,
but what they did obtain were very interesting. Among these
were the rare forest Kingfisher (Crittura cyanotis), a small new
species of Megapodius, and one specimen of the large and interesting Maleo (Megacephalon rubripes), to obtain which was
one of my chief reasons for visiting this district. Getting no
more, however, after ten days' search I removed to Licoupang,
at the extremity of the peninsula, a place celebrated for these
birds, as well as for the Babirúsa and Sapi-utan. I found here
Mr. Goldmann, the eldest son of the Governor of the Moluccas,
who was superintending the establishment of some Government salt-works. This was a better locality, and I obtained
some fine butterflies and very good birds, among which was one
more specimen of the rare ground dove (Phlegænas tristigmata),
which I had first obtained near the Máros waterfall in South
Celebes.

Hearing what I was particularly in search of, Mr. Goldmann
kindly offered to make a hunting-party to the place where the
"Maleos" are most abundant, a remote and uninhabited sea-
beach about twenty miles distant. The climate here was quite
different to that on the mountains, not a drop of rain having
fallen for four months ; so I made arrangements to stay on the
beach a week, in order to secure a good number of specimens.
We went partly by boat and partly through the forest, accompanied by the Major or head-man of Licoupang, with a dozen
natives and about twenty dogs. On the way they caught a
young Sapi-utan and five wild pigs. Of the former I preserved
the head. This animal is entirely confined to the remote
mountain forests of Celebes and one or two adjacent islands
which form part of the same group. In the adults the head is

black, with a white mark over each eye, one on each cheek, and another on the throat. The horns are very smooth and sharp when young, but become thicker and ridged at the bottom with age. Most naturalists consider this curious animal to be a small ox, but from the character of the horns, the fine coat of hair and the descending dewlap, it seemed closely to approach the antelopes.

Arrived at our destination we built a hut and prepared for a stay of some days, I to shoot and skin "Maleos," Mr. Goldmann and the Major to hunt wild pigs, Babirúsa, and Sapi-utan. The place is situated in the large bay between the islands of Limbé and Banca, and consists of a steep beach more than a mile in length, of deep loose and coarse black volcanic sand or rather gravel, very fatiguing to walk over. It is bounded at each extremity by a small river, with hilly ground beyond ; while the forest behind the beach itself is tolerably level and its growth stunted. We have here probably an ancient lava stream from the Klabat volcano, which has flowed down a valley into the sea, and the decomposition of which has formed the loose black sand. In confirmation of this view it may be mentioned, that the beaches beyond the small rivers in both directions are of white sand.

It is in this loose, hot black sand that those singular birds the "Maleos" deposit their eggs. In the months of August and September, when there is little or no rain, they come down in pairs from the interior to this or to one or two other favourite spots, and scratch holes three or four feet deep, just above high-water mark, where the female deposits a single large egg, which she covers over with about a foot of sand, and then returns to the forest. At the end of ten or twelve days she comes again to the same spot to lay another egg, and each female bird is supposed to lay six or eight eggs during the season. The male assists the female in making the hole, coming down and returning with her. The appearance of the bird when walking on the beach is very handsome. The glossy black and rosy white of the plumage, the helmeted head and elevated tail, like that of the common fowl, give a striking character, which their stately and somewhat sedate walk renders still more remarkable. There is hardly any difference between the sexes, except that the casque or bonnet at the back of the head and the tubercles at the nostrils are a little larger, and the beautiful rosy salmon colour a little deeper in the male bird, but the difference is so slight that it is not always possible to tell a male from a female without dissection. They run quickly, but when shot at or suddenly disturbed take wing with a heavy noisy flight to some neighbouring tree, where they settle on a low branch ; and they probably roost at night in a similar situation. Many birds lay in the same hole, for a dozen eggs are often found together ; and these are so large that it is not possible for the body of the bird to contain more than one fully-developed egg at the same time.

In all the female birds which I shot, none of the eggs besides the one large one exceeded the size of peas, and there were only eight or nine of these, which is probably the extreme number a bird can lay in one season.

Every year the natives come for fifty miles round to obtain these eggs, which are esteemed a great delicacy, and when quite fresh are indeed delicious. They are richer than hens' eggs and of a finer flavour, and each one completely fills an ordinary tea-cup, and forms with bread or rice a very good meal. The colour of the shell is a pale brick red, or very rarely pure white. They are elongate and very slightly smaller at one end, from four to four and a half inches long by two and a quarter or two and a half wide.

After the eggs are deposited in the sand they are no further cared for by the mother. The young birds on breaking the shell, work their way up through the sand and run off at once to the forest; and I was assured by Mr. Duivenboden of Ternate, that they can fly the very day they are hatched. He had taken some eggs on board his schooner which hatched during the night, and in the morning the little birds flew readily across the cabin. Considering the great distances the birds come to deposit the eggs in a proper situation (often ten or fifteen miles) it seems extraordinary that they should take no further care of them. It is, however, quite certain that they neither do nor can watch them. The eggs being deposited by a number of hens in succession in the same hole, would render it impossible for each to distinguish its own; and the food necessary for such large birds (consisting entirely of fallen fruits) can only be obtained by roaming over an extensive district, so that if the numbers of birds which come down to this single beach in the breeding season, amounting to many hundreds, were obliged to remain in the vicinity, many would perish of hunger.

In the structure of the feet of this bird, we may detect a cause for its departing from the habits of its nearest allies, the Megapodii and Talegalli, which heap up earth, leaves, stones, and sticks into a huge mound, in which they bury their eggs. The feet of the Maleo are not nearly so large or strong in pro-portion as in these birds, while its claws are short and straight instead of being long and much curved. The toes are, however, strongly webbed at the base, forming a broad powerful foot, which, with the rather long leg, is well adapted to scratch away the loose sand (which flies up in a perfect shower when the birds are at work), but which could not without much labour accumu-late the heaps of miscellaneous rubbish, which the large grasping feet of the Megapodius bring together with ease.

We may also, I think, see in the peculiar organization of the entire family of the Megapodidæ or Brush Turkeys, a reason why they depart so widely from the usual habits of the Class of birds. Each egg being so large as entirely to fill up the ab-dominal cavity and with difficulty pass the walls of the pelvis, a considerable interval is required before the successive eggs can be matured (the natives say about thirteen days). Each bird lays six or eight eggs, or even more each season, so that between the first and last there may be an interval of two or three months. Now, if these eggs were hatched in the ordinary way, either the parents must keep sitting continually for this long period, or if they only began to sit after the last egg was deposited, the first would be exposed to injury by the climate, or to destruction by the large lizards, snakes, or other animals which abound in the district ; because such large birds must roam about a good deal in search of food. Here then we seem to have a case in which the habits of a bird may be directly traced to its exceptional organization ; for it will hardly be maintained that this abnormal structure and peculiar food were given to the Megapodidæ in order that they might not exhibit that parental affection, or possess those domestic instincts so general in the Class of birds, and which so much excite our admiration.

It has generally been the custom of writers on Natural History to take the habits and instincts of animals as fixed points, and to consider their structure and organization as specially adapted to be in accordance with these. This assumption is however an arbitrary one, and has the bad effect of stifling inquiry into the nature and causes of "instincts and habits," treating them as directly due to a "first cause," and therefore incomprehensible to us. I believe that a careful consideration of the structure of a species, and of the peculiar physical and organic conditions by which it is surrounded, or has been surrounded in past ages, will often, as in this case, throw much light on the origin of its habits and instincts. These again, combined with changes in external conditions, react upon structure, and by means of "variation" and "natural selection" both are kept in harmony.

My friends remained three days, and got plenty of wild pigs and two Anóas, but the latter were much injured by the dogs, and I could only preserve the heads. A grand hunt which we attempted on the third day failed, owing to bad management in driving in the game, and we waited for five hours perched on platforms in trees without getting a shot, although we had been assured that pigs, Babirúsas, and Anóas would rush past us in dozens. I myself, with two men, stayed three days longer to get more specimens of the Maleos, and succeeded in preserving twenty-six very fine ones, the flesh and eggs of which supplied us with abundance of good food.

The Major sent a boat, as he had promised, to take home my baggage, while I walked through the forest with my two boys and a guide, about fourteen miles. For the first half of the distance there was no path, and we had often to cut our way through tangled rattans or thickets of bamboo. In some of our turnings to find the most practicable route I expressed my fear that we were losing our way, as the sun being vertical I could see no possible clue to the right direction. My conductors, however, laughed at the idea, which they seemed to consider quite ludicrous ; and sure enough, about half way, we suddenly encountered a little hut where people from Licoupang came to hunt and smoke wild pigs. My guide told me he had never before traversed the forest between these two points ; and this is what is considered by some travellers as one of the savage "instincts," whereas it is merely the result of wide general knowledge. The man knew the topography of the whole district : the slope of the land, the direction of the streams, the belts of bamboo or rattan, and many other indications of locality and direction ; and he was thus enabled to hit straight upon the hut, in the vicinity of which he had often hunted. In a forest of which he knew nothing, he would be quite as much at a loss as a European. Thus it is, I am convinced, with all the wonderful accounts of Indians finding their way through trackless forests to definite points. They may never have passed straight between the two particular points before, but they are well acquainted with the vicinity of both, and have such a general knowledge of the whole country, its water system, its soil and its vegetation, that as they approach the point they are to reach, many easily-recognized indications enable them to hit upon it with certainty.

The chief feature of this forest was the abundance of rattan palms, hanging from the trees, and turning and twisting about on the ground, often in inextricable confusion. One wonders at first how they can get into such queer shapes ; but it is evidently caused by the decay and fall of the trees up which they have first climbed, after which they grow along the ground till they meet with another trunk up which to ascend. A tangled mass of twisted living rattan is therefore a sign that at some former period a large tree has fallen there, though there may be not the slightest vestige of it left. The rattan seems to have unlimited powers of growth, and a single plant may mount up several trees in succession, and thus reach the enormous length they are said sometimes to attain. They much improve the appearance of a forest as seen from the coast ; for they vary the otherwise monotonous tree-tops with feathery crowns of leaves rising clear above them, and each terminated by an erect leafy spike like a lightning-conductor.

The other most interesting object in the forest was a beautiful palm, whose perfectly smooth and cylindrical stem rises erect to more than a hundred feet high, with a thickness of only eight or ten inches ; while the fan-shaped leaves which compose its crown are almost complete circles of six or eight feet diameter, borne aloft on long and slender petioles, and beautifully toothed round the edge by the extremities of the leaflets, which are separated only for a few inches from the circumference. It is probably the Livistona rotundifolia of botanists, and is the most complete and beautiful fan-leaf I have ever seen, serving admirably for folding into water-buckets and *impromptu* baskets, as well as for thatching and other purposes.

A few days afterwards I returned to Menado on horseback, sending my baggage round by sea ; and had just time to pack up all my collections to go by the next mail steamer to Amboyna. I will now devote a few pages to an account of the chief peculiarities of the Zoology of Celebes, and its relation to that of the surrounding countries.

CHAPTER XVIII.

NATURAL HISTORY OF CELEBES.

THE position of Celebes is the most central in the Archipelago. Immediately to the north are the Philippine islands ; on the west is Borneo ; on the east are the Molucca islands ; and on the south is the Timor group : and it is on all sides so connected with these islands by its own satellites, by small islets, and by coral reefs, that neither by inspection on the map nor by actual observation around its coast, is it possible to determine accurately which should be grouped with it, and which with the surrounding districts. Such being the case, we should naturally expect to find, that the productions of this central island in some degree represented the richness and variety of the whole Archipelago, while we should not expect much individuality in a country, so situated, that it would seem as if it were pre-eminently fitted to receive stragglers and immigrants from all around.

As so often happens in nature, however, the fact turns out to be just the reverse of what we should have expected ; and an examination of its animal productions shows Celebes to be at once the poorest in the number of its species, and the most isolated in the character of its productions, of all the great islands in the Archipelago. With its attendant islets it spreads over an extent of sea hardly inferior in length and breadth to that occupied by Borneo, while its actual land area is nearly double that of Java ; yet its Mammalia ~~and terrestrial birds~~ ~~number scarcely more than half the species~~ found in the last-named island. Its position is such that it could receive immigrants from every side more readily than Java, yet in proportion to the species which inhabit it far fewer seem derived from other islands, while far more are altogether peculiar to it ; and a considerable number of its animal forms are so remarkable, as to find no close allies in any other part of the world. I now propose to examine the best known groups of Celebesian animals in some detail, to study their relations to those of other islands, and to call attention to the many points of interest which they suggest.

We know far more of the birds of Celebes than we do of any other group of animals. No less than 205 species have been discovered, and though no doubt many more wading and swimming birds have to be added, yet the list of land birds, 144 in number, and which for our present purpose are much the most important, must be very nearly complete. I myself assiduously collected birds in Celebes for nearly ten months, and my assistant, Mr. Allen, spent two months in the Sula islands. The Dutch naturalist Forsten spent two years in Northern Celebes (twenty years before my visit), and collections of birds had also been sent to Holland from Macassar. The French ship of discovery, *L'Astrolabe*, also touched at Menado and procured collections. Since my return home, the Dutch naturalists Rosenberg and Bernstein have made extensive collections both in North Celebes and in the Sula islands ; yet all their researches combined have only added eight species of land birds to those forming part of my own collection—a fact which renders it almost certain that there are very few more to discover.[1]

Besides Salayer and Boutong on the south, with Peling and Bungay on the east, the three islands of the Sula (or Zula) Archipelago also belong zoologically to Celebes, although their position is such, that it would seem more natural to group them with the Moluccas. About 48 land birds are now known from the Sula group, and if we reject from these five species which have a wide range over the Archipelago, the remainder are much more characteristic of Celebes than of the Moluccas. Thirty-one species are identical with those of the former island, and four are representatives of Celebes forms, while only eleven are Moluccan species, and two more representatives.

But although the Sula islands belong to Celebes, they are so close to Bouru and the southern islands of the Gilolo group, that several purely Moluccan forms have migrated there, which are quite unknown to the island of Celebes itself ; the whole thirteen Moluccan species being in this category, thus adding to the productions of Celebes a foreign element which does not really belong to it. In studying the peculiarities of the Celebesian fauna, it will therefore be well to consider only the productions of the main island.

The number of land birds in the island of Celebes is 128, and from these we may, as before, strike out a small number of species which roam over the whole Archipelago (often from India to the Pacific), and which therefore only serve to disguise the peculiarities of individual islands. These are 20 in number, and leave 108 species which we may consider as more especially characteristic of the island. On accurately comparing these with the birds of all the surrounding countries, we find that only nine extend into the islands westward, and nineteen into the islands eastward, while no less than 80 are entirely confined to the Celebesian fauna—a degree of individuality which, considering

[1] Dr. B. Meyer and other naturalists have since explored the island and its surrounding islets, and have raised the total number of its birds to nearly 400, of which 288 are land birds.

the situation of the island, is hardly to be equalled in any other part of the world. If we still more closely examine these 80 species, we shall be struck by the many peculiarities of structure they present, and by the curious affinities with distant parts of the world which many of them seem to indicate. These points are of so much interest and importance that it will be necessary to pass in review all those species which are peculiar to the island, and to call attention to whatever is most worthy of remark.

Six species of the Hawk tribe are peculiar to Celebes ; three of these are very distinct from allied birds which range over all India to Java and Borneo, and which thus seem to be suddenly changed on entering Celebes. Another (Accipiter trinotatus), is a beautiful hawk, with elegant rows of large round white spots on the tail, rendering it very conspicuous and quite different from any other known bird of the family. Three owls are also peculiar ; and one, a barn owl (Strix rosenbergii), is very much larger and stronger than its ally Strix javanica, which ranges from India through all the islands as far as Lombock.

Of the ten Parrots found in Celebes, eight are peculiar. Among them are two species of the singular raquet-tailed parrots forming the genus Prioniturus, and which are characterized by possessing two long spoon-shaped feathers in the tail. Two allied species are found in the adjacent island of Mindanao, one of the Philippines, and this form of tail is found in no other parrots in the whole world. A small species of Lorikeet (Trichoglossus flavoviridis) seems to have its nearest ally in Australia.

The three Woodpeckers which inhabit the island are all peculiar, and are allied to species found in Java and Borneo, although very different from them all.

Among the three peculiar Cuckoos two are very remarkable. Phœnicophaus callirhynchus is the largest and handsomest species of its genus, and is distinguished by the three colours of its beak, bright yellow, red, and black. Eudynamis melanorynchus differs from all its allies in having a jet-black bill, whereas the other species of the genus always have it green, yellow, or reddish.

The Celebes Roller (Coracias temmincki) is an interesting example of one species of a genus being cut off from the rest. There are species of Coracias in Europe, Asia, and Africa, but none in the Malay peninsula, Sumatra, Java, or Borneo. The present species seems therefore quite out of place ; and what is still more curious is the fact, that it is not at all like any of the Asiatic species, but seems more to resemble those of Africa.

In the next family, the Bee-eaters, is another equally isolated bird, Meropogon forsteni, which combines the characters of African and Indian Bee-eaters, and whose only near ally, Meropogon breweri, was discovered by M. Du Chaillu in West Africa !

The two Celebes Hornbills have no close allies in those which abound in the surrounding countries. The only Thrush, Geocichla erythronota, is most nearly allied to a species peculiar to Timor. Two of the Flycatchers are closely allied to Indian species which are not found in the Malay islands. Two genera somewhat allied to the Magpies (Streptocitta and Charitornis), but whose affinities are so doubtful that Professor Schlegel places them among the Starlings, are entirely confined to Celebes. They are beautiful long-tailed birds, with black and white plumage, and with the feathers of the head somewhat rigid and scale-like.

Doubtfully allied to the Starlings are two other very isolated and beautiful birds. One, Enodes erythrophrys, has ashy and yellow plumage, but is ornamented with broad stripes of orange-red above the eyes. The other, Basilornis celebensis, is a blue-black bird with a white patch on each side of the breast, and the head ornamented with a beautiful compressed scaly crest of feathers, resembling in form that of the well-known Cock-of-the-rock of South America. The only ally to this bird is found in Ceram, and has the feathers of the crest elongated upwards into quite a different form.

A still more curious bird is the Scissirostrum pagei, which although it is at present classed in the Starling family, differs from all other species in the form of the bill and nostrils, and seems most nearly allied in its general structure to the Ox-peckers (Buphaga) of tropical Africa, next to which the celebrated ornithologist Prince Bonaparte finally placed it. It is almost entirely of a slaty colour, with yellow bill and feet, but the feathers of the rump and upper tail-coverts each terminate in a rigid glossy pencil or tuft of a vivid crimson. These pretty little birds take the place of the metallic-green starlings of the genus Calornis, which are found in most other islands of the Archipelago, but which are absent from Celebes.[1] They go in flocks, feeding upon grain and fruits, often frequenting dead trees, in holes of which they build their nests; and they cling to the trunks as easily as woodpeckers or creepers.

Out of eighteen Pigeons found in Celebes eleven are peculiar to it. Two of them, Ptilonopus gularis and Turacæna menadensis, have their nearest allies in Timor. Two others, Carpophaga forsteni and Phlægenas tristigmata, most resemble Philippine island species; and Carpophaga radiata belongs to a New Guinea group. Lastly, in the Gallinaceous tribe, the curious helmeted Maleo (Megacephalon rubripes) is quite isolated, having its nearest (but still distant) allies in the Brush-turkeys of Australia and New Guinea.

Judging, therefore, by the opinions of the eminent naturalists who have described and classified its birds, we find that many of the species have no near allies whatever in the countries

---

[1] Calornis neglecta, first found in the Sula Islands, has now been discovered in Celebes by Dr. Meyer.

which surround Celebes, but are either quite isolated, or indicate relations with such distant regions as New Guinea, Australia, India, or Africa. Other cases of similar remote affinities between the productions of distant countries no doubt exist, but in no spot upon the globe that I am yet acquainted with do so many of them occur together, or do they form so decided a feature in the natural history of the country.

The Mammalia of Celebes are ~~very~~ few in number, consisting of ~~fourteen~~ *forty* terrestrial species and ~~seven~~ *twenty three* bats. Of the former no less than ~~eleven~~ are peculiar, including two which there is reason to believe may have been recently carried into other islands by man. Three species which have a tolerably wide range in the Archipelago, are—1, The curious Lemur, Tarsius spectrum, which is found in all the islands westward as far as Malacca, and also in the Philippine Islands; 2, The common Malay Civet, Viverra tangalunga, which has a still wider range; and 3, a Deer, which seems to be the same as the Rusa hippelaphus of Java, and was probably introduced by man at an early period.

The more characteristic species are as follows :—

Cynopithecus nigrescens, a curious baboon-like monkey, if not a true baboon, which abounds all over Celebes, and is found nowhere else but in the one small island of Batchian, into which it has probably been introduced accidentally. An allied species is found in the Philippines, but in no other island of the Archipelago is there anything resembling them. These creatures are about the size of a spaniel, of a jet-black colour, and have the projecting dog-like muzzle and overhanging brows of the baboons. They have large red callosities and a short fleshy tail, scarcely an inch long, and hardly visible. They go in large bands, living chiefly in the trees, but often descending on the ground and robbing gardens and orchards.

Anoa depressicornis, the Sapi-utan, or wild cow of the Malays, is an animal which has been the cause of much controversy, as to whether it should be classed as ox, buffalo, or antelope. It is smaller than any other wild cattle, and in many respects seems to approach some of the ox-like antelopes of Africa. It is found only in the mountains, and is said never to inhabit places where there are deer. It is somewhat smaller than a small Highland cow, and has long straight horns, which are ringed at the base and slope backwards over the neck.

The wild pig seems to be of a species peculiar to the island; but a much more curious animal of this family is the Babirusa or Pig-deer, so named by the Malays from its long and slender legs, and curved tusks resembling horns. This extraordinary creature resembles a pig in general appearance, but it does not dig with its snout, as it feeds on fallen fruits. The tusks of the lower jaw are very long and sharp, but the upper ones instead of growing downwards in the usual way are completely reversed, growing upwards out of bony sockets through the skin on each side of the snout, curving backwards to near the eyes, and in old animals often reaching eight or ten inches in length. It is difficult to understand what can be the use of these extraordinary horn-like teeth. Some of the old writers supposed that they served as hooks, by which the creature could rest its head on a branch. But the way in which they usually diverge just over and in front of the eye has suggested the more probable idea, that they serve to guard these organs from thorns and spines while hunting for fallen fruits among the tangled

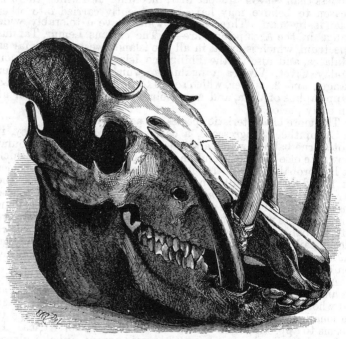

SKULL OF BABIRUSA.

thickets of rattans and other spiny plants. Even this, however, is not satisfactory, for the female, who must seek her food in the same way, does not possess them. I should be inclined to believe rather, that these tusks were once useful, and were then worn down as fast as they grew; but that changed conditions of life have rendered them unnecessary, and they now develop into a monstrous form, just as the incisors of the Beaver or Rabbit will go on growing, if the opposite teeth do not wear them away. In old animals they reach an enormous size, and are generally broken off as if by fighting.

Here again we have a resemblance to the Wart-hogs of Africa, whose upper canines grow outwards and curve up so as to form a transition from the usual mode of growth to that of the Babirusa. In other respects there seems no affinity between these animals, and the Babirusa stands completely isolated, having no resemblance to the pigs of any other part of the world. It is found all over Celebes and in the Sula islands, and also in Bouru, the only spot beyond the Celebes group to which it extends; and which island also shows some affinity to the Sula islands in its birds, indicating perhaps a closer connexion between them at some former period than now exists.

The other terrestrial mammals of Celebes are, five species of squirrels, which are all distinct from those of Java and Borneo, and mark the furthest eastward range of the genus in the tropics; and two of Eastern opossums (Cuscus), which are different from those of the Moluccas, and mark the furthest westward extension of this genus and of the Marsupial order. Thus we see that the Mammalia of Celebes are no less individual and remarkable than the birds, since three of the largest and most interesting species have no near allies in surrounding countries, but seem vaguely to indicate a relation to the African continent.

Many groups of insects appear to be especially subject to local influences, their forms and colours changing with each change of conditions, or even with a change of locality where the conditions seem almost identical. We should therefore anticipate that the individuality manifested in the higher animals would be still more prominent in these creatures with less stable organisms. On the other hand, however, we have to consider that the dispersion and migration of insects is much more easily effected than that of mammals or even of birds. They are much more likely to be carried away by violent winds; their eggs may be carried on leaves either by storms of wind or by floating trees, and their larvæ and pupæ often buried in trunks of trees or enclosed in waterproof cocoons, may be floated for days or weeks uninjured over the ocean. These facilities of distribution tend to assimilate the productions of adjacent lands in two ways: first, by direct mutual interchange of species; and secondly, by repeated immigrations of fresh individuals of a species common to other islands, which by intercrossing tend to obliterate the changes of form and colour, which differences of conditions might otherwise produce. Bearing these facts in mind, we shall find that the individuality of the insects of Celebes is even greater than we have any reason to expect.

For the purpose of insuring accuracy in comparisons with other islands, I shall confine myself to those groups which are best known, or which I have myself carefully studied. Beginning with the Papilionidæ or Swallow-tailed butterflies, Celebes possesses 24 species, of which the large number of 18 are not found in any other island. If we compare this with Borneo, which out of 29 species has only two not found elsewhere, the difference is as striking as anything can be. In the family of the Pieridæ, or white butterflies, the difference is not quite so great, owing perhaps to the more wandering habits of the group ; but it is still very remarkable. Out of 30 species inhabiting Celebes, 19 are peculiar, while Java (from which more species are known than from Sumatra or Borneo), out of 37 species has only 13 peculiar. The Danaidæ are large, but weak-flying butterflies, which frequent forests and gardens, and are plainly but often very richly coloured. Of these my own collection contains 16 species from Celebes and 15 from Borneo ; but whereas no less than 14 are confined to the former island, only two are peculiar to the latter. The Nymphalidæ are a very extensive group, of generally strong-winged and very bright-coloured butterflies, very abundant in the tropics, and represented in our own country by our Fritillaries, our Vanessas, and our Purple-emperor. Some months ago I drew up a list of the Eastern species of this group, including all the new ones discovered by myself, and arrived at the following comparative results :—

Species of Nymphalidæ.	Species peculiar to each island.	Percentage of peculiar Species.
Java 70	 23	 33
Borneo . . . 52	 15	 29
Celebes . . . 48	 35	 73

The Coleoptera are so extensive that few of the groups have yet been carefully worked out. I will therefore refer to one only, which I have myself recently studied—the Cetoniadæ or Rose-chafers,—a group of beetles which, owing to their extreme beauty, have been much sought after. From Java 37 species of these insects are known, and from Celebes only 30 ; yet only 13, or 35 per cent., are peculiar to the former island, and 19, or 63 per cent., to the latter.

The result of these comparisons is, that although Celebes is a single large island with only a few smaller ones closely grouped around it, we must really consider it as forming one of the great divisions of the Archipelago, equal in rank and importance to the whole of the Moluccan or Philippine groups, to the Papuan islands, or to the Indo-Malay islands (Java, Sumatra, Borneo, and the Malay peninsula). Taking those families of insects and birds which are best known, the following table shows the comparison of Celebes with the other groups of islands :—

	PAPILIONIDÆ AND PIERIDÆ. Per cent. of peculiar Species.	HAWKS, PARROTS, AND PIGEONS. Per cent. of peculiar Species.
Indo-Malay region 56	 54
Philippine group 66	 73
Celebes 69	 60
Moluccan group 52	 62
Timor group 42	 47
Papuan group 64	 47

These large and well-known families well represent the general character of the zoology of Celebes ; and they show that this island is really one of the most isolated portions of the Archipelago, although situated in its very centre.

But the insects of Celebes present us with other phenomena more curious and more difficult to explain than their specific individuality. The butterflies of that island are in many cases characterized by a peculiarity of outline, which distinguishes them at a glance from those of any other part of the world. It is most strongly manifested in the Papilios and the Pieridæ, and consists in the fore-wings being either strongly curved or abruptly bent near the base, or in the extremity being elongated and often somewhat hooked. Out of the 14 species of Papilio in Celebes, 13 exhibit this peculiarity in a greater or less degree, when compared with the most nearly allied species of the surrounding islands. Ten species of Pieridæ have the same character, and in four or five of the Nymphalidæ it is also very distinctly marked. In almost every case the species found in Celebes are much larger than those of the islands westward, and at least equal to those of the Moluccas, or even larger. The difference of form is however the most remarkable feature, as it is altogether a new thing for a whole set of species in one country to differ in exactly the same way from the corresponding sets in all the surrounding countries ; and it is so well marked, that without looking at the details of colouring, most Celebes Papilios and many Pieridæ can be at once distinguished from those of other islands by their form alone.

The outside figure of each pair here given shows the exact size and form of the fore-wing in a butterfly of Celebes, while the inner one represents the most closely allied species from one of the adjacent islands. Figure 1 shows the strongly curved margin of the Celebes species, Papilio gigon, compared with the much straighter margin of Papilio demolion from Singapore and Java. Figure 2 shows the abrupt bend over the base of the wing in Papilio miletus of Celebes compared with the slight curvature in the common Papilio sarpedon, which has almost exactly the same form from India to New Guinea and Australia. Figure 3 shows the elongated wing of Tachyris zarinda, a native of Celebes, compared with the much shorter wing of Tachyris nero, a very closely allied species found in all the western islands. The difference of form is in each case sufficiently obvious, but when the insects themselves are compared it is much more striking than in these partial outlines.

From the analogy of birds, we should suppose that the pointed wing gave increased rapidity of flight, since it is a character of terns, swallows, falcons, and of the swift-flying pigeons. A short and rounded wing, on the other hand, always accompanies a more feeble or more laborious flight, and one much less under command. We might suppose, therefore, that the butterflies which possess this peculiar form were better able to escape pursuit. But there seems no unusual abundance of insectivorous birds to render this necessary ; and as we cannot believe that such a curious peculiarity is without meaning, it seems probable that it is the result of a former condition of things, when the island possessed a much richer fauna, the relics of which we see in the isolated birds and Mammalia now inhabiting it ; and when the abundance of insectivorous creatures rendered

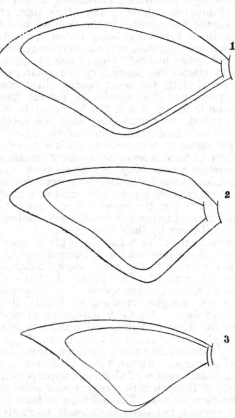

some unusual means of escape a necessity for the large-winged and showy butterflies. It is some confirmation of this view, that neither the very small nor the very obscurely coloured groups of butterflies have elongated wings, nor is any modification perceptible in those strong-winged groups which already possess great strength and rapidity of flight. These were already sufficiently protected from their enemies, and did not require increased power of escaping from them. It is not at all clear what effect the peculiar curvature of the wings has in modifying flight.

Another curious feature in the zoology of Celebes is also worthy of attention. I allude to the absence of several groups which are found on both sides of it, in the Indo-Malay islands as well as in the Moluccas ; and which thus seem to be unable, from some unknown cause, to obtain a footing in the intervening island. In Birds we have the two families of Podargidæ and Laniadæ, which range over the whole Archipelago and into Australia, and which yet have no representative in Celebes. The genera Ceyx among Kingfishers, Criniger among Thrushes, Rhipidura among Flycatchers, and Erythrura among Finches, are all found in the Moluccas as well as in Borneo and Java,— but not a single species belonging to any one of them is found in Celebes. Among insects, the large genus of Rose-chafers, Lomaptera, is found in every country and island between India and New Guinea, except Celebes. This unexpected absence of many groups, from one limited district in the very centre of their area of distribution, is a phenomenon not altogether unique, but, I believe, nowhere so well marked as in this case ; and it certainly adds considerably to the strange character of this remarkable island.

The anomalies and eccentricities in the natural history of Celebes which I have endeavoured to sketch in this chapter, all point to an origin in a remote antiquity. The history of extinct animals teaches us that their distribution in time and in space are strikingly similar. The rule is, that just as the productions of adjacent areas usually resemble each other closely, so do the productions of successive periods in the same area ; and as the productions of remote areas generally differ widely, so do the productions of the same area at remote epochs. We are therefore led irresistibly to the conclusion, that change of species, still more of generic and of family form, is a matter of time. But time may have led to a change of species in one country, while in another the forms have been more permanent, or the change may have gone on at an equal rate but in a different manner in both. In either case the amount of individuality in the productions of a district, will be to some extent a measure of the time that district has been isolated from those that surround it. Judged by this standard, Celebes must be one of the oldest parts of the Archipelago. It probably dates from a period not only anterior to that when Borneo, Java, and Sumatra were separated from the continent, but from that still more remote epoch when the land that now constitutes these islands had not risen above the ocean. Such an antiquity is necessary to account for the number of animal forms it possesses, which show no relation to those of India or Australia, but rather with those of Africa ; and we are led to speculate on the possibility of there having once existed a continent in the Indian Ocean which might serve as a bridge to connect these distant countries. Now it is a curious fact, that the existence of such a land has been already thought necessary to account for the distribution of the curious Quadrumana forming the family of the Lemurs. These have their metropolis in Madagascar, but are found also in Africa, in Ceylon, in the peninsula of India, and in the Malay Archipelago as far as Celebes, which is its furthest eastern limit. Dr. Sclater has proposed for the hypothetical continent connecting these distant points, and whose former existence is indicated by the Mascarene islands and the Maldive coral group, the name of Lemuria. Whether or no we believe in its existence in the exact form here indicated, the student of geographical distribution must see in the extraordinary and isolated productions of Celebes proofs of the former existence of some continent from whence the ancestors of these creatures, and of many other intermediate forms, could have been derived.[1]

In this short sketch of the most striking peculiarities of the Natural History of Celebes, I have been obliged to enter much into details that I fear will have been uninteresting to the general reader, but unless I had done so my exposition would have lost much of its force and value. It is by these details alone that I have been able to prove the unusual features that Celebes presents to us. Situated in the very midst of an Archipelago, and closely hemmed in on every side by islands teeming with varied forms of life, its productions have yet a surprising amount of individuality. While it is poor in the actual number of its species, it is yet wonderfully rich in peculiar forms, many of which are singular or beautiful, and are in some cases absolutely unique upon the globe. We behold here the curious phenomenon of groups of insects changing their outline in a similar manner when compared with those of surrounding islands, suggesting some common cause which never seems to have acted elsewhere in exactly the same way. Celebes, therefore, presents us with a most striking example of the interest that attaches to the study of the geographical distribution of animals. We can see that their present distribution upon the globe is the result of all the more recent changes the earth's surface has undergone; and by a careful study of the phenomena we are sometimes able to deduce approximately what those past changes must have been, in order to produce the distribution we find to exist. In the comparatively simple case of the Timor group, we were able to deduce these changes with some approach to certainty. In the much more complicated case of Celebes we can only indicate their general nature, since we now see the result, not of any single or recent change only, but of a whole series of the later revolutions which have resulted in the present distribution of land in the Eastern Hemisphere.

[1] I have since come to the conclusion that no such connecting land as Lemuria is required to explain the facts. (*See* my *Island Life*, pages 395 and 427.)

CHAPTER XIX.

BANDA.

(DECEMBER 1857, MAY 1859, APRIL 1861.)

THE Dutch mail steamer in which I travelled from Macassar
to Banda and Amboyna was a roomy and comfortable vessel,
although it would only go six miles an hour in the finest weather.
As there were but three passengers besides myself, we had abundance of room, and I was able to enjoy a voyage more than I had
ever done before. The arrangements are somewhat different
from those on board English or Indian steamers. There are no
cabin servants, as every cabin passenger invariably brings his
own, and the ship's stewards attend only to the saloon and the
eating department. At six A.M. a cup of tea or coffee is provided
for those who like it. At seven to eight there is a light breakfast
of tea, eggs, sardines, &c. At ten, Madeira, gin, and bitters are
brought on deck as a whet for the substantial eleven o'clock
breakfast, which differs from a dinner only in the absence of
soup. Cups of tea and coffee are brought round at three P.M. ;
bitters, &c, again at five ; a good dinner with beer and claret at
half-past six, concluded by tea and coffee at eight. Between
whiles, beer and sodawater are supplied when called for, so there
is no lack of little gastronomical excitements to while away the
tedium of a sea voyage.

Our first stopping place was Coupang, at the west end of the
large island of Timor. We then coasted along that island for
several hundred miles, having always a view of hilly ranges
covered with scanty vegetation, rising ridge behind ridge to the
height of six or seven thousand feet. Turning off towards
Banda we passed Pulo-Cambing, Wetter, and Roma, all of which
are desolate and barren volcanic islands, almost as uninviting as
Aden, and offering a strange contrast to the usual verdure and
luxuriance of the Archipelago. In two days more we reached
the volcanic group of Banda, covered with an unusually dense
and brilliant green vegetation, indicating that we had passed
beyond the range of the hot dry winds from the plains of Central
Australia. Banda is a lovely little spot, its three islands enclosing a secure harbour from whence no outlet is visible, and
with water so transparent, that living corals and even the
minutest objects are plainly seen on the volcanic sand at a
depth of seven or eight fathoms. The ever-smoking volcano
rears its bare cone on one side, while the two larger islands are
clothed with vegetation to the summit of the hills.

Going on shore, I walked up a pretty path which leads to the
highest point of the island on which the town is situated, where
there is a telegraph station and a magnificent view. Below lies

the little town, with its neat red-tiled white houses and the thatched cottages of the natives, bounded on one side by the old Portuguese fort. Beyond, about half a mile distant, lies the larger island in the shape of a horseshoe, formed of a range of abrupt hills covered with fine forest and nutmeg gardens ; while close opposite the town is the volcano, forming a nearly perfect cone, the lower part only covered with a light green bushy vegetation. On its north side the outline is more uneven, and there is a slight hollow or chasm about one-fifth of the way down, from which constantly issue two columns of smoke, which also rises less abundantly from the rugged surface around and from some spots nearer the summit. A white efflorescence, probably sulphur, is thickly spread over the upper part of the mountain, marked by the narrow black vertical lines of water gullies. The smoke unites as it rises, and forms a dense cloud, which in calm damp weather spreads out into a wide canopy hiding the top of the mountain. At night and early morning it often rises up straight and leaves the whole outline clear.

It is only when actually gazing on an active volcano that one can fully realize its awfulness and grandeur. Whence comes that inexhaustible fire whose dense and sulphureous smoke for ever issues from this bare and desolate peak? Whence the mighty forces that produced that peak, and still from time to time exhibit themselves in the earthquakes that always occur in the vicinity of volcanic vents? The knowledge from childhood, of the fact that volcanoes and earthquakes exist, has taken away somewhat of the strange and exceptional character that really belongs to them. The inhabitant of most parts of northern Europe sees in the earth the emblem of stability and repose. His whole life-experience, and that of all his age and generation, teaches him that the earth is solid and firm, that its massive rocks may contain water in abundance but never fire ; and these essential characteristics of the earth are manifest in every mountain his country contains. A volcano is a fact opposed to all this mass of experience, a fact of so awful a character that, if it were the rule instead of the exception, it would make the earth uninhabitable ; a fact so strange and unaccountable that we may be sure it would not be believed on any human testimony, if presented to us now for the first time, as a natural phenomenon happening in a distant country.

The summit of the small island is composed of a highly crystalline basalt ; lower down I found a hard stratified slaty sandstone, while on the beach are huge blocks of lava, and scattered masses of white coralline limestone. The larger island has coral rock to a height of three or four hundred feet, while above is lava and basalt. It seems probable, therefore, that this little group of four islands is the fragment of a larger district which was perhaps once connected with Ceram, but which was separated and broken up by the same forces which formed the volcanic cone. When I visited the larger island on another occasion, I saw a considerable tract covered with large forest trees, dead, but still standing. This was a record of the last great earthquake only two years ago, when the sea broke in over this part of the island and so flooded it as to destroy the vegetation on all the low lands. Almost every year there is an earthquake here, and at intervals of a few years very severe ones, which throw down houses and carry ships out of the harbour bodily into the streets.

Notwithstanding the losses incurred by these terrific visitations, and the small size and isolated position of these little islands, they have been and still are of considerable value to the Dutch Government, as the chief nutmeg-garden in the world. Almost the whole surface is planted with nutmegs, grown under the shade of lofty Kanary trees (Kanarium commune). The light volcanic soil, the shade, and the excessive moisture of these islands, where it rains more or less every month in the year, seem exactly to suit the nutmeg-tree, which requires no manure and scarcely any attention. All the year round flowers and ripe fruit are to be found, and none of those diseases occur which under a forced and unnatural system of cultivation have ruined the nutmeg planters of Singapore and Penang.

Few cultivated plants are more beautiful than nutmeg-trees. They are handsomely shaped and glossy-leaved, growing to the height of twenty or thirty feet, and bearing small yellowish flowers. The fruit is the size and colour of a peach, but rather oval. It is of a tough fleshy consistence, but when ripe splits open, and shows the dark-brown nut within, covered with the crimson mace, and is then a most beautiful object. Within the thin hard shell of the nut is the seed, which is the nutmeg of commerce. The nuts are eaten by the large pigeons of Banda, which digest the mace but cast up the nut with its seed uninjured.

The nutmeg trade has hitherto been a strict monopoly of the Dutch Government ; but since leaving the country I believe that this monopoly has been partially or wholly discontinued, a proceeding which appears exceedingly injudicious and quite unnecessary. There are cases in which monopolies are perfectly justifiable, and I believe this to be one of them. A small country like Holland cannot afford to keep distant and expensive colonies at a loss ; and having possession of a very small island where a valuable product, *not a necessary of life*, can be obtained at little cost, it is almost the duty of the state to monopolise it. No injury is done thereby to any one, but a great benefit is conferred on the whole population of Holland and its dependencies, since the produce of the state monopolies save them from the weight of a heavy taxation. Had the Government not kept the nutmeg trade of Banda in its own hands, it is probable that the whole of the islands would long ago have become the property of one or more large capitalists. The monopoly would have been almost the same, since no known spot on the globe can produce nutmegs so cheaply as Banda, but the profits of the monopoly would have gone to a few individuals instead of to the nation. As an illustration of how a state monopoly may become a state duty, let us suppose that no gold existed in Australia, but that it had been found in immense quantities by one of our ships in some small and barren island. In this case it would plainly become the duty of the State to keep and work the mines for the public benefit, since by doing so, the gain would be fairly divided among the whole population by decrease of taxation ; whereas by leaving it open to free trade while merely keeping the government of the island, we should certainly produce enormous evils during the first struggle for the precious metal, which would ultimately subside into the monopoly of some wealthy individual or great company, whose enormous revenue would not equally benefit the community. The nutmegs of Banda and the tin of Banca are to some extent parallel cases to this supposititious one, and I believe the Dutch Government will act most unwisely if they give up their monopoly.

Even the destruction of the nutmeg and clove trees in many islands, in order to restrict their cultivation to one or two where the monopoly could be easily guarded, usually made the theme of so much virtuous indignation against the Dutch, may be defended on similar principles, and is certainly not nearly so bad as many monopolies we ourselves have till very recently maintained. Nutmegs and cloves are not necessaries of life ; they are not even used as spices by the natives of the Moluccas, and no one was materially or permanently injured by the destruction of the trees, since there are a hundred other products that can be grown in the same islands, equally valuable and far more beneficial in a social point of view. It is a case exactly parallel to our prohibition of the growth of tobacco in England, for fiscal purposes, and is, morally and economically, neither better nor worse. The salt monopoly which we so long maintained in India was much worse. As long as we keep up a system of excise and customs on articles of daily use, which requires an elaborate array of officers and coastguards to carry into effect, and which creates a number of purely legal crimes, it is the height of absurdity for us to affect indignation at the conduct of the Dutch, who carried out a much more justifiable, less hurtful, and more profitable system in their Eastern possessions. I challenge objectors to point out any physical or moral evils that have actually resulted from the action of the Dutch Government in this matter ; whereas such evils are the admitted results of every one of our monopolies and restrictions.[1]

1 In the *Daily News* parliamentary report of March 28th, 1890, I read the following :— "Baron H. de Worms said it was true that in the Newara Eliya district of Ceylon land sales, affecting 10,283 men, women, and children, had taken place for non-payment of the paddy-tax, and that 981 persons had died of consequent want and disease, and 2,539 had been left destitute." Here is a dry official statement of the result of our taxing the people's food ; and it was such a very ordinary matter to our legislators that no further notice seems to have been taken of it. And we dare to abuse the Dutch of three centuries back for destroying spice trees !—for which they paid a fair compensation, and the results of which were probably beneficial rather than hurtful to the cultivators of the soil ! (*See* Chap. XXI.)

The conditions of the two experiments are totally different. The true "political economy" of a higher, when governing a lower race, has never yet been worked out. The application of our "political economy" to such cases invariably results in the extinction or degradation of the lower race; whence we may consider it probable that one of the necessary conditions of its truth is, the approximate mental and social unity of the society in which it is applied. I shall again refer to this subject in my chapter on Ternate, one of the most celebrated of the old spice-islands.

The natives of Banda are very much mixed, and it is probable that at least three-fourths of the population are mongrels, in various degrees of Malay, Papuan, Arab, Portuguese, and Dutch. The first two form the basis of the larger portion, and the dark skins, pronounced features, and more or less frizzly hair of the Papuans preponderate. There seems little doubt that the aborigines of Banda were Papuans, and a portion of them still exists in the Ké islands, where they emigrated when the Portuguese first took possession of their native island. It is such people as these that are often looked upon as transitional forms between two very distinct races, like the Malays and Papuans, whereas they are only examples of intermixture.

The animal productions of Banda, though very few, are interesting. The islands have perhaps no truly indigenous Mammalia but bats. The deer of the Moluccas and the pig have probably been introduced. A species of Cuscus or Eastern opossum is also found at Banda, and this may be truly indigenous in the sense of not having been introduced by man. Of birds, during my three visits of one or two days each, I collected eight kinds, and the Dutch collectors have added a few others. The most remarkable is a fine and very handsome fruit-pigeon, Carpophaga concinna, which feeds upon the nutmegs, or rather on the mace, and whose loud booming note is to be continually heard. This bird is found in the Ké and Matabello islands as well as Banda, but not in Ceran or any of the larger islands, which are inhabited by allied but very distinct species. A beautiful small fruit-dove, Ptilonopus diadematus, is also peculiar to Banda.

CHAPTER XX.

AMBOYNA.

(DECEMBER 1857, OCTOBER 1859, FEBRUARY 1860.)

TWENTY hours from Banda brought us to Amboyna, the capital of the Moluccas, and one of the oldest European settlements in the East. The island consists of two peninsulas, so nearly divided by inlets of the sea, as to leave only a sandy isthmus about a mile wide near their eastern extremity. The western inlet is several miles long and forms a fine harbour, on the southern side of which is situated the town of Amboyna. I had a letter of introduction to Dr. Mohnike, the chief medical officer of the Moluccas, a German and a naturalist. I found that he could write and read English, but could not speak it, being like myself a bad linguist ; so we had to use French as a medium of communication. He kindly offered me a room during my stay in Amboyna, and introduced me to his junior, Dr. Doleschall, a Hungarian, and also an entomologist. He was an intelligent and most amiable young man, but I was shocked to find that he was dying of consumption, though still able to perform the duties of his office. In the evening my host took me to the residence of the Governor, Mr. Goldmann, who received me in a most kind and cordial manner, and offered me every assistance. The town of Amboyna consists of a few business streets, and a number of roads set out at right angles to each other, bordered by hedges of flowering shrubs, and enclosing country houses and huts embosomed in palms and fruit trees. Hills and mountains form the background in almost every direction, and there are few places more enjoyable for a morning or evening stroll than these sandy roads and shady lanes in the suburbs of the ancient city of Amboyna.

There are no active volcanoes in the island, nor is it now subject to frequent earthquakes, although very severe ones have occurred and may be expected again. Mr. William Funnell, in his voyage with Dampier to the South Seas in 1705, says : "Whilst we were here [at Amboyna] we had a great earthquake, which continued two days, in which time it did a great deal of mischief ; for the ground burst open in many places, and swallowed up several houses and whole families. Several of the people were dug out again, but most of them dead, and many had their legs or arms broken by the fall of the houses. The castle walls were rent asunder in several places, and we thought that it and all the houses would have fallen down. The ground where we were swelled like a wave in the sea, but near us we had no hurt done." There are also numerous records of eruptions of a volcano on the west side of the island. In 1674 an eruption destroyed a village. In 1694 there was another eruption. In 1797 much vapour and heat was emitted. Other eruptions occurred in 1816 and 1820, and in 1824 a new crater is said to have been formed. Yet so capricious is the action of these subterranean fires, that since the last-named epoch all eruptive symptoms have so completely ceased, that I was assured by many of the most intelligent European inhabitants of Amboyna, that they had never heard of any such thing as a volcano on the island.

During the few days that elapsed before I could make arrangements to visit the interior, I enjoyed myself much in the society of the two doctors, both amiable and well-educated

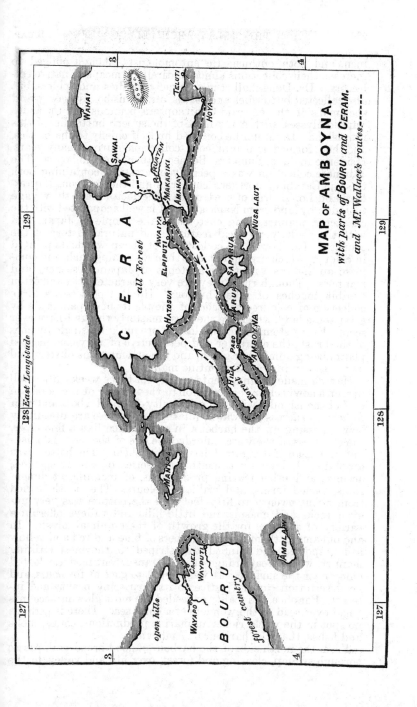

MAP of **AMBOYNA.**
with parts of Bouru and Ceram.
and Mr Wallace's routes------

128 *East Longitude* 129

129

127

128

3

4

WAHAI

SAWAI

CERAM
all Forest

R. RUATAN
MAKARIKI
AMAHAY
AWAIYA
ELIPUTI
HATOSUA

TELUTI
HOYA
AMANAY

NUSA LAUT
SAPARUA
HARUKU

PASO
AMBOYNA
Forest

MANIPA

AMBLAW

BOURU
open hills
WAYAPO
CAJELI
WAYPUTI
forest country

men, and both enthusiastic entomologists, though obliged to increase their collections almost entirely by means of native collectors. Dr. Doleschall studied chiefly the flies and spiders, but also collected butterflies and moths, and in his boxes I saw grand specimens of the emerald Ornithoptera priamus and the azure Papilio ulysses, with many more of the superb butterflies of this rich island. Dr. Mohnike confined himself chiefly to the beetles, and had formed a magnificent collection during many years' residence in Java, Sumatra, Borneo, Japan, and Amboyna. The Japanese collection was especially interesting, containing both the fine Carabi of northern countries and the gorgeous Buprestidæ and Longicorns of the tropics. The doctor made the voyage to Jeddo by land from Nagasaki, and is well acquainted with the character, manners, and customs of the people of Japan, and with the geology, physical features, and natural history of the country. He showed me collections of cheap woodcuts printed in colours, which are sold at less than a farthing each, and comprise an endless variety of sketches of Japanese scenery and manners. Though rude, they are very characteristic, and often exhibit touches of great humour. He also possesses a large collection of coloured sketches of the plants of Japan, made by a Japanese lady, which are the most masterly things I have ever seen. Every stem, twig, and leaf is produced by single touches of the brush, the character and perspective of very complicated plants being admirably given, and the articulations of stem and leaves shown in a most scientific manner.

Having made arrangements to stay for three weeks at a small hut, on a newly cleared plantation in the interior of the northern half of the island, I with some difficulty obtained a boat and men to take me across the water, for the Amboynese are dreadfully lazy. Passing up the harbour, in appearance like a fine river, the clearness of the water afforded me one of the most astonishing and beautiful sights I have ever beheld. The bottom was absolutely hidden by a continuous series of corals, sponges, actiniæ, and other marine productions, of magnificent dimensions, varied forms, and brilliant colours. The depth varied from about twenty to fifty feet, and the bottom was very uneven, rocks and chasms, and little hills and valleys, offering a variety of stations for the growth of these animal forests. In and out among them moved numbers of blue and red and yellow fishes, spotted and banded and striped in the most striking manner, while great orange or rosy transparent medusæ floated along near the surface. It was a sight to gaze at for hours, and no description can do justice to its surpassing beauty and interest. For once, the reality exceeded the most glowing accounts I had ever read of the wonders of a coral sea. There is perhaps no spot in the world richer in marine productions, corals, shells and fishes, than the harbour of Amboyna.

From the north side of the harbour, a good broad path passes through swamp, clearing and forest, over hill and valley, to the

farther side of the island; the coralline rock constantly protruding through the deep red earth which fills all the hollows,
and is more or less spread over the plains and hill-sides. The
forest vegetation is here of the most luxuriant character; ferns
and palms abound, and the climbing rattans were more
abundant than I had ever seen them, forming tangled festoons
over almost every large forest tree. The cottage I was to occupy
was situated in a large clearing of about a hundred acres, part
of which was already planted with young cacao-trees and
plantains to shade them, while the rest was covered with dead
and half-burnt forest trees; and on one side there was a tract
where the trees had been recently felled and were not yet burnt.
The path by which I had arrived continued along one side of
this clearing, and then again entering the virgin forest passed
over hill and dale to the northern side of the island.

My abode was merely a little thatched hut, consisting of an
open verandah in front and a small dark sleeping-room behind.
It was raised about five feet from the ground, and was reached
by rude steps to the centre of the verandah. The walls and
floor were of bamboo, and it contained a table, two bamboo
chairs, and a couch. Here I soon made myself comfortable, and
set to work hunting for insects among the more recently felled
timber, which swarmed with fine Curculionidæ, Longicorns, and
Buprestidæ, most of them remarkable for their elegant forms or
brilliant colours, and almost all entirely new to me. Only the
entomologist can appreciate the delight with which I hunted
about for hours in the hot sunshine, among the branches and
twigs and bark of the fallen trees, every few minutes securing
insects which were at that time almost all rare or new to
European collections.

In the shady forest paths were many fine butterflies, most
conspicuous among which was the shining blue Papilio ulysses,
one of the princes of the tribe. Though at that time so rare in
Europe, I found it absolutely common in Amboyna, though not
easy to obtain in fine condition, a large number of the specimens
being found when captured to have the wings torn or broken.
It flies with a rather weak undulating motion, and from its
large size, its tailed wings and brilliant colour, is one of the
most tropical-looking insects the naturalist can gaze upon.

There is a remarkable contrast between the beetles of Amboyna
and those of Macassar, the latter generally small and obscure,
the former large and brilliant. On the whole, the insects here
most resemble those of the Aru islands, but they are almost
always of distinct species, and when they are most nearly allied
to each other the species of Amboyna are of larger size and more
brilliant colours, so that one might be led to conclude that, in
passing east and west into a less favourable soil and climate,
they had degenerated into less striking forms.

Of an evening I generally sat reading in the verandah, ready
to capture any insects that were attracted to the light. One

night about nine o'clock I heard a curious noise and rustling overhead, as if some heavy animal were crawling slowly over the thatch. The noise soon ceased, and I thought no more about it and went to bed soon afterwards. The next afternoon just before dinner, being rather tired with my day's work, I was lying on the couch with a book in my hand, when gazing upwards I saw a large mass of something overhead which I had not noticed before. Looking more carefully I could see yellow and black marks, and thought it must be a tortoise-shell put up there out of the way between the ridge-pole and the roof. Continuing to gaze, it suddenly resolved itself into a large snake, compactly coiled up in a kind of knot; and I could detect his head and his bright eyes in the very centre of the folds. The noise of the evening before was now explained. A python had climbed up one of the posts of the house, and had made his way under the thatch within a yard of my head, and taken up a comfortable position in the roof—and I had slept soundly all night directly under him.

I called to my two boys who were skinning birds below and said, "Here's a big snake in the roof;" but as soon as I had shown it to them they rushed out of the house and begged me to come out directly. Finding they were too much afraid to do anything, we called some of the labourers in the plantation, and soon had half a dozen men in consultation outside. One of these, a native of Bouru, where there are a great many snakes, said he would get him out, and proceeded to work in a business-like manner. He made a strong noose of rattan, and with a long pole in the other hand poked at the snake, which then began slowly to uncoil itself. He then managed to slip the noose over its head, and getting it well on to the body, dragged the animal down. There was a great scuffle as the snake coiled round the chairs and posts to resist his enemy, but at length the man caught hold of its tail, rushed out of the house (running so quick that the creature seemed quite confounded), and tried to strike its head against a tree. He missed however, and let go, and the snake got under a dead trunk close by. It was again poked out, and again the Bouru man caught hold of its tail, and running away quickly dashed its head with a swing against a tree, and it was then easily killed with a hatchet. It was about twelve feet long and very thick, capable of doing much mischief and of swallowing a dog or a child.

I did not get a great many birds here. The most remarkable were the fine crimson lory, Eos rubra—a brush-tongued parroquet of a vivid crimson colour, which was very abundant. Large flocks of them came about the plantation, and formed a magnificent object when they settled down upon some flowering tree, on the nectar of which lories feed. I also obtained one or two specimens of the fine racquet-tailed kingfisher of Amboyna, Tanysiptera nais, one of the most singular and beautiful of that beautiful family. These birds differ from all other kingfishers

EJECTING AN INTRUDER.

(which have usually
short tails) by having
the two middle tail-
feathers immensely
lengthened and very
narrowly webbed, but
terminated by a spoon-shaped
enlargement, as in the mot-
mots and some of the hum-
ming-birds. They belong to
that division of the family
termed kinghunters, living
chiefly on insects and small
land-molluscs, which they
dart down upon and pick up
from the ground, just as a
kingfisher picks a fish out of
the water. They are confined
to a very limited area, comprising the Moluccas, New
Guinea, and Northern Aus-
tralia. About ten species of
these birds are now known,
all much resembling each
other, but yet sufficiently distinguishable in every locality.
The Amboynese species, of
which a very accurate representation is here given, is one
of the largest and handsomest.
It is full seventeen inches
long to the tips of the tail-
feathers ; the bill is coral red,
the under-surface pure white,
the back and wings deep
purple, while the shoulders,
head and nape, and some
spots on the upper part of
the back and wings, are pure
azure blue. The tail is white,
with the feathers narrowly
blue-edged, but the narrow
part of the long feathers is
rich blue. This was an entirely new species, and has
been well named after an
ocean goddess, by Mr. G. R.
Gray.

On Christmas eve I re-
turned to Amboyna, where I

RACQUET-TAILED KINGFISHER.

stayed about ten days with my kind friend Dr. Mohnike. Considering that I had been away only twenty days, and that on five or six of those I was prevented doing anything by wet weather and slight attacks of fever, I had made a very nice collection of insects, comprising a much larger proportion of large and brilliant species than I had ever before obtained in so short a time. Of the beautiful metallic Buprestidæ I had about a dozen handsome species, yet in the doctor's collection I observed four or five more very fine ones, so that Amboyna is unusually rich in this elegant group.

During my stay here I had a good opportunity of seeing how Europeans live in the Dutch colonies, where they have adopted customs far more in accordance with the climate than we have done in our tropical possessions. Almost all business is transacted in the morning between the hours of seven and twelve, the afternoon being given up to repose, and the evening to visiting. When in the house during the heat of the day, and even at dinner, they use a loose cotton dress, only putting on a suit of thin European-made clothes for out of doors and evening wear. They often walk about after sunset bareheaded, reserving the black hat for visits of ceremony. Life is thus made far more agreeable, and the fatigue and discomfort incident to the climate greatly diminished. Christmas day is not made much of, but on New Year's day official and complimentary visits are paid, and about sunset we went to the Governor's, where a large party of ladies and gentlemen were assembled. Tea and coffee were handed round, as is almost universal during a visit as well as cigars, for on no occasion is smoking prohibited in Dutch colonies, cigars being generally lighted before the cloth is withdrawn at dinner, even though half the company are ladies. I here saw for the first time the rare black lory from New Guinea, Chalcopsitta atra. The plumage is rather glossy and slightly tinged with yellowish and purple, the bill and feet being entirely black.

The native Amboynese who reside in the city are a strange, half-civilized, half-savage, lazy people, who seem to be a mixture of at least three races, Portuguese, Malay, and Papuan or Ceramese, with an occasional cross of Chinese or Dutch. The Portuguese element decidedly predominates in the old Christian population, as indicated by features, habits, and the retention of many Portuguese words in the Malay, which is now their language. They have a peculiar style of dress which they wear among themselves, a close-fitting white shirt with black trousers and a black frock or upper shirt. The women seem to prefer a dress entirely black. On festivals and holy days every man wears the swallow-tail coat, chimney-pot hat, and their accompaniments, displaying all the absurdity of our European fashionable dress. Though now Protestants, they preserve at feasts and weddings the processions and music of the Catholic Church, curiously mixed up with the gongs and dances of the aborigines of the country. Their language has still much more Portuguese than Dutch in it, although they have been in close communication with the latter nation for more than two hundred and fifty years; even many names of birds, trees and other natural objects, as well as many domestic terms, being plainly Portuguese.[1] This people seems to have had a marvellous power of colonization, and a capacity for impressing their national characteristics on every country they conquered, or in which they effected a merely temporary settlement. In a suburb of Amboyna there is a village of aboriginal Malays who are Mahometans, and who speak a peculiar language allied to those of Ceram, as well as Malay. They are chiefly fishermen, and are said to be both more industrious and more honest than the native Christians.

I went on Sunday, by invitation, to see a collection of shells and fish made by a gentleman of Amboyna. The fishes are perhaps unrivalled for variety and beauty by those of any one spot on the earth. The celebrated Dutch ichthyologist, Dr. Bleeker, has given a catalogue of seven hundred and eighty species found at Amboyna, a number almost equal to those of all the seas and rivers of Europe. A large proportion of them are of the most brilliant colours, being marked with bands and spots of the purest yellows, reds, and blues; while their forms present all that strange and endless variety so characteristic of the inhabitants of the ocean. The shells are also very numerous, and comprise a number of the finest species in the world. The Mactras and Ostreas in particular struck me by the variety and beauty of their colours. Shells have long been an object of traffic in Amboyna; many of the natives get their living by collecting and cleaning them, and almost every visitor takes away a small collection. The result is that many of the commoner sorts have lost all value in the eyes of the amateur, numbers of the handsome but very common cones, cowries, and olives sold in the streets of London for a penny each, being natives of the distant isle of Amboyna, where they cannot be bought so cheaply. The fishes in the collection were all well preserved in clear spirit in hundreds of glass jars, and the shells were arranged in large shallow pith boxes lined with paper, every specimen being fastened down with thread. I roughly estimated that there were nearly a thousand different kinds of shells, and perhaps ten thousand specimens, while the collection of Amboyna fishes was nearly perfect.

On the 4th of January I left Amboyna for Ternate; but two

---

1 The following are a few of the Portuguese words in common use by the Malay-speaking natives of Amboyna and the other Moluccan islands: Pombo (pigeon); milo (maize); testa (forehead); horas (hours); alfinete (pin); cadeira (chair); lenco (handkerchief); fresco (cool); trigo (flour); sono (sleep); familia (family); histori (talk); vosse (you); mesmo (even); cuñhado (brother-in-law); senhor (sir); nyora for signora (madam) —None of them, however, have the least notion that these words belong to a European language.

years later, in October 1859, I again visited it after my residence
in Menado, and stayed a month in the town in a small house
which I hired for the sake of assorting and packing up a large
and varied collection which I had brought with me from North
Celebes, Ternate, and Gilolo. I was obliged to do this because
the mail-steamer would have come the following month by way
of Amboyna to Ternate, and I should have been delayed two
months before I could have reached the former place. I then
paid my first visit to Ceram, and on returning to prepare for
my second more complete exploration of that island, I stayed
(much against my will) two months at Paso, on the isthmus
which connects the two portions of the island of Amboyna.
This village is situated on the eastern side of the isthmus, on
sandy ground, with a very pleasant view over the sea to the
island of Harúka. On the Amboyna side of the isthmus there
is a small river which has been continued by a shallow canal to
within thirty yards of high-water mark on the other side.
Across this small space, which is sandy and but slightly elevated,
all small boats and praus can be easily dragged, and all the
smaller traffic from Ceram and the islands of Saparúa and
Haruka, passes through Paso. The canal is not continued quite
through, merely because every spring-tide would throw up just
such a sand-bank as now exists.

I had been informed that the fine butterfly Ornithoptera
priamus was plentiful here, as well as the racquet-tailed kingfisher and the ring-necked lory. I found, however, that I had
missed the time for the former ; and birds of all kinds were
very scarce, although I obtained a few good ones, including one
or two of the above-mentioned rarities. I was much pleased
to get here the fine long-armed chafer, Euchirus longimanus.
This extraordinary insect is rarely or never captured except when
it comes to drink the sap of the sugar palms, where it is found by
the natives when they go early in the morning to take away the
bamboos which have been filled during the night. For some
time one or two were brought me every day, generally alive.
They are sluggish insects, and pull themselves lazily along by
means of their immense fore-legs. A figure of this and other
Moluccan beetles is given in the 27th chapter of this work.

I was kept at Paso by an inflammatory eruption, brought on
by the constant attacks of small acari like harvest-bugs, for
which the forests of Ceram are famous, and also by the want of
nourishing food while in that island. At one time I was
covered with severe boils. I had them on my eye, cheek,
armpits, elbows, back, thighs, knees, and ankles, so that I was
unable to sit or walk, and had great difficulty in finding a side
to lie upon without pain. These continued for some weeks,
fresh ones coming out as fast as others got well ; but good
living and sea baths ultimately cured them.

About the end of January Charles Allen, who had been my
assistant in Malacca and Borneo, again joined me on agreement

for three years ; and as soon as I got tolerably well, we had plenty to do laying in stores and making arrangements for our ensuing campaign. Our greatest difficulty was in obtaining men, but at last we succeeded in getting two each. An Amboyna Christian named Theodorus Matakena, who had been some time with me and had learnt to skin birds very well, agreed to go with Allen, as well as a very quiet and industrious lad named Cornelius, whom I had brought from Menado. I had two Amboynese, named Petrus Rehatta, and Mesach Matakena, the latter of whom had two brothers, named respectively Shadrach and Abednego, in accordance with the usual custom among these people of giving only Scripture names to their children.

During the time I resided in this place I enjoyed a luxury I have never met with either before or since—the true bread-fruit. A good deal of it has been planted about here and in the surrounding villages, and almost every day we had opportunities of purchasing some, as all the boats going to Amboyna were unloaded just opposite my door to be dragged across the isthmus. Though it grows in several other parts of the Archipelago, it is nowhere abundant, and the season for it only lasts a short time. It is baked entire in the hot embers, and the inside scooped out with a spoon. I compared it to Yorkshire pudding; Charles Allen said it was like mashed potatoes and milk. It is generally about the size of a melon, a little fibrous towards the centre, but everywhere else quite smooth and puddingy, something in consistence between yeast-dumplings and batter-pudding. We sometimes made curry or stew of it, or fried it in slices; but it is no way so good as simply baked. It may be eaten sweet or savory. With meat and gravy it is a vegetable superior to any I know, either in temperate or tropical countries. With sugar, milk, butter, or treacle, it is a delicious pudding, having a very slight and delicate but characteristic flavour, which, like that of good bread and potatoes, one never gets tired of. The reason why it is comparatively scarce is, that it is a fruit of which the seeds are entirely aborted by cultivation, and the tree can therefore only be propagated by cuttings. The seed-bearing variety is common all over the tropics, and though the seeds are very good eating, resembling chestnuts, the fruit is quite worthless as a vegetable. Now that steam and Ward's cases render the transport of young plants so easy, it is much to be wished that the best varieties of this unequalled vegetable should be introduced into our West India islands, and largely propagated there. As the fruit will keep some time after being gathered, we might then be able to obtain this tropical luxury in Covent Garden Market.

Although the few months I at various times spent in Amboyna were not altogether very profitable to me in the way of collections, yet it will always remain as a bright spot in the review of my Eastern travels, since it was there that I first made the acquaintance of those glorious birds and insects, which render the Moluccas classic ground in the eyes of the naturalist, and characterize its fauna as one of the most remarkable and beautiful upon the globe. On the 20th of February I finally quitted Amboyna for Ceram and Waigiou, leaving Charles Allen to go by a Government boat to Wahai on the north coast of Ceram, and thence to the unexplored island of Mysol.

CHAPTER XXI.

THE MOLUCCAS—TERNATE.

On the morning of the 8th of January, 1858, I arrived at Ternate, the fourth of a row of fine conical volcanic islands which skirt the west coast of the large and almost unknown island of Gilolo. The largest and most perfectly conical mountain is Tidore, which is over five thousand feet high—Ternate being very nearly the same height, but with a more rounded and irregular summit.[1] The town of Ternate is concealed from view till we enter between the two islands, when it is discovered stretching along the shore at the very base of the mountain. Its situation is fine, and there are grand views on every side. Close opposite is the rugged promontory and beautiful volcanic cone of Tidore; to the east is the long mountainous coast of Gilolo, terminated towards the north by a group of three lofty volcanic peaks, while immediately behind the town rises the huge mountain, sloping easily at first and covered with thick groves of fruit trees, but soon becoming steeper, and furrowed with deep gullies. Almost to the summit, whence issue perpetually faint wreaths of smoke, it is clothed with vegetation, and looks calm and beautiful, although beneath are hidden fires which occasionally burst forth in lava-streams, but more frequently make their existence known by the earthquakes which have many times devastated the town.

I brought letters of introduction to Mr. Duivenboden, a native of Ternate, of an ancient Dutch family, but who was educated in England and speaks our language perfectly. He was a very rich man, owned half the town, possessed many ships, and above a hundred slaves. He was, moreover, well educated, and fond of literature and science—a phenomenon in these regions. He was generally known as the king of Ternate, from his large property and great influence with the native Rajahs and their subjects. Through his assistance I obtained a house, rather ruinous, but well adapted to my purpose, being close to the town, yet with a free outlet to the country and the mountain. A few needful repairs were soon made, some bamboo furniture and other necessaries obtained, and after a visit to the Resident and Police Magistrate I found myself an inhabitant of the earthquake-tortured

[1] The officers of the *Challenger* found that Ternate was 5,600 feet high and Tidore 5,900 feet.

island of Ternate, and able to look about me and lay down the plan of my campaign for the ensuing year. I retained this house for three years, as I found it very convenient to have a place to return to after my voyages to the various islands of the Moluccas and New Guinea, where I could pack my collections, recruit my health, and make preparations for future journeys. To avoid repetitions, I will in this chapter combine what notes I have about Ternate.

A description of my house (the plan of which is here shown), will enable the reader to understand a very common mode of building in these islands. There is of course only one floor. The

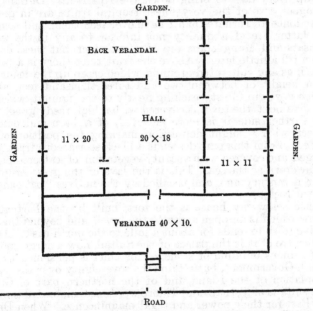

walls are of stone up to three feet high; on this are strong squared posts supporting the roof, everywhere except in the verandah filled in with the leaf-stems of the sago-palm, fitted neatly in wooden framing. The floor is of stucco, and the ceilings are like the walls. The house is forty feet square, consists of four rooms, a hall, and two verandahs, and is surrounded by a wilderness of fruit trees. A deep well supplied me with pure cold water, a great luxury in this climate. Five minutes' walk down the road brought me to the market and the beach, while in the opposite direction there were no more European houses between me and the mountain. In this house I spent many happy days. Returning to it after a three or four months' absence in some uncivilized region, I enjoyed the unwonted luxuries of milk and fresh bread, and regular supplies of fish and eggs, meat and vegetables, which were often sorely needed to restore my health and energy. I had ample space and convenience for unpacking, sorting, and arranging my treasures, and I had delightful walks in the suburbs of the town, or up the lower slopes, of the mountain, when I desired a little exercise, or had time for collecting.

The lower part of the mountain, behind the town of Ternate, is almost entirely covered with a forest of fruit trees, and during the season hundreds of men and women, boys and girls, go up every day to bring down the ripe fruit. Durians and Mangoes, two of the very finest tropical fruits, are in greater abundance at Ternate than I have ever seen them, and some of the latter are of a quality not inferior to any in the world. Lansats and Mangūstans are also abundant, but these do not ripen till a little later. Above the fruit trees there is a belt of clearings and cultivated grounds, which creep up the mountain to a height of between two and three thousand feet, above which is virgin forest, reaching nearly to the summit, which on the side next the town is covered with a high reedy grass. On the further side it is more elevated, of a bare and desolate aspect, with a slight depression marking the position of the crater. From this part descends a black scoriaceous tract, very rugged, and covered with a scanty vegetation of scattered bushes as far down as the sea. This is the lava of the great eruption, near a century ago, and is called by the natives "batu-angas" (burnt rock).

Just below my house is the fort, built by the Portuguese, below which is an open space to the beach, and beyond this the native town extends for about a mile to the north-east. About the centre of it is the palace of the Sultan, now a large, untidy, half-ruinous building of stone. This chief is pensioned by the Dutch Government, but retains the sovereignty over the native population of the island, and of the northern part of Gilolo. The sultans of Ternate and Tidore were once celebrated through the East for their power and regal magnificence. When Drake visited Ternate in 1579, the Portuguese had been driven out of the island, although they still had a settlement at Tidore. He gives a glowing account of the Sultan: "The King had a very rich canopy with embossings of gold borne over him, and was guarded with twelve lances. From the waist to the ground was all cloth of gold, and that very rich; in the attire of his head were finely wreathed in diverse rings of plaited gold, of an inch or more in breadth, which made a fair and princely show, somewhat resembling a crown in form; about his neck he had a chain of perfect gold, the links very great and one fold double; on his left hand was a diamond, an emerald, a ruby and a turky; on his right hand in one ring a big and perfect turky, and in another ring many diamonds of a smaller size."

All this glitter of barbaric gold was the produce of the spice trade, of which the Sultans kept the monopoly, and by which they became wealthy. Ternate, with the small islands in a line south of it, as far as Batchian, constitute the ancient Moluccas, the native country of the clove, as well as the only part in which it was cultivated. Nutmegs and mace were procured from the natives of New Guinea and the adjacent islands, where they grew wild ; and the profits on spice cargoes were so enormous, that the European traders were glad to give gold and jewels, and the finest manufactures of Europe or of India, in exchange. When the Dutch established their influence in these seas, and relieved the native princes from their Portuguese oppressors, they saw that the easiest way to repay themselves would be to get this spice trade into their own hands. For this purpose they adopted the wise principle of concentrating the culture of these valuable products in those spots only of which they could have complete control. To do this effectually it was necessary to abolish the culture and trade in all other places, which they succeeded in doing by treaty with the native rulers. These agreed to have all the spice trees in their possessions destroyed. They gave up large though fluctuating revenues, but they gained in return a fixed subsidy, freedom from the constant attacks and harsh oppressions of the Portuguese, and a continuance of their regal power and exclusive authority over their own subjects, which is maintained in all the islands except Ternate to this day.

It is no doubt supposed by most Englishmen, who have been accustomed to look upon this act of the Dutch with vague horror, as something utterly unprincipled and barbarous, that the native population suffered grievously by this destruction of such valuable property. But it is certain that this was not the case. The Sultans kept this lucrative trade entirely in their own hands as a rigid monopoly, and they would take care not to give their subjects more than would amount to their usual wages, while they would surely exact as large a quantity of spice as they could possibly obtain. Drake and other early voyagers always seem to have purchased their spice-cargoes from the Sultans and Rajahs, and not from the cultivators. Now the absorption of so much labour in the cultivation of this one product must necessarily have raised the price of food and other necessaries ; and when it was abolished, more rice would be grown, more sago made, more fish caught, and more tortoise-shell, rattan, gum-dammer, and other valuable products of the seas and forests would be obtained. I believe, therefore, that this abolition of the spice trade in the Moluccas was actually beneficial to the inhabitants, and that it was an act both wise in itself and morally and politically justifiable.[1]

In the selection of the places in which to carry on the cultiva-

1 See ante, p. 222, and footnote.

tion, the Dutch were not altogether fortunate or wise. Banda was chosen for nutmegs, and was eminently successful, since it continues to this day to produce a large supply of this spice, and to yield a considerable revenue. Amboyna was fixed upon for establishing the clove cultivation ; but the soil and climate, although apparently very similar to those of its native islands, are not favourable, and for some years the Government have actually been paying to the cultivators a higher rate than they could purchase cloves for elsewhere, owing to a great fall in the price since the rate of payment was fixed for a term of years by the Dutch Government, and which rate is still most honourably paid.

In walking about the suburbs of Ternate, we find everywhere the ruins of massive stone and brick buildings, gateways and arches, showing at once the superior wealth of the ancient town and the destructive effects of earthquakes. It was during my second stay in the town, after my return from New Guinea, that I first felt an earthquake. It was a very slight one, scarcely more than has been felt in this country, but occurring in a place that had been many times destroyed by them it was rather more exciting. I had just awoke at gun-fire (5 A.M.), when suddenly the thatch began to rustle and shake as if an army of cats were galloping over it, and immediately afterwards my bed shook too, so that for an instant I imagined myself back in New Guinea, in my fragile house, which shook when an old cock went to roost on the ridge ; but remembering that I was now on a solid earthen floor, I said to myself, " Why, it's an earthquake," and lay still in the pleasing expectation of another shock ; but none came, and this was the only earthquake I ever felt in Ternate.

The last great one was in February 1840, when almost every house in the place was destroyed. It began about midnight on the Chinese New Year's festival, at which time every one stays up nearly all night feasting at the Chinamen's houses and seeing the processions. This prevented any lives being lost, as every one ran out of doors at the first shock, which was not very severe. The second, a few minutes afterwards, threw down a great many houses, and others, which continued all night and part of the next day, completed the devastation. The line of disturbance was very narrow, so that the native town a mile to the east scarcely suffered at all. The wave passed from north to south, through the islands of Tidore and Makian, and terminated in Batchian, where it was not felt till four the following afternoon, thus taking no less than sixteen hours to travel a hundred miles, or about six miles an hour. It is singular that on this occasion there was no rushing up of the tide, or other commotion of the sea, as is usually the case during great earthquakes.

The people of Ternate are of three well-marked races : the Ternate Malays, the Orang Sirani, and the Dutch. The first are an intrusive Malay race somewhat allied to the Macassar people, who settled in the country at a very early epoch, drove out the indigenes, who were no doubt the same as those of the adjacent mainland of Gilolo, and established a monarchy. They perhaps obtained many of their wives from the natives, which will account for the extraordinary language they speak—in some respects closely allied to that of the natives of Gilolo, while it contains much that points to a Malayan origin. To most of these people the Malay language is quite unintelligible, although such as are engaged in trade are obliged to acquire it. "Orang Sirani," or Nazarenes, is the name given by the Malays to the Christian descendants of the Portuguese, who resemble those of Amboyna, and, like them, speak only Malay. There are also a number of Chinese merchants, many of them natives of the place, a few Arabs, and a number of half-breeds between all these races and native women. Besides these there are some Papuan slaves, and a few natives of other islands settled here, making up a motley and very puzzling population, till inquiry and observation have shown the distinct origin of its component parts.

Soon after my first arrival in Ternate I went to the island of Gilolo, accompanied by two sons of Mr. Duivenboden, and by a young Chinaman, a brother of my landlord, who lent us the boat and crew. These latter were all slaves, mostly Papuans, and at starting I saw something of the relation of master and slave in this part of the world. The crew had been ordered to be ready at three in the morning, instead of which none appeared till five, we having all been kept waiting in the dark and cold for two hours. When at length they came they were scolded by their master, but only in a bantering manner, and laughed and joked with him in reply. Then, just as we were starting, one of the strongest men refused to go at all, and his master had to beg and persuade him to go, and only succeeded by assuring him that I would give him something ; so with this promise, and knowing that there would be plenty to eat and drink and little to do, the black gentleman was induced to favour us with his company and assistance. In three hours' rowing and sailing we reached our destination, Sedingole, where there is a house belonging to the Sultan of Tidore, who sometimes goes there hunting. It was a dirty ruinous shed, with no furniture but a few bamboo bedsteads. On taking a walk into the country, I saw at once that it was no place for me. For many miles extends a plain covered with coarse high grass, thickly dotted here and there with trees, the forest country only commencing at the hills a good way in the interior. Such a place would produce few birds and no insects, and we therefore arranged to stay only two days, and then go on to Dodinga, at the narrow central isthmus of Gilolo, whence my friends would return to Ternate. We amused ourselves shooting parrots, lories, and pigeons, and trying to shoot deer, of which we saw plenty, but could not get one ; and our crew went out fishing with a net, so we did not want for provisions. When the time came for us to continue our journey, a fresh difficulty presented itself, for our gentlemen slaves refused in a body to go with us, saying very determinedly that they would return to Ternate. So their masters were obliged to submit, and I was left behind to get to Dodinga as I could. Luckily I succeeded in hiring a small boat, which took me there the same night, with my two men and my baggage.

Two or three years after this, and about the same length of time before I left the East, the Dutch emancipated all their slaves, paying their owners a small compensation. No ill results followed. Owing to the amicable relations which had always existed between them and their masters, due no doubt in part to the Government having long accorded them legal rights and protection against cruelty and ill-usage, many continued in the same service, and after a little temporary difficulty in some cases, almost all returned to work either for their old or for new masters. The Government took the very proper step of placing every emancipated slave under the surveillance of the police-magistrate. They were obliged to show that they were working for a living, and had some honestly-acquired means of existence. All who could not do so were placed upon public works at low wages, and thus were kept from the temptation to peculation or other crimes, which the excitement of newly-acquired freedom, and disinclination to labour, might have led them into.

CHAPTER XXII.

GILOLO.

(MARCH AND SEPTEMBER 1858.)

I MADE but few and comparatively short visits to this large and little known island, but obtained a considerable knowledge of its natural history by sending first my boy Ali, and then my assistant, Charles Allen, who stayed two or three months each in the northern peninsula, and brought me back large collections of birds and insects. In this chapter I propose to give a sketch of the parts which I myself visited. My first stay was at Dodinga, situated at the head of a deep bay exactly opposite Ternate, and a short distance up a little stream which penetrates a few miles inland. The village is a small one, and is completely shut in by low hills.

As soon as I arrived, I applied to the head man of the village for a house to live in, but all were occupied, and there was much difficulty in finding one. In the meantime I unloaded my baggage on the beach and made some tea, and afterwards discovered a small hut which the owner was willing to vacate if I would pay him five guilders for a month's rent. As this was something less than the fee-simple value of the dwelling, I agreed to give it him for the privilege of immediate occupation, only stipulating that he was to make the roof water-tight. This he agreed to do, and came every day to talk and look at me; and when I each time insisted upon his immediately mending the roof according to contract, all the answer I could get was, "Ea nanti" (Yes, wait a little). However, when I threatened to deduct a quarter guilder from the rent for every day it was not done, and a guilder extra if any of my things were wetted, he condescended to work for half an hour, which did all that was absolutely necessary.

On the top of a bank, of about a hundred feet ascent from the water, stands the very small but substantial fort erected by the Portuguese. Its battlements and turrets have long since been overthrown by earthquakes, by which its massive structure has also been rent; but it cannot well be thrown down, being a solid mass of stonework, forming a platform about ten feet high, and perhaps forty feet square. It is approached by narrow steps under an archway, and is now surmounted by a row of thatched hovels, in which live the small garrison, consisting of a Dutch corporal and four Javanese soldiers, the sole representatives of the Netherlands Government in the island. The village is occupied entirely by Ternate men. The true indigenes of Gilolo, "Alfuros" as they are here called, live on the eastern coast, or in the interior of the northern peninsula. The distance across the isthmus at this place is only two miles, and there is a good path, along which rice and sago are brought from the eastern villages. The whole isthmus is very rugged, though not high, being a succession of little abrupt hills and valleys, with angular masses of limestone rock everywhere projecting, and often almost blocking up the pathway. Most of it is virgin forest, very luxuriant and picturesque, and at this time having abundance of large scarlet Ixoras in flower, which made it exceptionally gay. I got some very nice insects here, though, owing to illness most of the time, my collection was a small one; and my boy Ali shot me a pair of one of the most beautiful birds of the East, Pitta gigas, a large ground-thrush, whose plumage of velvety black above is relieved by a breast of pure white, shoulders of azure blue, and belly of vivid crimson. It has very long and strong legs, and hops about with such activity in the dense tangled forest, bristling with rocks, as to make it very difficult to shoot.

In September 1858, after my return from New Guinea, I went to stay some time at the village of Djilolo, situated in a bay on the northern peninsula. Here I obtained a house through the kindness of the Resident of Ternate, who sent orders to prepare one for me. The first walk into the unexplored forests of a new locality is a moment of intense interest to the naturalist, as it is almost sure to furnish him with something curious or hitherto unknown. The first thing I saw here was a flock of small parroquets, of which I shot a pair, and was pleased to find a most beautiful little long-tailed bird, ornamented with green, red, and blue colours, and quite new to me. It was a variety of the Charmosyna placentis, one of the smallest and most elegant of the brush-tongued lories. My hunters soon shot me several other fine birds, and I myself found a specimen of the rare and beautiful day-flying moth, Cocytia d'Urvillei.

The village of Djilolo was formerly the chief residence of the Sultans of Ternate, till about eighty years ago, when at the request of the Dutch they removed to their present abode. The place was then no doubt much more populous, as is indicated by the wide extent of cleared land in the neighbourhood, now covered with coarse high grass, very disagreeable to walk through, and utterly barren to the naturalist. A few days' exploring showed me that only some small patches of forest remained for miles round, and the result was a scarcity of insects and a very limited variety of birds, which obliged me to change my locality. There was another village called Sahoe, to which there was a road of about twelve miles overland, and this had been recommended to me as a good place for birds, and as possessing a large population both of Mahometans and Alfuros, which latter race I much wished to see. I set off one morning to examine this place myself, expecting to pass through some extent of forest on my way. In this however I was much disappointed, as the whole road lies through grass and scrubby thickets, and it was only after reaching the village of Sahoe that some high forest land was perceived stretching towards the mountains to the north of it. About half-way we had to pass a deep river on a bamboo raft, which almost sunk beneath us. This stream was said to rise a long way off to the northward.

Although Sahoe did not at all appear what I expected, I determined to give it a trial, and a few days afterwards obtained a boat to carry my things by sea while I walked overland. A large house on the beach belonging to the Sultan was given me. It stood alone, and was quite open on every side, so that little privacy could be had, but as I only intended to stay a short time I made it do. A very few days dispelled all hopes I might have entertained of making good collections in this place. Nothing was to be found in every direction but interminable tracts of reedy grass, eight or ten feet high, traversed by narrow paths, often almost impassable. Here and there were clumps of fruit trees, patches of low wood, and abundance of plantations and rice grounds, all of which are, in tropical regions, a very desert for the entomologist. The virgin forest that I was in search of, existed only on the summits and on the steep rocky sides of the mountains a long way off, and in inaccessible situations. In the suburbs of the village I found a fair number of bees and wasps, and some small but interesting beetles. Two or three new birds were obtained by my hunters, and by incessant inquiries and promises I succeeded in getting the natives to bring me some land shells, among which was a very fine and handsome one, Helix pyrostoma. I was, however, completely wasting my time here compared with what I might be doing in a good locality, and after a week returned to Ternate, quite disappointed with my first attempts at collecting in Gilolo.

In the country round about Sahoe, and in the interior, there is a large population of indigenes, numbers of whom came daily into the village, bringing their produce for sale, while others were engaged as labourers by the Chinese and Ternate traders. A careful examination convinced me that these people are radically distinct from all the Malay races. There stature and their features, as well as their disposition and habits, are almost the same as those of the Papuans ; their hair is semi-Papuan— neither straight, smooth, and glossy, like all true Malays', nor so frizzly and woolly as the perfect Papuan type, but always crisp, waved, and rough, such as often occurs among the true Papuans, but never among the Malays. Their colour alone is often exactly that of the Malay, or even lighter. Of course there has been intermixture, and there occur occasionally individuals which it is difficult to classify ; but in most cases the large, somewhat aquiline nose, with elongated apex, the tall stature, the waved hair, the bearded face, and hairy body, as well as the less reserved manner and louder voice, unmistakably proclaim the Papuan type. Here then I had discovered the exact boundary line between the Malay and Papuan races, and at a spot where no other writer had expected it. I was very much pleased at this determination, as it gave me a clue to one of the most difficult problems in Ethnology, and enabled me in many other places to separate the two races, and to unravel their intermixtures.

On my return from Waigiou in 1860, I stayed some days on the southern extremity of Gilolo, but, beyond seeing something more of its structure and general character, obtained very little additional information. It is only in the northern peninsula that there are any indigenes, the whole of the rest of the island, with Batchian and the other islands westward, being exclusively inhabited by Malay tribes, allied to those of Ternate and Tidore. This would seem to indicate that the Alfuros were a comparatively recent immigration, and that they had come from the north or east, perhaps from some of the islands of the Pacific. It is otherwise difficult to understand how so many fertile districts should possess no true indigenes.

Gilolo, or Halmaheira as it is called by the Malays and Dutch, seems to have been recently modified by upheaval and subsidence. In 1673, a mountain is said to have been upheaved at Gamokonora on the northern peninsula. All the parts that I have seen have either been volcanic or coralline, and along the coast there are fringing coral reefs very dangerous to navigation. At the same time, the character of its natural history proves it to be a rather ancient land, since it possesses a number of animals peculiar to itself or common to the small islands around it, but almost always distinct from those of New Guinea on the east, of Ceram on the south, and of Celebes and the Sula islands on the west.

The island of Morty, close to the north-eastern extremity of Gilolo, was visited by my assistant Charles Allen, as well as by Dr. Bernstein ; and the collections obtained there present some curious differences from those of the main island. About fifty-six species of land-birds are known to inhabit this island, and of these a kingfisher (Tanysiptera doris), a honeysucker (Tropidorhynchus fuscicapillus), and a large crow-like starling (Lycocorax morotensis), are quite distinct from allied species found in Gilolo. The island is coralline and sandy, and we must therefore believe it to have been separated from Gilolo at a somewhat remote epoch ; while we learn from its natural history that an arm of the sea twenty-five miles wide serves to limit the range even of birds of considerable powers of flight.

CHAPTER XXIII.

TERNATE TO THE KAIÓA ISLANDS AND BATCHIAN.

(OCTOBER 1858.)

ON returning to Ternate from Sahoe, I at once began making preparations for a journey to Batchian, an island which I had been constantly recommended to visit since I had arrived in this part of the Moluccas. After all was ready I found that I should have to hire a boat, as no opportunity of obtaining a passage presented itself. I accordingly went into the native town, and could only find two boats for hire, one much larger than I required, and the other far smaller than I wished. I chose the smaller one, chiefly because it would not cost me one-third as much as the larger one, and also because in a coasting voyage a small vessel can be more easily managed, and more readily got into a place of safety during violent gales than a large one. I took with me my Bornean lad Ali, who was now very useful to me ; Lahagi, a native of Ternate, a very good steady man, and a fair shooter, who had been with me to New Guinea ; Lahi, a native of Gilolo, who could speak Malay, as woodcutter and general assistant ; and Garo, a boy who was to act as cook. As the boat was so small that we had hardly room to stow ourselves away when all my stores were on board, I only took one other man named Latchi, as pilot. He was a Papuan slave, a tall, strong black fellow, but very civil and careful. The boat I had hired from a Chinaman named Lau Keng Tong, for five guilders a month.

We started on the morning of October 9th, but had not got a hundred yards from land, when a strong head wind sprung up, against which we could not row, so we crept along shore to below the town, and waited till the turn of the tide should enable us to cross over to the coast of Tidore. About three in the afternoon we got off, and found that our boat sailed well, and would keep pretty close to the wind. We got on a good way before the wind fell and we had to take to our oars again. We landed on a nice sandy beach to cook our suppers, just as the sun set behind the rugged volcanic hills, to the south of the great cone of Tidore, and soon after beheld the planet Venus shining in the twilight with the brilliancy of a new moon, and casting a very distinct shadow. We left again a little before seven, and as we got out from the shadow of the mountain I observed a bright light over one part of the ridge, and soon after, what seemed a fire of remarkable whiteness on the very summit of the hill. I called the attention of my men to it, and they too thought it merely a fire ; but a few minutes afterwards, as we got farther off shore, the light rose clear up above the ridge of the hill, and some faint clouds clearing away from it, discovered the magnificent comet which was at the same time astonishing all Europe. The nucleus presented to the naked eye a distinct disc of brilliant white light, from which the tail rose at an angle of about 30° or 35° with the horizon, curving slightly downwards, and terminating in a broad brush of faint light, the curvature of which diminished till it was nearly straight at the end. The portion of the tail next the comet appeared three or four times as bright as the most luminous portion of the milky way, and what struck me as a singular feature was that its upper margin, from the nucleus to very near the extremity, was clearly and almost sharply defined, while the lower side gradually shaded off into obscurity. Directly it rose above the ridge of the hill, I said to my men, " See, it's not a fire, it's a bintang ber-ekor," (" tailed-star," the Malay idiom for a comet). " So it is," said they ; and all declared that they had often heard tell of such, but had never seen one till now. I had no telescope with me, nor any instrument at hand, but I estimated the length of the tail at about 20°, and the width, towards the extremity, about 4° or 5°.

The whole of the next day we were obliged to stop near the village of Tidore, owing to a strong wind right in our teeth. The country was all cultivated, and I in vain searched for any insects worth capturing. One of my men went out to shoot, but returned home without a single bird. At sunset, the wind having dropped, we quitted Tidore, and reached the next island, March, where we stayed till morning. The comet was again visible, but not nearly so brilliant, being partly obscured by clouds, and dimmed by the light of the new moon. We then rowed across

to the island of Motir, which is so surrounded with coral-reefs
that it is dangerous to approach. These are perfectly flat, and
are only covered at high water, ending in craggy vertical walls
of coral in very deep water. When there is a little wind, it is
dangerous to come near these rocks; but luckily it was quite
smooth, so we moored to their edge, while the men crawled over
the reef to the land, to make a fire and cook our dinner—the
boat having no accommodation for more than heating water for
my morning and evening coffee. We then rowed along the edge
of the reef to the end of the island, and were glad to get a nice
westerly breeze, which carried us over the strait to the island of
Makian, where we arrived about 8 P.M. The sky was quite clear,
and though the moon shone brightly, the comet appeared with
quite as much splendour as when we first saw it.

The coasts of these small islands are very different according
to their geological formation. The volcanoes, active or extinct,
have steep black beaches of volcanic sand, or are fringed with
rugged masses of lava and basalt. Coral is generally absent,
occurring only in small patches in quiet bays, and rarely or
never forming reefs. Ternate, Tidore, and Makian belong to
this class. Islands of volcanic origin, not themselves volcanoes,
but which have been probably recently upraised, are generally
more or less completely surrounded by fringing reefs of coral,
and have beaches of shining white coral sand. Their coasts
present volcanic conglomerates, basalt, and in some places a
foundation of stratified rocks, with patches of upraised coral.
Mareh and Motir are of this character, the outline of the latter
giving it the appearance of having been a true volcano, and it is
said by Forrest to have thrown out stones in 1778. The next
day (Oct. 12th), we coasted along the island of Makian, which
consists of a single grand volcano. It was now quiescent, but
about two centuries ago (in 1646) there was a terrible eruption,
which blew up the whole top of the mountain, leaving the
truncated jagged summit and vast gloomy crater valley which
at this time distinguished it. It was said to have been as lofty
as Tidore before this catastrophe.[1]

I stayed some time at a place where I saw a new clearing on
a very steep part of the mountain, and obtained a few interesting insects. In the evening we went on to the extreme southern
point, to be ready to pass across the fifteen-mile strait to the
island of Kaióa. At five the next morning we started, but the
wind, which had hitherto been westerly, now got to the south
and south-west, and we had to row almost all the way with a
burning sun overhead. As we approached land a fine breeze
sprang up, and we went along at a great pace; yet after an

[1] Soon after I left the Archipelago, on the 29th of December, 1862, another eruption
of this mountain suddenly took place, which caused great devastation in the island.
All the villages and crops were destroyed, and numbers of the inhabitants killed. The
sand and ashes fell so thick that the crops were partially destroyed fifty miles off, at
Ternate, where it was so dark the following day that lamps had to be lighted at noon.
For the position of this and the adjacent islands, see the map in Chapter XXXVII.

hour we were no nearer, and found we were in a violent current carrying us out to sea. At length we overcame it, and got on shore just as the sun set, having been exactly thirteen hours coming fifteen miles. We landed on a beach of hard coralline rock, with rugged cliffs of the same, resembling those of the Ké Islands (Chap. XXIX.) It was accompanied by a brilliancy and luxuriance of the vegetation very like what I had observed at those islands, which so much pleased me that I resolved to stay a few days at the chief village, and see if their animal productions were correspondingly interesting. While searching for a secure anchorage for the night we again saw the comet, still apparently as brilliant as at first, but the tail had now risen to a higher angle.

October 14*th*.—All this day we coasted along the Kaióa Islands, which have much the appearance and outline of Ké on a small scale, with the addition of flat swampy tracts along shore, and outlying coral reefs. Contrary winds and currents had prevented our taking the proper course to the west of them, and we had to go by a circuitous route round the southern extremity of one island, often having to go far out to sea on account of coral reefs. On trying to pass a channel through one of these reefs we were grounded, and all had to get out into the water, which in this shallow strait had been so heated by the sun as to be disagreeably warm, and drag our vessel a considerable distance among weeds and sponges, corals and prickly corallines. It was late at night when we reached the little village harbour, and we were all pretty well knocked up by hard work, and having had nothing but very brackish water to drink all day— the best we could find at our last stopping-place. There was a house close to the shore, built for the use of the Resident of Ternate when he made his official visits, but now occupied by several native travelling merchants, among whom I found a place to sleep.

The next morning early I went to the village to find the "Kapala," or head man. I informed him that I wanted to stay a few days in the house at the landing, and begged him to have it made ready for me. He was very civil, and came down at once to get it cleared, when we found that the traders had already left, on hearing that I required it. There were no doors to it, so I obtained the loan of a couple of hurdles to keep out dogs and other animals. The land here was evidently sinking rapidly, as shown by the number of trees standing in salt water dead and dying. After breakfast I started for a walk to the forest-covered hill above the village, with a couple of boys as guides. It was exceedingly hot and dry, no rain having fallen for two months. When we reached an elevation of about two hundred feet, the coralline rock which fringes the shore was succeeded by a hard crystalline rock, a kind of metamorphic sandstone. This would indicate that there had been a recent elevation of more than two hundred feet, which had still more recently changed into a movement of subsidence. The hill was very rugged, but among dry sticks and fallen trees I found some good insects, mostly of forms and species I was already acquainted with from Ternate and Gilolo. Finding no good paths I returned, and explored the lower ground eastward of the village, passing through a long range of plantain and tobacco grounds, encumbered with felled and burnt logs, on which I found quantities of beetles of the family Buprestidæ of six different species, one of which was new to me. I then reached a path in the swampy forest where I hoped to find some butterflies, but was disappointed. Being now pretty well exhausted by the intense heat, I thought it wise to return and reserve further exploration for the next day.

When I sat down in the afternoon to arrange my insects, the house was surrounded by men, women, and children, lost in amazement at my unaccountable proceedings ; and when, after pinning out the specimens, I proceeded to write the name of the place on small circular tickets, and attach one to each, even the old Kapala, the Mahometan priest, and some Malay traders could not repress signs of astonishment. If they had known a little more about the ways and opinions of white men, they would probably have looked upon me as a fool or a madman, but in their ignorance they accepted my operations as worthy of all respect, although utterly beyond their comprehension.

The next day (October 16th) I went beyond the swamp, and found a place where a new clearing was being made in the virgin forest. It was a long and hot walk, and the search among the fallen trunks and branches was very fatiguing, but I was rewarded by obtaining about seventy distinct species of beetles, of which at least a dozen were new to me, and many others rare and interesting. I have never in my life seen beetles so abundant as they were on this spot. Some dozen species of good-sized golden Buprestidæ, green rose-chafers (Lomaptera), and long-horned weevils (Anthribidæ), were so abundant that they rose up in swarms as I walked along, filling the air with a loud buzzing hum. Along with these, several fine Longicorns were almost equally common, forming such an assemblage as for once to realize that idea of tropical luxuriance which one obtains by looking over the drawers of a well-filled cabinet. On the under sides of the trunks clung numbers of smaller or more sluggish Longicorns, while on the branches at the edge of the clearing others could be detected sitting with outstretched antennæ ready to take flight at the least alarm. It was a glorious spot, and one which will always live in my memory as exhibiting the insect-life of the tropics in unexampled luxuriance. For the three following days I continued to visit this locality, adding each time many new species to my collection—the following notes of which may be interesting to entomologists. October 15th, 33 species of beetles ; 16th, 70 species ; 17th, 47 species ; 18th, 40 species ; 19th, 56 species—in

all about a hundred species, of which forty were new to me.
There were forty-four species of Longicorns among them, and
on the last day I took twenty-eight species of Longicorns, of
which five were new to me.

My boys were less fortunate in shooting. The only birds at
all common were the great red parrot (Eclectus grandis), found
in most of the Moluccas, a crow, and a Megapodius, or mound-
maker. A few of the pretty racquet-tailed kingfishers were
also obtained, but in very poor plumage. They proved, however, to be of a different species from those found in the other
islands, and come nearest to the bird originally described by
Linnæus under the name of Alcedo dea, and which came from
Ternate. This would indicate that the small chain of islands
parallel to Gilolo have a few peculiar species in common, a fact
which certainly occurs in insects.

The people of Kaióa interested me much. They are evidently
a mixed race, having Malay and Papuan affinities, and are
allied to the peoples of Ternate and of Gilolo. They possess a
peculiar language, somewhat resembling those of the surrounding islands, but quite distinct. They are now Mahometans,
and are subject to Ternate. The only fruits seen here were
papaws and pine-apples, the rocky soil and dry climate being
unfavourable. Rice, maize, and plantains flourish well, except
that they suffer from occasional dry seasons like the present
one. There is a little cotton grown, from which the women
weave sarongs (Malay petticoats). There is only one well of
good water on the islands, situated close to the landing-place,
to which all the inhabitants come for drinking water. The
men are good boat-builders, and they make a regular trade of
it and seem to be very well off.

After five days at Kaióa we continued our journey, and soon
got among the narrow straits and islands which lead down to
the town of Batchian. In the evening we stayed at a settlement
of Galéla men. These are natives of a district in the extreme
north of Gilolo, and are great wanderers over this part of the
Archipelago. They build large and roomy praus with out-
riggers, and settle on any coast or island they take a fancy for.
They hunt deer and wild pig, drying the meat ; they catch
turtle and tripang ; they cut down the forest and plant rice or
maize, and are altogether remarkably energetic and industrious.
They are very fine people, of light complexion, tall, and with
Papuan features, coming nearer to the drawings and descriptions
of the true Polynesians of Tahiti and Owyhee than any I have
seen.

During this voyage I had several times had an opportunity
of seeing my men get fire by friction. A sharp-edged piece of
bamboo is rubbed across the convex surface of another piece,
on which a small notch is first cut. The rubbing is slow at first
and gradually quicker, till it becomes very rapid, and the fine
powder rubbed off ignites and falls through the hole which the

rubbing has cut in the bamboo. This is done with great quickness and certainty. The Ternate people use bamboo in another way. They strike its flinty surface with a bit of broken china, and produce a spark, which they catch in some kind of tinder.

On the evening of October 21st we reached our destination, having been twelve days on the voyage. It had been fine weather all the time, and, although very hot, I had enjoyed myself exceedingly, and had besides obtained some experience in boat work among islands and coral reefs, which enabled me afterwards to undertake much longer voyages of the same kind. The village or town of Batchian is situated at the head of a wide and deep bay, where a low isthmus connects the northern and southern mountainous parts of the island. To the south is a fine range of mountains, and I had noticed at several of our landing-places that the geological formation of the island was very different from those around it. Whenever rock was visible it was either sandstone in thin layers, dipping south, or a pebbly conglomerate. Sometimes there was a little coralline limestone, but no volcanic rocks. The forest had a dense luxuriance and loftiness seldom found on the dry and porous lavas and raised coral reefs of Ternate and Gilolo ; and hoping for a corresponding richness in the birds and insects, it was with much satisfaction and with considerable expectation that I began my explorations in the hitherto unknown island of Batchian.

CHAPTER XXIV.

BATCHIAN.

(OCTOBER 1858 TO APRIL 1859.)

I LANDED opposite the house kept for the use of the Resident of Ternate, and was met by a respectable middle-aged Malay, who told me he was Secretary to the Sultan, and would receive the official letter with which I had been provided. On giving it him, he at once informed me I might have the use of the official residence which was empty. I soon got my things on shore, but on looking about me found that the house would never do to stay long in. There was no water except at a considerable distance, and one of my men would be almost entirely occupied getting water and firewood, and I should myself have to walk all through the village every day to the forest, and live almost in public, a thing I much dislike. The rooms were all boarded, and had ceilings, which are a great nuisance, as there are no means of hanging anything up except by driving nails, and not half the conveniences of a native bamboo and thatch cottage. I accordingly inquired for a house outside of the village on the road to the coal mines, and was informed by the Secretary that there was a small one belonging to the Sultan, and that he would go with me early next morning to see it.

We had to pass one large river, by a rude but substantial bridge, and to wade through another fine pebbly stream of clear water, just beyond which the little hut was situated. It was very small, not raised on posts, but with the earth for a floor, and was built almost entirely of the leaf-stems of the sago-palm, called here " gaba-gaba." Across the river behind rose a forest-clad bank, and a good road close in front of the house led through cultivated grounds to the forest about half a mile on, and thence to the coal mines four miles further. These advantages at once decided me, and I told the Secretary I would be very glad to occupy the house. I therefore sent my two men immediately to buy "ataps " (palm-leaf thatch) to repair the roof, and the next day, with the assistance of eight of the Sultan's men, got all my stores and furniture carried up and pretty comfortably arranged. A rough bamboo bedstead was soon constructed, and a table made of boards which I had brought with me, fixed under the window. Two bamboo chairs, an easy cane chair, and hanging shelves suspended with insulating oil cups, so as to be safe from ants, completed my furnishing arrangements.

In the afternoon succeeding my arrival, the Secretary accompanied me to visit the Sultan. We were kept waiting a few minutes in an outer gate-house, and then ushered to the door of a rude, half-fortified, whitewashed house. A small table and three chairs were placed in a large outer corridor, and an old dirty-faced man with grey hair and a grimy beard, dressed in a speckled blue cotton jacket and loose red trousers, came forward, shook hands, and asked me to be seated. After a quarter of an hour's conversation on my pursuits, in which his Majesty seemed to take great interest, tea and cakes—of rather better quality than usual on such occasions—were brought in. I thanked him for the house, and offered to show him my collections, which he promised to come and look at. He then asked me to teach him to take views—to make maps—to get him a small gun from England, and a milch-goat from Bengal ; all of which requests I evaded as skilfully as I was able, and we parted very good friends. He seemed a sensible old man, and lamented the small population of the island, which he assured me was rich in many valuable minerals, including gold ; but there were not people enough to look after them and work them. I described to him the great rush of population on the discovery of the Australian gold mines, and the huge nuggets found there, with which he was much interested, and exclaimed, "Oh ! if we had but people like that, my country would be quite as rich !"

The morning after I had got into my new house, I sent my

boys out to shoot, and went myself to explore the road to the
coal mines. In less than half a mile it entered the virgin forest,
at a place where some magnificent trees formed a kind of
natural avenue. The first part was flat and swampy, but it
soon rose a little, and ran alongside the fine stream which
passed behind my house, and which here rushed and gurgled
over a rocky or pebbly bed, sometimes leaving wide sandbanks
on its margins, and at other places flowing between high banks
crowned with a varied and magnificent forest vegetation. After
about two miles, the valley narrowed, and the road was carried
along the steep hill-side which rose abruptly from the water's
edge. In some places the rock had been cut away, but its
surface was already covered with elegant ferns and creepers.
Gigantic tree-ferns were abundant, and the whole forest had an
air of luxuriance and rich variety which it never attains in the
dry volcanic soil to which I had been lately accustomed. A
little further the road passed to the other side of the valley by
a bridge across the stream at a place where a great mass of
rock in the middle offered an excellent support for it, and two
miles more of most picturesque and interesting road brought
me to the mining establishment.

This is situated in a large open space, at a spot where two
tributaries fall into the main stream. Several forest-paths and
new clearings offered fine collecting grounds, and I captured
some new and interesting insects; but as it was getting late I
had to reserve a more thorough exploration for future occasions.
Coal had been discovered here some years before, and the road
was made in order to bring down a sufficient quantity for a fair
trial on the Dutch steamers. The quality, however, was not
thought sufficiently good, and the mines were abandoned.
Quite recently, works had been commenced in another spot, in
hopes of finding a better vein. There were about eighty men
employed, chiefly convicts; but this was far too small a number
for mining operations in such a country, where the mere
keeping a few miles of road in repair requires the constant
work of several men. If coal of sufficiently good quality should
be found, a tramroad would be made, and would be very easily
worked, owing to the regular descent of the valley.

Just as I got home I overtook Ali returning from shooting
with some birds hanging from his belt. He seemed much
pleased, and said, " Look here, sir, what a curious bird," holding
out what at first completely puzzled me. I saw a bird with a
mass of splendid green feathers on its breast, elongated into
two glittering tufts; but what I could not understand was a
pair of long white feathers, which stuck straight out from each
shoulder. Ali assured me that the bird stuck them out this
way itself, when fluttering its wings, and that they had remained so without his touching them. I now saw that I had
got a great prize, no less than a completely new form of the
Bird of Paradise, differing most remarkably from every other

WALLACE'S STANDARD WING, MALE AND FEMALE.

known bird. The general plumage is very sober, being a pure ashy olive, with a purplish tinge on the back ; the crown of the head is beautifully glossed with pale metallic violet, and the feathers of the front extend as much over the beak as in most of the family. The neck and breast are scaled with fine metallic green, and the feathers on the lower part are elongated on each side, so as to form a two-pointed gorget, which can be folded beneath the wings, or partially erected and spread out in the same way as the side plumes of most of the birds of paradise. The four long white plumes which give the bird its altogether unique character, spring from little tubercles close to the upper edge of the shoulder or bend of the wing ; they are narrow, gently curved, and equally webbed on both sides, of a pure creamy white colour. They are about six inches long, equalling the wing, and can be raised at right angles to it, or laid along the body at the pleasure of the bird. The bill is horn colour, the legs yellow, and the iris pale olive. This striking novelty has been named by Mr. G. R. Gray of the British Museum, Semioptera Wallacei, or "Wallace's Standard wing."

A few days later I obtained an exceedingly beautiful new butterfly, allied to the fine blue Papilio Ulysses, but differing from it in the colour being of a more intense tint, and in having a row of blue stripes; around the margin of the lower wings. This good beginning was, however, rather deceptive, and I soon found that insects, and especially butterflies, were somewhat scarce, and birds in far less variety than I had anticipated. Several of the fine Moluccan species were however obtained. The handsome red lory with green wings and a yellow spot in the back (Lorius garrulus) was not uncommon. When the Jambu, or rose apple (Eugenia sp.), was in flower in the village, flocks of the little lorikeet (Charmosyna placentis), already met with in Gilolo, came to feed upon the nectar, and I obtained as many specimens as I desired. Another beautiful bird of the parrot tribe was the Geoffroyus cyanicollis, a green parrot with a red bill and head, which colour shaded on the crown into azure blue, and thence into verditer blue and the green of the back. Two large and handsome fruit pigeons, with metallic green, ashy, and rufous plumage, were not uncommon ; and I was rewarded by finding a splendid deep blue roller (Eurystomus azureus), a lovely golden-capped sunbird (Nectarinea auriceps), and a fine racquet-tailed kingfisher (Tanysiptera isis), all of which were entirely new to ornitholgists. Of insects I obtained a considerable number of interesting beetles, including many fine longicorns, among which was the largest and handsomest species of the genus Glenea yet discovered. Among butterflies the beautiful little Danis sebæ was abundant, making the forests gay with its delicate wings of white and the richest metallic blue ; while showy Papilios, and pretty Pieridæ, and dark, rich Euplæas, many of them new, furnished a constant source of interest and pleasing occupation.

The island of Batchian possesses no really indigenous inhabitants, the interior being altogether uninhabited, and there are only a few small villages on various parts of the coast ; yet I found here four distinct races, which would wofully mislead an ethnological traveller unable to obtain information as to their origin. First there are the Batchian Malays, probably the earliest colonists, differing very little from those of Ternate. Their language, however, seems to have more of the Papuan element, with a mixture of pure Malay, showing that the settlement is one of stragglers of various races, although now sufficiently homogeneous. Then there are the "Orang Sirani," as at Ternate and Amboyna. Many of these have the Portuguese physiognomy strikingly preserved, but combined with a skin generally darker than the Malays. Some national customs are retained, and the Malay, which is their only language, contains a large number of Portuguese words and idioms. The third race consists of the Galela men from the north of Gilolo, a singular people, whom I have already described ; and the fourth is a colony from Tomōré, in the eastern peninsula of Celebes. These people were brought here at their own request a few years ago, to avoid extermination by another tribe. They have a very light complexion, open Tartar physiognomy, low stature, and a language of the Bugis type. They are an industrious agricultural people, and supply the town with vegetables. They make a good deal of bark cloth, similar to the tapa of the Polynesians, by cutting down the proper trees and taking off large cylinders of bark, which is beaten with mallets till it separates from the wood. It is then soaked, and so continuously and regularly beaten out that it becomes as thin and as tough as parchment. In this form it is much used for wrappers for clothes ; and they also make jackets of it, sewn neatly together and stained with the juice of another kind of bark, which gives it a dark red colour and renders it nearly waterproof.

Here are four very distinct kinds of people who may all be seen any day in and about the town of Batchian. Now if we suppose a traveller ignorant of Malay, picking up a word or two here and there of the "Batchian language," and noting down the "physical and moral peculiarities, manners, and customs of the Batchian people"—(for there are travellers who do all this in four-and-twenty hours)—what an accurate and instructive chapter we should have ! what transitions would be pointed out, what theories of the origin of races would be developed ! while the next traveller might flatly contradict every statement and arrive at exactly opposite conclusions.

Soon after I arrived here the Dutch Government introduced a new copper coinage of *cents* instead of *doits* (the 100th instead of the 120th part of a guilder), and all the old coins were ordered to be sent to Ternate to be changed. I sent a bag containing 6,000 doits, and duly received the new money by return of the boat. When Ali went to bring it, however, the captain required a written order ; so I waited to send again the next day, and it was lucky I did so, for that night my house was entered, all my boxes carried out and ransacked, and the various articles left on the road about twenty yards off, where we found them at five in the morning, when, on getting up and finding the house empty, we rushed out to discover tracks of the thieves. Not being able to find the copper money which they thought I had just received, they decamped, taking nothing but a few yards of cotton cloth, and a black coat and trousers, which latter were picked up a few days afterwards hidden in the grass. There was no doubt whatever who were the thieves. Convicts are employed to guard the Government stores when the boat arrives from Ternate. Two of them watch all night, and often take the opportunity to roam about and commit robberies.

The next day I received my money, and secured it well in a strong box fastened under my bed. I took out five or six hundred cents for daily expenses, and put them in a small japanned box, which always stood upon my table. In the afternoon I went for a short walk, and on my return this box and my keys, which I had carelessly left on the table, were gone. Two of my boys were in the house, but had heard nothing. I immediately gave information of the two robberies to the Director at the mines and to the Commandant at the fort, and got for answer, that if I caught the thief in the act I might shoot him. By inquiry in the village, we afterwards found that one of the convicts who was on duty at the Government rice-store in the village had quitted his guard, was seen to pass over the bridge towards my house, was seen again within two hundred yards of my house, and on returning over the bridge into the village carried something under his arm, carefully covered with his sarong. My box was stolen between the hours he was seen going and returning, and it was so small as to be easily carried in the way described. This seemed pretty clear circumstantial evidence. I accused the man and brought the witnesses to the Commandant. The man was examined, and confessed having gone to the river close to my house to bathe ; but said he had gone no further, having climbed up a cocoa-nut tree and brought home two nuts, which he had covered over, *because he was ashamed to be seen carrying them!* This explanation was thought satisfactory, and he was acquitted. I lost my cash and my box, a seal I much valued, with other small articles, and all my keys— the severest loss by far. Luckily my large cash-box was left locked, but so were others which I required to open immediately. There was, however, a very clever blacksmith employed to do ironwork for the mines, and he picked my locks for me when I required them, and in a few days made me new keys, which I used all the time I was abroad.

Towards the end of November the wet season set in, and we had daily and almost incessant rains, with only about one or two hours sunshine in the morning. The flat parts of the forest became flooded, the roads filled with mud, and insects and birds were scarcer than ever. On December 13th, in the afternoon, we had a sharp earthquake shock, which made the house and furniture shake and rattle for five minutes, and the trees and shrubs wave as if a gust of wind had passed over them. About the middle of December I removed to the village, in order more easily to explore the district to the west of it, and to be near the sea when I wished to return to Ternate. I obtained the use of a good-sized house in the Campong Sirani (or Christian village), and at Christmas and the New Year had to endure the incessant gun-firing, drum-beating, and fiddling of the inhabitants.

These people are very fond of music and dancing, and it would astonish a European to visit one of their assemblies. We enter a gloomy palm-leaf hut in which two or three very dim lamps barely render darkness visible. The floor is of black sandy earth, the roof hid in a smoky, impenetrable blackness; two or three benches stand against the walls, and the orchestra consists of a fiddle, a fife, a drum, and a triangle. There is plenty of company, consisting of young men and women, all very neatly dressed in white and black—a true Portuguese habit. Quadrilles, waltzes, polkas, and mazurkas are danced with great vigour and much skill. The refreshments are muddy coffee and a few sweetmeats. Dancing is kept up for hours, and all is conducted with much decorum and propriety. A party of this kind meets about once a week, the principal inhabitants taking it by turns, and all who please come in without much ceremony.

It is astonishing how little these people have altered in three hundred years, although in that time they have changed their language and lost all knowledge of their own nationality. They are still in manners and appearance almost pure Portuguese, very similar to those with whom I had become acquainted on the banks of the Amazon. They live very poorly as regards their house and furniture, but preserve a semi-European dress, and have almost all full suits of black for Sundays. They are nominally Protestants, but Sunday evening is their grand day for music and dancing. The men are often good hunters; and two or three times a week, deer or wild pigs are brought to the village, which, with fish and fowls, enables them to live well. They are almost the only people in the Archipelago who eat the great fruit-eating bats called by us "flying foxes." These ugly creatures are considered a great delicacy, and are much sought after. At about the beginning of the year they come in large flocks to eat fruit, and congregate during the day on some small islands in the bay, hanging by thousands on the trees, especially on dead ones. They can then be easily caught or knocked down with sticks, and are brought home by basketfuls. They require to be carefully prepared, as the skin and fur has a rank and powerful foxy odour; but they are generally cooked with abundance of spices and condiments, and are really very good eating, something like hare. The Orang Sirani are good cooks, having a much greater variety of savoury dishes than the Malays. Here, they live chiefly on sago as bread, with a little rice occasionally, and abundance of vegetables and fruit.

It is a curious fact that everywhere in the East where the Portuguese have mixed with the native races they have become darker in colour than either of the parent stocks. This is the case almost always with these "Orang Sirani" in the Moluccas, and with the Portuguese of Malacca. The reverse is the case in South America, where the mixture of the Portuguese or Brazilian with the Indian produces the "Mameluco," who is not unfrequently lighter than either parent, and always lighter than the Indian. The women at Batchian, although generally fairer than the men, are coarse in features, and very far inferior in beauty to the mixed Dutch-Malay girls, or even to many pure Malays.

The part of the village in which I resided was a grove of cocoa-nut trees, and at night, when the dead leaves were sometimes collected together and burnt, the effect was most magnificent—the tall stems, the fine crowns of foliage, and the immense fruit-clusters, being brilliantly illuminated against a dark sky, and appearing like a fairy palace supported on a hundred columns, and groined over with leafy arches. The cocoa-nut tree, when well grown, is certainly the prince of palms both for beauty and utility.

During my very first walk into the forest at Batchian, I had seen sitting on a leaf out of reach, an immense butterfly of a dark colour marked with white and yellow spots. I could not capture it as it flew away high up into the forest, but I at once saw that it was a female of a new species of Ornithoptera or "bird-winged butterfly," the pride of the Eastern tropics. I was very anxious to get it and to find the male, which in this genus is always of extreme beauty. During the two succeeding months I only saw it once again, and shortly afterwards I saw the male flying high in the air at the mining village. I had begun to despair of ever getting a specimen, as it seemed so rare and wild; till one day, about the beginning of January, I found a beautiful shrub with large white leafy bracts and yellow flowers, a species of Mussænda, and saw one of these noble insects hovering over it, but it was too quick for me, and flew away. The next day I went again to the same shrub and succeeded in catching a female, and the day after a fine male. I found it to be as I had expected, a perfectly new and most magnificent species, and one of the most gorgeously coloured butterflies in the world. Fine specimens of the male are more than seven inches across the wings, which are velvety black and fiery orange, the latter colour replacing the green of the allied species. The beauty and brilliancy of this insect are indescribable, and none but a naturalist can understand the intense excitement I experienced when I at length captured it. On taking it out of my net and opening the glorious wings, my heart began to beat violently, the blood rushed to my head, and I felt much more like fainting than I have done when in apprehension of immediate death. I had a headache the rest of the day, so great was the excitement produced by what will appear to most people a very inadequate cause.

I had decided to return to Ternate in a week or two more, but this grand capture determined me to stay on till I obtained a good series of the new butterfly, which I have since named Ornithoptera crœsus. The Mussænda bush was an admirable place, which I could visit every day on my way to the forest ; and as it was situated in a dense thicket of shrubs and creepers, I set my man Lahi to clear a space all round it, so that I could easily get at any insect that might visit it. Afterwards, finding that it was often necessary to wait some time there, I had a little seat put up under a tree by the side of it, where I came every day to eat my lunch, and thus had half an hour's watching about noon, besides a chance as I passed it in the morning. In this way I obtained on an average one specimen a day for a long time, but more than half of these were females, and more than half the remainder worn or broken specimens, so that I should not have obtained many perfect males had I not found another station for them.

As soon as I had seen them come to flowers, I sent my man Lahi with a net on purpose to search for them, as they had also been seen at some flowering trees on the beach, and I promised him half a day's wages extra for every good specimen he could catch. After a day or two he brought me two very fair specimens, and told me he had caught them in the bed of a large rocky stream that descends from the mountains to the sea about a mile below the village. They flew down this river, settling occasionally on stones and rocks in the water, and he was obliged to wade up it or jump from rock to rock to get at them. I went with him one day, but found that the stream was far too rapid and the stones too slippery for me to do anything, so I left it entirely to him, and all the rest of the time we stayed in Batchian he used to be out all day, generally bringing me one, and on good days two or three specimens. I was thus able to bring away with me more than a hundred of both sexes, including perhaps twenty very fine males, though not more than five or six that were absolutely perfect.

My daily walk now led me, first about half a mile along the sandy beach, then through a sago swamp over a causeway of very shaky poles to the village of the Tomōré people. Beyond this was the forest with patches of new clearing, shady paths, and a considerable quantity of felled timber. I found this a very fair collecting ground, especially for beetles. The fallen trunks in the clearings abounded with golden Buprestidæ and curious Brenthidæ and longicorns, while in the forest I found abundance of the smaller Curculionidæ, many longicorns, and some fine green Carabidæ.

Butterflies were not abundant, but I obtained a few more of the fine blue Papilio, and a number of beautiful little Lycænidæ, as well as a single specimen of the very rare Papilio Wallacei, of which I had taken the hitherto unique specimen in the Aru Islands.

The most interesting birds I obtained here were the beautiful blue kingfisher, Todiramphus diops; the fine green and purple doves, Ptilonopus superbus and P. iogaster, and several new birds of small size. My shooters still brought me in specimens of the Semioptera Wallacei, and I was greatly excited by the positive statements of several of the native hunters that another species of this bird existed, much handsomer and more remarkable. They declared that the plumage was glossy black, with metallic green breast as in my species, but that the white shoulder plumes were twice as long, and hung down far below the body of the bird. They declared that when hunting pigs or deer far in the forest they occasionally saw this bird, but that it was rare. I immediately offered twelve guilders (a pound) for a specimen; but all in vain, and I am to this day uncertain whether such a bird exists. Since I left, the German naturalist, Dr. Bernstein, stayed many months in the island with a large staff of hunters collecting for the Leyden Museum; and as he was not more successful than myself, we must consider either that the bird is very rare, or is altogether a myth.

Batchian is remarkable as being the most eastern point on the globe inhabited by any of the Quadrumana. A large black baboon-monkey (Cynopithecus nigrescens) is abundant in some parts of the forest. This animal has bare red callosities, and a rudimentary tail about an inch long—a mere fleshy tubercle, which may be very easily overlooked. It is the same species that is found all over the forests of Celebes, and as none of the other Mammalia of that island extend into Batchian I am inclined to suppose that this species has been accidentally introduced by the roaming Malays, who often carry about with them tame monkeys and other animals. This is rendered more probable by the fact that the animal is not found in Gilolo, which is only separated from Batchian by a very narrow strait. The introduction may have been very recent, as in a fertile and unoccupied island such an animal would multiply rapidly. The only other mammals obtained were an Eastern opossum, which Dr. Gray has described as Cuscus ornatus; the little flying opossum, Belideus ariel; a Civet cat, Viverra zebetha; and nine species of bats, most of the smaller ones being caught in the dusk with my butterfly net as they flew about before the house.

After much delay, owing to bad weather and the illness of one of my men, I determined to visit Kasserota (formerly the chief village), situated up a small stream, on an island close to the north coast of Batchian, where I was told that many rare birds were found. After my boat was loaded and everything ready, three days of heavy squalls prevented our starting, and

it was not till the 21st of March that we got away. Early next
morning we entered the little river, and in about an hour we
reached the Sultan's house, which I had obtained permission to
use. It was situated on the bank of the river, and surrounded
by a forest of fruit trees, among which were some of the very
loftiest and most graceful cocoa-nut palms I have ever seen.
It rained nearly all that day, and I could do little but unload
and unpack. Towards the afternoon it cleared up, and I attempted to explore in various directions, but found to my disgust
that the only path was a perfect mud swamp, along which it
was almost impossible to walk, and the surrounding forest so
damp and dark as to promise little in the way of insects. I
found too on inquiry that the people here made no clearings,
living entirely on sago, fruit, fish, and game ; and the path only
led to a steep rocky mountain equally impracticable and unproductive. The next day I sent my men to this hill, hoping it
might produce some good birds ; but they returned with only
two common species, and I myself had been able to get nothing,
every little track I had attempted to follow leading to a dense
sago swamp. I saw that I should waste time by staying here,
and determined to leave the following day.

This is one of those spots so hard for the European naturalist
to conceive, where with all the riches of a tropical vegetation,
and partly perhaps from the very luxuriance of that vegetation,
insects are as scarce as in the most barren parts of Europe, and
hardly more conspicuous. In temperate climates there is a
tolerable uniformity in the distribution of insects over those
parts of a country in which there is a similarity in the vegetation, any deficiency being easily accounted for by the absence
of wood or uniformity of surface. The traveller hastily passing
through such a country can at once pick out a collecting ground
which will afford him a fair notion of its entomology. Here the
case is different. There are certain requisites of a good collecting ground which can only be ascertained to exist by some days'
search in the vicinity of each village. In some places there is no
virgin forest, as at Djilolo and Sahoe ; in others there are no
open pathways or clearings, as here. At Batchian there are
only two tolerable collecting places,—the road to the coal mines,
and the new clearings made by the Tomōré people, the latter
being by far the most productive. I believe the fact to be that
insects are pretty uniformly distributed over these countries
(where the forests have not been cleared away), and are so
scarce in any one spot that searching for them is almost useless.
If the forest is all cleared away, almost all the insects disappear
with it ; but when small clearings and paths are made, the fallen
trees in various stages of drying and decay, the rotting leaves,
the loosening bark and the fungoid growths upon it, together
with the flowers that appear in much greater abundance where
the light is admitted, are so many attractions to the insects for
miles around, and cause a wonderful accumulation of species and

individuals. When the entomologist can discover such a spot, he does more in a month than he could possibly do by a year's search in the depths of the undisturbed forest.

The next morning we left early, and reached the mouth of the little river in about an hour. It flows through a perfectly flat alluvial plain, but there are hills which approach it near the mouth. Towards the lower part, in a swamp where the salt-water must enter at high tides, were a number of elegant tree-ferns from eight to fifteen feet high. These are generally considered to be mountain plants, and rarely to occur on the equator at an elevation of less than one or two thousand feet. In Borneo, in the Aru Islands, and on the banks of the Amazon, I have observed them at the level of the sea, and think it probable that the altitude supposed to be requisite for them may have been deduced from facts observed in countries where the plains and lowlands are largely cultivated, and most of the indigenous vegetation destroyed. Such is the case in most parts of Java, India, Jamaica, and Brazil, where the vegetation of the tropics has been most fully explored.

Coming out to sea we turned northwards, and in about two hours' sail reached a few huts, called Langundi, where some Galela men had established themselves as collectors of gum-dammar, with which they made torches for the supply of the Ternate market. About a hundred yards back rises a rather steep hill, and a short walk having shown me that there was a tolerable path up it, I determined to stay here for a few days. Opposite us, and all along this coast of Batchian, stretches a row of fine islands completely uninhabited. Whenever I asked the reason why no one goes to live in them, the answer always was, "For fear of the Magindano pirates." Every year these scourges of the Archipelago wander in one direction or another, making their rendezvous on some uninhabited island, and carrying devastation to all the small settlements around ; robbing, destroying, killing, or taking captive all they meet with. Their long well-manned praus escape from the pursuit of any sailing vessel by pulling away right in the wind's eye, and the warning smoke of a steamer generally enables them to hide in some shallow bay, or narrow river, or forest-covered inlet, till the danger is passed. The only effectual way to put a stop to their depredations would be to attack them in their strongholds and villages, and compel them to give up piracy, and submit to strict surveillance. Sir James Brooke did this with the pirates of the north-west coast of Borneo, and deserves the thanks of the whole population of the Archipelago for having rid them of half their enemies.

All along the beach here, and in the adjacent strip of sandy lowland, is a remarkable display of Pandaneæ or Screw-pines. Some are like huge branching candelabra, forty or fifty feet high, and bearing at the end of each branch a tuft of immense sword-shaped leaves, six or eight inches wide, and as many feet long.

Others have a single unbranched stem, six or seven feet high, the upper part clothed with the spirally arranged leaves, and bearing a single terminal fruit as large as a swan's egg. Others of intermediate size have irregular clusters of rough red fruits, and all have more or less spiny-edged leaves and ringed stems. The young plants of the larger species have smooth, glossy thick leaves, sometimes ten feet long and eight inches wide, which are used all over the Moluccas and New Guinea to make "cocoyas" or sleeping mats, which are often very prettily ornamented with coloured patterns. Higher up on the hill is a forest of immense trees, among which those producing the resin called dammar (Dammara sp.) are abundant. The inhabitants of several small villages in Batchian are entirely engaged in searching for this product, and making it into torches by pounding it and filling it into tubes of palm leaves about a yard long, which are the only lights used by many of the natives. Sometimes the dammar accumulates in large masses of ten or twenty pounds' weight, either attached to the trunk, or found buried in the ground at the foot of the trees. The most extraordinary trees of the forest are, however, a kind of fig, the aerial roots of which form a pyramid near a hundred feet high, terminating just where the tree branches out above, so that there is no real trunk. This pyramid or cone is formed of roots of every size, mostly descending in straight lines, but more or less obliquely—and so crossing each other, and connected by cross branches, which grow from one to another; as to form a dense and complicated network, to which nothing but a photograph could do justice (see illustration at page 64). The Kanary is also abundant in this forest, the nut of which has a very agreeable flavour, and produces an excellent oil. The fleshy outer covering of the nut is the favourite food of the great green pigeons of these islands (Carpophaga perspicillata), and their hoarse cooings and heavy flutterings among the branches can be almost continually heard.

After ten days at Langundi, finding it impossible to get the bird I was particularly in search of (the Nicobar pigeon, or a new species allied to it), and finding no new birds, and very few insects, I left early on the morning of April 1st, and in the evening entered a river on the main island of Batchian, (Langundi, like Kasserota, being on a distinct island), where some Malays and Galela men have a small village, and have made extensive rice-fields and plantain grounds. Here we found a good house near the river bank, where the water was fresh and clear, and the owner, a respectable Batchian Malay, offered me sleeping room and the use of the verandah if I liked to stay. Seeing forest all round within a short distance, I accepted his offer, and the next morning before breakfast walked out to explore, and on the skirts of the forest captured a few interesting insects.

Afterwards, I found a path which led for a mile or more through a very fine forest, richer in palms than any I had seen

in the Moluccas. One of these especially attracted my attention from its elegance. The stem was not thicker than my wrist,
yet it was very lofty, and bore clusters of bright red fruit. It
was apparently a species of Areca. Another of immense height
closely resembled in appearance the Euterpes of South America.
Here also grew the fan-leafed palm, whose small, nearly entire
leaves are used to make the dammar torches, and to form the
water-buckets in universal use. During this walk I saw near a
dozen species of palms, as well as two or three Pandani different
from those of Langundi. There were also some very fine climbing ferns and true wild Plantains (Musa), bearing an edible fruit
not so large as one's thumb, and consisting of a mass of seeds
just covered with pulp and skin. The people assured me they
had tried the experiment of sowing and cultivating this species,
but could not improve it. They probably did not grow it in
sufficient quantity, and did not persevere sufficiently long.

Batchian is an island that would perhaps repay the researches
of a botanist better than any other in the whole Archipelago.
It contains a great variety of surface and of soil, abundance of
large and small streams, many of which are navigable for some
distance, and there being no savage inhabitants, every part of
it can be visited with perfect safety. It possesses gold, copper,
and coal, hot springs and geysers, sedimentary and volcanic
rocks and coralline limestone, alluvial plains, abrupt hills and
lofty mountains, a moist climate, and a grand and luxuriant
forest vegetation.

The few days I stayed here produced me several new insects,
but scarcely any birds. Butterflies and birds are in fact remarkably scarce in these forests. One may walk a whole day and
not see more than two or three species of either. In everything but beetles these eastern islands are very deficient
compared with the western (Java, Borneo, &c.), and much more
so if compared with the forests of South America, where twenty
or thirty species of butterflies may be caught every day, and on
very good days a hundred—a number we can hardly reach here
in months of unremitting search. In birds there is the same
difference. In most parts of tropical America we may always
find some species of woodpecker tanager, bushshrike, chatterer,
trogon, toucan, cuckoo, and tyrant-fly-catcher; and a few days'
active search will produce more variety than can be here met
with in as many months. Yet, along with this poverty of
individuals and of species, there are in almost every class and
order some one or two species of such extreme beauty or
singularity, as to vie with, or even surpass, anything that even
South America can produce.

One afternoon when I was arranging my insects, and surrounded by a crowd of wondering spectators, I showed one of
them how to look at a small insect with a hand-lens, which
caused such evident wonder that all the rest wanted to see it
too. I therefore fixed the glass firmly to a piece of soft wood at

the proper focus, and put under it a little spiny beetle of the genus Hispa, and then passed it round for examination. The excitement was immense. Some declared it was a yard long; others were frightened, and instantly dropped it, and all were as much astonished, and made as much shouting and gesticulation as children at a pantomime, or at a Christmas exhibition of the oxyhydrogen microscope. And all this excitement was produced by a little pocket-lens, an inch and a half focus, and therefore magnifying only four or five times, but which to their unaccustomed eyes appeared to enlarge a hundredfold.

On the last day of my stay here, one of my hunters succeeded in finding and shooting the beautiful Nicobar pigeon, of which I had been so long in search. None of the residents had ever seen it, which shows that it is rare and shy. My specimen was a female in beautiful condition, and the glossy coppery and green of its plumage, the snow-white tail and beautiful pendent feathers of the neck, were greatly admired. I subsequently obtained a specimen in New Guinea, and once saw it in the Kaióa islands. It is found also in some small islands near Macassar, in others near Borneo, and in the Nicobar islands, whence it receives its name. It is a ground-feeder, only going upon trees to roost, and is a very heavy, fleshy bird. This may account for the fact of its being found chiefly on very small islands, while in the western half of the Archipelago it seems entirely absent from the larger ones. Being a ground feeder it is subject to the attacks of carnivorous quadrupeds, which are not found in the very small islands. Its wide distribution over the whole length of the Archipelago, from extreme west to east, is, however, very extraordinary, since, with the exception of a few of the birds of prey, not a single land bird has so wide a range. Ground-feeding birds are generally deficient in power of extended flight, and this species is so bulky and heavy that it appears at first sight quite unable to fly a mile. A closer examination shows, however, that its wings are remarkably large, perhaps in proportion to its size larger than those of any other pigeon, and its pectoral muscles are immense. A fact communicated to me by the son of my friend Mr. Duivenboden of Ternate, would show that, in accordance with these peculiarities of structure, it possesses the power of flying long distances. Mr. D. established an oil factory on a small coral island, a hundred miles north of New Guinea, with no intervening land. After the island had been settled a year, and traversed in every direction, his son paid it a visit; and just as the schooner was coming to an anchor, a bird was seen flying from seaward which fell into the water exhausted before it could reach the shore. A boat was sent to pick it up, and it was found to be a Nicobar pigeon, which must have come from New Guinea, and flown a hundred miles, since no such bird previously inhabited the island.

This is certainly a very curious case of adaptation to an unusual and exceptional necessity. The bird does not ordinarily require great powers of flight, since it lives in the forest, feeds on fallen fruits, and roosts in low trees like other ground pigeons. The majority of the individuals, therefore, can never make full use of their enormously powerful wings, till the exceptional case occurs of an individual being blown out to sea, or driven to emigrate by the incursion of some carnivorous animal, or the pressure of scarcity of food. A modification exactly opposite to that which produced the wingless birds (the Apteryx, Cassowary, and Dodo), appears to have here taken place ; and it is curious that in both cases an insular habitat should have been the moving cause. The explanation is probably the same as that applied by Mr. Darwin to the case of the Madeira beetles, many of which are wingless, while some of the winged ones have the wings better developed than the same species on the continent. It was advantageous to these insects either never to fly at all, and thus not run the risk of being blown out to sea, or to fly so well as to be able either to return to land, or to migrate safely to the continent. Bad flying was worse than not flying at all. So, while in such islands as New Zealand and Mauritius, far from all land, it was safer for a ground-feeding bird not to fly at all, and the short-winged individuals continually surviving, prepared the way for a wingless group of birds ; in a vast Archipelago thickly strewn with islands and islets it was advantageous to be able occasionally to migrate, and thus the long and strong-winged varieties maintained their existence longest, and ultimately supplanted all others, and spread the race over the whole Archipelago.

Besides this pigeon, the only new bird I obtained during the trip was a rare goat-sucker (Batrachostomus crinifrons), the only species of the genus yet found in the Moluccas. Among my insects the best were the rare Pieris aruna, of a rich chrome yellow colour, with a black border and remarkable white antennæ—perhaps the very finest butterfly of the genus ; and a large black wasp-like insect, with immense jaws like a stag-beetle, which has been named Megachile pluto by Mr. F. Smith. I collected about a hundred species of beetles quite new to me, but mostly very minute, and also many rare and handsome ones which I had already found in Batchian. On the whole I was tolerably satisfied with my seventeen days' excursion, which was a very agreeable one, and enabled me to see a good deal of the island. I had hired a roomy boat, and brought with me a small table and my rattan chair. These were great comforts, as, wherever there was a roof, I could immediately instal myself, and work and eat at ease. When I could not find accommodation on shore I slept in the boat, which was always drawn up on the beach if we stayed for a few days at one spot.

On my return to Batchian I packed up my collections, and prepared for my return to Ternate. When I first came I had sent back my boat by the pilot, with two or three other men

who had been glad of the opportunity. I now took advantage
of a Government boat which had just arrived with rice for the
troops, and obtained permission to return in her, and accordingly
started on the 13th of April, having resided only a week short
of six months on the island of Batchian. The boat was one of
the kind called "Kora-kora," quite open, very low, and about
four tons burthen. It had outriggers of bamboo about five feet
off each side, which supported a bamboo platform extending the
whole length of the vessel. On the extreme outside of this sat
the twenty rowers, while within was a convenient passage fore
and aft. The middle portion of the boat was covered with a
thatch-house, in which baggage and passengers are stowed; the
gunwale was not more than a foot above water, and from the
great top and side weight, and general clumsiness, these boats
are dangerous in heavy weather, and are not unfrequently lost.
A triangle mast and mat sail carried us on when the wind was
favourable, which (as usual) it never was, although, according
to the monsoon, it ought to have been. Our water, carried in
bamboos, would only last two days, and as the voyage occupied
seven, we had to touch at a great many places. The captain
was not very energetic, and the men rowed as little as they
pleased, or we might have reached Ternate in three days, having
had fine weather and little wind all the way.

There were several passengers besides myself: three or four
Javanese soldiers, two convicts whose time had expired (one,
curiously enough, being the man who had stolen my cash-box
and keys), the schoolmaster's wife and a servant going on a visit
to Ternate, and a Chinese trader going to buy goods. We had
to sleep all together in the cabin, packed pretty close; but they
very civilly allowed me plenty of room for my mattress, and we
got on very well together. There was a little cook-house in the
bows, where we could boil our rice and make our coffee, every
one of course bringing his own provisions, and arranging his
meal-times as he found most convenient. The passage would
have been agreeable enough but for the dreadful "tom-toms,"
or wooden drums, which are beaten incessantly while the men
are rowing. Two men were engaged constantly at them,
making a fearful din the whole voyage. The rowers are men
sent by the Sultan of Ternate. They get about threepence a
day, and find their own provisions. Each man had a strong
wooden "betel" box, on which he generally sat, a sleeping-mat,
and a change of clothes—rowing naked, with only a sarong or a
waist-cloth. They sleep in their places, covered with their mat,
which keeps out the rain pretty well. They chew betel or smoke
cigarettes incessantly; eat dry sago and a little salt fish; seldom
sing while rowing, except when excited and wanting to reach a
stopping-place, and do not talk a great deal. They are mostly
Malays, with a sprinkling of Alfuros from Gilolo, and Papuans
from Guebe or Waigiou.

One afternoon we stayed at Makian; many of the men went

on shore, and a great deal of plantains, bananas, and other fruits were brought on board. We then went on a little way, and in the evening anchored again. When going to bed for the night, I put out my candle, there being still a glimmering lamp burning, and, missing my handkerchief, thought I saw it on a box which formed one side of my bed, and put out my hand to take it. I quickly drew back on feeling something cool and very smooth, which moved as I touched it. "Bring the light, quick," I cried ; "here's a snake." And there he was, sure enough, nicely coiled up, with his head just raised to inquire who had disturbed him. It was now necessary to catch or kill him neatly, or he would escape among the piles of miscellaneous luggage, and we should hardly sleep comfortably. One of the ex-convicts volunteered to catch him with his hand wrapped up in a cloth, but from the way he went about it I saw he was nervous and would let the thing go, so I would not allow him to make the attempt. I then got a chopping-knife, and carefully moving my insect nets, which hung just over the snake and prevented me getting a free blow, I cut him quietly across the back, holding him down while my boy with another knife crushed his head. On examination, I found he had large poison fangs, and it is a wonder he did not bite me when I first touched him.

Thinking it very unlikely that two snakes had got on board at the same time, I turned in and went to sleep ; but having all the time a vague, dreamy idea that I might put my hand on another one, I lay wonderfully still, not turning over once all night, quite the reverse of my usual habits. The next day we reached Ternate, and I ensconced myself in my comfortable house, to examine all my treasures, and pack them securely for the voyage home.

CHAPTER XXV.

CERAM, GORAM, AND THE MATABELLO ISLANDS.

(OCTOBER 1859 TO JUNE 1860.)

I LEFT Amboyna for my first visit to Ceram at three o'clock in the morning of October 29th, after having been delayed several days by the boat's crew, who could not be got together. Captain Van der Beck, who gave me a passage in his boat, had been running after them all day, and at midnight we had to search for two of my men who had disappeared at the last moment. One we found at supper in his own house, and rather tipsy with his parting libations of arrack, but the other was gone across the bay, and we were obliged to leave without him. We stayed some hours at two villages near the east end of Amboyna, at one of which we had to discharge some wood for the missionaries' house, and on the third afternoon reached

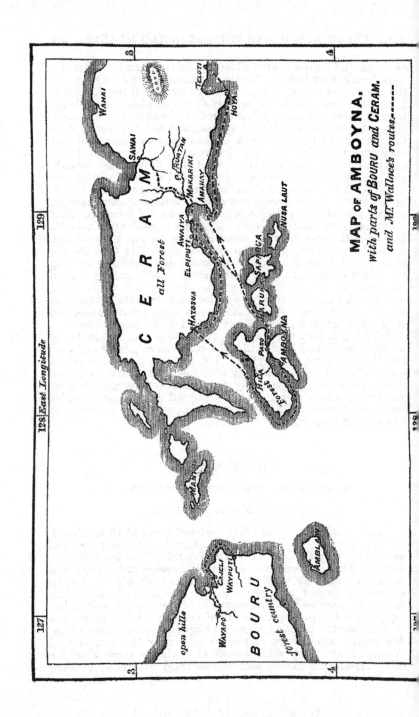

MAP of **AMBOYNA.**

with parts of *Bouru and* CERAM.

and Mr Wallace's routes------

Captain Van der Beck's plantation, situated at Hatosúa, in that part of Ceram opposite to the island of Amboyna. This was a clearing in flat and rather swampy forest, about twenty acres in extent, and mostly planted with cacao and tobacco. Besides a small cottage occupied by the workmen, there was a large shed for tobacco drying, a corner of which was offered me ; and thinking from the look of the place that I should find good collecting ground here, I fitted up temporary tables, benches, and beds, and made all preparations for some weeks' stay. A few days, however, served to show that I should be disappointed. Beetles were tolerably abundant, and I obtained plenty of fine long-horned Anthribidæ and pretty Longicorns, but they were mostly the same species as I had found during my first short visit to Amboyna. There were very few paths in the forest, which seemed poor in birds and butterflies, and day after day my men brought me nothing worth notice. I was therefore soon obliged to think about changing my locality, as I could evidently obtain no proper notion of the productions of the almost entirely unexplored island of Ceram by staying in this place.

I rather regretted leaving, because my host was one of the most remarkable men and most entertaining companions I had ever met with. He was a Fleming by birth, and, like so many of his countrymen, had a wonderful talent for languages. When quite a youth he had accompanied a Government official who was sent to report on the trade and commerce of the Mediterranean, and had acquired the colloquial language of every place they stayed a few weeks at. He had afterwards made voyages to St. Petersburg, and to other parts of Europe, including a few weeks in London, and had then come out to the East, where he had been for some years trading and speculating in the various islands. He now spoke Dutch, French, Malay, and Javanese, all equally well ; English with a very slight accent, but with perfect fluency, and a most complete knowledge of idiom, in which I often tried to puzzle him in vain. German and Italian were also quite familiar to him, and his acquaintance with European languages included Modern Greek, Turkish, Russian, and colloquial Hebrew and Latin. As a test of his power, I may mention that he had made a voyage to the out-of-the-way island of Salibaboo, and had stayed there trading a few weeks. As I was collecting vocabularies, he told me he thought he could remember some words, and dictated a considerable number. Some time after I met with a short list of words taken down in those islands, and in every case they agreed with those he had given me. He used to sing a Hebrew drinking-song, which he had learned from some Jews with whom he had once travelled, and astonished by joining in their conversation ; and he had a never-ending fund of tale and anecdote about the people he had met and the places he had visited.

In most of the villages of this part of Ceram are schools and native schoolmasters, and the inhabitants have been long converted to Christianity. In the larger villages there are European missionaries; but there is little or no external difference between the Christian and Alfuro villages, nor, as far as I have seen, in their inhabitants. The people seem more decidedly Papuan than those of Gilolo. They are darker in colour, and a number of them have the frizzly Papuan hair; their features also are harsh and prominent, and the women in particular are far less engaging than those of the Malay race. Captain Van der Beck was never tired of abusing the inhabitants of these Christian villages as thieves, liars, and drunkards, besides being incorrigibly lazy. In the city of Amboyna my friends Doctors Mohnike and Doleschall, as well as most of the European residents and traders, made exactly the same complaint, and would rather have Mahometans for servants, even if convicts, than any of the native Christians. One great cause of this is the fact that with the Mahometans temperance is a part of their religion, and has become so much a habit that practically the rule is never transgressed. One fertile source of want, and one great incentive to idleness and crime, is thus present with the one class, but absent in the other; but besides this the Christians look upon themselves as nearly the equals of the Europeans, who profess the same religion, and as far superior to the followers of Islam, and are therefore prone to despise work, and to endeavour to live by trade, or by cultivating their own land. It need hardly be said that with people in this low state of civilization religion is almost wholly ceremonial, and that neither are the doctrines of Christianity comprehended, nor its moral precepts obeyed. At the same time, as far as my own experience goes, I have found the better class of "Orang Sirani" as civil, obliging, and industrious as the Malays, and only inferior to them from their tendency to get intoxicated.

Having written to the Assistant Resident of Saparua (who has jurisdiction over the opposite part of the coast of Ceram) for a boat to pursue my journey, I received one rather larger than necessary with a crew of twenty men. I therefore bade adieu to my kind friend Captain Van der Beck, and left on the evening after its arrival for the village of Elpiputi, which we reached in two days. I had intended to stay here, but not liking the appearance of the place, which seemed to have no virgin forest near it, I determined to proceed about twelve miles further up the bay of Amahay, to a village recently formed, and inhabited by indigenes from the interior, and where some extensive cacao plantations were being made by some gentlemen of Amboyna. I reached the place (called Awaiya) the same afternoon, and with the assistance of Mr. Peters (the manager of the plantations) and the native chief, obtained a small house, got all my things on shore, and paid and discharged my twenty boatmen, two of whom had almost driven me to distraction by beating tom-toms the whole voyage.

I found the people here very nearly in a state of nature, and going almost naked. The men wear their frizzly hair gathered into a flat circular knot over the left temple, which has a very knowing look, and in their ears cylinders of wood as thick as one's finger, and coloured red at the ends. Armlets and anklets of woven grass or of silver, with necklaces of beads or of small fruits, complete their attire. The women wear similar ornaments, but have their hair loose. All are tall, with a dark brown skin, and well marked Papuan physiognomy. There is an Amboyna schoolmaster in the village, and a good number of children attend school every morning. Such of the inhabitants as have become Christians may be known by their wearing their hair loose, and adopting to some extent the native Christian dress—trousers and a loose shirt. Very few speak Malay, all these coast villages having been recently formed by inducing natives to leave the inaccessible interior. In all the central part of Ceram there now remains only one populous village in the mountains. Towards the east and the extreme west are a few others, with which exceptions all the inhabitants of Ceram are collected on the coast. In the northern and eastern districts they are mostly Mahometans, while on the south-west coast, nearest Amboyna, they are nominal Christians.

In all this part of the Archipelago, the Dutch make very praiseworthy efforts to improve the condition of the aborigines by establishing schoolmasters in every village (who are mostly natives of Amboyna or Saparua, who have been instructed by the resident missionaries), and by employing native vaccinators to prevent the ravages of smallpox. They also encourage the settlement of Europeans, and the formation of new plantations of cacao and coffee, one of the best means of raising the condition of the natives, who thus obtain work at fair wages, and have the opportunity of acquiring something of European tastes and habits.

My collections here did not progress much better than at my former station, except that butterflies were a little more plentiful, and some very fine species were to be found in the morning on the sea-beach, sitting so quietly on the wet sand that they could be caught with the fingers. In this way I had many fine specimens of Papilios brought me by the children. Beetles, however, were scarce, and birds still more so, and I began to think that the handsome species which I had so often heard were found in Ceram must be entirely confined to the eastern extremity of the island.

A few miles further north, at the head of the Bay of Amahay, is situated the village of Makariki, from whence there is a native path quite across the island to the north coast. My friend, Mr. Rosenberg, whose acquaintance I had made at New Guinea, and who was now the Government superintendent of all this part of Ceram, returned from Wahai, on the north coast, after I had been three weeks at Awaiya, and showed me some fine butterflies he had obtained on the mountain streams in the interior. He indicated a spot about the centre of the island where he thought I might advantageously stay a few days. I accordingly visited Makariki with him the next day, and he instructed the chief of the village to furnish me with men to carry my baggage, and accompany me on my excursion. As the people of the village wanted to be at home on Christmas-day, it was necessary to start as soon as possible; so we agreed that the men should be ready in two days, and I returned to make my arrangements.

I put up the smallest quantity of baggage possible for a six days' trip, and on the morning of December 18th we left Makariki, with six men carrying my baggage and their own pro-visions, and a lad from Awaiya, who was accustomed to catch butterflies for me. My two Amboyna hunters I left behind to shoot and skin what birds they could while I was away. Quitting the village, we first walked briskly for an hour through a dense tangled undergrowth, dripping wet from a storm of the previous night, and full of mud holes. After crossing several small streams we reached one of the largest rivers in Ceram, called Ruatan, which it was necessary to cross. It was both deep and rapid. The baggage was first taken over, parcel by parcel, on the men's heads, the water reaching nearly up to their armpits, and then two men returned to assist me. The water was above my waist, and so strong that I should certainly have been carried off my feet had I attempted to cross alone; and it was a matter of astonishment to me how the men could give me any assistance, since I found the greatest diffi-culty in getting my foot down again when I had once moved it off the bottom. The greater strength and grasping power of their feet, from going always barefoot, no doubt gave them a surer footing in the rapid water.

After well wringing out our wet clothes and putting them on, we again proceeded along a similar narrow forest track as before, choked with rotten leaves and dead trees, and in the more open parts overgrown with tangled vegetation. Another hour brought us to a smaller stream flowing in a wide gravelly bed, up which our road lay. Here we stayed half an hour to breakfast, and then went on, continually crossing the stream, or walking on its stony and gravelly banks, till about noon, when it became rocky and enclosed by low hills. A little further we entered a regular mountain-gorge, and had to clamber over rocks, and every moment cross and recross the water, or take short cuts through the forest. This was fatiguing work; and about three in the afternoon, the sky being over-cast, and thunder in the mountains indicating an approaching storm, we had to look out for a camping place, and soon after reached one of Mr. Rosenberg's old ones. The skeleton of his little sleeping-hut remained, and my men cut leaves and made a hasty roof just as the rain commenced. The baggage was covered over with leaves, and the men sheltered themselves as they could till the storm was over, by which time a flood came down the river, which effectually stopped our further march, even had we wished to proceed. We then lighted fires ; I made some coffee, and my men roasted their fish and plantains, and as soon as it was dark, we made ourselves comfortable for the night.

Starting at six the next morning, we had three hours of the same kind of walking, during which we crossed the river at least thirty or forty times, the water being generally knee-deep. This brought us to a place where the road left the stream, and here we stopped to breakfast. We then had a long walk over the mountain, by a tolerable path, which reached an elevation of about fifteen hundred feet above the sea. Here I noticed one of the smallest and most elegant tree ferns I had ever seen, the stem being scarcely thicker than my thumb, yet reaching a height of fifteen or twenty feet. I also caught a new butterfly of the genus Pieris, and a magnificent female specimen of Papilio gambrisius, of which I had hitherto only found the males, which are smaller and very different in colour. Descending the other side of the ridge, by a very steep path, we reached another river at a spot which is about the centre of the island, and which was to be our resting-place for two or three days.

In a couple of hours my men had built a little sleeping-shed for me, about eight feet by four, with a bench of split poles, they themselves occupying two or three smaller ones, which had been put up by former passengers.

The river here was about twenty yards wide, running over a pebbly and sometimes a rocky bed, and bordered by steep hills with occasionally flat swampy spots between their base and the stream. The whole country was one dense, unbroken, and very damp and gloomy virgin forest. Just at our resting-place there was a little bush-covered island in the middle of the channel, so that the opening in the forest made by the river was wider than usual, and allowed a few gleams of sunshine to penetrate. Here there were several handsome butterflies flying about, the finest of which, however, escaped me, and I never saw it again during my stay. In the two days and a half which we remained here, I wandered almost all day up and down the stream, searching after butterflies, of which I got, in all, fifty or sixty specimens, with several species quite new to me. There were many others which I saw only once, and did not capture, causing me to regret that there was no village in these interior valleys where I could stay a month. In the early part of each morning I went out with my gun in search of birds, and two of my men were out almost all day after deer ; but we were all equally unsuccessful, getting absolutely nothing the whole time we were in the forest. The only good bird seen was the fine Amboyna lory, but these were always too high to shoot ; besides this, the great Moluccan hornbill, which I did not want, was almost the only bird met with. I saw not a single ground-thrush, or kingfisher, or pigeon ;

and, in fact, have never been in a forest so utterly desert of animal life as this appeared to be. Even in all other groups of insects, except butterflies, there was the same poverty. I had hoped to find some rare tiger beetles, as I had done in similar situations in Celebes ; but, though I searched closely in forest, river-bed, and mountain-brook, I could find nothing but the two common Amboyna species. Other beetles there were absolutely none.

The constant walking in water, and over rocks and pebbles, quite destroyed the two pair of shoes I brought with me, so that, on my return, they actually fell to pieces, and the last day I had to walk in my stockings very painfully, and reached home quite lame. On our way back from Makariki, as on our way there, we had storm and rain at sea, and we arrived at Awaiya late in the evening, with all our baggage drenched, and ourselves thoroughly uncomfortable. All the time I had been in Ceram I had suffered much from the irritating bites of an invisible acarus, which is worse than mosquitoes, ants, and every other pest, because it is impossible to guard against them. This last journey in the forest left me covered from head to foot with inflamed lumps, which, after my return to Amboyna, produced a serious disease, confining me to the house for nearly two months, —a not very pleasant memento of my first visit to Ceram, which terminated with the year 1859.

It was not till the 24th of February, 1860, that I started again, intending to pass from village to village along the coast, staying where I found a suitable locality. I had a letter from the Governor of the Moluccas, requesting all the chiefs to supply me with boats and men to carry me on my journey. The first boat took me in two days to Amahay, on the opposite side of the bay to Awaiya. The chief here, wonderful to relate, did not make any excuses for delay, but immediately ordered out the boat which was to carry me on, put my baggage on board, set up mast and sails after dark, and had the men ready that night ; so that we were actually on our way at five the next morning— a display of energy and activity I scarcely ever saw before in a native chief on such an occasion. We touched at Cepa, and stayed for the night at Tamilan, the first two Mahometan villages on the south coast of Ceram. The next day, about noon, we reached Hoya, which was as far as my present boat and crew were going to take me. The anchorage is about a mile east of the village, which is faced by coral reefs, and we had to wait for the evening tide to move up and unload the boat into the strange rotten wooden pavilion kept for visitors.

There was no boat here large enough to take my baggage ; and although two would have done very well, the Rajah insisted upon sending four. The reason of this I found was, that there were four small villages under his rule, and by sending a boat from each he would avoid the difficult task of choosing two and letting off the others. I was told that at the next village of

Teluti there were plenty of Alfuros, and that I could get abundance of lories and other birds. The Rajah declared that black and yellow lories and black cockatoos were found there ; but I am inclined to think he knew very well he was telling me lies, and that it was only a scheme to satisfy me with his plan of taking me to that village, instead of a day's journey further on, as I desired. Here, as at most of the villages, I was asked for spirits, the people being mere nominal Mahometans, who confine their religion almost entirely to a disgust at pork, and a few other forbidden articles of food. The next morning, after much trouble, we got our cargoes loaded, and had a delightful row across the deep bay of Teluti, with a view of the grand central mountain-range of Ceram. Our four boats were rowed by sixty men, with flags flying and tom-toms beating, as well as very vigorous shouting and singing to keep up their spirits. The sea was smooth, the morning bright, and the whole scene very exhilarating. On landing, the Orang-kaya and several of the chief men, in gorgeous silk jackets, were waiting to receive us, and conducted me to a house prepared for my reception, where I determined to stay a few days, and see if the country round produced anything new.

My first inquiries were about the lories, but I could get very little satisfactory information. The only kinds known were the ring-necked lory and the common red and green lorikeet, both common at Amboyna. Black lories and cockatoos were quite unknown. The Alfuros resided in the mountains five or six days' journey away, and there were only one or two live birds to be found in the village, and these were worthless. My hunters could get nothing but a few common birds ; and notwithstanding fine mountains, luxuriant forests, and a locality a hundred miles eastward, I could find no new insects, and extremely few even of the common species of Amboyna and West Ceram. It was evidently no use stopping at such a place, and I was determined to move on as soon as possible.

The village of Teluti is populous, but straggling and very dirty. Sago trees here cover the mountain side, instead of growing as usual in low swamps but a closer examination shows that they grow in swampy patches, which have formed among the loose rocks that cover the ground, and which are kept constantly full of moisture by the rains, and by the abundance of rills which trickle down among them. This sago forms almost the whole subsistence of the inhabitants, who appear to cultivate nothing but a few small patches of maize and sweet potatoes. Hence, as before explained, the scarcity of insects. The Orang-kaya has fine clothes, handsome lamps, and other expensive European goods, yet lives every day on sago and fish as miserably as the rest.

After three days in this barren place I left on the morning of March 6th, in two boats of the same size as those which had brought me to Teluti. With some difficulty I had obtained

permission to take these boats on to Tobo, where I intended to
stay a while, and therefore got on pretty quickly, changing men
at the village of Laiemu, and arriving in a heavy rain at Ahtiago.
As there was a good deal of surf here, and likely to be more if
the wind blew hard during the night, our boats were pulled up
on the beach; and after supping at the Orang-kaya's house,
and writing down a vocabulary of the language of the Alfuros,
who live in the mountains inland, I returned to sleep in the
boat. Next morning we proceeded, changing men at Warenama,
and again at Hatometen, at both of which places there was
much surf and no harbour, so that the men had to go on shore
and come on board by swimming. Arriving in the evening of
March 7th at Batuassa, the first village belonging to the Rajah
of Tobo, and under the government of Banda, the surf was very
heavy, owing to a strong westward swell. We therefore rounded
the rocky point on which the village was situated, but found it
very little better on the other side. We were obliged, however,
to go on shore here; and waiting till the people on the beach
had made preparations, by placing a row of logs from the
water's edge on which to pull up our boats, we rowed as quickly
as we could straight on to them, after watching till the heaviest
surfs had passed. The moment we touched ground our men all
jumped out, and, assisted by those on shore, attempted to haul
up the boat high and dry, but not having sufficient hands, the
surf repeatedly broke into the stern. The steepness of the
beach, however, prevented any damage being done, and the
other boat having both crews to haul at it, was got up without
difficulty.

The next morning, the water being low, the breakers were at
some distance from shore, and we had to watch for a smooth
moment after bringing the boats to the water's edge, and so got
safely out to sea. At the two next villages, Tobo and Ossong,
we also took in fresh men, who came swimming through the
surf; and at the latter place the Rajah came on board and
accompanied me to Kissa-laut, where he has a house which he
lent me during my stay. Here again was a heavy surf, and it
was with great difficulty we got the boats safely hauled up. At
Amboyna I had been promised at this season a calm sea and the
wind off shore, but in this case, as in every other, I had been
unable to obtain any reliable information as to the winds and
seasons of places distant two or three days' journey. It appears,
however, that owing to the general direction of the island of
Ceram (E.S.E. and W.N.W.), there is a heavy surf and scarcely
any shelter on the south coast during the west monsoon, when
alone a journey to the eastward can be safely made; while
during the east monsoon, when I proposed to return along the
north coast to Wahai, I should probably find that equally
exposed and dangerous. But although the general direction of
the west monsoon in the Banda sea causes a heavy swell, with
bad surf on the coast, yet we had little advantage of the wind;

for, owing I suppose to the numerous bays and headlands, we had contrary south-east or even due east winds all the way, and had to make almost the whole distance from Amboyna by force of rowing. We had therefore all the disadvantages, and none of the advantages, of this west monsoon, which I was told would insure me a quick and pleasant journey.

I was delayed at Kissa-laut just four weeks, although after the first three days I saw that it would be quite useless for me to stay, and begged the Rajah to give me a prau and men to carry me on to Goram. But instead of getting one close at hand, he insisted on sending several miles off; and when after many delays it at length arrived, it was altogether unsuitable and too small to carry my baggage. Another was then ordered to be brought immediately, and was promised in three days, but double that time elapsed and none appeared, and we were obliged at length to get one at the adjoining village, where it might have been so much more easily obtained at first. Then came caulking and covering over, and quarrels between the owner and the Rajah's men, which occupied more than another ten days, during all which time I was getting absolutely nothing, finding this part of Ceram a perfect desert in zoology, although a most beautiful country, and with a very luxuriant vegetation. It was a complete puzzle, which to this day I have not been able to understand ; the only thing I obtained worth notice during my month's stay here being a few good land shells.

At length, on April 4th, we succeeded in getting away in our little boat of about four tons burthen, in which my numerous boxes were with difficulty packed so as to leave sleeping and cooking room. The craft could not boast an ounce of iron or a foot of rope in any part of its construction, nor a morsel of pitch or paint in its decoration. The planks were fastened together in the usual ingenious way with pegs and rattans. The mast was a bamboo triangle, requiring no shrouds, and carrying a long mat sail ; two rudders were hung on the quarters by rattans, the anchor was of wood, and a long and thick rattan served as a cable. Our crew consisted of four men, whose sole accommodation was about three feet by four in the bows and stern, with the sloping thatch roof to stretch themselves upon for a change. We had nearly a hundred miles to go, fully exposed to the swell of the Banda sea, which is sometimes very considerable ; but we luckily had it calm and smooth, so that we made the voyage in comparative comfort.

On the second day we passed the eastern extremity of Ceram, formed of a group of hummocky limestone hills ; and, sailing by the islands of Kwammer and Keffing, both thickly inhabited, came in sight of the little town of Kilwaru, which appears to rise out of the sea like a rustic Venice. This place has really a most extraordinary appearance, as not a particle of land or vegetation can be seen, but a long way out at sea a large village seems to float upon the water. There is of course a small island

of several acres in extent ; but the houses are built so closely all
round it upon piles in the water, that it is completely hidden.
It is a place of great traffic, being the emporium for much of the
produce of these Eastern seas, and is the residence of many
Bugis and Ceramese traders, and appears to have been chosen
on account of its being close to the only deep channel between
the extensive shoals of Ceram-laut and those bordering the east
end of Ceram. We now had contrary east winds, and were
obliged to pole over the shallow coral reefs of Ceram-laut for
nearly thirty miles. The only danger of our voyage was just at
its termination, for as we were rowing towards Manowolko, the
largest of the Goram group, we were carried out so rapidly by
a strong westerly current, that I was almost certain at one time
we should pass clear of the island ; in which case our situation
would have been both disagreeable and dangerous, as, with the
east wind which had just set in, we might have been unable to
return for many days, and we had not a day's water on board.
At the critical moment I served out some strong spirits to my
men, which put fresh vigour into their arms, and carried us out
of the influence of the current before it was too late.

MANOWOLKO, GORAM GROUP.

On arriving at Manowolko, we found the Rajah was at the
opposite island of Goram ; but he was immediately sent for, and
in the meantime a large shed was given for our accommodation.
At night the Rajah came, and the next day I had a visit from
him, and found, as I expected, that I had already made his
acquaintance three years before at Aru. He was very friendly,
and we had a long talk ; but when I begged for a boat and men
to take me on to Ké, he made a host of difficulties. There were
no praus, as all had gone to Ké or Aru ; and even if one were
found, there were no men, as it was the season when all were
away trading. But he promised to see about it, and I was
obliged to wait. For the next two or three days there was
more talking and more difficulties were raised, and I had time
to make an examination of the island and the people.

Manowolko is about fifteen miles long, and is a mere upraised
coral-reef. Two or three hundred yards inland rise cliffs of
coral rock, in many parts perpendicular, and one or two
hundred feet high ; and this, I was informed, is characteristic of
the whole island, in which there is no other kind of rock, and
no stream of water. A few cracks and chasms furnish paths to
the top of these cliffs, where there is an open undulating country,
in which the chief vegetable grounds of the inhabitants are
situated.

The people here—at least the chief men—were of a much
purer Malay race than the Mahometans of the mainland of
Ceram, which is perhaps due to there having been no indigenes
on these small islands when the first settlers arrived. In

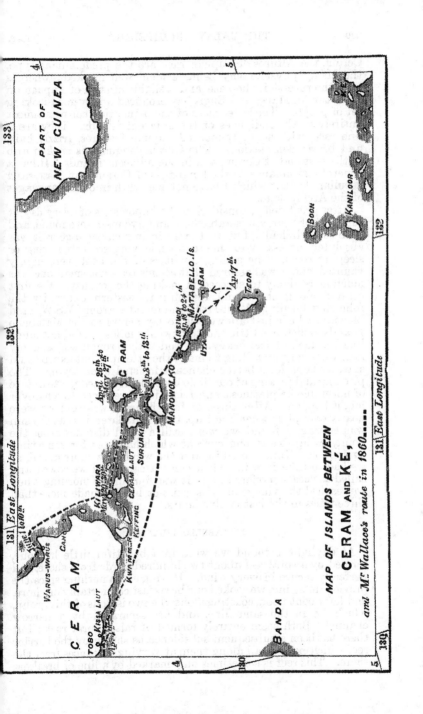

MAP OF ISLANDS BETWEEN
CERAM AND KÉ,
and Mr Wallace's route in 1860.------

Ceram, the Alfuros of Papuan race are the predominant type, the Malay physiognomy being seldom well marked; whereas here the reverse is the case, and a slight infusion of Papuan on a mixture of Malay and Bugis has produced a very good-looking set of people. The lower class of the population consists almost entirely of the indigenes of the adjacent islands. They are a fine race, with strongly-marked Papuan features, frizzly hair, and brown complexions. The Goram language is spoken also at the east end of Ceram, and in the adjacent islands. It has a general resemblance to the languages of Ceram, but possesses a peculiar element which I have not met with in other languages of the Archipelago.

After great delay, considering the importance of every day at this time of year, a miserable boat and five men were found, and with some difficulty I stowed away in it such baggage as it was absolutely necessary for me to take, leaving scarcely sitting or sleeping room. The sailing qualities of the boat were highly vaunted, and I was assured that at this season a small one was much more likely to succeed in making the journey. We first coasted along the island, reaching its eastern extremity the following morning (April 11th), and found a strong W.S.W. wind blowing, which just allowed us to lay across to the Matabello Islands, a distance little short of twenty miles. I did not much like the look of the heavy sky and rather rough sea, and my men were very unwilling to make the attempt; but as we could scarcely hope for a better chance, I insisted upon trying. The pitching and jerking of our little boat soon reduced me to a state of miserable helplessness, and I lay down, resigned to whatever might happen. After three or four hours, I was told we were nearly over; but when I got up, two hours later, just as the sun was setting, I found we were still a good distance from the point, owing to a strong current which had been for some time against us. Night closed in, and the wind drew more ahead, so we had to take in sail. Then came a calm, and we rowed and sailed as occasion offered; and it was four in the morning when we reached the village of Kissiwoi, not having made more than three miles in the last twelve hours.

MATABELLO ISLANDS.

At daylight I found we were in a beautiful little harbour, formed by a coral reef about two hundred yards from shore, and perfectly secure in every wind. Having eaten nothing since the previous morning, we cooked our breakfast comfortably on shore, and left about noon, coasting along the two islands of this group, which lie in the same line, and are separated by a narrow channel. Both seem entirely formed of raised coral rock; but there has been a subsequent subsidence, as shown by the barrier reef which extends all along them at varying distances from the shore. This reef is sometimes only marked by a line of breakers when there is a little swell on the sea ; in other places there is a ridge of dead coral above the water, which is here and there high enough to support a few low bushes. This was the first example I had met with of a true barrier reef due to subsidence, as has been so clearly shown by Mr. Darwin. In a sheltered archipelago they will seldom be distinguishable, from the absence of those huge rolling waves and breakers which in the wide ocean throw up a barrier of broken coral far above the usual high-water mark, while here they rarely rise to the surface.

On reaching the end of the southern island, called Uta, we were kept waiting two days for a wind that would enable us to pass over to the next island, Teor, and I began to despair of ever reaching Ké, and determined on returning. We left with a south wind, which suddenly changed to north-east, and induced me to turn again southward in the hopes that this was the commencement of a few days' favourable weather. We sailed on very well in the direction of Teor for about an hour, after which the wind shifted to W.S.W., and we were driven much out of our course, and at nightfall found ourselves in the open sea, and full ten miles to leeward of our destination. My men were now all very much frightened, for if we went on we might be a week at sea in our little open boat, laden almost to the water's edge ; or we might drift on to the coast of New Guinea, in which case we should most likely all be murdered. I could not deny these probabilities, and although I showed them that we could not get back to our starting-point with the wind as it was, they insisted upon returning. We accordingly put about, and found that we could lay no nearer to Uta than to Teor ; however, by great good luck, about ten o'clock we hit upon a little coral island, and lay under its lee till morning, when a favourable change of wind brought us back to Uta, and by evening (April 18th) we reached our first anchorage in Matabello, where I resolved to stay a few days, and then return to Goram. It was with much regret that I gave up my trip to Ké and the intervening islands, which I had looked forward to as likely to make up for my disappointment in Ceram, since my short visit on my voyage to Aru had produced me so many rare and beautiful insects.

The natives of Matabello are almost entirely occupied in making cocoa-nut oil, which they sell to the Bugis and Goram traders, who carry it to Banda and Amboyna. The rugged coral rock seems very favourable to the growth of the cocoa-nut palm, which abounds over the whole island to the very highest points, and produces fruit all the year round. Along with it are great numbers of the areca or betel-nut palm, the nuts of which are sliced, dried, and ground into a paste, which is much used by the betel-chewing Malays and Papuans. All the little children here, even such as can just run alone, carried between their lips a mass of the nasty-looking red paste, which is even more disgusting than to see them at the same age smoking cigars, which is very common even before they are weaned. Cocoa-nuts, sweet potatoes, an occasional sago cake, and the refuse nut after the oil has been extracted by boiling, form the chief sustenance of these people; and the effect of this poor and unwholesome diet is seen in the frequency of eruptions and scurvy skin diseases, and the numerous sores that disfigure the faces of the children.

The villages are situated on high and rugged coral peaks, only accessible by steep narrow paths, with ladders and bridges over yawning chasms. They are filthy with rotten husks and oil refuse, and the huts are dark, greasy, and dirty in the extreme. The people are wretched, ugly, dirty savages, clothed in un-changed rags, and living in the most miserable manner; and as every drop of fresh water has to be brought up from the beach, washing is never thought of; yet they are actually wealthy, and have the means of purchasing all the necessaries and luxuries of life. Fowls are abundant, and eggs were given me whenever I visited the villages, but these are never eaten, being looked upon as pets or as merchandise. Almost all of the women wear massive gold earrings, and in every village there are dozens of small bronze cannon lying about on the ground, although they have cost on the average perhaps 10l. apiece. The chief men of each village came to visit me, clothed in robes of silk and flowered satin, though their houses and their daily fare are no better than those of the other inhabitants. What a contrast between these people and such savages as the best tribes of hill Dyaks in Borneo, or the Indians of the Uaupes in South America, living on the banks of clear streams, clean in their persons and their houses, with abundance of wholesome food, and exhibiting its effect in healthy skins and beauty of form and feature! There is in fact almost as much difference between the various races of savage as of civilized peoples, and we may safely affirm that the better specimens of the former are much superior to the lower examples of the latter class.

One of the few luxuries of Matabello is the palm wine, which is the fermented sap from the flower stems of the cocoa-nut. It is really a very nice drink, more like cider than beer, though quite as intoxicating as the latter. Young cocoa-nuts are also very abundant, so that anywhere in the island it is only neces-sary to go a few yards to find a delicious beverage by climbing up a tree for it. It is the water of the young fruit that is drunk before the pulp has hardened; it is then more abundant, clear, and refreshing, and the thin coating of gelatinous pulp is thought a great luxury. The water of full-grown cocoa-nuts is always thrown away as undrinkable, although it is delicious in comparison with that of the old dry nuts which alone we obtain in this country. The cocoa-nut pulp I did not like at first; but fruits are so scarce, except at particular seasons, that one soon learns to appreciate anything of a fruity nature.

Many persons in Europe are under the impression that fruits

of delicious flavour abound in the tropical forests, and they will no doubt be surprised to learn that the truly wild fruits of this grand and luxuriant archipelago, the vegetation of which will vie with that of any part of the world, are in almost every island inferior in abundance and quality to those of Britain. Wild strawberries and raspberries are found in some places, but they are such poor tasteless things as to be hardly worth eating, and there is nothing to compare with our blackberries and whortleberries. The kanary-nut may be considered equal to a hazel-nut, but I have met with nothing else superior to our crabs, our haws, beech-nuts, wild plums, and acorns; fruits which would be highly esteemed by the natives of these islands, and would form an important part of their sustenance. All the fine tropical fruits are as much cultivated productions as our apples, peaches, and plums, and their wild prototypes, when found, are generally either tasteless or uneatable.

The people of Matabello, like those of most of the Mahometan villages of East Ceram and Goram, amused me much by their strange ideas concerning the Russian war. They believe that the Russians were not only most thoroughly beaten by the Turks, but were absolutely conquered, and all converted to Islamism! And they can hardly be convinced that such is not the case, and that had it not been for the assistance of France and England, the poor Sultan would have fared ill. Another of their notions is, that the Turks are the largest and strongest people in the world—in fact a race of giants; that they eat enormous quantities of meat, and are a most ferocious and irresistible nation. Whence such strangely incorrect opinions could have arisen it is difficult to understand, unless they are derived from Arab priests, or hadjis returned from Mecca, who may have heard of the ancient prowess of the Turkish armies when they made all Europe tremble, and suppose that their character and warlike capacity must be the same at the present time.

GORAM.

A steady south-east wind having set in, we returned to Manowolko on the 25th of April, and the day after crossed over to Ondor, the chief village of Goram.

Around this island extends, with few interruptions, an encircling coral reef about a quarter of a mile from the shore, visible as a stripe of pale green water, but only at very lowest ebb-tides showing any rock above the surface. There are several deep entrances through this reef, and inside it there is good anchorage in all weathers. The land rises gradually to a moderate height, and numerous small streams descend on all sides. The mere existence of these streams would prove that the island was not entirely coralline, as in that case all the water would sink through the porous rock as it does at Manowolko and Matabello; but we have more positive proof in the pebbles and stones of their beds, which exhibit a variety of stratified crystalline rocks. About a hundred yards from the beach rises a wall of coral rock, ten or twenty feet high, above which is an undulating surface of rugged coral, which slopes *downward* towards the interior, and then after a slight ascent is bounded by a second wall of coral. Similar walls occur higher up, and coral is found on the highest part of the island.

This peculiar structure teaches us that before the coral was formed land existed in this spot; that this land sunk gradually beneath the waters, but with intervals of rest, during which encircling reefs were formed around it at different elevations; that it then rose to above its present elevation, and is now again sinking. We infer this, because encircling reefs are a proof of subsidence; and if the island were again elevated about a hundred feet, what is now the reef and the shallow sea within it would form a wall of coral rock, and an undulating coralline plain, exactly similar to those that still exist at various altitudes up to the summit of the island. We learn also that these changes have taken place at a comparatively recent epoch, for the surface of the coral has scarcely suffered from the action of the weather, and hundreds of sea-shells, exactly resembling those still found upon the beach, and many of them retaining their gloss and even their colour, are scattered over the surface of the island to near its summit.

Whether the Goram group formed originally part of New Guinea or of Ceram it is scarcely possible to determine, and its productions will throw little light upon the question, if, as I suppose, the islands have been entirely submerged within the epoch of existing species of animals, as in that case it must owe its present fauna and flora to recent immigration from surrounding lands; and with this view its poverty in species very well agrees. It possesses much in common with East Ceram, but at the same time has a good deal of resemblance to the Ké Islands and Banda. The fine pigeon, Carpophaga concinna, inhabits Ké, Banda, Matabello, and Goram, and is replaced by a distinct species, C. neglecta, in Ceram. The insects of these four islands have also a common *facies*—facts which seem to indicate that some more extensive land has recently disappeared from the area they now occupy, and has supplied them with a few of its peculiar productions.

The Goram people (among whom I stayed a month) are a race of traders. Every year they visit the Tenimber, Ké, and Aru Islands, the whole north-west coast of New Guinea from Oetanata to Salwatty, and the islands of Waigiou and Mysol. They also extend their voyages to Tidore and Ternate, as well as to Banda and Amboyna. Their praus are all made by that wonderful race of boat-builders, the Ké islanders, who annually turn out some hundreds of boats, large and small, which can hardly be surpassed for beauty of form and goodness of workmanship. They trade chiefly in tripang, the medicinal mussoi bark, wild nutmegs, and tortoise-shell, which they sell to the Bugis traders at Ceram-laut or Aru, few of them caring to take their products to any other market. In other respects they are a lazy race, living very poorly, and much given to opium smoking. The only native manufactures are sail-matting, coarse cotton cloth, and pandanus-leaf boxes, prettily stained and ornamented with shell-work.

In the island of Goram, only eight or ten miles long, there are about a dozen Rajahs, scarcely better off than the rest of the inhabitants, and exercising a mere nominal sway, except when any order is received from the Dutch Government, when, being backed by a higher power, they show a little more strict authority. My friend the Rajah of Ammer (commonly called Rajah of Goram) told me that a few years ago, before the Dutch had interfered in the affairs of the island, the trade was not carried on so peaceably as at present, rival praus often fighting when on the way to the same locality, or trafficking in the same village. Now such a thing is never thought of—one of the good effects of the superintendence of a civilized government. Disputes between villages are still, however, sometimes settled by fighting, and I one day saw about fifty men, carrying long guns and heavy cartridge-belts, march through the village. They had come from the other side of the island on some question of trespass or boundary, and were prepared for war if peaceable negotiations should fail.

While at Manowolko I had purchased for 100 florins (9*l*.) a small prau, which was brought over the next day, as I was informed it was more easy to have the necessary alterations made in Goram, where several Ké workmen were settled.

As soon as we began getting my prau ready I was obliged to give up collecting, as I found that unless I was constantly on the spot myself very little work would be done. As I proposed making some long voyages in this boat, I determined to fit it up conveniently, and was obliged to do all the inside work myself, assisted by my two Amboynese boys. I had plenty of visitors, surprised to see a white man at work, and much astonished at the novel arrangements I was making in one of their native vessels. Luckily I had a few tools of my own, including a small saw and some chisels, and these were now severely tried, cutting and fitting heavy iron-wood planks for the flooring and the posts that support the triangular mast. Being of the best London make, they stood the work well, and without them it would have been impossible for me to have finished my boat with half the neatness, or in double the time. I had a Ké workman to put in new ribs, for which I bought nails of a Bugis trader, at 8*d*. a pound. My gimlets were, however, too small; and having no augers we were obliged to bore all the holes with hot irons, a most tedious and unsatisfactory operation.

Five men had engaged to work at the prau till finished, and then go with me to Mysol, Waigiou, and Ternate. Their ideas

of work were, however, very different from mine, and I had
immense difficulty with them ; seldom more than two or three
coming together, and a hundred excuses being given for working
only half a day when they did come. Yet they were constantly
begging advances of money, saying they had nothing to eat.
When I gave it them they were sure to stay away the next day,
and when I refused any further advances some of them declined
working any more. As the boat approached completion my
difficulties with the men increased. The uncle of one had
commenced a war, or sort of faction fight, and wanted his
assistance ; another's wife was ill, and would not let him come ;
a third had fever and ague, and pains in his head and back ;
and a fourth had an inexorable creditor who would not let him
go out of his sight. They had all received a month's wages in
advance ; and though the amount was not large, it was necessary to make them pay it back, or I should get no men at all. I
therefore sent the village constable after two, and kept them in
custody a day, when they returned about three-fourths of what
they owed me. The sick man also paid, and the steersman
found a substitute who was willing to take his debt, and receive
only the balance of his wages.

About this time we had a striking proof of the dangers of
New Guinea trading. Six men arrived at the village in a small
boat almost starved, having escaped one of two praus, the remainder of whose crews (fourteen in number) had been murdered
by the natives of New Guinea. The praus had left this village
a few months before, and among the murdered men were the
Rajah's son, and the relations or slaves of many of the inhabitants. The cry of lamentation that arose when the news arrived
was most distressing. A score of women, who had lost husbands,
brothers, sons, or more distant relatives, set up at once the most
dismal shrieks, and groans, and wailings, which continued at
intervals till late at night ; and as the chief houses in the village
were crowded together round that which I occupied, our situation
was anything but agreeable.

It seems that the village where the attack took place (nearly
opposite the small island of Lakahia) is known to be dangerous,
and the vessels had only gone there a few days before to buy
some tripang. The crew were living on shore, the praus being
in a small river close by, and they were attacked and murdered
in the day-time while bargaining with the Papuans. The six
men who survived were on board the praus, and escaped by at
once getting into the small boat and rowing out to sea.

This south-west part of New Guinea, known to the native
traders as "Papua Kowiyee" and "Papua Onen," is inhabited
by the most treacherous and bloodthirsty tribes. It is in these
districts that the commanders and portions of the crews of many
of the early discovery ships were murdered, and scarcely a year
now passes but some lives are lost. The Goram and Ceram
traders are themselves generally inoffensive ; they are well ac-

quainted with the character of these natives, and are not likely
to provoke an attack by any insults or open attempt at robbery
or imposition. They are accustomed to visit the same places
every year, and the natives can have no fear of them, as may be
alleged in excuse for their attacks on Europeans. In other extensive districts inhabited by the same Papuan races, such as
Mysol, Salwatty, Waigiou, and some parts of the adjacent coast,
the people have taken the first step in civilization, owing probably to the settlement of traders of mixed breed among them,
and for many years no such attacks have taken place. On the
south-west coast, and in the large island of Jobie, however, the
natives are in a very barbarous condition, and take every opportunity of robbery and murder—a habit which is confirmed by
the impunity they experience, owing to the vast extent of wild
mountain and forest country forbidding all pursuit or attempt
at punishment. In the very same village, four years before,
more than fifty Goram men were murdered ; and as these
savages obtain an immense booty in the praus and all their
appurtenances, it is to be feared that such attacks will continue
to be made at intervals as long as traders visit the same spots
and attempt no retaliation. Punishment could only be inflicted
on these people by very arbitrary measures, such as by obtaining possession of some of the chiefs by stratagem, and rendering
them responsible for the capture of the murderers at the peril
of their own heads. But anything of this kind would be quite
contrary to the system adopted by the Dutch Government in its
dealings with natives.

GORAM TO WAHAI IN CERAM.

When my boat was at length launched and loaded, I got my
men together, and actually set sail the next day (May 27th),
much to the astonishment of the Goram people, to whom such
punctuality was a novelty. I had a crew of three men and a
boy, besides my two Amboyna lads, which was sufficient for
sailing, though rather too few if obliged to row much. The next
day was very wet, with squalls, calms, and contrary winds, and
with some difficulty we reached Kilwaru, the metropolis of the
Bugis traders in the far East. As I wanted to make some purchases, I stayed here two days, and sent two of my boxes of
specimens by a Macassar prau to be forwarded to Ternate,
thus relieving myself of a considerable incumbrance. I bought
knives, basins, and handkerchiefs for barter, which with the
choppers, cloth, and beads I had brought with me, made a
pretty good assortment. I also bought two tower muskets to
satisfy my crew, who insisted on the necessity of being armed
against attacks of pirates ; and with spices and a few articles
of food for the voyage nearly my last doit was expended.
The little island of Kilwaru is a mere sandbank, just large
enough to contain a small village, and situated between the

islands of Ceram-laut, and Kissa—straits about a third of a mile wide separating it from each of them. It is surrounded by coral reefs, and offers good anchorage in both monsoons. Though not more than fifty yards across, and not elevated more than three or four feet above the highest tides, it has wells of excellent drinking water—a singular phenomenon, which would seem to imply deep-seated subterranean channels connecting it with other islands. These advantages, with its situation in the centre of the Papuan trading district, lead to its being so much frequented by the Bugis traders. Here the Goram men bring the produce of their little voyages, which they exchange for cloth, sago cakes, and opium ; and the inhabitants of all the surrounding islands visit it with the same object. It is the rendezvous of the praus trading to various parts of New Guinea, which here assort and dry their cargoes, and refit for the voyage home. Trìpang and mussoi bark are the most bulky articles of produce brought here, with wild nutmegs, tortoise-shell, pearls, and birds of paradise, in smaller quantities. The villagers of the mainland of Ceram bring their sago, which is thus distributed to the islands farther east, while rice from Bali and Macassar can also be purchased at a moderate price. The Goram men come here for their supplies of opium, both for their own consumption and for barter in Mysol and Waigiou, where they have introduced it, and where the chiefs and wealthy men are passionately fond of it. Schooners from Bali come to buy Papuan slaves, while the sea-wandering Bugis arrive from distant Singapore, in their lumbering praus, bringing thence the produce of the Chinamen's workshops and Kling's bazaar, as well as of the looms of Lancashire and Massachusetts.

One of the Bugis traders who had arrived a few days before from Mysol, brought me news of my assistant, Charles Allen, with whom he was well acquainted, and who, he assured me, was making large collections of birds and insects, although he had not obtained any birds of paradise, Silinta, where he was staying, not being a good place for them. This was on the whole satisfactory, and I was anxious to reach him as soon as possible.

Leaving Kilwaru early in the morning of June 1st, with a strong east wind we doubled the point of Ceram about noon, the heavy sea causing my prau to roll about a good deal, to the damage of our crockery. As bad weather seemed coming on, we got inside the reefs and anchored opposite the village of Warus-warus to wait for a change. The night was very squally, and though in a good harbour we rolled and jerked uneasily ; but in the morning I had greater cause for uneasiness in the discovery that our entire Goram crew had decamped, taking with them all they possessed and a little more, and leaving us without any small boat in which to land. I immediately told my Amboyna men to load and fire the muskets as a signal of distress, which was soon answered by the village chief sending off a boat, which took me on shore. I requested that messengers should be immediately sent to the neighbouring villages in quest of the fugitives, which was promptly done. My prau was brought into a small creek, where it could securely rest in the mud at low water, and part of a house was given me in which I could stay for a while. I now found my progress again suddenly checked, just when I thought I had overcome my chief difficulties. As I had treated my men with the greatest kindness, and had given them almost everything they had asked for, I can impute their running away only to their being totally unaccustomed to the restraint of a European master, and to some undefined dread of my ultimate intentions regarding them. The oldest man was an opium smoker, and a reputed thief, but I had been obliged to take him at the last moment as a substitute for another. I feel sure it was he who induced the others to run away, and as they knew the country well, and had several hours' start of us, there was little chance of catching them.

We were here in the great sago district of East Ceram, which supplies most of the surrounding islands with their daily bread, and during our week's delay I had an opportunity of seeing the whole process of making it, and obtaining some interesting statistics. The sago tree is a palm, thicker and larger than the cocoa-nut tree, although rarely so tall, and having immense pinnate spiny leaves, which completely cover the trunk till it is many years old. It has a creeping root-stem like the Nipa palm, and when about ten or fifteen years of age sends up an immense terminal spike of flowers, after which the tree dies. It grows in swamps, or in swampy hollows on the rocky slopes of hills, where it seems to thrive equally well as when exposed to the influx of salt or brackish water. The midribs of the immense leaves form one of the most useful articles in these lands, supplying the place of bamboo, to which for many purposes they are superior. They are twelve or fifteen feet long, and, when very fine, as thick in the lower part as a man's leg. They are very light, consisting entirely of a firm pith covered with a hard thin rind or bark. Entire houses are built of these; they form admirable roofing-poles for thatch; split and well-supported, they do for flooring; and when chosen of equal size, and pegged together side by side to fill up the panels of framed wooden houses, they have a very neat appearance, and make better walls and partitions than boards, as they do not shrink, require no paint or varnish, and are not a quarter the expense. When carefully split and shaved smooth they are formed into light boards with pegs of the bark itself, and are the foundation of the leaf-covered boxes of Goram. All the insect-boxes I used in the Moluccas were thus made at Amboyna, and when covered with stout paper inside and out, are strong, light, and secure the insect-pins remarkably well. The leaflets of the sago folded and tied side by side on the smaller midribs form the "atap"

or thatch in universal use, while the product of the trunk is the
staple food of some hundred thousands of men.

When sago is to be made, a full-grown tree is selected just
before it is going to flower. It is cut down close to the ground,
the leaves and leaf-stalks cleared away, and a broad strip of

SAGO CLUB.

the bark taken off the upper side of the trunk. This exposes
the pithy matter, which is of a rusty colour near the bottom of
the tree, but higher up pure white, about as hard as a dry
apple, but with woody fibres running through it about a quarter
of an inch apart. This pith is cut or broken down into a coarse

SAGO WASHING.

powder by means of a tool constructed for the purpose—a club
of hard and heavy wood, having a piece of sharp quartz rock
firmly embedded into its blunt end, and projecting about half
an inch. By successive blows of this, narrow strips of the pith
are cut away, and fall down into the cylinder formed by the
bark. Proceeding steadily on, the whole trunk is cleared out

leaving a skin not more than half an inch in thickness. This material is carried away (in baskets made of the sheathing bases of the leaves) to the nearest water, where a washing-machine is put up, which is composed almost entirely of the sago tree itself. The large sheathing bases of the leaves form the troughs, and the fibrous covering from the leaf-stalks of the young cocoa-nut the strainer. Water is poured on the mass of pith, which is kneaded and pressed against the strainer till the starch is all dissolved and has passed through, when the fibrous refuse is thrown away, and a fresh basketful put in its place. The water charged with sago starch passes on to a trough, with a depression in the centre, where the sediment is deposited, the surplus water trickling off by a shallow outlet. When the trough is nearly full, the mass of starch, which has a slight reddish tinge, is made into cylinders of about thirty pounds' weight, and neatly covered with sago leaves, and in this state is sold as raw sago.

Boiled with water this forms a thick glutinous mass, with a rather astringent taste, and is eaten with salt, limes, and chillies. Sago-bread is made in large quantities, by baking it into cakes in a small clay oven containing six or eight slits side by side, each about three-quarters of an inch wide, and six or eight inches square. The raw sago is broken up, dried in the sun, powdered, and finely sifted. The oven is heated over a clear fire of embers, and is lightly filled with the sago-powder. The openings are then covered with a flat piece of sago bark,

SAGO OVEN.

and in about five minutes the cakes are turned out sufficiently baked. The hot cakes are very nice with butter, and when made with the addition of a little sugar and grated cocoa-nut are quite a delicacy. They are soft, and something like corn-flour cakes, but have a slight characteristic flavour which is lost in the refined sago we use in this country. When not wanted for immediate use, they are dried for several days in the sun, and tied up in bundles of twenty. They will then keep for years; they are very hard, and very rough and dry, but the people are used to them from infancy, and little children may be seen gnawing at them as contentedly as ours with their bread-and-butter. If dipped in water and then toasted, they become almost as good as when fresh baked; and thus treated they were my daily substitute for bread with my coffee. Soaked and boiled they make a very good pudding or vegetable, and served well to economize our rice, which is sometimes difficult to get so far east.

It is truly an extraordinary sight to witness a whole tree-trunk, perhaps twenty feet long and four or five in circumference, converted into food with so little labour and preparation. A good-sized tree will produce thirty tomans or bundles of thirty pounds each, and each toman will make sixty cakes of three to the pound. Two of these cakes are as much as a man can eat at one meal, and five are considered a full day's allowance ; so that, reckoning a tree to produce 1,800 cakes, weighing 600 pounds, it will supply a man with food for a whole year. The labour to produce this is very moderate. Two men will finish a tree in five days, and two women will bake the whole into cakes in five days more ; but the raw sago will keep very well, and can be baked as wanted, so that we may estimate that in ten days a man may produce food for the whole year. This is on the supposition that he possesses sago trees of his own, for they are now all private property. If he does not, he has to pay about seven and sixpence for one ; and as labour here is fivepence a day, the total cost of a year's food for one man is about twelve shillings. The effect of this cheapness of food is decidedly prejudicial, for the inhabitants of the sago countries are never so well off as those where rice is cultivated. Many of the people here have neither vegetables nor fruit, but live almost entirely on sago and a little fish. Having few occupations at home, they wander about on petty trading or fishing expeditions to the neighbouring islands ; and as far as the comforts of life are concerned, are much inferior to the wild hill-Dyaks of Borneo, or to many of the more barbarous tribes of the Archipelago.

The country round Warus-warus is low and swampy, and owing to the absence of cultivation there were scarcely any paths leading into the forest. I was therefore unable to collect much during my enforced stay, and found no rare birds or insects to improve my opinion of Ceram as a collecting ground. Finding it quite impossible to get men here to accompany me on the whole voyage, I was obliged to be content with a crew to take me as far as Wahai, on the middle of the north coast of Ceram, and the chief Dutch station in the island. The journey took us five days, owing to calms and light winds, and no incident of any interest occurred on it, nor did I obtain at our stopping places a single addition to my collections worth naming. At Wahai, which I reached on the 15th of June, I was hospitably received by the Commandant and my old friend Herr Rosenberg, who was now on an official visit here. He lent me some money to pay my men, and I was lucky enough to obtain three others willing to make the voyage with me to Ternate, and one more who was to return from Mysol. One of my Amboyna lads, however, left me, so that I was still rather short of hands.

I found here a letter from Charles Allen, who was at Silinta in Mysol, anxiously expecting me, as he was out of rice and other necessaries, and was short of insect-pins. He was also ill, and if I did not soon come would return to Wahai.

As my voyage from this place to Waigiou was among islands

inhabited by the Papuan race, and was an eventful and dis-
astrous one, I will narrate its chief incidents in a separate chapter
in that division of my work devoted to the Papuan Islands. I
now have to pass over a year spent in Waigiou and Timor, in
order to describe my visit to the island of Bouru, which concluded my explorations of the Moluccas.

CHAPTER XXVI

BOURU.

(MAY AND JUNE 1861. *Map*, p. 279.)

I HAD long wished to visit the large island of Bouru, which
lies due west of Ceram, and of which scarcely anything appeared
to be known to naturalists, except that it contained a babirusa
very like that of Celebes. I therefore made arrangements for
staying there two months after leaving Timor Delli in 1861.
This I could conveniently do by means of the Dutch mail-
steamers, which make a monthly round of the Moluccas.

We arrived at the harbour of Cajeli on the 4th of May ; a gun
was fired, the Commandant of the fort came alongside in a
native boat to receive the post-packet, and took me and my
baggage on shore, the steamer going off again without coming
to an anchor. We went to the house of the Opzeiner, or over-
seer, a native of Amboyna—Bouru being too poor a place to
deserve even an Assistant Resident ; yet the appearance of the
village was very far superior to that of Delli, which possesses
"His Excellency the Governor," and the little fort, in perfect
order, surrounded by neat grass-plots and straight walks,
although manned by only a dozen Javanese soldiers with an
Adjutant for commander, was a very Sebastopol in comparison
with the miserable mud enclosure at Delli, with its numerous
staff of Lieutenants, Captain, and Major. Yet this, as well as
most of the forts in the Moluccas, was originally built by the
Portuguese themselves. Oh! Lusitania, how art thou fallen !

While the Opzeiner was reading his letters, I took a walk
round the village with a guide in search of a house. The whole
place was dreadfully damp and muddy, being built in a swamp
with not a spot of ground raised a foot above it, and surrounded
by swamps on every side. The houses were mostly well built,
of wooden framework filled in with gaba-gaba (leaf-stems of the
sago-palm), but as they had no whitewash, and the floors were
of bare black earth like the roads, and generally on the same
level, they were extremely damp and gloomy. At length I found
one with the floor raised about a foot, and succeeded in making
a bargain with the owner to turn out immediately, so that by
night I had installed myself comfortably. The chairs and tables

were left for me ; and as the whole of the remaining furniture
in the house consisted of a little crockery and a few clothes-
boxes, it was not much trouble for the owners to move into the
house of some relatives, and thus obtain a few silver rupees very
easily. Every foot of ground between the houses throughout the
village is crammed with fruit trees, so that the sun and air have
no chance of penetrating. This must be very cool and pleasant
in the dry season, but makes it damp and unhealthy at other
times of the year. Unfortunately I had come two months too
soon, for the rains were not yet over, and mud and water were
the prominent features of the country. About a mile behind
and to the east of the village the hills commence, but they are
very barren, being covered with scanty coarse grass and scattered trees of the Melaleuca cajuputi, from the leaves of which
the celebrated cajeput oil is made. Such districts are absolutely
destitute of interest for the zoologist. A few miles further on
rose higher mountains, apparently well covered with forest, but
they were entirely uninhabited and trackless, and practically inaccessible to a traveller with limited time and means. It became
evident, therefore, that I must leave Cajeli for some better collecting ground, and finding a man who was going a few miles
eastward to a village on the coast where he said there were hills
and forest, I sent my boy Ali with him to explore and report on
the capabilities of the district. At the same time I arranged to
go myself on a little excursion up a river which flows into the
bay about five miles north of the town, to a village of the Alfuros,
or indigenes, where I thought I might perhaps find a good
collecting ground.

The Rajah of Cajeli, a good-tempered old man, offered to
accompany me, as the village was under his government ; and
we started one morning early, in a long narrow boat with eight
rowers. In about two hours we entered the river and commenced our inland journey against a very powerful current.
The stream was about a hundred yards wide, and was generally
bordered with high grass, and occasionally bushes and palm-
trees. The country round was flat and more or less swampy,
with scattered trees and shrubs. At every bend we crossed the
river to avoid the strength of the current, and arrived at our
landing-place about four o'clock, in a torrent of rain. Here we
waited for an hour, crouching under a leaky mat till the Alfuros
arrived who had been sent for from the village to carry
my baggage, when we set off along a path of whose extreme
muddiness I had been warned before starting.

I turned up my trousers as high as possible, grasped a stout
stick to prevent awkward falls, and then boldly plunged into
the first mud-hole, which was immediately succeeded by another
and another. The mud or mud and water was knee-deep, with
little intervals of firmer ground between, making progression
exceedingly difficult. The path was bordered with high rigid
grass, growing in dense clumps separated by water, so that

nothing was to be gained by leaving the beaten track, and we were obliged to go floundering on, never knowing where our feet would rest, as the mud was now a few inches, now two feet, deep, and the bottom very uneven, so that the foot slid down to the lowest part, and made it difficult to keep one's balance. One step would be upon a concealed stick or log, almost dislocating the ankle, while the next would plunge into soft mud above the knee. It rained all the way, and the long grass, six feet high, met over the path ; so that we could not see a step of the way ahead, and received a double drenching. Before we got to the village it was dark, and we had to cross over a small but deep and swollen stream by a narrow log of wood, which was more than a foot under water. There was a slender shaking stick for a handrail, and it was nervous work feeling in the dark in the rushing water for a safe place on which to place the advanced foot. After an hour of this most disagreeable and fatiguing walk we reached the village, followed by the men with our guns, ammunition, boxes, and bedding, all more or less soaked. We consoled ourselves with some hot tea and cold fowl, and went early to bed.

The next morning was clear and fine, and I set out soon after sunrise to explore the neighbourhood   The village had evidently been newly formed, and consisted of a single straight street of very miserable huts totally deficient in every comfort, and as bare and cheerless inside as out. It was situated on a little elevated patch of coarse gravelly soil, covered with the usual high rigid grass, which came up close to the backs of the houses. At a short distance in several directions were patches of forest, but all on low and swampy ground. I made one attempt along the only path I could find, but soon came upon a deep mud-hole, and found that I must walk barefoot if at all ; so I returned and deferred further exploration till after breakfast. I then went on into the jungle and found patches of sago-palms and a low forest vegetation, but the paths were everywhere full of mud-holes, and intersected by muddy streams and tracts of swamp, so that walking was not pleasurable, and too much attention to one's steps was not favourable to insect catching, which requires above everything freedom of motion. I shot a few birds, and caught a few butterflies, but all were the same as I had already obtained about Cajeli.

On my return to the village I was told that the same kind of ground extended for many miles in every direction, and I at once decided that Wayapo was not a suitable place to stay at. The next morning early we waded back again through the mud and long wet grass to our boat, and by mid-day reached Cajeli, where I waited Ali's return to decide on my future movements. He came the following day, and gave a very bad account of Pelah, where he had been. There was a little brush and trees along the beach, and hills inland covered with high grass and cajuputi trees—my dread and abhorrence. On inquiring who could give me trustworthy information, I was referred to the Lieutenant of the Burghers, who had travelled all round the island, and was a very intelligent fellow. I asked him to tell me if he knew of any part of Bouru where there was no "kusu-kusu," as the coarse grass of the country is called. He assured me that a good deal of the south coast was forest land, while along the north was almost entirely swamp and grassy hills. After minute inquiries, I found that the forest country commenced at a place called Waypoti, only a few miles beyond Pelah, but that, as the coast beyond that place was exposed to the east monsoon and dangerous for praus, it was necessary to walk. I immediately went to the Opzeiner, and he called the Rajah. We had a consultation, and arranged for a boat to take me the next evening but one to Pelah, whence I was to proceed on foot, the Orang-kaya going the day before to call the Alfuros to carry my baggage.

The journey was made as arranged, and on May 19th we arrived at Waypoti, having walked about ten miles along the beach, and through stony forest bordering the sea, with occasional plunges of a mile or two into the interior. We found no village, but scattered houses and plantations, with hilly country pretty well covered with forest, and looking rather promising. A low hut with a very rotten roof, showing the sky through in several places, was the only one I could obtain. Luckily it did not rain that night, and the next day we pulled down some of the walls to repair the roof, which was of immediate importance, especially over our beds and table.

About half a mile from the house was a fine mountain stream, running swiftly over a bed of rocks and pebbles, and beyond this was a hill covered with fine forest. By carefully picking my way I could wade across this river without getting much above my knees, although I would sometimes slip off a rock and go into a hole up to my waist, and about twice a week I went across it in order to explore the forest. Unfortunately there were no paths here of any extent, and it did not prove very productive either in insects or birds. To add to my difficulties, I had stupidly left my only pair of strong boots on board the steamer, and my others were by this time all dropping to pieces, so that I was obliged to walk about barefooted, and in constant fear of hurting my feet, and causing a wound which might lay me up for weeks, as had happened in Borneo, Aru, and Dorey. Although there were numerous plantations of maize and plantains, there were no new clearings; and as without these it is almost impossible to find many of the best kinds of insects, I determined to make one myself, and with much difficulty engaged two men to clear a patch of forest, from which I hoped to obtain many fine beetles before I left.

During the whole of my stay, however, insects never became plentiful. My clearing produced me a few fine Longicorns and Buprestidæ, different from any I had before seen, together with several of the Amboyna species, but by no means so numerous or so beautiful as I had found in that small island. For example, I collected only 210 different kinds of beetles during my two months' stay at Bouru, while in three weeks at Amboyna, in 1857, I found more than 300 species. One of the finest insects found at Bouru was a large Cerambyx, of a deep shining chestnut colour, and with very long antennæ. It varied greatly in size, the largest specimens being three inches long, while the smallest were only an inch, the antennæ varying from one and a half to five inches.

One day my boy Ali came home with a story of a big snake. He was walking through some high grass, and stepped on something which he took for a small fallen tree, but it felt cold and yielding to his feet, and far to the right and left there was a waving and rustling of the herbage. He jumped back in affright and prepared to shoot, but could not get a good view of the creature, and it passed away, he said, like a tree being dragged along through the grass. As he had several times already shot large snakes, which he declared were all as nothing compared with this, I am inclined to believe it must really have been a monster. Such creatures are rather plentiful here, for a man living close by showed me on his thigh the marks where he had been seized by one close to his house. It was big enough to take the man's thigh in its mouth, and he would probably have been killed and devoured by it had not his cries brought out his neighbours, who destroyed it with their choppers. As far as I could make out it was about twenty feet long, but Ali's was probably much larger.

It sometimes amuses me to observe how, a few days after I have taken possession of it, a native hut seems quite a comfortable home. My house at Waypoti was a bare shed, with a large bamboo platform at one side. At one end of this platform, which was elevated about three feet, I fixed up my mosquito curtain, and partly enclosed it with a large Scotch plaid, making a comfortable little sleeping apartment. I put up a rude table on legs buried in the earthen floor, and had my comfortable rattan-chair for a seat. A line across one corner carried my daily-washed cotton clothing, and on a bamboo shelf was arranged my small stock of crockery and hardware. Boxes were ranged against the thatch walls, and hanging shelves, to preserve my collections from ants while drying, were suspended both without and within the house. On my table lay books, penknives, scissors, pliers, and pins, with insect and bird labels, all of which were unsolved mysteries to the native mind.

Most of the people here had never seen a pin, and the better informed took a pride in teaching their more ignorant companions the peculiarities and uses of that strange European production—a needle with a head, but no eye! Even paper, which we throw away hourly as rubbish, was to them a curiosity; and I often saw them picking up little scraps which had been swept out of the house, and carefully putting them away in their betel-pouch. Then when I took my morning coffee and evening tea, how many were the strange things displayed to them! Teapot, teacups, teaspoons, were all more or less curious in their eyes; tea, sugar, biscuit, and butter, were articles of human consumption seen by many of them for the first time. One asks if that whitish powder is "gula passir" (sand-sugar), so called to distinguish it from the coarse lump palm-sugar or molasses of native manufacture; and the biscuit is considered a sort of European sago-cake, which the inhabitants of those remote regions are obliged to use in the absence of the genuine article. My pursuits were of course utterly beyond their comprehension. They continually asked me what white people did with the birds and insects I took so much care to preserve. If I only kept what was beautiful, they might perhaps comprehend it; but to see ants, and flies, and small, ugly insects put away so carefully was a great puzzle to them, and they were convinced that there must be some medical or magical use for them which I kept a profound secret. These people were in fact as completely unacquainted with civilized life as the Indians of the Rocky Mountains, or the savages of Central Africa—yet a steamship, that highest triumph of human ingenuity, with its little floating epitome of European civilization, touches monthly at Cajeli, twenty miles off; while at Amboyna, only sixty miles distant, a European population and government have been established for more than three hundred years.

Having seen a good many of the natives of Bouru from different villages, and from distant parts of the island, I feel convinced that they consist of two distinct races now partially amalgamated. The larger portion are Malays of the Celebes type, often exactly similar to the Tomóre people of East Celebes, whom I found settled in Batchian; while others altogether resemble the Alfuros of Ceram. The influx of two races can easily be accounted for. The Sula Islands, which are closely connected with East Celebes, approach to within forty miles of the north coast of Bouru, while the island of Manipa offers an easy point of departure for the people of Ceram. I was confirmed in this view by finding that the languages of Bouru possessed distinct resemblances to that of Sula, as well as to those of Ceram.

Soon after we had arrived at Waypoti, Ali had seen a beautiful little bird of the genus Pitta, which I was very anxious to obtain, as in almost every island the species are different, and none were yet known from Bouru. He and my other hunter continued to see it two or three times a week, and to hear its peculiar note much oftener, but could never get a specimen, owing to its always frequenting the most dense thorny thickets, where only hasty glimpses of it could be obtained, and at so short a distance that it would be difficult to avoid blowing the bird to pieces. Ali was very much annoyed that he could not get a specimen of this bird, in going after which he had already severely wounded his feet with thorns ; and when we had only two days more to stay, he went of his own accord one evening to sleep at a little hut in the forest some miles off, in order to have a last try for it at daybreak, when many birds come out to feed, and are very intent on their morning meal. The next evening he brought me home two specimens, one with the head blown completely off, and otherwise too much injured to preserve, the other in very good order, and which I at once saw to be a new species, very like the Pitta celebensis, but ornamented with a square patch of bright red on the nape of the neck.

The next day after securing this prize we returned to Cajeli, and packing up my collections left Bouru by the steamer. During our two days' stay at Ternate, I took on board what baggage I had left there, and bade adieu to all my friends. We then crossed over to Menado, on our way to Macassar and Java, and I finally quitted the Moluccas, among whose luxuriant and beautiful islands I had wandered for more than three years.

My collections in Bouru, though not extensive, were of considerable interest ; for out of sixty-six species of birds which I collected there, no less than seventeen were new, or had not been previously found in any island of the Moluccas. Among these were two kingfishers, Tanysiptera acis and Ceyx Cajeli ; a beautiful sunbird, Nectarinea proserpina ; a handsome little black and white flycatcher, Monarcha loricata, whose swelling throat was beautifully scaled with metallic blue ; and several of less interest. I also obtained a skull of the babirusa, one specimen of which was killed by native hunters during my residence at Cajeli.

CHAPTER XXVII.

THE NATURAL HISTORY OF THE MOLUCCAS.

THE Moluccas consists of three large islands, Gilolo, Ceram, and Bouru, the two former being each about two hundred miles long ; and a great number of smaller isles and islets, the most important of which are Batchian, Morty, Obi, Ké, Timor-laut, and Amboyna ; and among the smaller ones, Ternate, Tidore, Kaióa, and Banda. These occupy a space of ten degrees of latitude by eight of longitude, and they are connected by groups of small islets to New Guinea on the east, the Philippines on the north, Celebes on the west, and Timor on the south. It will be as well to bear in mind these main features of extent and geographical position while we survey their animal productions and discuss their relations to the countries which surround them on every side in almost equal proximity.

We will first consider the Mammalia, or warm-blooded quadrupeds, which present us with some singular anomalies. The land mammals are exceedingly few in number, only ten being yet known from the entire group. The bats or aërial mammals, on the other hand, are numerous—not less than twenty-five species being already known. But even this exceeding poverty of terrestrial mammals does not at all represent the real poverty of the Moluccas in this class of animals ; for, as we shall soon see, there is good reason to believe that several of the species have been introduced by man, either purposely or by accident.

The only quadrumanous animal in the group is the curious baboon-monkey, Cynopithecus nigrescens, already described as being one of the characteristic animals of Celebes. This is found only in the island of Batchian ; and it seems so much out of place there—as it is difficult to imagine how it could have reached the island by any natural means of dispersal, and yet not have passed by the same means over the narrow strait to Gilolo—that it seems more likely to have originated from some individuals which had escaped from confinement, these and similar animals being often kept as pets by the Malays, and carried about in their praus.

Of all the carnivorous animals of the Archipelago the only one found in the Moluccas is the Viverra tangalunga, which inhabits both Batchian and Bouru, and probably some of the other islands. I am inclined to think that this also may have been introduced accidentally, for it is often made captive by the Malays, who procure civet from it, and it is an animal very restless and untamable, and therefore likely to escape. This view is rendered still more probable by what Antonio de Morga tells us was the custom in the Philippines in 1602. He says that " the natives of Mindanao carry about civet-cats in cages, and sell them in the islands ; and they take the civet from them, and let them go again." The same species is common in the Philippines and in all the large islands of the Indo-Malay region.

The only Moluccan ruminant is a deer, which was once supposed to be a distinct species, but is now generally considered to be a slight variety of the Rusa hippelaphus of Java. Deer are often tamed and petted, and their flesh is so much esteemed by all Malays, that it is very natural they should endeavour to introduce them into the remote islands in which they settled, and whose luxuriant forests seem so well adapted for their subsistence.

The strange babirusa of Celebes is also found in Bouru, but in no other Moluccan island, and it is somewhat difficult to imagine how it got there. It is true that there is some approximation between the birds of the Sula Islands (where the babirusa is also found) and those of Bouru, which seems to indicate that these islands have recently been closer together, or that some intervening land has disappeared. At this time the babirusa may have entered Bouru, since it probably swims as well as its allies the pigs. These are spread all over the Archipelago, even to several of the smaller islands, and in many cases the species are peculiar. It is evident, therefore, that they have some natural means of dispersal. There is a popular idea that pigs cannot swim, but Sir Charles Lyell has shown that this is a mistake. In his *Principles of Geology* (10th edit. vol. ii. p. 355) he adduces evidence to show that pigs have swum many miles at sea, and are able to swim with great ease and swiftness. I have myself seen a wild pig swimming across the arm of the sea that separates Singapore from the Peninsula of Malacca, and we thus have explained the curious fact, that of all the large mammals of the Indian region, pigs alone extend beyond the Moluccas and as far as New Guinea, although it is somewhat curious that they have not found their way to Australia.

The little shrew, Sorex myosurus, which is common in Sumatra, Borneo, and Java, is also found in the larger islands of the Moluccas, to which it may have been accidentally conveyed in native praus.

This completes the list of the placental mammals which are so characteristic of the Indian region ; and we see that, with the single exception of the pig, all may very probably have been introduced by man, since all except the pig are of species identical with those now abounding in the great Malay islands, or in Celebes.

The four remaining mammals are Marsupials, an order of the class Mammalia, which is very characteristic of the Australian fauna ; and these are probably true natives of the Moluccas, since they are either of peculiar species, or if found elsewhere are natives only of New Guinea or North Australia. The first is the small flying opossum, Belideus ariel, a beautiful little animal, exactly like a small flying squirrel in appearance, but belonging to the marsupial order. The other three are species of the curious genus Cuscus, which is peculiar to the Austro-Malayan region. These are opossum-like animals, with a long prehensile tail, of which the terminal half is generally bare. They have small heads, large eyes, and a dense covering of woolly fur, which is often pure white with irregular black spots or blotches, or sometimes ashy brown with or without white spots. They live in trees, feeding upon the leaves, of which they devour large quantities. They move about slowly, and are difficult to kill, owing to the thickness of their fur, and their tenacity of life. A heavy charge of shot will often lodge in the skin and do them no harm, and even breaking the spine or piercing the brain will not kill them for some hours. The natives everywhere eat their flesh, and as their motions are so slow, easily catch them by climbing ; so that it is wonderful they have not been exterminated. It may be, however, that their dense woolly fur protects them from birds of prey, and the islands they live in are too thinly inhabited for man to be able to exterminate them. The figure represents Cuscus ornatus, a new species discovered by me in Batchian, and which also inhabits Ternate. It is

peculiar to the Moluccas, while the two other species which
inhabit Ceram are found also in New Guinea and Waigiou.

In place of the excessive poverty of mammals which char-
acterizes the Moluccas, we have a very rich display of the
feathered tribes. The number of species of birds at present
known from the various islands of the Moluccan group is 265,
but of these only 70 belong to the usually abundant tribes of
the waders and swimmers, indicating that these are very im-

CUSCUS ORNATUS.

perfectly known. As they are also pre-eminently wanderers,
and are thus little fitted for illustrating the geographical distribution of life in a limited area, we will here leave them out
of consideration and confine our attention only to the 195 land-
birds.

When we consider that all Europe, with its varied climate and
vegetation, with every mile of its surface explored, and with
the immense extent of temperate Asia and Africa, which serve
as storehouses, from which it is continually recruited, only

supports 257 species of land-birds as residents or regular immigrants, we must look upon the numbers already procured in the
small and comparatively unknown islands of the Moluccas as
indicating a fauna of fully average richness in this department.
But when we come to examine the family groups which go to
make up this number, we find the most curious deficiencies in
some, balanced by equally striking redundancy in others. Thus
if we compare the birds of the Moluccas with those of India, as
given in Mr. Jerdon's work, we find that the three groups of the
parrots, kingfishers, and pigeons, form nearly *one-third* of the
whole land-birds in the former, while they amount to only *one-
twentieth* in the latter country. On the other hand, such wide-
spread groups as the thrushes, warblers, and finches, which in
India form nearly *one-third* of all the land-birds, dwindle down
in the Moluccas to *one-fourteenth*.

The reason of these peculiarities appears to be, that the
Moluccan fauna has been almost entirely derived from that of
New Guinea, in which country the same deficiency and the same
luxuriance is to be observed. Out of the seventy-eight
genera in which the Moluccan land-birds may be classed,
no less than seventy are characteristic of New Guinea, while
only six belong specially to the Indo-Malay islands. But
this close resemblance to New Guinea genera does not extend to
the species, for no less than 140 out of the 195 land-birds are
peculiar to the Moluccan islands, while 32 are found also in New
Guinea, and 15 in the Indo-Malay islands.[1] These facts teach
us, that though the birds of this group have evidently been derived mainly from New Guinea, yet the immigration has not
been a recent one, since there has been time for the greater
portion of the species to have become changed. We find, also,
that many very characteristic New Guinea forms have not
entered the Moluccas at all, while others found in Ceram and
Gilolo do not extend so far west as Bouru. Considering, further,
the absence of most of the New Guinea mammals from the
Moluccas, we are led to the conclusion that these islands are not
fragments which have been separated from New Guinea, but
form a distinct insular region, which has been upheaved independently at a rather remote epoch, and during all the muta-
tions it has undergone has been constantly receiving immigrants
from that great and productive island. The considerable length
of time the Moluccas have remained isolated is further indicated
by the occurrence of two peculiar genera of birds, Semioptera
and Lycocorax, which are found nowhere else.

We are able to divide this small archipelago into two well-
marked groups—that of Ceram, including also Bouru, Amboyna,
Banda, and Ké ; and that of Gilolo, including Morty, Batchian,

---

[1] A few species have been added in Bouru, Obi, Batchian, and other of the less
known islands, by Mr. H. O. Forbes, Dr. Guillemard, and the Dutch and German
naturalists, but they only slightly alter the figures, and do not at all affect the
conclusions here drawn.

Obi, Ternate, and other small islands. These divisions have
each a considerable number of peculiar species, no less than
fifty-five being found in the Ceram group only ; and besides
this, most of the separate islands have some species peculiar to
themselves.   Thus Morty Island has a peculiar kingfisher,
honey-sucker, and starling ; Ternate has a ground-thrush (Pitta)
and a flycatcher ; Banda has a pigeon, a shrike, and a Pitta ;
Ké has two flycatchers, a Zosterops, a shrike, a king-crow, and
a cuckoo ; and the remote Timor-laut, which should probably
come into the Moluccan group, has a cockatoo and lory as its
only known birds, and both are of peculiar species.[1]

The Moluccas are especially rich in the parrot tribe, no less
than twenty-two species, belonging to ten genera, inhabiting
them.   Among these is the large red-crested cockatoo, so commonly seen alive in Europe, two handsome red parrots of the
genus Eclectus, and five of the beautiful crimson lories, which
are almost exclusively confined to these islands and the New
Guinea group.   The pigeons are hardly less abundant or beautiful, twenty-one species being known, including twelve of the
beautiful green fruit pigeons, the smaller kinds of which are
ornamented with the most brilliant patches of colour on the
head and the under-surface.   Next to these come the kingfishers,
including sixteen species, almost all of which are beautiful, and
many are among the most brilliantly-coloured birds that exist.

One of the most curious groups of birds, the Megapodii, or
mound-makers, is very abundant in the Moluccas.   They are
gallinaceous birds, about the size of a small fowl, and generally
of a dark ashy or sooty colour, and they have remarkably large
and strong feet and long claws.   They are allied to the "Maleo"
of Celebes, of which an account has already been given, but
they differ in habits, most of these birds frequenting the scrubby
jungles along the sea-shore, where the soil is sandy, and there
is a considerable quantity of *débris*, consisting of sticks, shells,
seaweed, leaves, &c.   Of this rubbish the Megapodius forms
immense mounds, often six or eight feet high and twenty or
thirty feet in diameter, which they are enabled to do with comparative ease by means of their large feet, with which they can
grasp and throw backwards a quantity of material.   In the
centre of this mound, at a depth of two or three feet, the eggs
are deposited, and are hatched by the gentle heat produced by
the fermentation of the vegetable matter of the mound.   When
I first saw these mounds in the island of Lombock, I could
hardly believe that they were made by such small birds, but I

[1] Mr. H. O. Forbes visited these islands in 1882, and obtained a fine collection of
birds which now amount to eighty species.   Of these sixty-two are land-birds, and
twenty-six of these are peculiar to the island.   Their affinities are chiefly with the
Moluccas and New Guinea, but to some extent also with Timor and Australia.   (*See*
Forbes' *Naturalist's Wanderings in the Eastern Archipelago*, p. 355.)   The butterflies
collected by Mr. Forbes show similar affinities, but tending more towards Timor and
Australia, due probably to the more immediate dependence of butterflies on vegetation.

afterwards met with them frequently, and have once or twice come upon the birds engaged in making them. They run a few steps backwards, grasping a quantity of loose material in one foot, and throw it a long way behind them. When once properly buried the eggs seem to be no more cared for, the young birds working their way up through the heap of rubbish, and running off at once into the forest. They come out of the egg covered with thick downy feathers, and have no tail, although the wings are fully developed.

I was so fortunate as to discover a new species (Megapodius wallacei), which inhabits Gilolo, Ternate, and Bouru. It is the handsomest bird of the genus, being richly banded with reddish brown on the back and wings ; and it differs from the other species in its habits. It frequents the forests of the interior, and comes down to the sea-beach to deposit its eggs, but instead of making a mound, or scratching a hole to receive them, it burrows into the sand to the depth of about three feet obliquely downwards, and deposits its eggs at the bottom. It then loosely covers up the mouth of the hole, and is said by the natives to obliterate and disguise its own footmarks leading to and from the hole, by making many other tracks and scratches in the neighbourhood. It lays its eggs only at night, and at Bouru a bird was caught early one morning as it was coming out of its hole, in which several eggs were found. All these birds seem to be semi-nocturnal, for their loud wailing cries may be constantly heard late into the night and long before daybreak in the morning. The eggs are all of a rusty red colour, and very large for the size of the bird, being generally three or three and a quarter inches long, by two or two and a quarter wide. They are very good eating, and are much sought after by the natives.

Another large and extraordinary bird is the Cassowary, which inhabits the island of Ceram only. It is a stout and strong bird, standing five or six feet high, and covered with long coarse black hair-like feathers. The head is ornamented with a large horny casque or helmet, and the bare skin of the neck is conspicuous with bright blue and red colours. The wings are quite absent, and are replaced by a group of horny black spines like blunt porcupine quills. These birds wander about the vast mountainous forests that cover the island of Ceram, feeding chiefly on fallen fruits, and on insects or crustacea. The female lays from three to five large and beautifully shagreened green eggs upon a bed of leaves, the male and female sitting upon them alternately for about a month. This bird is the helmeted cassowary (Casuarius galeatus) of naturalists, and was for a long time the only species known. Others have since been discovered in New Guinea, New Britain, and North Australia.

It was in the Moluccas that I first discovered undoubted cases of "mimicry" among birds, and these are so curious that I must briefly describe them. It will be as well, however, first to explain what is meant by mimicry in natural history. At page

100, I have described a butterfly which, when at rest, so closely resembles a dead leaf, that it thereby escapes the attacks of its enemies. This is termed a " protective resemblance." If however the butterfly, being itself a savoury morsel to birds, had closely resembled another butterfly which was disagreeable to birds, and therefore never eaten by them, it would be as well protected as if it resembled a leaf ; and this is what has been happily termed "mimicry" by Mr. Bates, who first discovered the object of these curious external imitations of one insect by another belonging to a distinct genus or family, and sometimes even to a distinct order. The clear-winged moths which resemble wasps and hornets are the best examples of "mimicry" in our own country.

For a long time all the known cases of exact resemblance of one creature to quite a different one were confined to insects, and it was therefore with great pleasure that I discovered in the island of Bouru two birds which I constantly mistook for each other, and which yet belonged to two distinct and somewhat distant families. One of these is a honeysucker named Tropidorhynchus bouruensis, and the other a kind of oriole, which has been called Mimeta bouruensis. The oriole resembles the honeysucker in the following particulars : the upper and under surfaces of the two birds are exactly of the same tints of dark and light brown ; the Tropidorhynchus has a large bare black patch round the eyes ; this is copied in the Mimeta by a patch of black feathers. The top of the head of the Tropidorhynchus has a scaly appearance from the narrow scale-formed feathers, which are imitated by the broader feathers of the Mimeta having a dusky line down each. The Tropidorhynchus has a pale ruff formed of curious recurved feathers on the nape (which has given the whole genus the name of Friar birds) ; this is represented in the Mimeta by a pale band in the same position. Lastly, the bill of the Tropidorhynchus is raised into a protuberant keel at the base, and the Mimeta has the same character, although it is not a common one in the genus. The result is, that on a superficial examination the birds are identical, although they have important structural differences, and cannot be placed near each other in any natural arrangement.

In the adjacent island of Ceram we find very distinct species of both these genera, and, strange to say, these resemble each other quite as closely as do those of Bouru. The Tropidorhynchus subcornutus is of an earthy brown colour, washed with ochreish yellow, with bare orbits, dusky cheeks, and the usual recurved nape-ruff. The Mimeta forsteni which accompanies it, is absolutely identical in the tints of every part of the body, and the details are copied just as minutely as in the former species.

We have two kinds of evidence to tell us which bird in this case is the model, and which the copy. The honeysuckers are coloured in a manner which is very general in the whole family to which they belong, while the orioles seem to have departed

Euplolus (new species).
Euchirus longimanus, male.

Arachnobas
(new species).

Xenocerus semilunctuosus, fem.
Xenocerus (new species), male.

MOLUCCAN BEETLES

from the gay yellow tints so common among their allies. We should therefore conclude that it is the latter who mimic the former. If so, however, they must derive some advantage from the imitation, and as they are certainly weak birds, with small feet and claws, they may require it. Now the Tropidorhynchi are very strong and active birds, having powerful grasping claws, and long, curved, sharp beaks. They assemble together in groups and small flocks, and they have a very loud bawling note which can be heard at a great distance, and serves to collect a number together in time of danger. They are very plentiful and very pugnacious, frequently driving away crows and even hawks, which perch on a tree where a few of them are assembled. It is very probable, therefore, that the smaller birds of prey have learnt to respect these birds and leave them alone, and it may thus be a great advantage for the weaker and less courageous Mimetas to be mistaken for them. This being the case, the laws of Variation and Survival of the Fittest, will suffice to explain how the resemblance has been brought about, without supposing any voluntary action on the part of the birds themselves; and those who have read Mr. Darwin's *Origin of Species* will have no difficulty in comprehending the whole process.

The insects of the Moluccas are pre-eminently beautiful, even when compared with the varied and beautiful productions of other parts of the Archipelago. The grand bird-winged butterflies (Ornithoptera) here reach their maximum of size and beauty, and many of the Papilios, Pieridæ, Danaidæ, and Nymphalidæ are equally pre-eminent. There is, perhaps, no island in the world so small as Amboyna where so many grand insects are to be found. Here are three of the very finest Ornithopteræ—priamus, helena, and remus; three of the handsomest and largest Papilios—ulysses, deiphobus, and gambrisius; one of the handsomest Pieridæ, Iphias leucippe; the largest of the Danaidæ, Hestia idea; and two unusually large and handsome Nymphalidæ—Diadema pandarus, and Charaxes euryalus. Among its beetles are the extraordinary Euchirus longimanus, whose enormous legs spread over a space of eight inches, and an unusual number of large and handsome Longicorns, Anthribidæ, and Buprestidæ.

The beetles figured on the plate as characteristic of the Moluccas are: 1. A small specimen of the Euchirus longimanus, or Long-armed Chafer, which has been already mentioned in the account of my residence at Amboyna (Chapter XX.). The female has the fore legs of moderate length. 2. A fine weevil, (an undescribed species of Eupholus,) of rich blue and emerald green colours, banded with black. It is a native of Ceram and Goram, and is found on foliage. 3. A female of Xenocerus semiluctuosus, one of the Anthribidæ of delicate silky white and black colours. It is abundant on fallen trunks and stumps in Ceram and Amboyna. 4. An undescribed species of Xenocerus;

a male, with very long and curious antennæ, and elegant black and white markings. It is found on fallen trunks in Batchian. 5. An undescribed species of Arachnobas, a curious genus of weevils peculiar to the Moluccas and New Guinea, and remarkable for their long legs, and their habit of often sitting on leaves, and turning rapidly round the edge to the under-surface when disturbed. It was found in Gilolo. All these insects are represented of the natural size.

Like the birds, the insects of the Moluccas show a decided affinity with those of New Guinea rather than with the productions of the great western islands of the Archipelago, but the difference in form and structure between the productions of the east and west is not nearly so marked here as in birds. This is probably due to the more immediate dependence of insects on climate and vegetation, and the greater facilities for their distribution in the varied stages of egg, pupa, and perfect insect. This has led to a general uniformity in the insect-life of the whole Archipelago, in accordance with the general uniformity of its climate and vegetation ; while on the other hand the great susceptibility of the insect organization to the action of external conditions has led to infinite detailed modifications of form and colour, which have in many cases given a considerable diversity to the productions of adjacent islands.

Owing to the great preponderance among the birds of parrots, pigeons, kingfishers, and sunbirds, almost all of gay or delicate colours and many adorned with the most gorgeous plumage, and to the numbers of very large and showy butterflies which are almost everywhere to be met with, the forests of the Moluccas offer to the naturalist a very striking example of the luxuriance and beauty of animal life in the tropics. Yet the almost entire absence of Mammalia, and of such widespread groups of birds as woodpeckers, thrushes, jays, tits, and pheasants, must convince him that he is in a part of the world which has in reality but little in common with the great Asiatic continent, although an unbroken chain of islands seem to link them to it.

CHAPTER XXVIII.

MACASSAR TO THE ARU ISLANDS IN A NATIVE PRAU.

(DECEMBER 1856.)

IT was the beginning of December, and the rainy season at Macassar had just set in. For nearly three months I had beheld the sun rise daily above the palm-groves, mount to the zenith, and descend like a globe of fire into the ocean, unobscured for a single moment of his course : now dark leaden clouds had gathered over the whole heavens, and seemed to have rendered him permanently invisible. The strong east winds, warm and dry and dust-laden, which had hitherto blown as certainly as the sun had risen, were now replaced by variable gusty breezes and heavy rains, often continuous for three days and nights together; and the parched and fissured rice stubbles which during the dry weather had extended in every direction for miles around the town, were already so flooded as to be only passable by boats, or by means of a labyrinth of paths on the top of the narrow banks which divided the separate properties.

Five months of this kind of weather might be expected in Southern Celebes, and I therefore determined to seek some more favourable climate for collecting in during that period, and to return in the next dry season to complete my exploration of the district. Fortunately for me I was in one of the great emporiums of the native trade of the Archipelago. Rattans from Borneo, sandal-wood and bees'-wax from Flores and Timor, tripang from the Gulf of Carpentaria, cajuputi-oil from Bouru, wild nutmegs and mussoi-bark from New Guinea, are all to be found in the stores of the Chinese and Bugis merchants of Macassar, along with the rice and coffee which are the chief products of the surrounding country. More important than all these however is the trade to Aru, a group of islands situated on the south-west coast of New Guinea, and of which almost the whole produce comes to Macassar in native vessels. These islands are quite out of the track of all European trade, and are inhabited only by black mop-headed savages, who yet contribute to the luxurious tastes of the most civilized races. Pearls, mother-of-pearl, and tortoise-shell, find their way to Europe, while edible birds' nests and "tripang" or sea-slug are obtained by shiploads for the gastronomic enjoyment of the Chinese.

The trade to these islands has existed from very early times, and it is from them that Birds of Paradise, of the two kinds known to Linnæus, were first brought. The native vessels can only make the voyage once a year, owing to the monsoons. They leave Macassar in December or January, at the beginning of the west monsoon, and return in July or August with the full strength of the east monsoon. Even by the Macassar people themselves, the voyage to the Aru Islands is looked upon as a rather wild and romantic expedition, full of novel sights and strange adventures. He who has made it is looked up to as an authority, and it remains with many the unachieved ambition of their lives. I myself had hoped rather than expected ever to reach this "Ultima Thule" of the East; and when I found that I really could do so now, had I but courage to trust myself for a thousand miles' voyage in a Bugis prau, and for six or seven months among lawless traders and ferocious savages—I felt somewhat as I did when, a schoolboy, I was for the first time allowed to travel outside the stage-coach, to visit that scene of

all that is strange and new and wonderful to young imaginations
—London !

By the help of some kind friends I was introduced to the
owner of one of the large praus which was to sail in a few days.
He was a Javanese half-caste, intelligent, mild, and gentlemanly
in his manners, and had a young and pretty Dutch wife, whom
he was going to leave behind during his absence. When we
talked about passage money he would fix no sum, but insisted
on leaving it entirely to me to pay on my return exactly what
I liked. "And then," said he, "whether you give me one dollar
or a hundred, I shall be satisfied, and shall ask no more."

The remainder of my stay was fully occupied in laying in
stores, engaging servants, and making every other preparation
for an absence of seven months from even the outskirts of
civilization. On the morning of December 13th, when we went
on board at daybreak, it was raining hard. We set sail and it
came on to blow. Our boat was lost astern, our sails damaged,
and the evening found us back again in Macassar harbour. We
remained there four days longer, owing to its raining all the
time, thus rendering it impossible to dry and repair the huge
mat sails. All these dreary days I remained on board, and
during the rare intervals when it didn't rain, made myself
acquainted with our outlandish craft, some of the peculiarities
of which I will now endeavour to describe.

It was a vessel of about seventy tons burthen, and shaped
something like a Chinese junk. The deck sloped considerably
downward to the bows, which are thus the lowest part of the
ship. There were two large rudders, but instead of being placed
astern they were hung on the quarters from strong cross beams,
which projected out two or three feet on each side, and to which
extent the deck overhung the sides of the vessel amidships. The
rudders were not hinged but hung with slings of rattan, the
friction of which keeps them in any position in which they are
placed, and thus perhaps facilitates steering. The tillers were
not on deck, but entered the vessel through two square openings
into a lower or half deck about three feet high, in which sit the
two steersmen. In the after part of the vessel was a low poop,
about three and a half feet high, which forms the captain's
cabin, its furniture consisting of boxes, mats, and pillows. In
front of the poop and mainmast was a little thatched house on
deck, about four feet high to the ridge ; and one compartment
of this, forming a cabin six and a half feet long by five and a
half wide, I had all to myself, and it was the snuggest and most
comfortable little place I ever enjoyed at sea. It was entered
by a low sliding door of thatch on one side, and had a very small
window on the other. The floor was of split bamboo, pleasantly
elastic, raised six inches above the deck, so as to be quite dry.
It was covered with fine cane mats, for the manufacture of which
Macassar is celebrated ; against the further wall were arranged
my gun-case, insect-boxes, clothes, and books; my mattress

occupied the middle, and next the door were my canteen, lamp, and little store of luxuries for the voyage ; while guns, revolver, and hunting knife hung conveniently from the roof. During these four miserable days I was quite jolly in this little snuggery —more so than I should have been if confined the same time to the gilded and uncomfortable saloon of a first-class steamer. Then, how comparatively sweet was everything on board—no paint, no tar, no new rope, (vilest of smells to the qualmish !) no grease, or oil, or varnish ; but instead of these, bamboo and rattan, and coir rope and palm thatch ; pure vegetable fibres, which smell pleasantly if they smell at all, and recall quiet scenes in the green and shady forest.

Our ship had two masts, if masts they can be called, which were great movable triangles. If in an ordinary ship you replace the shrouds and backstay by strong timbers, and take away the mast altogether, you have the arrangement adopted on board a prau. Above my cabin, and resting on cross-beams attached to the masts, was a wilderness of yards and spars, mostly formed of bamboo. The mainyard, an immense affair nearly a hundred feet long, was formed of many pieces of wood and bamboo bound together with rattans in an ingenious manner. The sail carried by this was of an oblong shape, and was hung out of the centre, so that when the short end was hauled down on deck the long end mounted high in the air, making up for the lowness of the mast itself. The foresail was of the same shape, but smaller. Both these were of matting, and with two jibs and a fore and aft sail astern of cotton canvas, completed our rig.

The crew consisted of about thirty men, natives of Macassar and the adjacent coasts and islands. They were mostly young, and were short, broad-faced, good-humoured-looking fellows. Their dress consisted generally of a pair of trousers only when at work, and a handkerchief twisted round the head, to which in the evening they would add a thin cotton jacket. Four of the elder men were " jurumudis," or steersmen, who had to squat (two at a time) in the little steerage before described, changing every six hours. Then there was an old man, the " juragan," or captain, but who was really what we should call the first mate ; he occupied the other half of the little house on deck. There were about ten respectable men, Chinese or Bugis, whom our owner used to call " his own people." He treated them very well, shared his meals with them, and spoke to them always with perfect politeness ; yet they were most of them a kind of slave debtors, bound over by the police magistrate to work for him at mere nominal wages for a term of years till their debts were liquidated. This is a Dutch institution in this part of the world, and seems to work well. It is a great boon to traders, who can do nothing in these thinly-populated regions without trusting goods to agents and petty dealers, who frequently squander them away in gambling and debauchery. The lower classes are almost all in a chronic state of debt. The merchant trusts them again and again, till the amount is something serious, when he brings them to court and has their services allotted to him for its liquidation. The debtors seem to think this no disgrace, but rather enjoy their freedom from responsibility, and the dignity of their position under a wealthy and well-known merchant. They trade a little on their own account, and both parties seem to get on very well together. The plan seems a more sensible one than that which we adopt, of effectually preventing a man from earning anything towards paying his debts by shutting him up in a jail.

My own servants were three in number. Ali, the Malay boy whom I had picked up in Borneo, was my head man. He had already been with me a year, could turn his hand to anything, and was quite attentive and trustworthy. He was a good shot, and fond of shooting, and I had taught him to skin birds very well. The second, named Baderoon, was a Macassar lad, also a pretty good boy, but a desperate gambler. Under pretence of buying a house for his mother, and clothes for himself, he had received four months' wages about a week before we sailed, and in a day or two gambled away every dollar of it. He had come on board with no clothes, no betel, or tobacco, or salt fish, all which necessary articles I was obliged to send Ali to buy for him. These two lads were about sixteen, I should suppose; the third was younger, a sharp little rascal named Baso, who had been with me a month or two, and had learnt to cook tolerably. He was to fulfil the important office of cook and housekeeper, for I could not get any regular servants to go to such a terribly remote country; one might as well ask a *chef de cuisine* to go to Patagonia.

On the fifth day that I had spent on board (Dec. 15th) the rain ceased, and final preparations were made for starting. Sails were dried and furled, boats were constantly coming and going, and stores for the voyage, fruit, vegetables, fish, and palm-sugar, were taken on board. In the afternoon two women arrived with a large party of friends and relations, and at parting there was a general nose-rubbing (the Malay kiss), and some tears shed. These were promising symptoms for our getting off the next day; and accordingly, at three in the morning, the owner came on board, the anchor was immediately weighed, and by four we set sail. Just as we were fairly off and clear of the other praus, the old juragan repeated some prayers, all around responding with "Allah il Allah," and a few strokes on a gong as an accompaniment, concluding with all wishing each other "Salaamat jalan" (a safe and happy journey). We had a light breeze, a calm sea, and a fine morning, a prosperous commencement of our voyage of about a thousand miles to the far-famed Aru Islands.

The wind continued light and variable all day, with a calm in the evening before the land breeze sprang up. We were then passing the island of "Tanakaki" (foot of the land), at the extreme south of this part of Celebes. There are some dangerous

rocks here, and as I was standing by the bulwarks, I happened
to spit over the side ; one of the men begged I would not do so
just now, but spit on deck, as they were much afraid of this
place. Not quite comprehending I made him repeat his request,
when, seeing he was in earnest, I said, "Very well, I suppose
there are 'hantus' (spirits) here." "Yes," said he, "and they
don't like anything to be thrown overboard ; many a prau has
been lost by doing it." Upon which I promised to be very
careful. At sunset the good Mohammedans on board all repeated
a few words of prayer with a general chorus, reminding me of
the pleasing and impressive *Ave Maria* of Catholic countries.

Dec 20th.—At sunrise we were opposite the Bontyne mountain,
said to be one of the highest in Celebes. In the afternoon,
we passed the Salayer Straits and had a little squall, which
obliged us to lower our huge mast, sails, and heavy yards. The
rest of the evening we had a fine west wind, which carried us on
at near five knots an hour, as much as our lumbering old tub
can possibly go.

Dec. 21st.—A heavy swell from the south-west rolling us about
most uncomfortably. A steady wind was blowing, however,
and we got on very well.

Dec. 22nd.—The swell had gone down. We passed Boutong, a
large island, high, woody, and populous, the native place of some
of our crew. A small prau returning from Bali to the island
of Goram overtook us. The nakoda (captain) was known to our
owner. They had been two years away, but were full of people,
with several black Papuans on board. At 6 P.M. we passed
Wangiwangi, low but not flat, inhabited and subject to Boutong.
We had now fairly entered the Molucca Sea. After dark it was
a beautiful sight to look down on our rudders, from which rushed
eddying streams of phosphoric light gemmed with whirling
sparks of fire. It resembled (more nearly than anything else to
which I can compare it) one of the large irregular nebulous
star-clusters seen through a good telescope, with the additional
attraction of ever-changing form and dancing motion.

Dec. 23rd.— Fine red sunrise ; the island we left last evening
barely visible behind us. The Goram prau about a mile south
of us. They have no compass, yet they have kept a very true
course during the night. Our owner tells me they do it by the
swell of the sea, the direction of which they notice at sunset,
and sail by it during the night. In these seas they are never
(in fine weather) more than two days without seeing land. Of
course adverse winds or currents sometimes carry them away,
but they soon fall in with some island, and there are always some
old sailors on board who know it, and thence take a new course.
Last night a shark about five feet long was caught, and this
morning it was cut up and cooked. In the afternoon they got
another, and I had a little fried, and found it firm and dry, but
very palatable. In the evening the sun set in a heavy bank of
clouds, which, as darkness came on, assumed a fearfully black

appearance. According to custom, when strong wind or rain is expected, our large sails were furled, and with their yards let down on deck, and a small square foresail alone kept up. The great mat sails are most awkward things to manage in rough weather. The yards which support them are seventy feet long, and of course very heavy ; and the only way to furl them being to roll up the sail on the boom, it is a very dangerous thing to have them standing when overtaken by a squall. Our crew, though numerous enough for a vessel of 700 instead of one of 70 tons, have it very much their own way, and there seems to be seldom more than a dozen at work at a time. When anything important is to be done, however, all start up willingly enough, but then all think themselves at liberty to give their opinion, and half a dozen voices are heard giving orders, and there is such a shrieking and confusion that it seems wonderful anything gets done at all.

Considering we have fifty men of several tribes and tongues on board, wild, half-savage-looking fellows, and few of them feeling any of the restraints of morality or education, we get on wonderfully well. There is no fighting or quarrelling, as there would certainly be among the same number of Europeans with as little restraint upon their actions, and there is scarcely any of that noise and excitement which might be expected. In fine weather the greater part of them are quietly enjoying them-selves—some are sleeping under the shadow of the sails ; others, in little groups of three or four, are talking or chewing betel ; one is making a new handle to his chopping-knife, another is stitching away at a new pair of trousers or a shirt, and all are as quiet and well-conducted as on board the best-ordered English merchantman. Two or three take it by turns to watch in the bows and see after the braces and halyards of the great sails ; the two steersmen are below in the steerage ; our captain, or the juragan, gives the course, guided partly by the compass and partly by the direction of the wind, and a watch of two or three on the poop look after the trimming of the sails and call out the hours by the water-clock. This is a very ingenious contrivance, which measures time well in both rough weather and fine. It is simply a bucket half filled with water, in which floats the half of a well-scraped cocoa-nut shell. In the bottom of this shell is a very small hole, so that when placed to float in the bucket a fine thread of water squirts up into it. This gradually fills the shell, and the size of the hole is so adjusted to the capacity of the vessel that, exactly at the end of an hour, plump it goes to the bottom. The watch then cries out the number of hours from sunrise, and sets the shell afloat again empty. This is a very good measurer of time. I tested it with my watch and found that it hardly varied a minute from one hour to another, nor did the motion of the vessel have any effect upon it, as the water in the bucket of course kept level. It has a great advantage for a rude people in being easily understood, in being

rather bulky and easy to see, and in the final submergence being
accompanied with a little bubbling and commotion of the water,
which calls the attention to it. It is also quickly replaced if
lost while in harbour.

Our captain and owner I find to be a quiet, good-tempered
man, who seems to get on very well with all about him. When
at sea he drinks no wine or spirits, but indulges only in coffee
and cakes, morning and afternoon, in company with his supercargo and assistants. He is a man of some little education, can
read and write well both Dutch and Malay, uses a compass, and
has a chart. He has been a trader to Aru for many years, and
is well known to both Europeans and natives in this part of the
world.

Dec. 24th.—Fine, and little wind. No land in sight for the
first time since we left Macassar. At noon calm, with heavy
showers, in which our crew wash their clothes, and in the afternoon the prau is covered with shirts, trousers, and sarongs of
various gay colours. I made a discovery to-day which at first
rather alarmed me. The two ports, or openings, through which
the tillers enter from the lateral rudders are not more than three
or four feet above the surface of the water, which thus has a
free entrance into the vessel. I of course had imagined that
this open space from one side to the other was separated from
the hold by a water-tight bulkhead, so that a sea entering
might wash out at the further side, and do no more harm than
give the steersmen a drenching. To my surprise and dismay,
however, I find that it is completely open to the hold, so that
half-a-dozen seas rolling in on a stormy night would nearly, or
quite, swamp us. Think of a vessel going to sea for a month
with two holes, each a yard square, into the hold, at three feet
above the water-line—holes, too, which cannot possibly be
closed ! But our captain says all praus are so ; and though he
acknowledges the danger, " he does not know how to alter it—
the people are used to it ; he does not understand praus so well
as they do, and if such a great alteration were made, he should
be sure to have difficulty in getting a crew ! " This proves at all
events that praus must be good sea-boats, for the captain has been
continually making voyages in them for the last ten years, and
says he has never known water enough enter to do any harm.

Dec. 25th.—Christmas-day dawned upon us with gusts of
wind, driving rain, thunder and lightning, added to which a short
confused sea made our queer vessel pitch and roll very uncomfortably. About nine o'clock, however, it cleared up, and we
then saw ahead of us the fine island of Bouru, perhaps forty or
fifty miles distant, its mountains wreathed with clouds, while
its lower lands were still invisible. The afternoon was fine, and
the wind got round again to the west ; but although this is
really the west monsoon, there is no regularity or steadiness
about it, calms and breezes from every point of the compass
continually occurring. The captain, though nominally a Protest-

ant, seemed to have no idea of Christmas-day as a festival.
Our dinner was of rice and curry as usual, and an extra glass of
wine was all I could do to celebrate it.

Dec. 26th.—Fine view of the mountains of Bouru, which we
have now approached considerably. Our crew seem rather a
clumsy lot. They do not walk the deck with the easy swing of
English sailors, but hesitate and stagger like landsmen. In the
night the lower boom of our mainsail broke, and they were all
the morning repairing it. It consisted of two bamboos lashed
together thick end to thin, and was about seventy feet long.
The rigging and arrangement of these praus contrasts strangely
with that of European vessels, in which the various ropes and
spars, though much more numerous, are placed so as not to
interfere with each other's action. Here the case is quite different; for though there are no shrouds or stays to complicate
the matter, yet scarcely anything can be done without first
clearing something else out of the way. The large sails cannot
be shifted round to go on the other tack without first hauling
down the jibs, and the booms of the fore and aft sails have
to be lowered and completely detached to perform the same
operation. Then there are always a lot of ropes foul of each
other, and all the sails can never be set (though they are so few)
without a good part of their surface having the wind kept
out of them by others. Yet praus are much liked even by those
who have had European vessels, because of their cheapness both
in first cost and in keeping up; almost all repairs can be done
by the crew, and very few European stores are required.

Dec. 28th.—This day we saw the Banda group, the volcano
first appearing—a perfect cone, having very much the outline
of the Egyptian pyramids, and looking almost as regular. In
the evening the smoke rested over its summit like a small
stationary cloud. This was my first view of an active volcano,
but pictures and panoramas have so impressed such things on
one's mind, that when we at length behold them they seem
nothing extraordinary.

Dec. 30th.—Passed the island of Teor, and a group near it,
which are very incorrectly marked on the charts. Flying-fish
were numerous to-day. It is a smaller species than that of the
Atlantic, and more active and elegant in its motions. As they
skim along the surface they turn on their sides, so as fully to
display their beautiful fins, taking a flight of about a hundred
yards, rising and falling in a most graceful manner. At a
little distance they exactly resemble swallows, and no one who
sees them can doubt that they really do fly, not merely descend
in an oblique direction from the height they gain by their first
spring. In the evening an aquatic bird, a species of booby
(Sula fiber.) rested on our hen-coop, and was caught by the
neck by one of my boys.

Dec. 31st.—At daybreak the Ké Islands (pronounced kay)
were in sight, where we are to stay a few days. About noon

we rounded the northern point, and endeavoured to coast along
to the anchorage ; but being now on the leeward side of the
island, the wind came in violent irregular gusts, and then
leaving us altogether, we were carried back by a strong current.
Just then two boats-load of natives appeared, and our owner
having agreed with them to tow us into harbour, they tried
to do so, assisted by our own boat, but could make no way.   We
were therefore obliged to anchor in a very dangerous place on a
rocky bottom, and we were engaged till nearly dark getting
hawsers secured to some rocks under water.   The coast of Ké
along which we had passed was very picturesque.   Light-
coloured limestone rocks rose abruptly from the water to the
height of several hundred feet, everywhere broken into jutting
peaks and pinnacles, weather-worn into sharp points and honeycombed surfaces, and clothed throughout with a most varied and
luxuriant vegetation.   The cliffs above the sea offered to our
view screw-pines and arborescent Liliaceæ of strange forms,
mingled with shrubs and creepers ; while the higher slopes supported a dense growth of forest trees.   Here and there little bays
and inlets presented beaches of dazzling whiteness.   The water
was transparent as crystal, and tinged the rock-strewn slope
which plunged steeply into its unfathomable depths with colours
varying from emerald to lapis-lazuli.   The sea was calm as a
lake, and the glorious sun of the tropics threw a flood of golden
light over all.   The scene was to me inexpressibly delightful.   I
was in a new world, and could dream of the wonderful productions hid in those rocky forests, and in those azure abysses.
But few European feet had ever trodden the shores I gazed upon ;
its plants, and animals, and men were alike almost unknown,
and I could not help speculating on what my wanderings there
for a few days might bring to light.

CHAPTER XXIX.

THE KÉ ISLANDS.

(JANUARY 1857.)

THE native boats that had come to meet us were three or four
in number, containing in all about fifty men.   They were long
canoes, with the bow and stern rising up into a peak six or eight
feet high, decorated with shells and waving plumes of cassowaries'
hair.   I now had my first view of Papuans in their own country,
and in less than five minutes was convinced that the opinion
already arrived at by the examination of a few Timor and New
Guinea slaves was substantially correct, and that the people I
now had an opportunity of comparing side by side belonged to
two of the most distinct and strongly marked races that the

earth contains. Had I been blind, I could have been certain that these islanders were not Malays The loud, rapid, eager tones, the incessant motion, the intense vital activity manifested in speech and action, are the very antipodes of the quiet, unimpulsive, unanimated Malay. These Ké men came up singing and shouting, dipping their paddles deep in the water and throwing up clouds of spray ; as they approached nearer they stood up in their canoes and increased their noise and gesticulations ; and on coming alongside, without asking leave, and without a moment's hesitation, the greater part of them scrambled up on our deck just as if they were come to take possession of a captured vessel. Then commenced a scene of indescribable confusion. These forty black, naked, mop-headed savages seemed intoxicated with joy and excitement. Not one of them could remain still for a moment. Every individual of our crew was in turn surrounded and examined, asked for tobacco or arrack, grinned at and deserted for another. All talked at once, and our captain was regularly mobbed by the chief men, who wanted to be employed to tow us in, and who begged vociferously to be paid in advance. A few presents of tobacco made their eyes glisten ; they would express their satisfaction by grins and shouts, by rolling on deck, or by a headlong leap overboard. School-boys on an unexpected holiday, Irishmen at a fair, or midshipmen on shore, would give but faint idea of the exuberant animal enjoyment of these people.

Under similar circumstances Malays *could* not behave as these Papuans did. If they came on board a vessel (after asking permission), not a word would be at first spoken, except a few compliments, and only after some time, and very cautiously, would any approach be made to business. One would speak at a time, with a low voice and great deliberation, and the mode of making a bargain would be by quietly refusing all your offers, or even going away without saying another word about the matter, unless you advanced your price to what they were willing to accept. Our crew, many of whom had not made the voyage before, seemed quite scandalized at such unprecedented bad manners, and only very gradually made any approach to fraternization with the black fellows. They reminded me of a party of demure and well-behaved children suddenly broken in upon by a lot of wild, romping, riotous boys, whose conduct seems most extraordinary and very naughty !

These moral features are more striking and more conclusive of absolute diversity than even the physical contrast presented by the two races, though that is sufficiently remarkable. The sooty blackness of the skin, the mop-like head of frizzly hair, and, most important of all, the marked form of countenance of quite a different type from that of the Malay, are what we cannot believe to result from mere climatal or other modifying influences on one and the same race. The Malay face is of the

Mongolian type, broad and somewhat flat. The brows are depressed, the mouth wide, but not projecting, and the nose small and well formed but for the great dilatation of the nostrils. The face is smooth, and rarely develops the trace of a beard ; the hair black, coarse, and perfectly straight. The Papuan, on the other hand, has a face which we may say is compressed and projecting. The brows are protuberant and overhanging, the mouth large and prominent, while the nose is very large, the apex elongated downwards, the ridge thick, and the nostrils large. It is an obtrusive and remarkable feature in the countenance, the very reverse of what obtains in the Malay face. The twisted beard and frizzly hair complete this remarkable contrast. Here then I had reached a new world, inhabited by a strange people. Between the Malayan tribes, among whom I had for some years been living, and the Papuan races, whose country I had now entered, we may farly say that there is as much difference, both moral and physical, as between the red Indians of South America and the negroes of Guinea on the opposite side of the Atlantic.

Jan. 1st, 1857.—This has been a day of thorough enjoyment. I have wandered in the forests of an island rarely seen by Europeans. Before daybreak we left our anchorage, and in an hour reached the village of Har, where we were to stay three or four days. The range of hills here receded so as to form a small bay, and they were broken up into peaks and hummocks with intervening flats and hollows. A broad beach of the whitest sand lined the inner part of the bay, backed by a mass of cocoanut palms, among which the huts were concealed, and surmounted by a dense and varied growth of timber. Canoes and boats of various sizes were drawn up on the beach, and one or two idlers, with a few children and a dog, gazed at our prau as we came to an anchor.

When we went on shore the first thing that attracted us was a large and well-constructed shed, under which a long boat was being built, while others in various stages of completion were placed at intervals along the beach. Our captain, who wanted two of moderate size for the trade among the islands at Aru, immediately began bargaining for them, and in a short time had arranged the number of brass guns, gongs, sarongs, handkerchiefs, axes, white plates, tobacco and arrack which he was to give for a pair which could be got ready in four days. We then went to the village, which consisted only of three or four huts, situated immediately above the beach on an irregular rocky piece of ground overshadowed with cocoa-nuts, palms, bananas, and other fruit trees. The houses were very rude, black and half rotten, raised a few feet on posts with low sides of bamboo or planks and high thatched roofs. They had small doors and no windows, an opening under the projecting gables letting the smoke out and a little light in. The floors were of strips of bamboo, thin, slippery, and elastic, and so weak, that my feet were in danger of plunging through at every step. Native boxes of pandanus-leaves and slabs of palm pith, very neatly constructed, mats of the same, jars and cooking pots of native pottery, and a few European plates and basins, were the whole furniture, and the interior was throughout dark and smoke-blackened, and dismal in the extreme.

Accompanied by Ali and Baderoon, I now attempted to make some explorations, and we were followed by a train of boys eager to see what we were going to do. The most trodden path from the beach led us into a shady hollow, where the trees were of immense height and the undergrowth scanty. From the summits of these trees came at intervals a deep booming sound, which at first puzzled us, but which we soon found to proceed from some large pigeons. My boys shot at them, and after one or two misses, brought one down. It was a magnificent bird twenty inches long, of a bluish white colour, with the back wings and tail intense metallic green, with golden, blue, and violet reflexions, the feet coral red, and the eyes golden yellow. It is a rare species, which I have named Carpophaga concinna, and is found only in a few small islands, where, however, it abounds. It is the same species which in the island of Banda is called the nutmeg-pigeon, from its habit of devouring the fruits, the seed or nutmeg being thrown up entire and uninjured. Though these pigeons have a narrow beak, yet their jaws and throat are so extensible that they can swallow fruits of very large size. I had before shot a species much smaller than this one, which had a number of hard gobular palm-fruits in its crop, each more than an inch in diameter.

A little further the path divided into two, one leading along the beach, and across mangrove and sago swamps, the other rising to cultivated grounds. We therefore returned and taking a fresh departure from the village, endeavoured to ascend the hills and penetrate into the interior. The path, however, was a most trying one. Where there was earth, it was a deposit of reddish clay overlying the rock, and was worn so smooth by the attrition of naked feet that my shoes could obtain no hold on the sloping surface. A little farther we came to the bare rock, and this was worse, for it was so rugged and broken, and so honeycombed and weatherworn into sharp points and angles, that my boys, who had gone barefooted all their lives, could not stand it. Their feet began to bleed, and I saw that if I did not want them completely lamed it would be wise to turn back. My own shoes, which were rather thin, were but a poor protection, and would soon have been cut to pieces; yet our little naked guides tripped along with the greatest ease and unconcern, and seemed much astonished at our effeminacy in not being able to take a walk which to them was a perfectly agreeable one. During the rest of our stay in the island we were obliged to confine ourselves to the vicinity of the shore and the cultivated grounds, and those more level portions of the forest where a little soil had accumulated and the rock had been less exposed to atmospheric action.

The island of Ké (pronounced exactly as the letter K, but erroneously spelt in our maps Key or Ki) is long and narrow, running in a north and south direction, and consists almost entirely of rock and mountain. It is everywhere covered with luxuriant forests, and in its bays and inlets the sand is of dazzling whiteness, resulting from the decomposition of the coralline limestone of which it is entirely composed. In all the little swampy inlets and valleys sago trees abound, and these supply the main subsistence of the natives, who grow no rice, and have scarcely any other cultivated products but cocoa-nuts, plantains, and yams. From the cocoa-nuts, which surround every hut, and which thrive exceedingly on the porous limestone soil and under the influence of salt breezes, oil is made which is sold at a good price to the Aru traders, who all touch here to lay in their stock of this article, as well as to purchase boats and native crockery. Wooden bowls, pans, and trays, are also largely made here, hewn out of solid blocks of wood with knife and adze ; and these are carried to all parts of the Moluccas. But the art in which the natives of Ké pre-eminently excel is that of boat-building. Their forests supply abundance of fine timber though probably not more so than many other islands, and from some unknown causes these remote savages have come to excel in what seems a very difficult art. Their small canoes are beautifully formed, broad and low in the centre, but rising at each end, where they terminate in high-pointed beaks more or less carved, and ornamented with a plume of feathers. They are not hollowed out of a tree, but are regularly built of planks running from end to end, and so accurately fitted that it is often difficult to find a place where a knife-blade can be inserted between the joints. The larger ones are from 20 to 30 tons burthen, and are finished ready for sea without a nail or particle of iron being used, and with no other tools than axe, adze, and auger. These vessels are handsome to look at, good sailers, and admirable sea-boats, and will make long voyages with perfect safety, traversing the whole Archipelago from New Guinea to Singapore in seas, which, as every one who has sailed much in them can testify, are not so smooth and tempest free as word-painting travellers love to represent them.

The forests of Ké produce magnificent timber, tall, straight, and durable, of various qualities, some of which are said to be superior to the best Indian teak. To make each pair of planks used in the construction of the larger boats an entire tree is consumed. It is felled, often miles away from the shore, cut across to the proper length, and then hewn longitudinally into two equal portions. Each of these forms a plank by cutting down with the axe to a uniform thickness of three or four inches, leaving at first a solid block at each end to prevent splitting. Along the centre of each plank a series of projecting pieces are

left, standing up three or four inches, about the same width, and
a foot long ; these are of great importance in the construction
of the vessel. When a sufficient number of planks have been
made, they are laboriously dragged through' the forest by three
or four men each to the beach, where the boat is to be built.
A foundation piece, broad in the middle and rising considerably
at each end, is first laid on blocks and properly shored up. The
edges of this are worked true and smooth with the adze, and a
plank, properly curved and tapering at each end, is held firmly
up against it, while a line is struck along it which allows it to
be cut so as to fit exactly. A series of auger holes, about as
large as one's finger, are then bored along the opposite edges,
and pins of very hard wood are fitted to these, so that the two
planks are held firmly, and can be driven into the closest contact ; and difficult as this seems to do without any other aid
than rude practical skill in forming each edge to the true
corresponding curves, and in boring the holes so as exactly to
match both in position and direction, yet so well is it done that
the best European shipwright cannot produce sounder or closer-
fitting joints. The boat is built up in this way by fitting plank
to plank till the proper height and width are obtained. We
have now a skin held together entirely by the hard-wood pins
connecting the edges of the planks, very strong and elastic, but
having nothing but the adhesion of these pins to prevent the
planks gaping. In the smaller boats seats, in the larger ones
cross-beams are now fixed. They are sprung into slight notches
cut to receive them, and are further secured to the projecting
pieces of the plank below by a strong lashing of rattan. Ribs
are now formed of single pieces of tough wood chosen and
trimmed so as exactly to fit on to the projections from each
plank, being slightly notched to receive them, and securely
bound to them by rattans passed through a hole in each projecting piece close to the surface of the plank. The ends are
closed against the vertical prow and stern posts, and further
secured with pegs and rattans, and then the boat is complete ;
and when fitted with rudders, masts, and thatched covering, is
ready to do battle with the waves. A careful consideration of
the principle of this mode of construction, and allowing for the
strength and binding qualities of rattan (which resembles in
these respects wire rather than cordage), makes me believe that
a vessel carefully built in this manner is actually stronger and
safer than one fastened in the ordinary way with nails.

During our stay here we were all very busy. Our captain was
daily superintending the completion of his two small praus. All
day long native boats were coming with fish, cocoa-nuts, parrots
and lories, earthen pans, sirih leaf, wooden bowls, and trays, &c.
&c., which every one of the fifty inhabitants of our prau seemed
to be buying on his own account, till all available and most un-
available space of our vessel was occupied with these miscellaneous articles : for every man on board a prau considers

himself at liberty to trade, and to carry with him whatever he can afford to buy.

Money is unknown and valueless here—knives, cloth, and arrack forming the only medium of exchange, with tobacco for small coin. Every transaction is the subject of a special bargain, and the cause of much talking. It is absolutely necessary to offer very little, as the natives are never satisfied till you add a little more. They are then far better pleased than if you had given them twice the amount at first and refused to increase it.

I, too, was doing a little business, having persuaded some of the natives to collect insects for me ; and when they really found that I gave them most fragrant tobacco for worthless black and green beetles, I soon had scores of visitors, men, women, and children, bringing bamboos full of creeping things, which, alas ! too frequently had eaten each other into fragments during the tedium of a day's confinement. Of one grand new beetle, glittering with ruby and emerald tints, I got a large quantity, having first detected one of its wing-cases ornamenting the outside of a native's tobacco pouch. It was quite a new species, and had not been found elsewhere than on this little island. It is one of the Buprestidæ, and has been named Cyphogastra calepyga.

Each morning after an early breakfast I wandered by myself into the forest, where I found delightful occupation in capturing the large and handsome butterflies, which were tolerably abundant, and most of them new to me ; for I was now upon the confines of the Moluccas and New Guinea—a region the productions of which were then among the most precious and rare in the cabinets of Europe. Here my eyes were feasted for the first time with splendid scarlet lories on the wing, as well as by the sight of that most imperial butterfly, the " Priamus" of collectors, or a closely allied species, but flying so high that I did not succeed in capturing a specimen. One of them was brought me in a bamboo, boxed up with a lot of beetles, and of course torn to pieces. The principal drawback of the place for a collector is the want of good paths, and the dreadfully rugged character of the surface, requiring the attention to be so continually directed to securing a footing, as to make it very difficult to capture active winged things, who pass out of reach while one is glancing to see that the next step may not plunge one into a chasm or over a precipice. Another inconvenience is that there are no running streams, the rock being of so porous a nature that the surface-water everywhere penetrates its fissures ; at least such is the character of the neighbourhood we visited, the only water being small springs trickling out close to the sea-beach.

In the forests of Ké, arboreal Liliaceæ and Pandanaceæ abound, and give a character to the vegetation in the more exposed rocky places. Flowers were scarce, and there were not many orchids, but I noticed the fine white butterfly-orchis,

Phalænopsis grandiflora, or a species closely allied to it. The freshness and vigour of the vegetation was very pleasing, and on such an arid, rocky surface was a sure indication of a perpetually humid climate. Tall clean trunks, many of them buttressed, and immense trees of the fig family, with aërial roots stretching out and interlacing and matted together for fifty or a hundred feet above the ground, were the characteristic features ; and there was an absence of thorny shrubs and prickly rattans, which would have made these wilds very pleasant to roam in, had it not been for the sharp honeycombed rocks already alluded to. In damp places a fine undergrowth of broad-leaved herbaceous plants was found, about which swarmed little green lizards, with tails of the most "heavenly blue," twisting in and out among the stalks and foliage so actively that I often caught glimpses of their tails only, when they startled me by their resemblance to small snakes. Almost the only sounds in these primæval woods proceeded from two birds, the red lories, who utter shrill screams like most of the parrot tribe, and the large green nutmeg-pigeon, whose voice is either a loud and deep boom, like two notes struck upon a very large gong, or sometimes a harsh toad-like croak, altogether peculiar and remarkable. Only two quadrupeds are said by the natives to inhabit the island—a wild pig and a Cuscus, or Eastern opossum, of neither of which could I obtain specimens.

The insects were more abundant, and very interesting. Of butterflies I caught thirty-five species, most of them new to me, and many quite unknown in European collections. Among them was the fine yellow and black Papilio euchenor, of which but few specimens had been previously captured, and several other handsome butterflies of large size, as well as some beautiful little "blues," and some brilliant day-flying moths. The beetle tribe were less abundant, yet I obtained some very fine and rare species. On the leaves of a slender shrub in an old clearing I found several fine blue and black beetles of the genus Eupholus, which almost rival in beauty the diamond beetles of South America. Some cocoa-nut palms in blossom on the beach were frequented by a fine green floral beetle (Lomaptera papua), which, when the flowers were shaken, flew off like a small swarm of bees. I got one of our crew to climb up the tree, and he brought me a good number in his hand ; and seeing they were valuable, I sent him up again with my net to shake the flowers into, and thus secured a large quantity. My best capture, however, was the superb insect of the Buprestis family, already mentioned as having been obtained from the natives, who told me they found it in rotten trees in the mountains.

In the forest itself the only common and conspicuous coleoptera were two tiger beetles. One, Therates labiata, was much larger than our green tiger beetle, of a purple black colour, with green metallic glosses, and the broad upper lip of a bright yellow. It was always found upon foliage, generally of broad-leaved herbaceous plants, and in damp and gloomy situations, taking frequent short flights from leaf to leaf, and preserving an alert attitude, as if always looking out for its prey. Its vicinity could be immediately ascertained, often before it was seen, by a very pleasant odour, like otto of roses, which it seems to emit continually, and which may probably be attractive to the small insects on which it feeds. The other, Tricondyla aptera, is one of the most curious forms in the family of the Cicindelidæ, and is almost exclusively confined to the Malay Islands. In shape it resembles a very large ant, more than an inch long, and of a purple black colour. Like an ant also it is wingless, and is generally found ascending trees, passing around the trunks in a spiral direction when approached, to avoid capture, so that it requires a sudden run and active fingers to secure a specimen. This species emits the usual fetid odour of the ground beetles. My collections during our four days' stay at Ké were as follow :— Birds, 13 species ; insects, 194 species ; and 3 kinds of land-shells.

There are two kinds of people inhabiting these islands—the indigenes, who have the Papuan characters strongly marked, and who are pagans ; and a mixed race, who are nominally Mahometans, and wear cotton clothing, while the former use only a waist cloth of cotton or bark. These Mahometans are said to have been driven out of Banda by the early European settlers. They were probably a brown race, more allied to the Malays, and their mixed descendants here exhibit great variations of colour, hair, and features, graduating between the Malay and Papuan types. It is interesting to observe the influence of the early Portuguese trade with these countries in the words of their language, which still remain in use even among these remote and savage islanders. "Lenco" for handkerchief, and "faca" for knife, are here used to the exclusion of the proper Malay terms. The Portuguese and Spaniards were truly wonderful conquerors and colonizers. They effected more rapid changes in the countries they conquered than any other nations of modern times, resembling the Romans in their power of impressing their own language, religion, and manners on rude and barbarous tribes.

The striking contrast of character between these people and the Malays is exemplified in many little traits. One day when I was rambling in the forest, an old man stopped to look at me catching an insect. He stood very quiet till I had pinned and put it away in my collecting box, when he could contain himself no longer, but bent almost double, and enjoyed a hearty roar of laughter. Every one will recognize this as a true negro trait. A Malay would have stared, and asked with a tone of bewilderment what I was doing, for it is but little in his nature to laugh, never heartily, and still less at or in the presence of a stranger, to whom, however, his disdainful glances or whispered remarks are less agreeable than the most boisterous open expression of merriment. The women here were not so much frightened at strangers, or made to keep themselves so much secluded as among the Malay races ; the children were more merry and had the "nigger grin," while the noisy confusion of tongues among the men, and their excitement on very ordinary occasions, are altogether removed from the general taciturnity and reserve of the Malay.

The language of the Ké people consists of words of one, two, or three syllables in about equal proportions, and has many aspirated and a few guttural sounds. The different villages have slight differences of dialect, but they are mutually intelligible, and, except in words that have evidently been introduced during a long-continued commercial intercourse, seem to have no affinity whatever with the Malay languages.

Jan. 6th.—The small boats being finished, we sailed for Aru at 4 P.M., and as we left the shores of Ké had a fine view of its rugged and mountainous character ; ranges of hills, three or four thousand feet high, stretching southwards as far as the eye could reach, everywhere covered with a lofty, dense, and unbroken forest. We had very light winds, and it therefore took us thirty hours to make the passage of sixty miles to the low, or flat, but equally forest-covered Aru Islands, where we anchored in the harbour of Dobbo at nine in the evening of the next day.

My first voyage in a prau being thus satisfactorily terminated, I must, before taking leave of it for some months, bear testimony to the merits of the queer old-world vessel. Setting aside all ideas of danger, which is probably, after all, not more than in any other craft, I must declare that I have never, either before or since, made a twenty days' voyage so pleasantly, or perhaps, more correctly speaking, with so little discomfort. This I attribute chiefly to having my small cabin on deck, and entirely to myself, to having my own servants to wait upon me, and to the absence of all those marine-store smells of paint, pitch, tallow, and new cordage, which are to me insupportable. Something is also to be put down to freedom from all restraint of dress, hours of meals, &c., and to the civility and obliging disposition of the captain. I had agreed to have my meals with him, but whenever I wished it I had them in my own berth, and at what hours I felt inclined. The crew were all civil and good-tempered, and with very little discipline everything went on smoothly, and the vessel was kept very clean and in pretty good order, so that on the whole I was much delighted with the trip, and was inclined to rate the luxuries of the semi-barbarous prau as surpassing those of the most magnificent screw-steamer, that highest product of our civilization.

CHAPTER XXX.

THE ARU ISLANDS.—RESIDENCE IN DOBBO.

(JANUARY TO MARCH 1857.)

On the 8th of January, 1857, I landed at Dobbo, the trading settlement of the Bugis and Chinese, who annually visit the Aru Islands. It is situated on the small island of Wamma, upon a spit of sand which projects out to the north, and is just wide enough to contain three rows of houses. Though at first sight a most strange and desolate-looking place to build a village on, it has many advantages. There is a clear entrance from the west among the coral reefs that border the land, and there is good anchorage for vessels, on one side of the village or the other, in both the east and west monsoons. Being fully exposed to the sea-breezes in three directions it is healthy, and the soft sandy beach offers great facilities for hauling up the praus, in order to secure them from sea-worms and prepare them for the homeward voyage. At its southern extremity the sand-bank merges in the beach of the island, and is backed by a luxuriant growth of lofty forest. The houses are of various sizes, but are all built after one pattern, being merely large thatched sheds, a small portion of which, next the entrance, is used as a dwelling, while the rest is parted off, and often divided by one or two floors, in order better to stow away merchandise and native produce.

As we had arrived early in the season, most of the houses were empty, and the place looked desolate in the extreme—the whole of the inhabitants who received us on our landing amounting to about half-a-dozen Bugis and Chinese. Our captain, Herr Warzbergen, had promised to obtain a house for me, but unforeseen difficulties presented themselves. One which was to let had no roof, and the owner, who was building it on speculation, could not promise to finish it in less than a month. Another, of which the owner was dead, and of which I might therefore take undisputed possession as the first comer, wanted considerable repairs, and no one could be found to do the work, although about four times its value was offered. The captain, therefore, recommended me to take possession of a pretty good house near his own, whose owner was not expected for some weeks ; and as I was anxious to be on shore, I immediately had it cleared out, and by evening had all my things housed, and was regularly installed as an inhabitant of Dobbo. I had brought with me a cane chair, and a few light boards, which were soon rigged up into a table and shelves. A broad bamboo bench served as sofa and bedstead, my boxes were conveniently arranged, my mats

spread on the floor, a window cut in the palm-leaf wall to light
my table, and though the place was as miserable and gloomy
a shed as could be imagined, I felt as contented as if I had
obtained a well-furnished mansion, and looked forward to a
month's residence in it with unmixed satisfaction.

The next morning, after an early breakfast, I set off to explore
the virgin forests of Aru, anxious to set my mind at rest as to
the treasures they were likely to yield, and the probable success
of my long-meditated expedition. A little native imp was our
guide, seduced by the gift of a German knife, value three-
halfpence, and my Macassar boy Baderoon brought his chopper
to clear the path if necessary.

We had to walk about half a mile along the beach, the ground
behind the village being mostly swampy, and then turned into
the forest along a path which leads to the native village of
Wamma, about three miles off on the other side of the island.
The path was a narrow one, and very little used, often swampy
and obstructed by fallen trees, so that after about a mile we
lost it altogether, our guide having turned back, and we were
obliged to follow his example. In the meantime, however, I had
not been idle, and my day's captures determined the success of
my journey in an entomological point of view. I had taken
about thirty species of butterflies, more than I had ever captured
in a day since leaving the prolific banks of the Amazon, and
among them were many most rare and beautiful insects, hitherto
only known by a few specimens from New Guinea. The large
and handsome spectre-butterfly, Hestia durvillei; the pale-
winged peacock butterfly, Drusilla catops; and the most brilliant
and wonderful of the clear-winged moths, Cocytia d'Urvillei,
were especially interesting, as well as several little "blues,"
equalling in brilliancy and beauty anything the butterfly world
can produce. In the other groups of insects I was not so successful, but this was not to be wondered at in a mere exploring
ramble, when only what is most conspicuous and novel attracts
the attention. Several pretty beetles, a superb "bug," and a few
nice land-shells were obtained, and I returned in the afternoon
well satisfied with my first trial of the promised land.

The next two days were so wet and windy that there was no
going out; but on the succeeding one the sun shone brightly,
and I had the good fortune to capture one of the most magnificent insects the world contains, the great bird-winged butterfly, Ornithoptera poseidon. I trembled with excitement as I
saw it coming majestically towards me, and could hardly believe
I had really succeeded in my stroke till I had taken it out of
the net and was gazing, lost in admiration, at the velvet black
and brilliant green of its wings, seven inches across, its golden
body, and crimson breast. It is true I had seen similar insects
in cabinets at home, but it is quite another thing to capture
such one's self—to feel it struggling between one's fingers,
and to gaze upon its fresh and living beauty, a bright gem shin-

ing out amid the silent gloom of a dark and tangled forest. The village of Dobbo held that evening at least one contented man.

Jan. 26th.—Having now been here a fortnight, I began to understand a little of the place and its peculiarities. Praus continually arrived, and the merchant population increased almost daily. Every two or three days a fresh house was opened, and the necessary repairs made. In every direction men were bringing in poles, bamboos, rattans, and the leaves of the nipa palm to construct or repair the walls, thatch, doors, and shutters of their houses, which they do with great celerity. Some of the arrivals were Macassar men or Bugis, but more from the small island of Goram, at the east end of Ceram, whose inhabitants are the petty traders of the far East. Then the natives of Aru come in from the other side of the islands (called here "blakang tana," or "back of the country") with the produce they have collected during the preceding six months, and which they now sell to the traders, to some of whom they are most likely in debt. Almost all, or I may safely say all, the new arrivals pay me a visit, to see with their own eyes the unheard-of phenomenon of a person come to stay at Dobbo who does not trade! They have their own ideas of the uses that may possibly be made of stuffed birds, beetles, and shells which are not the right shells—that is, "mother-of-pearl." They every day bring me dead and broken shells, such as I can pick up by hundreds on the beach, and seem quite puzzled and distressed when I decline them. If, however, there are any snail shells among the lot, I take them, and ask for more—a principle of selection so utterly unintelligible to them, that they give it up in despair, or solve the problem by imputing hidden medical virtue to those which they see me preserve so carefully.

These traders are all of the Malay race, or a mixture of which Malay is the chief ingredient, with the exception of a few Chinese. The natives of Aru, on the other hand, are Papuans, with black or sooty brown skins, woolly or frizzly hair, thick-ridged prominent noses, and rather slender limbs. Most of them wear nothing but a waist-cloth, and a few of them may be seen all day long wandering about the half-deserted streets of Dobbo offering their little bit of merchandise for sale.

Living in a trader's house everything is brought to me as well as to the rest—bundles of smoked tripang, or *bêche de mer*, looking like sausages which have been rolled in mud and then thrown up the chimney; dried sharks' fins, mother-of-pearl shells, as well as Birds of Paradise, which, however, are so dirty and so badly preserved that I have as yet found no specimens worth purchasing. When I hardly look at the articles, and make no offer for them, they seem incredulous, and, as if fearing they have misunderstood me, again offer them, and declare what they want in return—knives, or tobacco, or sago, or handkerchiefs. I then have to endeavour to explain, through any interpreter who may be at hand, that neither tripang nor pearl oyster shells have any charms for me, and that I even decline to speculate in tortoiseshell, but that anything eatable I will buy—fish, or turtle, or vegetables of any sort. Almost the only food, however, that we can obtain with any regularity, are fish and cockles of very good quality, and to supply our daily wants it is absolutely necessary to be always provided with four articles —tobacco, knives, sago-cakes, and Dutch copper doits—because when the particular thing asked for is not forthcoming, the fish pass on to the next house, and we may go that day without a dinner. It is curious to see the baskets and buckets used here. The cockles are brought in large volute shells, probably the Cymbium ducale, while gigantic helmet-shells, a species of Cassis, suspended by a rattan handle, form the vessels in which fresh water is daily carried past my door. It is painful to a naturalist to see these splendid shells with their inner whorls ruthlessly broken away to fit them for their ignoble use.

My collections, however, got on but slowly, owing to the unexpectedly bad weather, violent winds with heavy showers having been so continuous as only to give me four good collecting days out of the first sixteen I spent here. Yet enough had been collected to show me that with time and fine weather I might expect to do something good. From the natives I obtained some very fine insects and a few pretty land-shells ; and of the small number of birds yet shot more than half were known New Guinea species, and therefore certainly rare in European collections, while the remainder were probably new. In one respect my hopes seemed doomed to be disappointed. I had anticipated the pleasure of myself preparing fine specimens of the Birds of Paradise, but I now learnt that they are all at this season out of plumage, and that it is in September and October that they have the long plumes of yellow silky feathers in full perfection. As all the praus return in July, I should not be able to spend that season in Aru without remaining another whole year, which was out of the question. I was informed, however, that the small red species, the "King Bird of Paradise," retains its plumage at all seasons, and this I might therefore hope to get.

As I became familiar with the forest scenery of the island, I perceived it to possess some characteristic features that distinguished it from that of Borneo and Malacca, while, what is very singular and interesting, it recalled to my mind the half-forgotten impressions of the forests of Equatorial America. For example, the palms were much more abundant than I had generally found them in the East, more generally mingled with the other vegetation, more varied in form and aspect, and presenting some of those lofty and majestic smooth-stemmed, pinnate-leaved species which recall the Uauassú (Attalea speciosa) of the Amazon, but which I had hitherto rarely met with in the Malayan islands.

In animal life the immense number and variety of spiders

and of lizards were circumstances that recalled the prolific regions of South America, more especially the abundance and varied colours of the little jumping spiders which abound on flowers and foliage, and are often perfect gems of beauty. The web-spinning species were also more numerous than I had ever seen them, and were a great annoyance, stretching their nets across the footpaths just about the height of my face ; and the threads composing these are so strong and glutinous as to require much trouble to free one's self from them. Then their inhabitants, great yellow-spotted monsters with bodies two inches long, and legs in proportion, are not pleasant things to run one's nose against while pursuing some gorgeous butterfly, or gazing aloft in search of some strange-voiced bird. I soon found it necessary not only to brush away the web, but also to destroy the spinner ; for at first, having cleared the path one day, I found the next morning that the industrious insects had spread their nets again in the very same places.

The lizards were equally striking by their numbers, variety, and the situations in which they were found. The beautiful blue-tailed species so abundant in Ké was not seen here. The Aru lizards are more varied but more sombre in their colours— shades of green, grey, brown, and even black, being very frequently seen. Every shrub and herbaceous plant was alive with them ; every rotten trunk or dead branch served as a station for some of these active little insect-hunters, who, I fear, to satisfy their gross appetites, destroy many gems of the insect world, which would feast the eyes and delight the hearts of our more discriminating entomologists. Another curious feature of the jungle here was the multitude of sea-shells everywhere met with on the ground and high up on the branches and foliage, all inhabited by hermit-crabs, who forsake the beach to wander in the forest. I have actually seen a spider carrying away a good-sized shell and devouring its (probably juvenile) tenant. On the beach, which I had to walk along every morning to reach the forest, these creatures swarmed by thousands. Every dead shell, from the largest to the most minute, was appropriated by them. They formed small social parties of ten or twenty around bits of stick or seaweed, but dispersed hurriedly at the sound of approaching footsteps. After a windy night, that nasty-looking Chinese delicacy the sea-slug was sometimes thrown up on the beach, which was at such times thickly strewn with some of the most beautiful shells that adorn our cabinets, along with fragments and masses of coral and strange sponges, of which I picked up more than twenty different sorts. In many cases sponge and coral are so much alike that it is only on touching them that they can be distinguished. Quantities of seaweed, too, are thrown up ; but strange as it may seem, these are far less beautiful and less varied than may be found on any favourable part of our own coasts.

The natives here, even those who seem to be of pure Papuan

race, were much more reserved and taciturn than those of Ké. This is probably because I only saw them as yet among strangers and in small parties. One must see the savage at home to know what he really is. Even here, however, the Papuan character sometimes breaks out. Little boys sing cheerfully as they walk along, or talk aloud to themselves (quite a negro characteristic); and, try all they can, the men cannot conceal their emotions in the true Malay fashion. A number of them were one day in my house, and having a fancy to try what sort of eating tripang would be, I bought a couple, paying for them with such an extravagant quantity of tobacco that the seller saw I was a green customer. He could not, however, conceal his delight, but as he smelt the fragrant weed, and exhibited the large handful to his companions, he grinned and twisted and gave silent chuckles in a most expressive pantomime. I had often before made the same mistake in paying a Malay for some trifle. In no case, however, was his pleasure visible on his countenance—a dull and stupid hesitation only showing his surprise, which would be exhibited exactly in the same way whether he was over or under paid. These little moral traits are of the greatest interest when taken in connexion with physical features. They do not admit of the same ready explanation by external causes which is so frequently applied to the latter. Writers on the races of mankind have too often to trust to the information of travellers who pass rapidly from country to country, and thus have few opportunities of becoming acquainted with peculiarities of national character, or even of ascertaining what is really the average physical conformation of the people. Such are exceedingly apt to be deceived in places where two races have long intermingled, by looking on intermediate forms and mixed habits as evidences of a natural transition from one race to the other, instead of an artificial mixture of two distinct peoples; and they will be the more readily led into this error if, as in the present case, writers on the subject should have been in the habit of classing these races as mere varieties of one stock, as closely related in physical conformation as from their geographical proximity one might suppose they ought to be. So far as I have yet seen, the Malay and Papuan appear to be as widely separated as any two human races that exist, being distinguished by physical, mental, and moral characteristics, all of the most marked and striking kind.

Feb. 5th.—I took advantage of a very fine calm day to pay a visit to the island of Wokan, which is about a mile from us, and forms part of the "tanna busar," or mainland of Aru. This is a large island, extending from north to south about a hundred miles, but so low in many parts as to be intersected by several creeks, which run completely through it, offering a passage for good-sized vessels. On the west side, where we are, there are only a few outlying islands, of which ours (Wamma) is the principal; but on the east coast are a great number of islands, extending some miles beyond the mainland, and forming the "blakang tana," or "back country," of the traders, being the principal seat of the pearl, tripang, and tortoiseshell fisheries. To the mainland many of the birds and animals of the country are altogether confined ; the Birds of Paradise, the black cockatoo, the great brush-turkey, and the cassowary, are none of them found on Wamma, or any of the detached islands. I did not, however, expect in this excursion to see any decided difference in the forest or its productions, and was therefore agreeably surprised. The beach was overhung with the drooping branches of large trees, loaded with Orchideæ, ferns, and other epiphytal plants. In the forest there was more variety, some parts being dry, and with trees of a lower growth, while in others there were some of the most beautiful palms I have ever seen, with a perfectly straight, smooth, slender stem, a hundred feet high, and a crown of handsome drooping leaves. But the greatest novelty and most striking feature to my eyes were the tree-ferns, which, after seven years spent in the tropics, I now saw in perfection for the first time. All I had hitherto met with were slender species, not more than twelve feet high, and they gave not the least idea of the supreme beauty of trees bearing their elegant heads of fronds more than thirty feet in the air, like those which were plentifully scattered about this forest. There is nothing in tropical vegetation so perfectly beautiful.

My boys shot five sorts of birds, none of which we had obtained during a month's shooting in Wamma. Two were very pretty flycatchers, already known from New Guinea ; one of them (Monarcha chrysomela), of brilliant black and bright orange colours, is by some authors considered to be the most beautiful of all flycatchers ; the other is pure white and velvety black, with a broad fleshy ring round the eye of an azure blue colour ; it is named the "spectacled flycatcher" (Monarcha telescopthalma), and was first found in New Guinea, along with the other, by the French naturalists during the voyage of the discovery-ship *Coquille*.

Feb. 18th.—Before leaving Macassar, I had written to the Governor of Amboyna requesting him to assist me with the native chiefs of Aru. I now received by a vessel which had arrived from Amboyna a very polite answer, informing me that orders had been sent to give me every assistance that I might require ; and I was just congratulating myself on being at length able to get a boat and men to go to the mainland and explore the interior, when a sudden check came in the form of a piratical incursion. A small prau arrived which had been attacked by pirates and had a man wounded. They were said to have five boats, but more were expected to be behind, and the traders were all in consternation, fearing that their small vessels sent trading to the "blakang tana" would be plundered. The Aru natives were of course dreadfully alarmed, as these marauders attack their villages, burn and murder, and carry away women

and children for slaves. Not a man will stir from his village for
some time, and I must remain still a prisoner in Dobbo. The
Governor of Amboyna, out of pure kindness, has told the chiefs
that they are to be responsible for my safety, so that they have
an excellent excuse for refusing to stir.

Several praus went out in search of the pirates, sentinels were
appointed, and watch-fires lighted on the beach to guard against
the possibility of a night attack, though it was hardly thought
they would be bold enough to attempt to plunder Dobbo. The
next day the praus returned, and we had positive information
that these scourges of the Eastern seas were really among us.
One of Herr Warzbergen's small praus also arrived in a sad
plight. It had been attacked six days before, just as it was returning from the "blakang tana." The crew escaped in their
small boat and hid in the jungle, while the pirates came up and
plundered the vessel. They took away everything but the cargo
of mother-of-pearl shell, which was too bulky for them. All the
clothes and boxes of the men, and the sails and cordage of the
prau, were cleared off. They had four large war boats, and fired
a volley of musketry as they came up, and sent off their small
boats to the attack. After they had left, our men observed from
their concealment that three had stayed behind with a small
boat; and being driven to desperation by the sight of the
plundering, one brave fellow swam off armed only with his
parang, or chopping-knife, and coming on them unawares made
a desperate attack, killing one and wounding the other two, receiving himself numbers of slight wounds, and then swimming
off again when almost exhausted. Two other praus were also
plundered, and the crew of one of them murdered to a man.
They are said to be Sooloo pirates, but have Bugis among them.
On their way here they have devastated one of the small islands
east of Ceram. It is now eleven years since they have visited
Aru, and by thus making their attacks at long and uncertain
intervals the alarm dies away, and they find a population for
the most part unarmed and unsuspicious of danger. None of the
small trading vessels now carry arms, though they did so for a
year or two after the last attack, which was just the time when
there was the least occasion for it. A week later one of the
smaller pirate boats was captured in the "blakang tana." Seven
men were killed and three taken prisoners. The larger vessels
have been often seen but cannot be caught, as they have very
strong crews, and can always escape by rowing out to sea in the
eye of the wind, returning at night. They will thus remain
among the innumerable islands and channels till the change of
the monsoon enables them to sail westward.

March 9th.—For four or five days we have had a continual gale
of wind, with occasional gusts of great fury, which seem as if
they would send Dobbo into the sea. Rain accompanies it
almost every alternate hour, so that it is not a pleasant time.
During such weather I can do little, but am busy getting ready

a boat I have purchased, for an excursion into the interior. There is immense difficulty about men, but I believe the "Orang-kaya," or head man of Wamma, will accompany me to see that I don't run into danger.

Having become quite an old inhabitant of Dobbo, I will endeavour to sketch the sights and sounds that pervade it, and the manners and customs of its inhabitants. The place is now pretty full, and the streets present a far more cheerful aspect than when we first arrived. Every house is a store, where the natives barter their produce for what they are most in need of. Knives, choppers, swords, guns, tobacco, gambier, plates, basins, handkerchiefs, sarongs, calicoes, and arrack, are the principal articles wanted by the natives; but some of the stores contain also tea, coffee, sugar, wine, biscuits, &c., for the supply of the traders; and others are full of fancy goods, china ornaments, looking-glasses, razors, umbrellas, pipes, and purses, which take the fancy of the wealthier natives. Every fine day mats are spread before the doors and the tripang is put out to dry, as well as sugar, salt, biscuit, tea, cloths, and other things that get injured by an excessively moist atmosphere. In the morning and evening spruce Chinamen stroll about or chat at each other's doors, in blue trousers, white jacket, and a queue into which red silk is plaited till it reaches almost to their heels. An old Bugis hadji regularly takes an evening stroll in all the dignity of flowing green silk robe and gay turban, followed by two small boys carrying his sirih and betel boxes.

In every vacant space new houses are being built, and all sorts of odd little cooking-sheds are erected against the old ones, while in some out-of-the-way corners, massive log pigsties are tenanted by growing porkers; for how could the Chinamen exist six months without one feast of pig? Here and there are stalls where bananas are sold, and every morning two little boys go about with trays of sweet rice and grated cocoa-nut, fried fish, or fried plantains; and whichever it may be, they have but one cry, and that is—"Chocolat—t—t!" This must be a Spanish or Portuguese cry, handed down for centuries, while its meaning has been lost. The Bugis sailors, while hoisting the mainsail, cry out, "Véla à véla,—véla, véla, véla!" repeated in an everlasting chorus. As "véla" is Portuguese for a sail, I supposed I had discovered the origin of this, but I found afterwards they used the same cry when heaving anchor, and often change it to "hela," which is so much an universal expression of exertion and hard breathing that it is most probably a mere interjectional cry.

I dare say there are now near five hundred people in Dobbo of various races, all met in this remote corner of the East, as they express it, "to look for their fortune;" to get money any way they can. They are most of them people who have the very worst reputation for honesty as well as every other form of morality—Chinese, Bugis, Ceramese, and half-caste Javanese, with a sprinkling of half-wild Papuans from Timor, Babber, and other islands—yet all goes on as yet very quietly. This motley, ignorant, bloodthirsty, thievish population live here without the shadow of a government, with no police, no courts, and no lawyers; yet they do not cut each other's throats; do not plunder each other day and night; do not fall into the anarchy such a state of things might be supposed to lead to. It is very extraordinary! It puts strange thoughts into one's head about the mountain-load of government under which people exist in Europe, and suggests the idea that we may be overgoverned. Think of the hundred Acts of Parliament annually enacted to prevent us, the people of England, from cutting each other's throats, or from doing to our neighbours as we would *not* be done by. Think of the thousands of lawyers and barristers whose whole lives are spent in telling us what the hundred Acts of Parliament mean, and one would be led to infer that if Dobbo has too little law England has too much.

Here we may behold in its simplest form the genius of Commerce at the work of Civilization. Trade is the magic that keeps all at peace, and unites these discordant elements into a well-behaved community. All are traders, and all know that peace and order are essential to successful trade, and thus a public opinion is created which puts down all lawlessness. Often in former years, when strolling along the Campong Glam in Singapore, I have thought how wild and ferocious the Bugis sailors looked, and how little I should like to trust myself among them. But now I find them to be very decent, well-behaved fellows; I walk daily unarmed in the jungle, where I meet them continually; I sleep in a palm-leaf hut, which any one may enter, with as little fear and as little danger of thieves or murder as if I were under the protection of the Metropolitan police. It is true the Dutch influence is felt here. The islands are nominally under the government of the Moluccas, which the native chiefs acknowledge; and in most years a commissioner arrives from Amboyna, who makes the tour of the islands, hears complaints, settles disputes, and carries away prisoner any heinous offender. This year he is not expected to come, as no orders have yet been received to prepare for him; so the people of Dobbo will probably be left to their own devices. One day a man was caught in the act of stealing a piece of iron from Herr Warzbergen's house, which he had entered by making a hole through the thatch wall. In the evening the chief traders of the place, Bugis and Chinese, assembled, the offender was tried and found guilty, and sentenced to receive twenty lashes on the spot. They were given with a small rattan in the middle of the street, not very severely, as the executioner appeared to sympathize a little with the culprit. The disgrace seemed to be thought as much of as the pain; for though any amount of clever cheating is thought rather meritorious than otherwise, open robbery and housebreaking meet with universal reprobation.

NATIVES OF ARU SHOOTING THE GREAT BIRD OF PARADISE.

CHAPTER XXXI.

THE ARU ISLANDS.—JOURNEY AND RESIDENCE IN THE INTERIOR.

(MARCH TO MAY 1857.)

MY boat was at length ready, and having obtained two men besides my own servants, after an enormous amount of talk and trouble, we left Dobbo on the morning of March 13th for the mainland of Aru. By noon we reached the mouth of a small river or creek, which we ascended, winding among mangrove swamps, with here and there a glimpse of dry land. In two hours we reached a house, or rather small shed, of the most miserable description, which our steersman, the " Orang-kaya " of Wamma, said was the place we were to stay at, and where he had assured me we could get every kind of bird and beast to be found in Aru. The shed was occupied by about a dozen men, women, and children ; two cooking fires were burning in it, and there seemed little prospect of my obtaining any accommodation. I however deferred inquiry till I had seen the neighbouring forest, and immediately started off with two men, net, and guns, along a path at the back of the house. In an hour's walk I saw enough to make me determine to give the place a trial, and on my return, finding the " Orang-kaya " was in a strong fever-fit and unable to do anything, I entered into negotiations with the owner of the house for the use of a slip at one end of it about five feet wide, for a week, and agreed to pay as rent one "parang," or chopping-knife. I then immediately got my boxes and bedding out of the boat, hung up a shelf for my bird-skins and insects, and got all ready for work next morning. My own boys slept in the boat to guard the remainder of my property ; a cooking place sheltered by a few mats was arranged under a tree close by, and I felt that degree of satisfaction and enjoyment which I always experience when, after much trouble and delay, I am on the point of beginning work in a new locality.

One of my first objects was to inquire for the people who are accustomed to shoot the paradise birds. They lived at some distance in the jungle, and a man was sent to call them. When they arrived, we had a talk by means of the "Orang-kaya " as interpreter, and they said they thought they could get some. They explained that they shoot the birds with a bow and arrow, the arrow having a conical wooden cap fitted to the end as large as a teacup, so as to kill the bird by the violence of the blow without making any wound or shedding any blood. The trees frequented by the birds are very lofty ; it is therefore necessary to erect a small leafy covering or hut among the branches, to which the hunter mounts before daylight in the morning and remains the whole day, and whenever a bird alights they are almost sure of securing it. (See Illustration.) They returned to their homes the same evening, and I never saw anything more of them, owing, as I afterwards found, to its being too early to obtain birds in good plumage.

The first two or three days of our stay here were very wet, and I obtained but few insects or birds, but at length, when I was beginning to despair, my boy Baderoon returned one day with a specimen which repaid me for months of delay and

MAP
OF THE
ARU ISLANDS.
Mr Wallace's Routes. ....

expectation. It was a small bird, a little less than a thrush. The greater part of its plumage was of an intense cinnabar red, with a gloss as of spun glass. On the head the feathers became short and velvety, and shaded into rich orange. Beneath, from the breast downwards, was pure white, with the softness and gloss of silk, and across the breast a band of deep metallic green separated this colour from the red of the throat. Above each eye was a round spot of the same metallic green ; the bill was yellow, and the feet and legs were of a fine cobalt blue, strikingly contrasting with all the other parts of the body. Merely in arrangement of colours and texture of plumage this little bird was a gem of the first water ; yet these comprised only half its strange beauty. Springing from each side of the breast, and ordinarily lying concealed under the wings, were little tufts of greyish feathers about two inches long, and each terminated by a broad band of intense emerald green. These plumes can be raised at the will of the bird, and spread out into a pair of elegant fans when the wings are elevated. But this is not the only ornament. The two middle feathers of the tail are in the form of slender wires about five inches long, and which diverge in a beautiful double curve. About half an inch of the end of this wire is webbed on the outer side only, and coloured of a fine metallic green, and being curled spirally inwards form a pair of elegant glittering buttons, hanging five inches below the body, and the same distance apart. These two ornaments, the breast fans and the spiral tipped tail wires, are altogether unique, not occurring on any other species of the eight thousand different birds that are known to exist upon the earth ; and, combined with the most exquisite beauty of plumage, render this one of the most perfectly lovely of the many lovely productions of nature. My transports of admiration and delight quite amused my Aru hosts, who saw nothing more in the "Burong raja" than we do in the robin or the goldfinch.[1]

Thus one of my objects in coming to the far East was accomplished. I had obtained a specimen of the King Bird of Paradise (Paradisea regia), which had been described by Linnæus from skins preserved in a mutilated state by the natives. I knew how few Europeans had ever beheld the perfect little organism I now gazed upon, and how very imperfectly it was still known in Europe. The emotions excited in the mind of a naturalist, who has long desired to see the actual thing which he has hitherto known only by description, drawing, or badly-preserved external covering—especially when that thing is of surpassing rarity and beauty—require the poetic faculty fully to express them. The remote island in which I found myself situated, in an almost unvisited sea, far from the tracks of merchant fleets and navies ; the wild, luxuriant tropical forest, which stretched far away on every side ; the rude, uncultured

[1] See the upper figure on Plate at commencement of Chapter XXXVIII.

savages who gathered round me—all had their influence in determining the emotions with which I gazed upon this "thing of beauty." I thought of the long ages of the past, during which the successive generations of this little creature had run their course—year by year being born, and living and dying amid these dark and gloomy woods, with no intelligent eye to gaze upon their loveliness; to all appearance such a wanton waste of beauty. Such ideas excite a feeling of melancholy. It seems sad that on the one hand such exquisite creatures should live out their lives and exhibit their charms only in these wild, inhospitable regions, doomed for ages yet to come to hopeless barbarism; while on the other hand, should civilized man ever reach these distant lands, and bring moral, intellectual, and physical light into the recesses of these virgin forests, we may be sure that he will so disturb the nicely-balanced relations of organic and inorganic nature as to cause the disappearance, and finally the extinction, of these very beings whose wonderful structure and beauty he alone is fitted to appreciate and enjoy. This consideration must surely tell us that all living things were *not* made for man. Many of them have no relation to him. The cycle of their existence has gone on independently of his, and is disturbed or broken by every advance in man's intellectual development; and their happiness and enjoyments, their loves and hates, their struggles for existence, their vigorous life and early death, would seem to be immediately related to their own well-being and perpetuation alone, limited only by the equal well-being and perpetuation of the numberless other organisms with which each is more or less intimately connected.

After the first king-bird was obtained, I went with my men into the forest, and we were not only rewarded with another in equally perfect plumage, but I was enabled to see a little of the habits of both it and the larger species. It frequents the lower trees of the less dense forests, and is very active, flying strongly with a whirring sound, and continually hopping or flying from branch to branch. It eats hard stone-bearing fruits as large as a gooseberry, and often flutters its wings after the manner of the South American manakins, at which time it elevates and expands the beautiful fans with which its breast is adorned. The natives of Aru call it "Goby-goby."

One day I got under a tree where a number of the Great Paradise birds were assembled, but they were high up in the thickest of the foliage, and flying and jumping about so continually that I could get no good view of them. At length I shot one, but it was a young specimen, and was entirely of a rich chocolate-brown colour, without either the metallic green throat or yellow plumes of the full-grown bird. All that I had yet seen resembled this, and the natives told me that it would be about two months before any would be found in full plumage. I still hoped, therefore, to get some. Their voice is most extraordinary. At early morn, before the sun has risen, we hear a loud cry of "Wawk—wawk—wawk, wŏk—wŏk—wŏk," which resounds through the forest, changing its direction continually. This is the Great Bird of Paradise going to seek his breakfast. Others soon follow his example; lories and parroquets cry shrilly; cockatoos scream; king-hunters croak and bark; and the various smaller birds chirp and whistle their morning song. As I lie listening to these interesting sounds, I realize my position as the first European who has ever lived for months together in the Aru Islands, a place which I had hoped rather than expected ever to visit. I think how many besides myself have longed to reach these almost fairy realms, and to see with their own eyes the many wonderful and beautiful things which I am daily encountering. But now Ali and Baderoon are up and getting ready their guns and ammunition, and little Baso has his fire lighted and is boiling my coffee, and I remember that I had a black cockatoo brought in late last night, which I must skin immediately, and so I jump up and begin my day's work very happily.

This cockatoo is the first I have ever seen, and is a great prize. It has a rather small and weak body, long weak legs, large wings, and an enormously developed head, ornamented with a magnificent crest, and armed with a sharp-pointed hooked bill of immense size and strength. The plumage is entirely black, but has all over it the curious powdery white secretion characteristic of cockatoos. The cheeks are bare, and of an intense blood-red colour. Instead of the harsh scream of the white cockatoos, its voice is a somewhat plaintive whistle. The tongue is a curious organ, being a slender fleshy cylinder of a deep red colour, terminated by a horny black plate, furrowed across and somewhat prehensile. The whole tongue has a considerable extensile power. I will here relate something of the habits of this bird, with which I have since become acquainted. It frequents the lower parts of the forest, and is seen singly, or at most two or three together. It flies slowly and noiselessly, and may be killed by a comparatively slight wound. It eats various fruits and seeds, but seems more particularly attached to the kernel of the kanary-nut, which grows on a lofty forest tree (Canarium commune), abundant in the islands where this bird is found; and the manner in which it gets at these seeds shows a correlation of structure and habits, which would point out the "kanary" as its special food. The shell of this nut is so excessively hard that only a heavy hammer will crack it; it is somewhat triangular, and the outside is quite smooth. The manner in which the bird opens these nuts is very curious. Taking one endways in its bill and keeping it firm by a pressure of the tongue, it cuts a transverse notch by a lateral sawing motion of the sharp-edged lower mandible. This done, it takes hold of the nut with its foot, and biting off a piece of leaf retains it in the deep notch of the upper mandible, and again seizing the nut, which is prevented from slipping by the elastic tissue of the leaf, fixes the edge of the lower mandible in the notch, and by a powerful nip breaks off a piece of the shell. Again taking the nut in its claws, it inserts the very long and sharp point of the bill and picks out the kernel, which is seized hold of, morsel

HEAD OF BLACK COCKATOO.

by morsel, by the extensible tongue. Thus every detail of form and structure in the extraordinary bill of this bird seems to have its use, and we may easily conceive that the black cockatoos have maintained themselves in competition with their more active and more numerous white allies, by their power of existing on a kind of food which no other bird is able to extract from its stony shell. The species is the Microglossum aterrimum of naturalists.

During the two weeks which I spent in this little settlement, I had good opportunities of observing the natives at their own home, and living in their usual manner. There is a great monotony and uniformity in every-day savage life, and it seemed to me a more miserable existence than when it had the charm of novelty. To begin with the most important fact in the existence of uncivilized peoples—their food—the Aru men have no regular supply, no staff of life, such as bread, rice, mandiocca, maize, or sago, which are the daily food of a large proportion of mankind. They have, however, many sorts of vegetables, plantains, yams, sweet potatoes, and raw sago ; and they chew up vast quantities of sugar-cane, as well as betel-nuts, gambir, and tobacco. Those who live on the coast have plenty of fish ; but when inland, as we are here, they only go to the sea occasionally, and then bring home cockles and other shell-fish by the boatload. Now and then they get wild pig or kangaroo, but too rarely to form anything like a regular part of their diet, which is essentially vegetable ; and what is of more importance, as affecting their health, green, watery vegetables, imperfectly cooked, and even these in varying and often insufficient quantities. To this diet may be attributed the prevalence of skin diseases, and ulcers on the legs and joints. The scurfy skin disease so common among savages has a close connexion with the poorness and irregularity of their living. The Malays, who are never without their daily rice, are generally free from it ; the hill-Dyaks of Borneo, who grow rice and live well, are clean skinned, while the less industrious and less cleanly tribes, who live for a portion of the year on fruits and vegetables only, are very subject to this malady. It seems clear that in this, as in other respects, man is not able to make a beast of himself with impunity, feeding like the cattle on the herbs and fruits of the earth, and taking no thought of the morrow. To maintain his health and beauty he must labour to prepare some farinaceous product capable of being stored and accumulated, so as to give him a regular supply of wholesome food. When this is obtained, he may add vegetables, fruits, and meat with advantage.

The chief luxury of the Aru people, besides betel and tobacco, is arrack (Java rum), which the traders bring in great quantities and sell very cheap. A day's fishing or rattan cutting will purchase at least a half-gallon bottle ; and when the tripang or birds' nests collected during a season are sold, they get whole boxes, each containing fifteen such bottles, which the inmates of a house will sit round day and night till they have finished. They themselves tell me that at such bouts they often tear to pieces the house they are in, break and destroy everything they can lay their hands on, and make such an infernal riot as is alarming to behold.

The houses and furniture are on a par with the food. A rude shed supported on rough and slender sticks rather than posts, no walls, but the floor raised to within a foot of the eaves, is the style of architecture they usually adopt. Inside there are partitioned walls of thatch, forming little boxes or sleeping places, to accommodate the two or three separate families that usually live together. A few mats, baskets, and cooking vessels, with plates and basins purchased from the Macassar traders, constitute their whole furniture ; spears and bows are their weapons ; a sarong or mat forms the clothing of the women, a waist-cloth of the men. For hours or even for days they sit idle in their houses, the women bringing in the vegetables or sago which form their food. Sometimes they hunt or fish a little, or work at their houses or canoes, but they seem to enjoy pure idleness, and work as little as they can. They have little to vary the monotony of life, little that can be called pleasure, except idleness and conversation. And they certainly do talk ! Every evening there is a little Babel around me : but as I understand not a word of it, I go on with my book or work undisturbed. Now and then they scream and shout, or laugh frantically for variety ; and this goes on alternately with vociferous talking of men, women, and children, till long after I am in my mosquito curtain and sound asleep.

At this place I obtained some light on the complicated mixture of races in Aru, which would utterly confound an ethnologist. Many of the natives, though equally dark with the others, have little of the Papuan physiognomy, but have more delicate features of the European type, with more glossy, curling hair. These at first quite puzzled me, for they have no more resemblance to Malay than to Papuan, and the darkness of skin and hair would forbid the idea of Dutch intermixture. Listening to their conversation, however, I detected some words that were familiar to me. " Accabó " was one ; and to be sure that it was not an accidental resemblance, I asked the speaker in Malay what "accabó" meant, and was told it meant "done or finished," a true Portuguese word, with its meaning retained. Again, I heard the word "jafui" often repeated, and could see, without inquiry, that its meaning was "he's gone," as in Portuguese. "Porco," too, seems a common name, though the people have no idea of its European meaning. This cleared up the difficulty. I at once understood that some early Portuguese traders had penetrated to these islands, and mixed with the natives, influencing their language, and leaving in their descendants for many generations the visible characteristics of their race. If to this we add the occasional mixture of Malay, Dutch, and Chinese with the indigenous Papuans, we have no reason to wonder at the curious varieties of form and feature occasionally to be met with in Aru. In this very house there was a Macassar man, with an Aru wife and a family of mixed children. In Dobbo I saw a Javanese and an Amboyna man, each with an

Aru wife and family ; and as this kind of mixture has been going on for at least three hundred years, and probably much longer, it has produced a decided effect on the physical characteristics of a considerable portion of the population of the islands, more especially in Dobbo and the parts nearest to it.

March 28th.—The "Orang-kaya" being very ill with fever had begged to go home, and had arranged with one of the men of the house to go on with me as his substitute. Now that I wanted to move, the bugbear of the pirates was brought up, and it was pronounced unsafe to go further than the next small river. This would not suit me, as I had determined to traverse the channel called Watelai to the "blakang-tana"; but my guide was firm in his dread of pirates, of which I knew there was now no danger, as several vessels had gone in search of them, as well as a Dutch gunboat which had arrived since I left Dobbo. I had, fortunately, by this time heard that the Dutch "Commissie" had really arrived, and therefore threatened that if my guide did not go with me immediately, I would appeal to the authorities, and he would certainly be obliged to give back the cloth which the "Orang-kaya" had transferred to him in prepayment. This had the desired effect ; matters were soon arranged, and we started the next morning. The wind, however, was dead against us, and after rowing hard till midday we put in to a small river where there were a few huts, to cook our dinners. The place did not look very promising, but as we could not reach our destination, the Watelai river, owing to the contrary wind, I thought we might as well wait here a day or two. I therefore paid a chopper for the use of a small shed, and got my bed and some boxes on shore. In the evening, after dark, we were suddenly alarmed by the cry of "Bajak ! bajak !" (Pirates !) The men all seized their bows and spears, and rushed down to the beach ; we got hold of our guns and prepared for action, but in a few minutes all came back laughing and chattering, for it had proved to be only a small boat and some of their own comrades returned from fishing. When all was quiet again, one of the men, who could speak a little Malay, came to me and begged me not to sleep too hard. "Why ?" said I. "Perhaps the pirates may really come," said he very seriously, which made me laugh and assure him I should sleep as hard as I could.

Two days were spent here, but the place was unproductive of insects or birds of interest, so we made another attempt to get on. As soon as we got a little away from the land we had a fair wind, and in six hours' sailing reached the entrance of the Watelai channel, which divides the most northerly from the middle portion of Aru. At its mouth this was about half a mile wide, but soon narrowed, and a mile or two on it assumed entirely the aspect of a river about the width of the Thames at London, winding among low but undulating and often hilly country. The scene was exactly such as might be expected in the interior of a continent. The channel continued of a uniform average width, with reaches and sinuous bends, one bank being often precipitous, or even forming vertical cliffs, while the other was flat and apparently alluvial; and it was only the pure salt-water, and the absence of any stream but the slight flux and reflux of the tide, that would enable a person to tell that he was navigating a strait and not a river. The wind was fair, and carried us along, with occasional assistance from our oars, till about three in the afternoon, when we landed where a little brook formed two or three basins in the coral rock, and then fell in a miniature cascade into the salt-water river. Here we bathed and cooked our dinner, and enjoyed ourselves lazily till sunset, when we pursued our way for two hours more, and then moored our little vessel to an overhanging tree for the night.

At five the next morning we started again, and in an hour overtook four large praus containing the "Commissie," who had come from Dobbo to make their official tour round the islands, and had passed us in the night. I paid a visit to the Dutchmen, one of whom spoke a little English, but we found that we could get on much better with Malay. They told me that they had been delayed going after the pirates to one of the northern islands, and had seen three of their vessels but could not catch them, because on being pursued they rowed out in the wind's eye, which they are enabled to do by having about fifty oars to each boat. Having had some tea with them, I bade them adieu, and turned up a narrow channel which our pilot said would take us to the village of Watelai, on the east side of Aru. After going some miles we found the channel nearly blocked up with coral, so that our boat grated along the bottom, crunching what may truly be called the living rock. Sometimes all hands had to get out and wade, to lighten the vessel and lift it over the shallowest places; but at length we overcame all obstacles and reached a wide bay or estuary studded with little rocks and islets, and opening to the eastern sea and the numerous islands of the "blakang-tana." I now found that the village we were going to was miles away; that we should have to go out to sea, and round a rocky point. A squall seemed coming on, and as I have a horror of small boats at sea, and from all I could learn Watelai village was not a place to stop at (no Birds of Paradise being found there), I determined to return and go to a village I had heard of up a tributary of the Watelai river, and situated nearly in the centre of the mainland of Aru. The people there were said to be good, and to be accustomed to hunting and bird-catching, being too far inland to get any part of their food from the sea. While I was deciding this point the squall burst upon us, and soon raised a rolling sea in the shallow water, which upset an oil bottle and a lamp, broke some of my crockery, and threw us all into confusion. Rowing hard we managed to get back into the main river by dusk, and looked out for a place to cook our suppers. It happened to be high water, and a very high tide, so that every piece of sand or beach was covered, and it was with the greatest difficulty, and after much groping in the dark, that we discovered a little sloping piece of rock about two feet square on which to make a fire and cook some rice. The next day we continued our way back, and on the following day entered a stream on the south side of the Watelai river, and ascending to where navigation ceased found the little village of Wanumbai, consisting of two large houses surrounded by plantations, amid the virgin forests of Aru.

As I liked the look of the place, and was desirous of staying some time, I sent my pilot to try and make a bargain for house accommodation. The owner and chief man of the place made many excuses. First, he was afraid I would not like his house, and then was doubtful whether his son, who was away, would like his admitting me. I had a long talk with him myself, and tried to explain what I was doing, and how many things I would buy of them, and showed him my stock of beads, and knives, and cloth, and tobacco, all of which I would spend with his family and friends if he would give me house-room. He seemed a little staggered at this, and said he would talk to his wife, and in the meantime I went for a little walk to see the neighbourhood. When I came back, I again sent my pilot, saying that I would go away if he would not give me part of his house. In about half an hour he returned with a demand for about half the cost of building a house, for the rent of a small portion of it for a few weeks. As the only difficulty now was a pecuniary one, I got out about ten yards of cloth, an axe, with a few beads and some tobacco, and sent them as my final offer for the part of the house which I had before pointed out. This was accepted after a little more talk, and I immediately proceeded to take possession.

The house was a good large one, raised as usual about seven feet on posts, the walls about three or four feet more, with a high-pitched roof. The floor was of bamboo laths, and in the sloping roof was an immense shutter, which could be lifted and propped up to admit light and air. At the end where this was situated the floor was raised about a foot, and this piece, about ten feet wide by twenty long, quite open to the rest of the house, was the portion I was to occupy. At one end of this piece, separated by a thatch partition, was a cooking place, with a clay floor and shelves for crockery. At the opposite end I had my mosquito curtain hung, and round the walls we arranged my boxes and other stores, fitted up a table and seat, and with a little cleaning and dusting made the place look quite comfortable. My boat was then hauled up on shore, and covered with palm-leaves, the sails and oars brought indoors, a hanging-stage for drying my specimens erected outside the house, and another inside, and my boys were set to clean their guns and get all ready for beginning work.

The next day I occupied myself in exploring the paths in the immediate neighbourhood. The small river up which we had ascended ceases to be navigable at this point, above which it is a little rocky brook, which quite dries up in the hot season. There was now, however, a fair stream of water in it; and a path which was partly in and partly by the side of the water, promised well for insects, as I here saw the magnificent blue butterfly, Papilio ulysses, as well as several other fine species, flopping lazily along, sometimes resting high up on the foliage which drooped over the water, at others settling down on the damp rock or on the edges of muddy pools. A little way on several paths branched off through patches of second-growth forest to cane-fields, gardens, and scattered houses, beyond which again the dark wall of verdure striped with tree-trunks, marked out the limits of the primæval forests. The voices of many birds promised good shooting, and on my return I found that my boys had already obtained two or three kinds I had not seen before; and in the evening a native brought me a rare and beautiful species of ground-thrush (Pitta novæ-guineæ) hitherto only known from New Guinea.

As I improved my acquaintance with them I became much interested in these people, who are a fair sample of the true savage inhabitants of the Aru Islands, tolerably free from foreign admixture. The house I lived in contained four or five families, and there were generally from six to a dozen visitors besides. They kept up a continual row from morning till night —talking, laughing, shouting, without intermission—not very pleasant, but interesting as a study of national character. My boy Ali said to me, "Banyak quot bitchara Orang Aru" (The Aru people are very strong talkers), never having been accustomed to such eloquence either in his own or any other country he had hitherto visited. Of an evening the men, having got over their first shyness, began to talk to me a little, asking about my country, &c., and in return I questioned them about any traditions they had of their own origin. I had, however, very little success, for I could not possibly make them understand the simple question of where the Aru people first came from. I put it in every possible way to them, but it was a subject quite beyond their speculations; they had evidently never thought of anything of the kind, and were unable to conceive a thing so remote and so unnecessary to be thought about as their own origin. Finding this hopeless, I asked if they knew when the trade with Aru first began, when the Bugis and Chinese and Macassar men first came in their praus to buy tripang and tortoise-shell, and birds' nests, and Paradise birds? This they comprehended, but replied that there had always been the same trade as long as they or their fathers recollected, but that this was the first time a real white man had come among them, and, said they, "You see how the people come every day from all the villages round to look at you." This was very flattering, and accounted for the great concourse of visitors which I had at first imagined was accidental. A few years before I had been one of the gazers at the Zulus and the Aztecs in London. Now the tables were turned upon me, for I was to these people a new and strange variety of man, and had the honour of affording to them, in my own person, an attractive exhibition, gratis.

All the men and boys of Aru are expert archers, never stirring without their bows and arrows. They shoot all sorts of birds, as well as pigs and kangaroos occasionally, and thus have a tolerably good supply of meat to eat with their vegetables. The result of this better living is superior healthiness, well-made bodies, and generally clear skins. They brought me numbers of small birds in exchange for beads or tobacco, but mauled them terribly, notwithstanding my repeated instructions. When they got a bird alive they would often tie a string to its leg, and keep it a day or two, till its plumage was so draggled and dirtied as to be almost worthless. One of the first things I got from them was a living specimen of the curious and beautiful racquet-tailed kingfisher. Seeing how much I admired it, they after-wards brought me several more, which were all caught before daybreak, sleeping in cavities of the rocky banks of the stream. My hunters also shot a few specimens, and almost all of them had the red bill more or less clogged with mud and earth. This indicates the habits of the bird, which, though popularly a king-fisher, never catches fish, but lives on insects and minute shells, which it picks up in the forest, darting down upon them from its perch on some low branch. The genus Tanysiptera, to which this bird belongs, is remarkable for the enormously lengthened tail, which in all other kingfishers is small and short. Linnæus named the species known to him "the goddess kingfisher" (Alcedo dea), from its extreme grace and beauty, the plumage being brilliant blue and white, with the bill red, like coral. Several species of these interesting birds are now known, all confined within the very limited area which comprises the Moluccas, New Guinea, and the extreme north of Australia. They resemble each other so closely that several of them can only be distinguished by careful comparison. One of the rarest, however, which inhabits New Guinea, is very distinct from the rest, being bright red beneath instead of white. That which I now obtained was a new one, and has been named Tanysiptera hydrocharis, but in general form and coloration it is exactly similar to the larger species found in Amboyna, and figured at page 229.

New and interesting birds were continually brought in, either by my own boys or by the natives, and at the end of a week Ali arrived triumphant one afternoon with a fine specimen of the Great Bird of Paradise. The ornamental plumes had not yet attained their full growth, but the richness of their glossy orange colouring, and the exquisite delicacy of the loosely waving feathers, were unsurpassable. At the same time a great black cockatoo was brought in, as well as a fine fruit-pigeon and several small birds, so that we were all kept hard at work skinning till sunset. Just as we had cleared away and packed up for the night, a strange beast was brought, which had been shot by the natives. It resembled in size, and in its white woolly covering, a small fat lamb, but had short legs, hand-like feet with large claws, and a long prehensile tail. It was a Cuscus (C. maculatus), one of the curious marsupial animals of the Papuan region, and I was very desirous to obtain the skin. The owners, however, said they wanted to eat it; and though I offered them a good price, and promised to give them all the meat, there was great hesitation. Suspecting the reason, I offered, though it was night, to set to work immediately and get out the body for them, to which they agreed. The creature was much hacked about, and the two hind feet almost cut off, but it was the largest and finest specimen of the kind I had seen; and after an hour's hard work I handed over the body to the owners, who immediately cut it up and roasted it for supper.

As this was a very good place for birds, I determined to remain a month longer, and took the opportunity of a native boat going to Dobbo to send Ali for a fresh supply of ammunition and provisions. They started on the 10th of April, and the house was crowded with about a hundred men, boys, women, and girls, bringing their loads of sugar-cane, plantains, sirih-leaf, yams, &c.; one lad going from each house to sell the produce and make purchases. The noise was indescribable. At least fifty of the hundred were always talking at once, and that not in the low, measured tones of the apathetically polite Malay, but with loud voices, shouts, and screaming laughter, in which the women and children were even more conspicuous than the men. It was only while gazing at me that their tongues were moderately quiet, because their eyes were fully occupied. The black vegetable soil here overlying the coral rock is very rich, and the sugar-cane was finer than any I had ever seen. The canes brought to the boat were often ten and even twelve feet long, and thick in proportion, with short joints throughout, swelling between the knots with the abundance of the rich juice. At Dobbo they get a high price for it, 1d. to 3d. a stick, and there is an insatiable demand among the crews of the praus and the Baba fishermen. Here they eat it continually. They half live on it, and sometimes feed their pigs with it. Near every house are great heaps of the refuse cane; and large wicker-baskets to contain this refuse as it is produced form a regular part of the furniture of a house. Whatever time of the day you enter, you are sure to find three or four people with a yard of cane in one hand, a knife in the other, and a basket between their legs, hacking, paring, chewing, and basket-filling, with a persevering assiduity which reminds one of a hungry cow grazing, or of a caterpillar eating up a leaf.

After five days' absence the boats returned from Dobbo,

bringing Ali and all the things I had sent for quite safe. A large party had assembled to be ready to carry home the goods brought, among which were a good many cocoa-nuts, which are a great luxury here. It seems strange that they should never plant them; but the reason simply is, that they cannot bring their hearts to bury a good nut for the prospective advantage of a crop twelve years hence. There is also the chance of the fruits being dug up and eaten unless watched night and day. Among the things I had sent for was a box of arrack, and I was now of course besieged with requests for a little drop. I gave them a flask (about two bottles), which was very soon finished, and I was assured that there were many present who had not had a taste. As I feared my box would very soon be emptied if I supplied all their demands, I told them I had given them one, but the second they must pay for, and that afterwards I must have a Paradise Bird for each flask. They immediately sent round to all the neighbouring houses, and mustered up a rupee in Dutch copper money, got their second flask, and drunk it as quickly as the first, and were then very talkative, but less noisy and importunate than I had expected. Two or three of them got round me and begged me for the twentieth time to tell them the name of my country. Then, as they could not pronounce it satisfactorily, they insisted that I was deceiving them, and that it was a name of my own invention. One funny old man, who bore a ludicrous resemblance to a friend of mine at home, was almost indignant. "Ung-lung!" said he, "who ever heard of such a name?—Ang-lang—Anger-lang—that can't be the name of your country; you are playing with us." Then he tried to give a convincing illustration. "My country is Wanumbai—anybody can say Wanumbai. I'm an orang-Wanumbai; but, N-glung! who ever heard of such a name? Do tell us the real name of your country, and then when you are gone we shall know how to talk about you." To this luminous argument and remonstrance I could oppose nothing but assertion, and the whole party remained firmly convinced that I was for some reason or other deceiving them. They then attacked me on another point—what all the animals and birds and insects and shells were preserved so carefully for. They had often asked me this before, and I had tried to explain to them that they would be stuffed, and made to look as if alive, and people in my country would go to look at them. But this was not satisfying; in my country there must be many better things to look at, and they could not believe I would take so much trouble with their birds and beasts just for people to look at. They did not want to look at them; and we, who made calico and glass and knives, and all sorts of wonderful things, could not want things from Aru to look at. They had evidently been thinking about it, and had at length got what seemed a very satisfactory theory; for the same old man said to me, in a low mysterious voice, "What becomes of them when you go on to the sea?" "Why, they are all packed up in boxes," said I. "What did you think became of them?" "They all come to life again, don't they?" said he; and though I tried to joke it off, and said if they did we should have plenty to eat at sea, he stuck to his opinion, and kept repeating, with an air of deep conviction, "Yes, they all come to life again, that's what they do—they all come to life again."

After a little while, and a good deal of talking among themselves, he began again—"I know all about it—oh, yes! Before you came we had rain every day—very wet indeed; now, ever since you have been here, it is fine hot weather. Oh, yes! I know all about it; you can't deceive me." And so I was set down as a conjurer, and was unable to repel the charge. But the conjurer was completely puzzled by the next question: "What," said the old man, "is the great ship, where the Bugis and Chinamen go to sell their things? It is always in the great sea—its name is Jong; tell us all about it." In vain I inquired what they knew about it; they knew nothing but that it was called "Jong," and was always in the sea, and was a very great ship, and concluded with, "Perhaps that is your country?" Finding that I could not or would not tell them anything about "Jong," there came more regrets that I would not tell them the real name of my country; and then a long string of compliments, to the effect that I was a much better sort of a person than the Bugis and Chinese, who sometimes came to trade with them, for I gave them things for nothing, and did not try to cheat them. How long would I stop? was the next earnest inquiry. Would I stay two or three months? They would get me plenty of birds and animals, and I might soon finish all the goods I had brought, and then, said the old spokesman, "Don't go away, but send for more things from Dobbo, and stay here a year or two." And then again the old story, "Do tell us the name of your country. We know the Bugis men, and the Macassar men, and the Java men, and the China men; only you, we don't know from what country you come. Ung-lung! it can't be; I know that is not the name of your country." Seeing no end to this long talk, I said I was tired, and wanted to go to sleep; so after begging—one a little bit of dry fish for his supper, and another a little salt to eat with his sago—they went off very quietly, and I went outside and took a stroll round the house by moonlight, thinking of the simple people and the strange productions of Aru, and then turned in under my mosquito curtain, to sleep with a sense of perfect security in the midst of these good-natured savages.

We now had seven or eight days of hot and dry weather, which reduced the little river to a succession of shallow pools connected by the smallest possible thread of trickling water. If there were a dry season like that of Macassar, the Aru Islands would be uninhabitable, as there is no part of them much above a hundred feet high; and the whole being a mass of porous coralline rock, allows the surface water rapidly to escape. The only dry season they have is for a month or two about September or October, and there is then an excessive scarcity of water, so that sometimes hundreds of birds and other animals die of drought. The natives then remove to houses near the sources of the small streams, where, in the shady depths of the forest, a small quantity of water still remains. Even then many of them have to go miles for their water, which they keep in large bamboos and use very sparingly. They assure me that they catch and kill game of all kinds, by watching at the water holes or setting snares around them. That would be the time for me to make my collections ; but the want of water would be a terrible annoyance, and the impossibility of getting away before another whole year had passed made it out of the question.

Ever since leaving Dobbo I had suffered terribly from insects, who seemed here bent upon revenging my long-continued persecution of their race. At our first stopping-place sand-flies were very abundant at night, penetrating to every part of the body, and producing a more lasting irritation than mosquitoes. My feet and ankles especially suffered, and were completely covered with little red swollen specks, which tormented me horribly. On arriving here we were delighted to find the house free from sand-flies or mosquitoes, but in the plantations where my daily walks led me, the day-biting mosquitoes swarmed, and seemed especially to delight in attacking my poor feet. After a month's incessant punishment, those useful members rebelled against such treatment and broke into open insurrection, throwing out numerous inflamed ulcers, which were very painful, and stopped me from walking. So I found myself confined to the house, and with no immediate prospect of leaving it. Wounds or sores in the feet are especially difficult to heal in hot climates, and I therefore dreaded them more than any other illness. The confinement was very annoying, as the fine hot weather was excellent for insects, of which I had every promise of obtaining a fine collection ; and it is only by daily and unremitting search that the smaller kinds, and the rarer and more interesting specimens, can be obtained. When I crawled down to the river-side to bathe, I often saw the blue-winged Papilio ulysses, or some other equally rare and beautiful insect ; but there was nothing for it but patience, and to return quietly to my bird-skinning, or whatever other work I had indoors. The stings and bites and ceaseless irritation caused by these pests of the tropical forests would be borne uncomplainingly ; but to be kept prisoner by them in so rich and unexplored a country, where rare and beautiful creatures are to be met with in every forest ramble—a country reached by such a long and tedious voyage, and which might not in the present century be again visited for the same purpose—is a punishment too severe for a naturalist to pass over in silence.

I had, however, some consolation in the birds my boys brought home daily, more especially the Paradiseas, which they at length obtained in full plumage. It was quite a relief to my mind to get these, for I could hardly have torn myself away from Aru had I not obtained specimens. But what I valued almost as much as the birds themselves, was the knowledge of their habits, which I was daily obtaining both from the accounts of my hunters, and from the conversation of the natives. The birds had now commenced what the people here call their "sácaleli," or dancing-parties, in certain trees in the forest, which are not fruit trees as I at first imagined, but which have an immense head of spreading branches and large but scattered leaves, giving a clear space for the birds to play and exhibit their plumes. On one of these trees a dozen or twenty full-plumaged male birds assemble together, raise up their wings, stretch out their necks, and elevate their exquisite plumes, keeping them in a continual vibration. Between whiles they fly across from branch to branch in great excitement, so that the whole tree is filled with waving plumes in every variety of attitude and motion. (See Illustration facing p. 337.) The bird itself is nearly as large as a crow, and is of a rich coffee brown colour. The head and neck is of a pure straw yellow above, and rich metallic green beneath. The long plumy tufts of golden orange feathers spring from the sides beneath each wing, and when the bird is in repose are partly concealed by them. At the time of its excitement, however, the wings are raised vertically over the back, the head is bent down and stretched out, and the long plumes are raised up and expanded till they form two magnificent golden fans striped with deep red at the base, and fading off into the pale brown tint of the finely divided and softly waving points. The whole bird is then overshadowed by them, the crouching body, yellow head, and emerald green throat forming but the foundation and setting to the golden glory which waves above. When seen in this attitude, the Bird of Paradise really deserves its name, and must be ranked as one of the most beautiful and most wonderful of living things. I continued also to get specimens of the lovely little king-bird occasionally, as well as numbers of brilliant pigeons, sweet little parroquets, and many curious small birds, most nearly resembling those of Australia and New Guinea.

Here, as among most savage people I have dwelt among, I was delighted with the beauty of the human form—a beauty of which stay-at-home civilized people can scarcely have any conception. What are the finest Grecian statues to the living, moving, breathing men I saw daily around me? The unrestrained grace of the naked savage as he goes about his daily occupations, or lounges at his ease, must be seen to be understood; and a youth bending his bow is the perfection of manly beauty. The women, however, except in extreme youth, are by no means so pleasant to look at as the men. Their strongly-marked features are very unfeminine, and hard work, privations, and very early marriages soon destroy whatever of beauty or grace they may for a short time possess. Their toilet is very simple, but also, I am sorry to say, very coarse and disgusting. It consists solely of a mat of plaited strips of palm leaves, worn tight round the body, and reaching from the hips to the knees. It seems not to be changed till worn out, is seldom washed, and is generally very dirty. This is the universal dress, except in a few cases, where Malay "sarongs" have come into use. Their frizzly hair is tied in a bunch at the back of the head. They delight in combing, or rather forking it, using for that purpose a large wooden fork with four diverging prongs, which answers the purpose of separating and arranging the long tangled, frizzly mass of cranial vegetation much better than any comb could do. The only ornaments of the women are earrings and necklaces, which they arrange in various tasteful ways. The ends of a necklace are often attached to the earrings, and then looped on to the hair-knot behind. This has really an elegant appearance, the beads hanging gracefully on each side of the head, and by establishing a connexion with the earrings give an appearance of utility to those barbarous ornaments. We recommend this style to the consideration of those of the fair sex who still bore holes in their ears and hang rings thereto. Another style of necklace among these Papuan belles is to wear two, each hanging on one side of the neck and under the opposite arm, so as to cross each other. This has a very pretty appearance, in part due to the contrast of the white beads or kangaroo teeth of which they are composed with the dark glossy skin. The earrings themselves are formed of a bar of copper or silver twisted so that the ends cross. The men, as usual among savages, adorn themselves more than the women. They wear necklaces, earrings, and finger rings, and delight in a band of plaited grass tight round the arm just below the shoulder, to which they attach a bunch of hair or bright coloured feathers by way of ornament. The teeth of small animals, either alone, or alternately with black or white beads, form their necklaces, and sometimes bracelets also. For these latter, however, they prefer brass wire, or the black, horny, wing-spines of the cassowary, which they consider a charm. Anklets of brass or shell, and tight plaited garters below the knee, complete their ordinary decorations.

Some natives of Kobror from further south, and who are reckoned the worst and least civilized of the Aru tribes, came one day to visit us. They have a rather more than usually savage appearance, owing to the greater amount of ornaments they use—the most conspicuous being a large horseshoe-shaped comb which they wear over the forehead, the ends resting on the temples. The back of the comb is fastened into a piece of wood, which is plated with tin in front, and above is attached a plume of feathers from a cock's tail. In other respects they scarcely differed from the people I was living with. They brought me a couple of birds, some shells and insects, showing that the report of the white man and his doings had reached their country. There was probably hardly a man in Aru who had not by this time heard of me.

Besides the domestic utensils already mentioned, the movable property of a native is very scanty. He has a good supply of spears and bows and arrows for hunting, a parang, or chopping-knife, and an axe—for the stone age has passed away here, owing to the commercial enterprise of the Bugis and other Malay races. Attached to a belt, or hung across his shoulder, he carries a little skin pouch and an ornamented bamboo, containing betel-nut, tobacco, and lime, and a small German wooden-handled knife is generally stuck between his waist-cloth of bark and his bare skin. Each man also possesses a "cadjan," or sleeping-mat, made of the broad leaves of a pandanus neatly sewn together in three layers. This mat is about four feet square, and when folded has one end sewn up, so that it forms a kind of sack open at one side. In the closed corner the head or feet can be placed, or by carrying it on the head in a shower it forms both coat and umbrella. It doubles up in a small compass for convenient carriage, and then forms a light and elastic cushion, so that on a journey it becomes clothing, house, bedding, and furniture, all in one.

The only ornaments in an Aru house are trophies of the chase —jaws of wild pigs, the heads and backbones of cassowaries, and plumes made from the feathers of the Bird of Paradise, cassowary, and domestic fowl. The spears, shields, knife-handles, and other utensils are more or less carved in fanciful designs, and the mats and leaf boxes are painted or plaited in neat patterns of red, black, and yellow colours. I must not forget these boxes, which are most ingeniously made of the pith of a palm leaf pegged together, lined inside with pandanus leaves, and outside with the same, or with plaited grass. All the joints and angles are covered with strips of split rattan sewn neatly on. The lid is covered with the brown leathery spathe of the Areca palm, which is impervious to water, and the whole box is neat, strong, and well finished. They are made from a few inches to two or three feet long, and being much esteemed by the Malays as clothes-boxes, are a regular article of export from Aru. The natives use the smaller ones for tobacco or betel-nut, but seldom have clothes enough to require the larger ones, which are only made for sale.

Among the domestic animals which may generally be seen in native houses are gaudy parrots, green, red, and blue, a few domestic fowls, which have baskets hung for them to lay in under the eaves, and who sleep on the ridge, and several half-starved, wolfish-looking dogs. Instead of rats and mice there are curious little marsupial animals about the same size, which run about at night and nibble anything eatable that may be left

uncovered. Four or five different kinds of ants attack everything not isolated by water, and one kind even swims across
that ; great spiders lurk in baskets and boxes, or hide in the
folds of my mosquito curtain ; centipedes and millepedes are
found everywhere. I have caught them under my pillow and
on my head ; while in every box, and under every board which
has lain for some days undisturbed, little scorpions are sure to
be found snugly ensconced, with their formidable tails quickly
turned up ready for attack or defence. Such companions seem
very alarming and dangerous, but all combined are not so bad
as the irritation of mosquitoes, or of the insect pests often found
at home. These latter are a constant and unceasing source of
torment and disgust, whereas you may live a long time among
scorpions, spiders, and centipedes, ugly and venomous though
they are, and get no harm from them. After living twelve
years in the tropics, I have never yet been bitten or stung by
either.

The lean and hungry dogs before mentioned were my greatest
enemies, and kept me constantly on the watch. If my boys
left the bird they were skinning for an instant, it was sure to
be carried off. Everything eatable had to be hung up to the
roof, to be out of their reach. Ali had just finished skinning a
fine King Bird of Paradise one day, when he dropped the skin.
Before he could stoop to pick it up, one of this famished race
had seized upon it, and he only succeeded in rescuing it from its
fangs after it was torn to tatters. Two skins of the large
Paradisea, which were quite dry and ready to pack away, were
incautiously left on my table for the night, wrapped up in
paper. The next morning they were gone, and only a few
scattered feathers indicated their fate. My hanging shelf was
out of their reach ; but having stupidly left a box which served
as a step, a full-plumaged Paradise Bird was next morning
missing ; and a dog below the house was to be seen still
mumbling over the fragments, with the fine golden plumes all
trampled in the mud. Every night, as soon as I was in bed, I
could hear them searching about for what they could devour,
under my table, and all about my boxes and baskets, keeping
me in a state of suspense till morning, lest something of value
might incautiously have been left within their reach. They
would drink the oil of my floating lamp and eat the wick, and
upset or break my crockery if my lazy boys had neglected to
wash away even the smell of anything eatable. Bad, however,
as they are here, they were worse in a Dyak's house in Borneo
where I was once staying, for there they gnawed off the tops of
my waterproof boots, ate a large piece out of an old leather
game-bag, besides devouring a portion of my mosquito curtain !

April 28th.—Last evening we had a grand consultation, which
had evidently been arranged and discussed beforehand. A
number of the natives gathered round me, and said they wanted
to talk. Two of the best Malay scholars helped each other, the

rest putting in hints and ideas in their own language. They told me a long rambling story ; but, partly owing to their imperfect knowledge of Malay, partly through my ignorance of local terms, and partly through the incoherence of their narrative, I could not make it out very clearly. It was, however, a tradition, and I was glad to find they had anything of the kind. A long time ago, they said, some strangers came to Aru, and came here to Wanumbai, and the chief of the Wanumbai people did not like them, and wanted them to go away, but they would not go, and so it came to fighting, and many Aru men were killed, and some, along with the chief, were taken prisoners, and carried away by the strangers. Some of the speakers, however, said that he was not carried away, but went away in his own boat to escape from the foreigners, and went to the sea and never came back again. But they all believe that the chief and the people that went with him still live in some foreign country ; and if they could but find out where, they would send for them to come back again. Now having some vague idea that white men must know every country beyond the sea, they wanted to know if I had met their people in my country or in the sea. They thought they must be there, for they could not imagine where else they could be. They had sought for them everywhere, they said—on the land and in the sea, in the forest and on the mountains, in the air and in the sky, and could not find them ; therefore, they must be in my country, and they begged me to tell them, for I must surely know, as I came from across the great sea. I tried to explain to them that their friends could not have reached my country in small boats ; and that there were plenty of islands like Aru all about the sea, which they would be sure to find. Besides, as it was so long ago, the chief and all the people must be dead. But they quite laughed at this idea, and said they were sure they were alive, for they had proof of it. And then they told me that a good many years ago, when the speakers were boys, some Wokan men who were out fishing met these lost people in the sea, and spoke to them ; and the chief gave the Wokan men a hundred fathoms of cloth to bring to the men of Wanumbai, to show that they were alive and would soon come back to them ; but the Wokan men were thieves, and kept the cloth, and they only heard of it afterwards ; and when they spoke about it, the Wokan men denied it, and pretended they had not received the cloth ;—so they were quite sure their friends were at that time alive and somewhere in the sea. And again, not many years ago, a report came to them that some Bugis traders had brought some children of their lost people ; so they went to Dobbo to see about it, and the owner of the house, who was now speaking to me, was one who went ; but the Bugis man would not let them see the children, and threatened to kill them if they came into his house. He kept the children shut up in a large box, and when he went away he took them with him. And at the end of each of these stories, they begged me in an imploring tone to tell them if I knew where their chief and their people now were.

By dint of questioning, I got some account of the strangers who had taken away their people. They said they were wonderfully strong, and each one could kill a great many Aru men; and when they were wounded, however badly, they spit upon the place, and it immediately became well. And they made a great net of rattans, and entangled their prisoners in it, and sunk them in the water; and the next day, when they pulled the net up on shore, they made the drowned men come to life again, and carried them away.

Much more of the same kind was told me, but in so confused and rambling a manner that I could make nothing out of it, till I inquired how long ago it was that all this happened, when they told me that after their people were taken away the Bugis came in their praus to trade in Aru, and to buy tripang and birds' nests. It is not impossible that something similar to what they related to me really happened when the early Portuguese discoverers first came to Aru, and has formed the foundation for a continually increasing accumulation of legend and fable. I have no doubt that to the next generation, or even before, I myself shall be transformed into a magician or a demigod, a worker of miracles, and a being of supernatural knowledge. They already believe that all the animals I preserve will come to life again; and to their children it will be related that they actually did so. An unusual spell of fine weather setting in just at my arrival has made them believe I can control the seasons; and the simple circumstance of my always walking alone in the forest is a wonder and a mystery to them, as well as my asking them about birds and animals I have not yet seen, and showing an acquaintance with their forms, colours, and habits. These facts are brought against me when I disclaim knowledge of what they wish me to tell them. "You must know," say they; "you know everything: you make the fine weather for your men to shoot; and you know all about our birds and our animals as well as we do; and you go alone into the forest and are not afraid." Therefore every confession of ignorance on my part is thought to be a blind, a mere excuse to avoid telling them too much. My very writing materials and books are to them weird things; and were I to choose to mystify them by a few simple experiments with lens and magnet, miracles without end would in a few years cluster about me; and future travellers, penetrating to Wanumbai, would hardly believe that a poor English naturalist, who had resided a few months among them, could have been the original of the supernatural being to whom so many marvels were attributed.

For some days I had noticed a good deal of excitement, and many strangers came and went armed with spears and cutlasses, bows and shields. I now found there was war near us—two neighbouring villages having a quarrel about some matter of local politics that I could not understand. They told me it was quite a common thing, and that they are rarely without fighting somewhere near. Individual quarrels are taken up by villages and tribes, and the nonpayment of the stipulated price for a wife is one of the most frequent causes of bitterness and bloodshed. One of the war shields was brought me to look at. It was made of rattans and covered with cotton twist, so as to be both light, strong, and very tough. I should think it would resist any ordinary bullet. About the middle there was an arm-hole with a shutter or flap over it. This enables the arm to be put through and the bow drawn, while the body and face, up to the eyes, remain protected, which cannot be done if the shield is carried on the arm by loops attached at the back in the ordinary way. A few of the young men from our house went to help their friends, but I could not hear that any of them were hurt, or that there was much hard fighting.

May 8th.—I had now been six weeks at Wanumbai, but for more than half the time was laid up in the house with ulcerated feet. My stores being nearly exhausted, and my bird and insect boxes full, and having no immediate prospect of getting the use of my legs again, I determined on returning to Dobbo. Birds had lately become rather scarce, and the Paradise birds had not yet become as plentiful as the natives assured me they would be in another month. The Wanumbai people seemed very sorry at my departure; and well they might be, for the shells and insects they picked up on the way to and from their plantations, and the birds the little boys shot with their bows and arrows, kept them all well supplied with tobacco and gambir, besides enabling them to accumulate a stock of beads and coppers for future expenses. The owner of the house was supplied gratis with a little rice, fish, or salt, whenever he asked for it, which I must say was not very often. On parting, I distributed among them my remnant stock of salt and tobacco, and gave my host a flask of arrack, and believe that on the whole my stay with these simple and good-natured people was productive of pleasure and profit to both parties. I fully intended to come back; and had I known that circumstances would have prevented my doing so, should have felt some sorrow in leaving a place where I had first seen so many rare and beautiful living things, and had so fully enjoyed the pleasure which fills the heart of the naturalist when he is so fortunate as to discover a district hitherto unexplored, and where every day brings forth new and unexpected treasures. We loaded our boat in the afternoon, and, starting before daybreak, by the help of a fair wind reached Dobbo late the same evening.

DOBBO, IN THE TRADING SEASON.

CHAPTER XXXII.

THE ARU ISLANDS.—SECOND RESIDENCE AT DOBBO.

(MAY AND JUNE 1857.)

DOBBO was full to overflowing, and I was obliged to occupy the court-house where the Commissioners hold their sittings. They had now left the island, and I found the situation agreeable, as it was at the end of the village, with a view down the principal street. It was a mere shed, but half of it had a roughly boarded floor, and by putting up a partition and opening a window I made it a very pleasant abode. In one of the boxes I had left in charge of Herr Warzbergen, a colony of small ants had settled and deposited millions of eggs. It was luckily a fine hot day, and by carrying the box some distance from the house, and placing every article in the sunshine for an hour or two, I got rid of them without damage, as they were fortunately a harmless species.

Dobbo now presented an animated appearance. Five or six new houses had been added to the street; the praus were all brought round to the western side of the point, where they were hauled up on the beach, and were being caulked and covered with a thick white lime-plaster for the homeward voyage, making them the brightest and cleanest looking things in the place. Most of the small boats had returned from the "blakang-tana" (back country), as the side of the islands towards New Guinea is called. Piles of firewood were being heaped up behind the houses; sail-makers and carpenters were busy at work; mother-of-pearl shell was being tied up in bundles; and the black and ugly smoked tripang was having a last exposure to the sun before loading. The spare portion of the crews were employed cutting and squaring timber, and boats from Ceram and Goram were constantly unloading their cargoes of sago-cake for the traders' homeward voyage. The fowls, ducks, and goats all looked fat and thriving on the refuse food of a dense population, and the Chinamen's pigs were in a state of obesity that foreboded early death. Parrots and lories and cockatoos, of a dozen different kinds, were suspended on bamboo perches at the doors of the houses, with metallic green or white fruit-pigeons which cooed musically at noon and eventide. Young cassowaries, strangely striped with black and brown, wandered about the houses or gambolled with the playfulness of kittens in the hot sunshine, with sometimes a pretty little kangaroo, caught in the Aru forests, but already tame and graceful as a petted fawn.

Of an evening there were more signs of life than at the time of my former residence. Tom-toms, jews'-harps, and even fiddles were to be heard, and the melancholy Malay songs sounded not unpleasantly far into the night. Almost every day there was a cock-fight in the street. The spectators make a ring, and after the long steel spurs are tied on, and the poor animals are set down to gash and kill each other, the excitement is immense. Those who have made bets scream and yell and jump frantically, if they think they are going to win or lose. But in a very few minutes it is all over ; there is a hurrah from the winners, the owners seize their cocks, the winning bird is caressed and admired, the loser is generally dead or very badly wounded, and his master may often be seen plucking out his feathers as he walks away, preparing him for the cooking pot while the poor bird is still alive.

A game at foot-ball, which generally took place at sunset, was, however, much more interesting to me. The ball used is a rather small one, and is made of rattan, hollow, light, and elastic. The player keeps it dancing a little while on his foot, then occasionally on his arm or thigh, till suddenly he gives it a good blow with the hollow of the foot, and sends it flying high in the air. Another player runs to meet it, and at its first bound catches it on his foot and plays in his turn. The ball must never be touched with the hand ; but the arm, shoulder, knee, or thigh are used at pleasure to rest the foot. Two or three played very skilfully, keeping the ball continually flying about, but the place was too confined to show off the game to advantage. One evening a quarrel arose from some dispute in the game, and there was a great row, and it was feared there would be a fight about it—not two men only, but a party of a dozen or twenty on each side, a regular battle with knives and krisses ; but after a large amount of talk it passed off quietly, and we heard nothing about it afterwards.

Most Europeans being gifted by nature with a luxuriant growth of hair upon their faces think it disfigures them, and keep up a continual struggle against her by mowing down every morning the crop which has sprouted up during the preceding twenty-four hours. Now the men of Mongolian race are, naturally, just as many of us want to be. They mostly pass their lives with faces as smooth and beardless as an infant's. But shaving seems an instinct of the human race ; for many of these people, having no hair to take off their faces, shave their heads. Others, however, set resolutely to work to force nature to give them a beard. One of the chief cock-fighters at Dobbo was a Javanese, a sort of master of the ceremonies of the ring, who tied on the spurs and acted as backer-up to one of the combatants. This man had succeeded, by assiduous cultivation, in raising a pair of moustaches which were a triumph of art, for they each contained about a dozen hairs more than three inches long, and which, being well greased and twisted, were distinctly visible (when not too far off) as a black thread hanging down on each side of his mouth. But the beard to match was the difficulty, for nature had cruelly refused to give him a rudiment of hair on his chin, and the most talented gardener could not do much if he had nothing to cultivate. But true genius triumphs over difficulties. Although there was no hair proper on the chin, there happened to be, rather on one side of it, a small mole or freckle which contained (as such things frequently do) a few stray hairs. These had been made the most of. They had reached four or five inches in length, and formed another black thread dangling from the left angle of the chin. The owner carried this as if it were something remarkable (as it certainly was); he often felt it affectionately, passed it between his fingers, and was evidently extremely proud of his moustaches and beard !

One of the most surprising things connected with Aru was the excessive cheapness of all articles of European or native manufacture. We were here two thousand miles beyond Singapore and Batavia, which are themselves emporiums of the " far East," in a place unvisited by, and almost unknown to, European traders; everything reached us through at least two or three hands, often many more; yet English calicoes and American cotton cloths could be bought for 8s. the piece; muskets for 15s.; common scissors and German knives at three-halfpence each; and other cutlery, cotton goods, and earthenware in the same proportion. The natives of this out-of-the-way country can, in fact, buy all these things at about the same money price as our workmen at home, but in reality very much cheaper, for the produce of a few hours' labour enables the savage to purchase in abundance what are to him luxuries, while to the European they are necessaries of life. The barbarian is no happier and no better off for this cheapness. On the contrary, it has a most injurious effect on him. He wants the stimulus of necessity to force him to labour; and if iron were as dear as silver, and calico as costly as satin, the effect would be beneficial to him. As it is, he has more idle hours, gets a more constant supply of tobacco, and can intoxicate himself with arrack more frequently and more thoroughly; for your Aru man scorns to get half drunk—a tumblerful of arrack is but a slight stimulus, and nothing less than half a gallon of spirit will make him tipsy to his own satisfaction.

It is not agreeable to reflect on this state of things. At least half of the vast multitudes of uncivilized peoples, on whom our gigantic manufacturing system, enormous capital, and intense competition force the produce of our looms and workshops, would be not a whit worse off physically, and would certainly be improved morally, if all the articles with which we supply them were double or treble their present prices. If at the same time the difference of cost, or a large portion of it, could find its way into the pockets of the manufacturing workmen, thousands would be raised from want to comfort, from starvation to health, and would be removed from one of the chief incentives to crime. It is difficult for an Englishman to avoid contemplating with pride our gigantic and ever-increasing manufactures and commerce, and thinking everything good that renders their progress still more rapid, either by lowering the price at which the articles can be produced, or by discovering new markets to which they may be sent. If, however, the question that is so frequently asked of the votaries of the less popular sciences were put here—" *Cui bono ?*"—it would be found more difficult to answer than had been imagined. The advantages, even to the few who reap them, would be seen to be mostly physical, while the widespread moral and intellectual evils resulting from unceasing labour, low wages, crowded dwellings, and monotonous occupations, to perhaps as large a number as those who gain any real advantage, might be held to show a balance of evil so great, as to lead the greatest admirers of our manufactures and commerce to doubt the advisability of their further development. It will be said: "We cannot stop it; capital must be employed; our population must be kept at work; if we hesitate a moment, other nations now hard pressing us will get ahead, and national ruin will follow." Some of this is true, some fallacious. It is undoubtedly a difficult problem which we have to solve; and I am inclined to think it is this difficulty that makes men conclude that what seems a necessary and unalterable state of things must be good—that its benefits must be greater than its evils. This was the feeling of the American advocates of slavery; they could not see an easy, comfortable way out of it. In our own case, however, it is to be hoped, that if a fair consideration of the matter in all its bearings shows that a preponderance of evil arises from the immensity of our manufactures and commerce—evil which must go on increasing with their increase—there is enough both of political wisdom and true philanthropy in Englishmen, to induce them to turn their superabundant wealth into other channels. The fact that has led to these remarks is surely a striking one: that in one of the most remote corners of the earth savages can buy clothing cheaper than the people of the country where it is made; that the weaver's child should shiver in the wintry wind, unable to purchase articles attainable by the wild natives of a tropical climate, where clothing is mere ornament or luxury, should make us pause ere we regard with unmixed admiration the system which has led to such a result, and cause us to look with ome suspicion on the further extension of that system. It must be remembered too that our commerce is not a purely natural growth. It has been ever fostered by the legislature, and forced to an unnatural luxuriance by the protection of our fleets and armies. The wisdom and the justice of this policy have been already doubted. So soon, therefore, as it is seen that the further extension of our manufactures and commerce would be an evil, the remedy is not far to seek.

After six weeks' confinement to the house I was at length

well, and could resume my daily walks in the forest. I did not, however, find it so productive as when I had first arrived at Dobbo. There was a damp stagnation about the paths, and insects were very scarce. In some of my best collecting places I now found a mass of rotting wood, mingled with young shoots, and overgrown with climbers, yet I always managed to add something daily to my extensive collections. I one day met with a curious example of failure of instinct, which, by showing it to be fallible, renders it very doubtful whether it is anything more than hereditary habit, dependent on delicate modifications of sensation. Some sailors cut down a good-sized tree, and, as is always my practice, I visited it daily for some time in search of insects. Among other beetles came swarms of the little cylindrical wood-borers (Platypus, Tesserocerus, &c.), and commenced making holes in the bark. After a day or two I was surprised to find hundreds of them sticking in the holes they had bored, and on examination discovered that the milky sap of the tree was of the nature of gutta-percha, hardening rapidly on exposure to the air, and gluing the little animals in self-dug graves. The habit of boring holes in trees in which to deposit their eggs was not accompanied by a sufficient instinctive knowledge of which trees were suitable, and which destructive to them. If, as is very probable, these trees have an attractive odour to certain species of borers, it might very likely lead to their becoming extinct ; while other species to whom the same odour was disagreeable, and who therefore avoided the dangerous trees, would survive, and would be credited by us with an instinct, whereas they would really be guided by a simple sensation.

Those curious little beetles, the Brenthidæ, were very abundant in Aru. The females have a pointed rostrum, with which they bore deep holes in the bark of dead trees, often burying the rostrum up to the eyes, and in these holes deposit their eggs. The males are larger, and have the rostrum dilated at the end, and sometimes terminating in a good-sized pair of jaws. I once saw two males fighting together ; each had a fore-leg laid across the neck of the other, and the rostrum bent quite in an attitude of defiance, and looking most ridiculous. Another time, two were fighting for a female, who stood close by busy at her boring. They pushed at each other with their rostra, and clawed and thumped, apparently in the greatest rage, although their coats of mail must have saved both from injury. The small one, however, soon ran away, acknowledging himself vanquished, In most Coleoptera the female is larger than the male, and it is therefore interesting, as bearing on the question of sexual selection, that in this case, as in the stag-beetles where the males fight together, they should be not only better armed, but also much larger than the females.

Just as we were going away, a handsome tree, allied to Erythrina, was in blossom, showing its masses of large crimson flowers scattered here and there about the forest. Could it have been seen from an elevation, it would have had a fine effect; from below I could only catch sight of masses of gorgeous colour in clusters and festoons overhead, about which flocks of blue and orange lories were fluttering and screaming.

A good many people died at Dobbo this season; I believe about twenty. They were buried in a little grove of Casuarinas behind my house. Among the traders was a Mahometan priest, who superintended the funerals, which were very simple. The body was wrapped up in new white cotton cloth, and was carried on a bier to the grave. All the spectators sat down on the ground, and the priest chanted some verses from the Koran. The graves were fenced round with a slight bamboo railing, and a little carved wooden head-post was put to mark the spot. There was

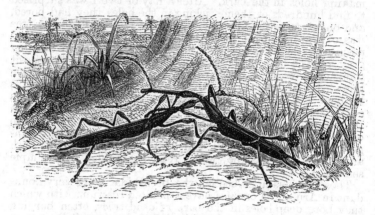

MALE BRENTHIDÆ (*Leptorhynchus angustatus*) FIGHTING.

also in the village a small mosque, where every Friday the faithful went to pray. This is probably more remote from Mecca than any other mosque in the world, and marks the farthest eastern extension of the Mahometan religion. The Chinese here, as elsewhere, showed their superior wealth and civilization by tombstones of solid granite brought from Singapore, with deeply-cut inscriptions, the characters of which are painted in red, blue, and gold. No people have more respect for the graves of their relations and friends than this strange, ubiquitous, money-getting people.

Soon after we had returned to Dobbo, my Macassar boy, Baderoon, took his wages and left me, because I scolded him for laziness. He then occupied himself in gambling, and at first had some luck, and bought ornaments, and had plenty of money. Then his luck turned; he lost everything, borrowed money and lost that, and was obliged to become the slave of his creditor till he had worked out the debt. He was a quick and active lad when he pleased, but was apt to be idle, and had such an incorrigible propensity for gambling, that it will very likely lead to his becoming a slave for life.

The end of June was now approaching, the east monsoon had set in steadily, and in another week or two Dobbo would be deserted. Preparations for departure were everywhere visible, and every sunny day (rather rare now) the streets were as crowded and as busy as beehives. Heaps of tripang were finally dried and packed up in sacks; mother-of-pearl shell, tied up with rattans into convenient bundles, was all day long being carried to the beach to be loaded; water-casks were filled, and cloths and mat-sails mended and strengthened for the run home before the strong east wind. Almost every day groups of natives arrived from the most distant parts of the islands, with cargoes of bananas and sugar-cane to exchange for tobacco, sago-bread, and other luxuries, before the general departure. The Chinamen killed their fat pig and made their parting feast, and kindly sent me some pork, and a basin of birds'-nest stew, which had very little more taste than a dish of vermicelli. My boy Ali returned from Wanumbai, where I had sent him alone for a fortnight to buy Paradise Birds and prepare the skins; he brought me sixteen glorious specimens, and had he not been very ill with fever and ague might have obtained twice the number. He had lived with the people whose house I had occupied, and it is a proof of their goodness, if fairly treated, that although he took with him a quantity of silver dollars to pay for the birds they caught, no attempt was made to rob him, which might have been done with the most perfect impunity. He was kindly treated when ill, and was brought back to me with the balance of the dollars he had not spent.

The Wanumbai people, like almost all the inhabitants of the Aru Islands, are perfect savages, and I saw no signs of any religion. There are, however, three or four villages on the coast where schoolmasters from Amboyna reside, and the people are nominally Christians, and are to some extent educated and civilized. I could not get much real knowledge of the customs of the Aru people during the short time I was among them, but they have evidently been considerably influenced by their long association with Mahometan traders. They often bury their dead, although the national custom is to expose the body on a raised stage till it decomposes. Though there is no limit to the number of wives a man may have, they seldom exceed one or two. A wife is regularly purchased from the parents, the price being a large assortment of articles, always including gongs, crockery, and cloth. They told me that some of the tribes kill the old men and women when they can no longer work, but I saw many very old and decrepit people, who seemed pretty well attended to. No doubt all who have much intercourse with the Bugis and Ceramese traders gradually lose many of their native customs, especially as these people often settle in their villages and marry native women.

The trade carried on at Dobbo is very considerable. This year there were fifteen large praus from Macassar, and perhaps a hundred small boats from Ceram, Goram, and Ké. The Macassar cargoes are worth about 1,000*l*. each, and the other boats take away perhaps about 3,000*l*. worth, so that the whole exports may be estimated at 18,000*l*. per annum. The largest and most bulky items are pearl-shell and tripang, or *bêche-de-mer*, with smaller quantities of tortoise-shell, edible birds' nests, pearls, ornamental woods, timber, and Birds of Paradise. These are purchased with a variety of goods. Of arrack, about equal in strength to ordinary West India rum, 3,000 boxes, each containing fifteen half-gallon bottles, are consumed annually. Native cloth from Celebes is much esteemed for its durability, and large quantities are sold, as well as white English calico and American unbleached cottons, common crockery, coarse cutlery, muskets, gunpowder, gongs, small brass cannon, and elephants' tusks. These three last articles constitute the wealth of the Aru people, with which they pay for their wives, or which they hoard up as " real property." Tobacco is in immense demand for chewing, and it must be very strong, or an Aru man will not look at it. Knowing how little these people generally work, the mass of produce obtained annually shows that the islands must be pretty thickly inhabited, especially along the coasts, as nine-tenths of the whole are marine productions.

It was on the 2nd of July that we left Aru, followed by all the Macassar praus, fifteen in number, who had agreed to sail in company. We passed south of Banda, and then steered due west, not seeing land for three days, till we sighted some low islands west of Bouton. We had a strong and steady south-east wind day and night, which carried us on at about five knots an hour, where a clipper ship would have made twelve. The sky was continually cloudy, dark, and threatening, with occasional drizzling showers, till we were west of Bouru, when it cleared up and we enjoyed the bright sunny skies of the dry season for the rest of our voyage. It is about here, therefore, that the seasons of the eastern and western regions of the Archipelago are divided. West of this line from June to December is generally fine, and often very dry, the rest of the year being the wet season. East of it the weather is exceedingly uncertain, each island, and each side of an island, having its own peculiarities. The difference seems to consist not so much in the distribution of the rainfall as in that of the clouds and the moistness of the atmosphere. In Aru, for example, when we left, the little streams were all dried up, although the weather was gloomy ; while in January, February, and March, when we had the hottest sunshine and the finest days, they were always flowing. The driest time of all the year in Aru occurs in September and October, just as it does in Java and Celebes.

The rainy seasons agree, therefore, with those of the western islands, although the weather is very different. The Molucca sea is of a very deep blue colour, quite distinct from the clear light blue of the Atlantic. In cloudy and dull weather it looks absolutely black, and when crested with foam has a stern and angry aspect. The wind continued fair and strong during our whole voyage, and we reached Macassar in perfect safety on the evening of the 11th of July, having made the passage from Aru (more than a thousand miles) in nine and a half days.

My expedition to the Aru Islands had been eminently successful. Although I had been for months confined to the house by illness, and had lost much time by the want of the means of locomotion, and by missing the right season at the right place, I brought away with me more than nine thousand specimens of natural objects, of about sixteen hundred distinct species. I had made the acquaintance of a strange and little-known race of men ; I had become familiar with the traders of the far East ; I had revelled in the delights of exploring a new fauna and flora, one of the most remarkable and most beautiful and least-known in the world ; and I had succeeded in the main object for which I had undertaken the journey—namely, to obtain fine specimens of the magnificent Birds of Paradise, and to be enabled to observe them in their native forests. By this success I was stimulated to continue my researches in the Moluccas and New Guinea for nearly five years longer, and it is still the portion of my travels to which I look back with the most complete satisfaction.

CHAPTER XXXIII.

THE ARU ISLANDS.—PHYSICAL GEOGRAPHY AND ASPECTS OF NATURE.

IN this chapter I propose to give a general sketch of the physical geography of the Aru Islands, and of their relation to the surrounding countries ; and shall thus be able to incorporate the information obtained from traders, and from the works of other naturalists, with my own observations in these exceedingly interesting and little-known regions.

The Aru group may be said to consist of one very large central island with a number of small ones scattered round it. The great island is called by the natives and traders " Tana-busar " (great or mainland), to distinguish it as a whole from Dobbo, or any of the detached islands. It is of an irregular oblong form, about eighty miles from north to south, and forty or fifty miles from east to west, in which direction it is traversed by three narrow channels, dividing it into four portions. These channels are always called rivers by the traders, which puzzled me much till I passed through one of them, and saw how exceedingly ap-

plicable the name was. The northern channel, called the river
of Watelai, is about a quarter of a mile wide at its entrance, but
soon narrows to about the eighth of a mile, which width it retains, with little variation, during its whole length of nearly
fifty miles, till it again widens at its eastern mouth. Its course
is moderately winding, and the banks are generally dry and
somewhat elevated. In many places there are low cliffs of hard
coralline limestone, more or less worn by the action of water;
while sometimes level spaces extend from the banks to low
ranges of hills a little inland. A few small streams enter it
from right and left, at the mouths of which are some little
rocky islands. The depth is very regular, being from ten to
fifteen fathoms, and it has thus every feature of a true river,
but for the salt water and the absence of a current. The other
two rivers, whose names are Vorkai and Maykor, are said to be
very similar in general character; but they are rather near together, and have a number of cross channels intersecting the
flat tract between them. On the south side of Maykor the
banks are very rocky, and from thence to the southern extremity of Aru is an uninterrupted extent of rather elevated
and very rocky country, penetrated by numerous small streams,
in the high limestone cliffs bordering which the edible birds'
nests of Aru are chiefly obtained. All my informants stated
that the two southern rivers are larger than Watelai.

The whole of Aru is low, but by no means so flat as it has
been represented, or as it appears from the sea. Most of it is
dry rocky ground, with a somewhat undulating surface, rising
here and there into abrupt hillocks, or cut into steep and narrow
ravines. Except the patches of swamp which are found at the
mouths of most of the small rivers, there is no absolutely level
ground, although the greatest elevation is probably not more
than two hundred feet. The rock which everywhere appears
in the ravines and brooks is a coralline limestone, in some places
soft and friable, in others so hard and crystalline as to resemble
our mountain limestone.

The small islands which surround the central mass are very
numerous; but most of them are on the east side, where they
form a fringe, often extending ten or fifteen miles from the main
islands. On the west there are very few, Wamma and Pulo Babi
being the chief, with Ougia and Wassia at the north-west
extremity. On the east side the sea is everywhere shallow, and
full of coral; and it is here that the pearl-shells are found which
form one of the chief staples of Aru trade. All the islands are
covered with a dense and very lofty forest.

The physical features here described are of peculiar interest,
and, as far as I am aware, are to some extent unique; for I have
been unable to find any other record of an island of the size of
Aru crossed by channels which exactly resemble true rivers.
How these channels originated were a complete puzzle to me,
till, after a long consideration of the whole of the natural

phenomena presented by these islands, I arrived at a conclusion which I will now endeavour to explain. There are three ways in which we may conceive islands which are not volcanic to have been formed, or to have been reduced to their present condition, —by elevation, by subsidence, or by separation from a continent or larger island. The existence of coral rock, or of raised beaches far inland, indicates recent elevation ; lagoon coral-islands, and such as have barrier or encircling reefs, have suffered subsidence ; while our own islands, whose productions are entirely those of the adjacent continent, have been separated from it. Now the Aru Islands are all coral rock, and the adjacent sea is shallow and full of coral ; it is therefore evident that they have been elevated from beneath the ocean at a not very distant epoch. But if we suppose that elevation to be the first and only cause of their present condition, we shall find ourselves quite unable to explain the curious river-channels which divide them. Fissures during upheaval would not pro-duce the regular width, the regular depth, or the winding curves which characterize them ; and the action of tides and currents during their elevation might form straits of irregular width and depth, but not the river-like channels which actually exist. If, again, we suppose the last movement to have been one of subsidence, reducing the size of the islands, these channels are quite as inexplicable ; for subsidence would necessarily lead to the flooding of all low tracts on the banks of the old rivers, and thus obliterate their courses ; whereas these remain perfect, and of nearly uniform width from end to end.

Now if these channels have ever been rivers they must have flowed from some higher regions, and this must have been to the east, because on the north and west the sea-bottom sinks down at a short distance from the shore to an unfathomable depth ; whereas on the east a shallow sea, nowhere exceeding fifty fathoms, extends quite across to New Guinea, a distance of about a hundred and fifty miles. An elevation of only three hundred feet would convert the whole of this sea into moderately high land, and make the Aru Islands a portion of New Guinea ; and the rivers which have their mouths àt Utanata and Wamuka might then have flowed on across Aru, in the channels which are now occupied by salt water. When the intervening land sunk down, we must suppose the land that now constitutes Aru to have remained nearly stationary—a not very improbable supposition, when we consider the great extent of the shallow sea, and the very small amount of depression the land need have undergone to produce it.

But the fact of the Aru Islands having once been connected with New Guinea does not rest on this evidence alone. There is such a striking resemblance between the productions of the two countries as only exists between portions of a common territory. I collected one hundred species of land-birds in the Aru Islands, and about eighty of them have been found on the mainland of New Guinea. Among these are the great wingless cassowary, two species of heavy brush-turkeys, and two of short winged thrushes, which could certainly not have passed over the 150 miles of open sea to the coast of New Guinea. This barrier is equally effectual in the case of many other birds which live only in the depths of the forest, as the kinghunters (Dacelo gaudichaudi), the fly-catching wrens (Todopsis), the great crown pigeon (Goura coronata), and the small wood doves (Ptilonopus perlatus, P. aurantiifrons, and P. coronulatus). Now, to show the real effect of such a barrier, let us take the island of Ceram, which is exactly the same distance from New Guinea, but separated from it by a deep sea. Out of about seventy land-birds inhabiting Ceram, only fifteen are found in New Guinea, and none of these are terrestrial or forest-haunting species. The cassowary is distinct ; the kingfishers, parrots, pigeons, fly-catchers, honeysuckers, thrushes, and cuckoos, are almost always quite distinct species. More than this, at least twenty genera, which are common to New Guinea and Aru, do not extend into Ceram, indicating with a force which every naturalist will appreciate, that the two latter countries have received their faunas in a radically different manner. Again, a true kangaroo is found in Aru, and the same species occurs in Mysol, which is equally Papuan in its productions, while either the same, or one closely allied to it, inhabits New Guinea ; but no such animal is found in Ceram, which is only sixty miles from Mysol. Another small marsupial animal (Perameles doreyanus) is common to Aru and New Guinea. The insects show exactly the same results. The butterflies of Aru are all either New Guinea species, or very slightly modified forms ; whereas those of Ceram are more distinct than are the birds of the two countries.

It is now generally admitted that we may safely reason on such facts as these, which supply a link in the defective geo-logical record. The upward and downward movements which any country has undergone, and the succession of such move-ments, can be determined with much accuracy ; but geology alone can tell us nothing of lands which have entirely dis-appeared beneath the ocean. Here physical geography and the distribution of animals and plants are of the greatest service. By ascertaining the depth of the seas separating one country from another, we can form some judgment of the changes which are taking place. If there are other evidences of sub-sidence, a shallow sea implies a former connexion of the adjacent lands ; but if this evidence is wanting, or if there is reason to suspect a rising of the land, then the shallow sea may be the result of that rising, and may indicate that the two countries will be joined at some future time, but not that they have previously been so. The nature of the animals and plants inhabiting these countries will, however, almost always enable us to determine this question. Mr. Darwin has shown us how we may determine in almost every case whether an island has ever been connected with a continent or larger land, by the presence or absence of terrestrial Mammalia and reptiles. What he terms "oceanic islands" possess neither of these groups of animals, though they may have a luxuriant vegetation, and a fair number of birds, insects, and land-shells ; and we therefore conclude that they have originated in mid-ocean, and have never been connected with the nearest masses of land. St. Helena, Madeira, and New Zealand are examples of oceanic islands. They possess all other classes of life, because these have means of dispersion over wide spaces of sea, which terrestrial mammals and birds have not, as is fully explained in Sir Charles Lyell's *Principles of Geology* and Mr. Darwin's *Origin of Species.* On the other hand, an island may never have been actually connected with the adjacent continents or islands, and yet may possess representatives of all classes of animals, because many terrestrial mammals and some reptiles have the means of passing over short distances of sea. But in these cases the number of species that have thus migrated will be very small, and there will be great deficiencies even in birds and flying insects, which we should imagine could easily cross over. The island of Timor (as I have already shown in Chapter XIII.) bears this relation to Australia ; for while it contains several birds and insects of Australian forms, no Australian mammal or reptile is found in it, and a great number of the most abundant and characteristic forms of Australian birds and insects are entirely absent. Contrast this with the British Islands, in which a large proportion of the plants, insects, reptiles, and Mammalia of the adjacent parts of the continent are fully represented, while there are no remarkable deficiencies of extensive groups, such as always occur when there is reason to believe there has been no such connexion. The case of Sumatra, Borneo, and Java, and the Asiatic continent is equally clear ; many large Mammalia, terrestrial birds, and reptiles being common to all, while a large number more are of closely allied forms. Now, geology has taught us that this representation by allied forms in the same locality implies lapse of time, and we therefore infer that in Great Britain, where almost every species is absolutely identical with those on the Continent, the separation has been very recent ; while in Sumatra and Java, where a considerable number of the continental species are represented by allied forms, the separation was more remote.

From these examples we may see how important a supplement to geological evidence is the study of the geographical distribution of animals and plants in determining the former condition of the earth's surface ; and how impossible it is to understand the former without taking the latter into account. The productions of the Aru Islands offer the strongest evidence that at no very distant epoch they formed a part of New Guinea ; and the peculiar physical features which I have described, indicate that they must have stood at very nearly the same level then as they do now, having been separated by the subsidence of the great plain which formerly connected them with it.

Persons who have formed the usual ideas of the vegetation of the tropics—who picture to themselves the abundance and brilliancy of the flowers, and the magnificent appearance of hundreds of forest trees covered with masses of coloured blossoms, will be surprised to hear, that though vegetation in Aru is highly luxuriant and varied, and would afford abundance of fine and curious plants to adorn our hothouses, yet bright and showy flowers are, as a general rule, altogether absent, or so very scarce as to produce no effect whatever on the general scenery. To give particulars : I have visited five distinct localities in the islands, I have wandered daily in the forests, and have passed along upwards of a hundred miles of coast and river during a period of six months, much of it very fine weather, and till just as I was about to leave, I never saw a single plant of striking brilliancy or beauty, hardly a shrub equal to a hawthorn, or a climber equal to a honeysuckle! It cannot be said that the flowering season had not arrived, for I saw many herbs, shrubs, and forest trees in flower, but all had blossoms of a green or greenish-white tint, not superior to our lime-trees. Here and there on the river banks and coasts are a few Convolvulaceæ, not equal to our garden Ipomæas, and in the deepest shades of the forest some fine scarlet and purple Zingiberaceæ, but so few and scattered as to be nothing amid the mass of green and flowerless vegetation. Yet the noble Cycadaceæ and screw-pines, thirty or forty feet high, the elegant tree ferns, the lofty palms, and the variety of beautiful and curious plants which everywhere meet the eye, attest the warmth and moisture of the tropics, and the fertility of the soil. It is true that Aru seemed to me exceptionally poor in flowers, but this is only an exaggeration of a general tropical feature ; for my whole experience in the equatorial regions of the west and the east has convinced me, that in the most luxuriant parts of the tropics, flowers are less abundant, on the average less showy, and are far less effective in adding colour to the landscape than in temperate climates. I have never seen in the tropics such brilliant masses of colour as even England can show in her furze-clad commons, her heathery mountain-sides, her glades of wild hyacinths, her fields of poppies, her meadows of buttercups and orchises—carpets of yellow, purple, azure-blue, and fiery crimson, which the tropics can rarely exhibit. We have smaller masses of colour in our hawthorn and crab trees, our holly and mountain-ash, our broom, foxgloves, primroses, and purple vetches, which clothe with gay colours the whole length and breadth of our land. These beauties are all common. They are characteristic of the country and the climate ; they have not to be sought for, but they gladden the eye at every

step. In the regions of the equator, on the other hand, whether
it be forest or savannah, a sombre green clothes universal nature.
You may journey for hours, and even for days, and meet with
nothing to break the monotony. Flowers are everywhere rare,
and anything at all striking is only to be met with at very
distant intervals.

The idea that nature exhibits gay colours in the tropics, and
that the general aspect of nature is there more bright and varied
in hue than with us, has even been made the foundation of
theories of art, and we have been forbidden to use bright
colours in our garments, and in the decorations of our dwellings, because it was supposed that we should be thereby acting
in opposition to the teachings of nature. The argument itself
is a very poor one, since it might with equal justice be maintained, that as we possess faculties for the appreciation of
colours, we should make up for the deficiencies of nature and
use the gayest tints in those regions where the landscape is
most monotonous. But the assumption on which the argument
is founded is totally false, so that even if the reasoning were
valid, we need not be afraid of outraging nature by decorating
our houses and our persons with all those gay hues which are
so lavishly spread over our fields and mountains, our hedges,
woods, and meadows.

It is very easy to see what has led to this erroneous view of
the nature of tropical vegetation. In our hothouses and at our
flower-shows we gather together the finest flowering plants from
the most distant regions of the earth, and exhibit them in a
proximity to each other which never occurs in nature. A
hundred distinct plants, all with bright, or strange, or gorgeous
flowers, make a wonderful show when brought together ; but
perhaps no two of these plants could ever be seen together in a
state of nature, each inhabiting a distant region, or a different
station. Again, all moderately warm extra-European countries
are mixed up with the tropics in general estimation, and a vague
idea is formed that whatever is pre-eminently beautiful *must*
come from the hottest parts of the earth. But the fact is quite
the contrary. Rhododendrons and azaleas are plants of temperate regions, the grandest lilies are from temperate Japan,
and a large proportion of our most showy flowering plants are
natives of the Himalayas, of the Cape, of the United States, of
Chili, or of China and Japan, all temperate regions. True, there
are a great number of grand and gorgeous flowers in the tropics,
but the proportion they bear to the mass of the vegetation is
exceedingly small ; so that what appears an anomaly is never-
theless a fact, and the effect of flowers on the general aspect of
nature is far less in the equatorial than in the temperate regions
of the earth.

CHAPTER XXXIV.

NEW GUINEA.—DOREY.

(MARCH TO JULY 1858.)

AFTER my return from Gilolo to Ternate, in March 1858, I made arrangements for my long-wished-for voyage to the mainland of New Guinea, where I anticipated that my collections would surpass those which I had formed at the Aru Islands. The poverty of Ternate in articles used by Europeans was shown by my searching in vain through all the stores for such common things as flour, metal spoons, wide-mouthed phials, bees'-wax, a pen-knife, and a stone or metal pestle and mortar. I took with me four servants : my head man Ali, and a Ternate lad named Jumaat (Friday), to shoot ; Lahagi, a steady, middle-aged man, to cut timber and assist me in insect-collecting ; and Loisa, a Javanese cook. As I knew I should have to build a house at Dorey, where I was going, I took with me eighty cadjans, or waterproof mats, made of pandanus leaves, to cover over my baggage on first landing, and to help to roof my house afterwards.

We started on the 25th of March in the schooner *Hester Helena*, belonging to my friend Mr. Duivenboden, and bound on a trading voyage along the north coast of New Guinea. Having calms and light airs, we were three days reaching Gané, near the south end of Gilolo, where we stayed to fill up our water-casks, and buy a few provisions. We obtained fowls, eggs, sago, plantains, sweet potatoes, yellow pumpkins, chilies, fish, and dried deers' meat ; and on the afternoon of the 29th proceeded on our voyage to Dorey harbour. We found it, however, by no means easy to get along ; for so near to the equator the monsoons entirely fail of their regularity, and after passing the southern point of Gilolo we had calms, light puffs of wind, and contrary currents, which kept us for five days in sight of the same islands, between it and Poppa. A squall then brought us on to the entrance of Dampier's Straits, where we were again becalmed, and were three more days creeping through them. Several native canoes now came off to us from Waigiou on one side, and Batanta on the other, bringing a few common shells, palm-leaf mats, cocoa-nuts, and pumpkins. They were very extravagant in their demands, being accustomed to sell their trifles to whalers and China ships, whose crews will purchase anything at ten times its value. My only purchases were a float belonging to a turtle-spear carved to resemble a bird, and a very well-made palm-leaf box, for which articles I gave a copper ring and a yard of calico. The canoes were very narrow and furnished with an outrigger, and in some of them there was only one man, who seemed to think nothing of coming out alone eight or ten miles from shore. The people were Papuans, much resembling the natives of Aru.

When we had got out of the Straits, and were fairly in the great Pacific Ocean, we had a steady wind for the first time since leaving Ternate, but unfortunately it was dead ahead, and we had to beat against it, tacking on and off the coast of New Guinea. I looked with intense interest on those rugged mountains, retreating ridge behind ridge into the interior, where the foot of civilized man had never trod. There was the country of the cassowary and the tree-kangaroo, and those dark forests produced the most extraordinary and the most beautiful of the feathered inhabitants of the earth—the varied species of Birds of Paradise. A few days more, and I hoped to be in pursuit of these, and of the scarcely less beautiful insects which accompany them. We had still, however, for several days only calms and light head-winds, and it was not till the 10th of April that a fine westerly breeze set in, followed by a squally night, which kept us off the entrance of Dorey harbour. The next morning we entered, and came to anchor off the small island of Mansinam, on which dwelt two German missionaries, Messrs. Otto and Geisler. The former immediately came on board to give us welcome, and invited us to go on shore and breakfast with him. We were then introduced to his companion—who was suffering dreadfully from an abscess on the heel, which had confined him to the house for six months—and to his wife, a young German woman, who had been out only three months. Unfortunately she could speak no Malay or English, and had to guess at our compliments on her excellent breakfast by the justice we did to it.

These missionaries were working men, and had been sent out as being more useful among savages than persons of a higher class. They had been here about two years, and Mr. Otto had already learnt to speak the Papuan language with fluency, and had begun translating some portions of the Bible. The language, however, is so poor that a considerable number of Malay words have to be used ; and it is very questionable whether it is possible to convey any idea of such a book to a people in so low a state of civilization. The only nominal converts yet made are a few of the women ; and some few of the children attend school, and are being taught to read, but they make little progress. There is one feature of this mission which I believe will materially interfere with its moral effect. The missionaries are allowed to trade to eke out the very small salaries granted them from Europe, and of course are obliged to carry out the trade principle of buying cheap and selling dear, in order to make a profit. Like all savages the natives are quite careless of the future, and when their small rice crops are gathered they bring a large portion of it to the missionaries, and sell it for knives, beads, axes, tobacco, or any other articles they may require. A few months later, in the wet season, when food is scarce, they come to buy it back again, and give in exchange tortoiseshell, tripang, wild nutmegs, or other produce. Of course the rice is sold at a much higher rate than it was bought, as is perfectly fair and just ; and the operation is on the whole thoroughly beneficial to the natives, who would otherwise consume and waste their food when it was abundant, and then starve. Yet I cannot imagine that the natives see it in this light. They must look upon the trading missionaries with some suspicion, and cannot feel so sure of their teachings being disinterested, as would be the case if they acted like the Jesuits in Singapore. The first thing to be done by the missionary in attempting to improve savages, is to convince them by his actions that he comes among them for their benefit only, and not for any private ends of his own. To do this he must act in a different way from other men, not trading and taking advantage of the necessities of those who want to sell, but rather giving to those who are in distress. It would be well if he conformed himself in some degree to native customs, and then endeavoured to show how these customs might be gradually modified, so as to be more healthful and more agreeable. A few energetic and devoted men acting in this way might probably effect a decided moral improvement on the lowest savage tribes, whereas trading missionaries, teaching what Jesus said, but not doing as He did, can scarcely be expected to do more than give them a very little of the superficial varnish of religion.

Dorey harbour is in a fine bay, at one extremity of which an elevated point juts out, and, with two or three small islands, forms a sheltered anchorage. The only vessel it contained when we arrived was a Dutch brig, laden with coals for the use of a war-steamer, which was expected daily, on an exploring expedition along the coasts of New Guinea, for the purpose of fixing on a locality for a colony. In the evening we paid it a visit, and landed at the village of Dorey, to look out for a place where I could build my house. Mr. Otto also made arrangements for me with some of the native chiefs, to send men to cut wood, rattans, and bamboo the next day.

The villages of Mansinam and Dorey presented some features quite new to me. The houses all stand completely in the water, and are reached by long rude bridges. They are very low, with the roof shaped like a large boat, bottom upwards. The posts which support the houses, bridges, and platforms, are small crooked sticks, placed without any regularity, and looking as if they were tumbling down. The floors are also formed of sticks, equally irregular, and so loose and far apart that I found it almost impossible to walk on them. The walls consist of bits of boards, old boats, rotten mats, attaps, and palm-leaves, stuck in anyhow here and there, and having altogether the most wretched and dilapidated appearance it is possible to conceive. Under the eaves of many of the houses hang human skulls, the trophies of their battles with the savage Arfaks of the interior, who often come to attack them. A large boat-shaped council-house is supported on larger posts, each of which is grossly carved to represent a naked male or female human figure, and other carvings still more revolting are placed upon the platform before the entrance. The view of an ancient lake-dweller's village, given as the frontispiece of Sir Charles Lyell's *Antiquity of Man*, is chiefly founded on a sketch of this very village of Dorey; but the extreme regularity of the structures there depicted has no

PAPUAN, NEW GUINEA.

place in the original, any more than it probably had in the actual lake-villages.

The people who inhabit these miserable huts are very similar to the Ké and Aru islanders, and many of them are very handsome, being tall and well-made, with well-cut features and large aquiline noses. Their colour is a deep brown, often approaching closely to black, and the fine mop-like heads of frizzly hair appear to be more common than elsewhere, and are considered a great ornament, a long six-pronged bamboo fork being kept stuck in them to serve the purpose of a comb; and this is assiduously used at idle moments to keep the densely growing mass from becoming matted and tangled. The majority have short woolly hair, which does not seem capable of an equally luxuriant development. A growth of hair somewhat similar to this, and almost as abundant, is found among the half-breeds between the Indian and Negro in South America. Can this be an indication that the Papuans are a mixed race ?

For the first three days after our arrival I was fully occupied from morning to night building a house, with the assistance of a dozen Papuans and my own men. It was immense trouble to get our labourers to work, as scarcely one of them could speak a word of Malay ; and it was only by the most energetic gesticulations, and going through a regular pantomime of what was wanted, that we could get them to do anything. If we made them understand that a few more poles were required, which two could have easily cut, six or eight would insist upon going together, although we needed their assistance in other things. One morning ten of them came to work, bringing only one chopper among them, although they knew I had none ready for use. I chose a place about two hundred yards from the beach, on an elevated ground, by the side of the chief path from the village of Dorey to the provision-grounds and the forest. Within twenty yards was a little stream, which furnished us with excellent water and a nice place to bathe. There was only low underwood to clear away, while some fine forest trees stood at a short distance, and we cut down the wood for about twenty yards round to give us light and air. The house, about twenty feet by fifteen, was built entirely of wood, with a bamboo floor, a single door of thatch, and a large window, looking over the sea, at which I fixed my table, and close beside it my bed, within a little partition. I bought a number of very large palm-leaf mats of the natives, which made excellent walls ; while the mats I had brought myself were used on the roof, and were covered over with attaps as soon as we could get them made. Outside, and rather behind, was a little hut, used for cooking, and a bench, roofed over, where my men could sit to skin birds and animals. When all was finished, I had my goods and stores brought up, arranged them conveniently inside, and then paid my Papuans with knives and choppers, and sent them away. The next day our schooner left for the more eastern islands, and I found myself fairly established as the only European inhabitant of the vast island of New Guinea.

As we had some doubt about the natives, we slept at first with loaded guns beside us and a watch set ; but after a few days, finding the people friendly, and feeling sure that they would not venture to attack five well-armed men, we took no further precautions. We had still a day or two's work in finishing up the house, stopping leaks, putting up our hanging shelves for drying specimens inside and out, and making the path down to the water, and a clear dry space in front of the house.

On the 17th, the steamer not having arrived, the coal-ship

left, having lain here a month, according to her contract; and on the same day my hunters went out to shoot for the first time, and brought home a magnificent crown pigeon and a few common birds. The next day they were more successful, and I was delighted to see them return with a Bird of Paradise in full plumage, a pair of the fine Papuan lories (Lorius domicella), four other lories and parroquets, a grackle (Gracula dumonti), a king-hunter (Dacelo gaudichaudi), a racquet-tailed kingfisher (Tanysiptera galatea), and two or three other birds of less beauty. I went myself to visit the native village on the hill behind Dorey, and took with me a small present of cloth, knives, and beads, to secure the good-will of the chief, and get him to send some men to catch or shoot birds for me. The houses were scattered about among rudely cultivated clearings. Two which I visited consisted of a central passage, on each side of which opened short passages, admitting to two rooms, each of which was a house accommodating a separate family. They were elevated at least fifteen feet above the ground, on a complete forest of poles, and were so rude and dilapidated that some of the small passages had openings in the floor of loose sticks, through which a child might fall. The inhabitants seemed rather uglier than those at Dorey village. They are, no doubt, the true indigenes of this part of New Guinea, living in the interior, and subsisting by cultivation and hunting. The Dorey men, on the other hand, are shore-dwellers, fishers and traders in a small way, and have thus the character of a colony who have migrated from another district. These hillmen or "Arfaks" differed much in physical features. They were generally black, but some were brown like Malays. Their hair, though always more or less frizzly, was sometimes short and matted, instead of being long, loose, and woolly; and this seemed to be a constitutional difference, not the effect of care and cultivation. Nearly half of them were afflicted with the scurfy skin-disease. The old chief seemed much pleased with his present, and promised (through an interpreter I brought with me) to protect

PAPUAN PIPE.

my men when they came there shooting, and also to procure me some birds and animals. While conversing, they smoked

tobacco of their own growing, in pipes cut from a single piece
of wood with a long upright handle.

We had arrived at Dorey about the end of the wet season,
when the whole country was soaked with moisture. The native
paths were so neglected as to be often mere tunnels closed over
with vegetation, and in such places there was always a fearful
accumulation of mud. To the naked Papuan this is no obstruction. He wades through it, and the next watercourse
makes him clean again ; but to myself, wearing boots and
trousers, it was a most disagreeable thing to have to go up to
my knees in a mud-hole every morning. The man I brought
with me to cut wood fell ill soon after we arrived, or I would
have set him to clear fresh paths in the worst places. For the
first ten days it generally rained every afternoon and all night ;
but by going out every hour of fine weather, I managed to get
on tolerably with my collections of birds and insects, finding
most of those collected by Lesson during his visit in the *Coquille*,
as well as many new ones. It appears, however, that Dorey is
not the place for Birds of Paradise, none of the natives being
accustomed to preserve them. Those sold here are all brought
from Amberbaki, about a hundred miles west, where the
Doreyans go to trade.

The islands in the bay, with the low lands near the coast,
seem to have been formed by recently raised coral reefs and are
much strewn with masses of coral but little altered. The ridge
behind my house, which runs out to the point, is also entirely
coral rock, although there are signs of a stratified foundation in
the ravines, and the rock itself is more compact and crystalline.
It is, therefore, probably older, a more recent elevation having
exposed the low grounds and islands. On the other side of the
bay rise the great mass of the Arfak mountains, said by the
French navigators to be about ten thousand feet high, and
inhabited by savage tribes. These are held in great dread by
the Dorey people, who have often been attacked and plundered
by them, and have some of their skulls hanging outside their
houses. If I was seen going into the forest anywhere in the
direction of the mountains, the little boys of the village would
shout after me, "Arfaki ! Arfaki ! " just as they did after Lesson
nearly forty years before.

On the 15th of May the Dutch war-steamer *Etna* arrived ; but,
as the coals had gone, it was obliged to stay till they came back.
The captain knew when the coalship was to arrive and how long
it was chartered to stay at Dorey, and could have been back in
time, but supposed it would wait for him, and so did not hurry
himself. The steamer lay at anchor just opposite my house,
and I had the advantage of hearing the half-hourly bells struck,
which was very pleasant after the monotonous silence of the
forest. The captain, doctor, engineer, and some other of the
officers paid me visits ; the servants came to the brook to wash
clothes, and the son of the Prince of Tidore, with one or two

companions, to bathe; otherwise I saw little of them, and was not disturbed by visitors so much as I had expected to be. About this time the weather set in pretty fine, but neither birds nor insects became much more abundant, and new birds were very scarce. None of the Birds of Paradise except the common one were ever met with, and we were still searching in vain for several of the fine birds which Lesson had obtained here. Insects were tolerably abundant, but were not on the average so fine as those of Amboyna, and I reluctantly came to the conclusion that Dorey was not a good collecting locality. Butterflies were very scarce, and were mostly the same as those which I had obtained at Aru.

Among the insects of other orders, the most curious and novel were a group of horned flies, of which I obtained four distinct species, settling on fallen trees and decaying trunks. These

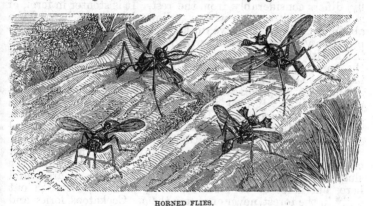

HORNED FLIES.

Elaphomia cervicornis.　　　　Elaphomia wallacei.
E. brevicornis.　　　　　　　　E. alcicornis.

remarkable insects, which have been described by Mr. W. W. Saunders as a new genus, under the name of Elaphomia or deer-flies, are about half an inch long, slender bodied, and with very long legs, which they draw together so as to elevate their bodies high above the surface they are standing upon. The front pair of legs are much shorter, and these are often stretched directly forwards, so as to resemble antennæ. The horns spring from beneath the eye, and seem to be a prolongation of the lower part of the orbit. In the largest and most singular species, named Elaphomia cervicornis or the stag-horned deer-fly, these horns are nearly as long as the body, having two branches, with two small snags near their bifurcation, so as to resemble the horns of a stag. They are black, with the tips pale, while the body and legs are yellowish brown, and the eyes (when alive) violet and green. The next species (Elaphomia wallacei) is of a dark brown colour, banded and spotted with yellow. The horns are about one-third the length of the insect, broad, flat, and of an elongated triangular form. They are of a beautiful pink colour, edged with black, and with a pale central stripe. The front part of the head is also pink, and the eyes violet pink, with a green stripe across them, giving the insect a very elegant and singular appearance. The third species (Elaphomia alcicornis, the elk-horned deer-fly) is a little smaller than the two already described, but resembling in colour Elaphomia wallacei. The horns are very remarkable, being suddenly dilated into a flat plate, strongly toothed round the outer margin, and strikingly resembling the horns of the elk, after which it has been named. They are of a yellowish colour margined with brown, and tipped with black on the three upper teeth. The fourth species (Elaphomia brevicornis, the short-horned deer-fly) differs considerably from the rest. It is stouter in form, of a nearly black colour, with a yellow ring at the base of the abdomen ; the wings have dusky stripes, and the head is compressed and dilated laterally, with very small flat horns, which are black with a pale centre, and look exactly like the rudiment of the horns of the two preceding species. None of the females have any trace of the horns, and Mr. Saunders places in the same genus a species which has no horns in either sex (Elaphomia polita). It is of a shining black colour, and resembles Elaphomia cervicornis in form, size, and general appearance. The figures above given represent these insects of their natural size and in characteristic attitudes.

The natives seldom brought me anything. They are poor creatures, and rarely shoot a bird, pig, or kangaroo, or even the sluggish opossum-like Cuscus. The tree-kangaroos are found here, but must be very scarce, as my hunters, although out daily in the forest, never once saw them. Cockatoos, lories, and parroquets were really the only common birds. Even pigeons were scarce, and in little variety, although we occasionally got the fine crown pigeon, which was always welcome as an addition to our scantily furnished larder.

Just before the steamer arrived I had wounded my ankle by clambering among the trunks and branches of fallen trees (which formed my best hunting grounds for insects), and, as usual with foot wounds in this climate, it turned into an obstinate ulcer, keeping me in the house for several days. When it healed up it was followed by an internal inflammation of the foot, which by the doctor's advice I poulticed incessantly for four or five days, bringing out a severe inflamed swelling on the tendon above the heel. This had to be leeched, and lanced, and doctored with ointments and poultices for several weeks, till I was almost driven to despair—for the weather was at length fine, and I was tantalized by seeing grand butterflies flying past my door, and thinking of the twenty or thirty new species of insects that I ought to be getting every day. And this, too, in

New Guinea!—a country which I might never visit again,—a country which no naturalist had ever resided in before—a country which contained more strange and new and beautiful natural objects than any other part of the globe. The naturalist will be able to appreciate my feelings, sitting from morning to night in my little hut, unable to move without a crutch, and my only solace the birds my hunters brought in every afternoon, and the few insects caught by my Ternate man, Lahagi, who now went out daily in my place, but who of course did not get a fourth part of what I should have obtained. To add to my troubles all my men were more or less ill, some with fever, others with dysentery or ague; at one time there were three of them besides myself all helpless, the cook alone being well, and having enough to do to wait upon us. The Prince of Tidore and the Resident of Banda were both on board the steamer, and were seeking Birds of Paradise, sending men round in every direction, so that there was no chance of my getting even native skins of the rarer kinds; and any birds, insects, or animals the Dorey people had to sell were taken on board the steamer, where purchasers were found for everything, and where a larger variety of articles were offered in exchange than I had to show.

After a month's close confinement in the house I was at length able to go out a little, and about the same time I succeeded in getting a boat and six natives to take Ali and Lahagi to Amberbaki, and to bring them back at the end of a month. Ali was charged to buy all the Birds of Paradise he could get, and to shoot and skin all other rare or new birds; and Lahagi was to collect insects, which I hoped might be more abundant than at Dorey. When I recommenced my daily walks in search of insects, I found a great change in the neighbourhood, and one very agreeable to me. All the time I had been laid up the ship's crew and the Javanese soldiers who had been brought in a tender (a sailing ship which had arrived soon after the *Etna*), had been employed cutting down, sawing, and splitting large trees for firewood, to enable the steamer to get back to Amboyna if the coal-ship did not return; and they had also cleared a number of wide, straight paths through the forest in various directions, greatly to the astonishment of the natives, who could not make out what it all meant. I had now a variety of walks, and a good deal of dead wood on which to search for insects; but notwithstanding these advantages, they were not nearly so plentiful as I had found them at Saráwak, or Amboyna, or Batchian, confirming my opinion that Dorey was not a good locality. It is quite probable, however, that at a station a few miles in the interior, away from the recently elevated coralline rocks and the influence of the sea air, a much more abundant harvest might be obtained.

One afternoon I went on board the steamer to return the captain's visit, and was shown some very nice sketches (by one of the lieutenants), made on the south coast, and also at the

Arfak mountain, to which they had made an excursion. From these and the captain's description, it appeared that the people of Arfak were similar to those of Dorey, and I could hear nothing of the straight-haired race which Lesson says inhabits the interior, but which no one has ever seen, and the account of which I suspect has originated in some mistake. The captain told me he had made a detailed survey of part of the south coast, and if the coal arrived should go away at once to Humboldt Bay, in longitude 141° east, which is the line up to which the Dutch claim New Guinea. On board the tender I found a brother naturalist, a German named Rosenberg, who was draughtsman to the surveying staff. He had brought two men with him to shoot and skin birds, and had been able to purchase a few rare skins from the natives. Among these was a pair of the superb Paradise Pie (Astrapia nigra) in tolerable preservation. They were brought from the island of Jobie, which may be its native country, as it certainly is of the rarer species of crown pigeon (Goura steursii), one of which was brought alive and sold on board. Jobie, however, is a very dangerous place, and sailors are often murdered there when on shore ; sometimes the vessels themselves being attacked. Wandammen, on the mainland opposite Jobie, where there are said to be plenty of birds, is even worse, and at either of these places my life would not have been worth a week's purchase had I ventured to live alone and unprotected as at Dorey. On board the steamer they had a pair of tree-kangaroos alive. They differ chiefly from the ground-kangaroo in having a more hairy tail, not thickened at the base, and not used as a prop ; and by the powerful claws on the fore-feet, by which they grasp the bark and branches, and seize the leaves on which they feed. They move along by short jumps on their hind-feet, which do not seem particularly well adapted for climbing trees. It has been supposed that these tree-kangaroos are a special adaptation to the swampy, half-drowned forests of New Guinea, in place of the usual form of the group, which is adapted only to dry ground. Mr. Windsor Earl makes much of this theory, but, unfortunately for it, the tree-kangaroos are chiefly found in the northern peninsula of New Guinea, which is entirely composed of hills and mountains with very little flat land, while the kangaroo of the low flat Aru Islands (Dorcopsis asiaticus) is a ground species. A more probable supposition seems to be, that the tree-kangaroo has been modified to enable it to feed on foliage in the vast forests of New Guinea, as these form the great natural feature which distinguishes that country from Australia.

On June 5th, the coal-ship arrived, having been sent back from Amboyna, with the addition of some fresh stores for the steamer. The wood, which had been almost all taken on board, was now unladen again, the coal taken in, and on the 17th both steamer and tender left for Humboldt Bay. We were then a little quiet again, and got something to eat; for while the vessels were here every bit of fish or vegetable was taken on board, and I had often to make a small parroquet serve for two meals. My men now returned from Amberbaki, but, alas! brought me almost nothing. They had visited several villages, and even went two days' journey into the interior, but could find no skins of Birds of Paradise to purchase, except the common kind, and very few even of those. The birds found were the same as at Dorey, but were still scarcer. None of the natives anywhere near the coast shoot or prepare Birds of Paradise, which come from far in the interior over two or three ranges of mountains, passing by barter from village to village till they reach the sea. There the natives of Dorey buy them, and on their return home sell them to the Bugis or Ternate traders. It is therefore hopeless for a traveller to go to any particular place on the coast of New Guinea where rare Paradise Birds may have been bought, in hopes of obtaining freshly killed specimens from the natives; and it also shows the scarcity of these birds in any one locality, since from the Amberbaki district, a celebrated place, where at least five or six species have been procured, not one of the rarer ones has been obtained this year. The Prince of Tidore, who would certainly have got them if any were to be had, was obliged to put up with a few of the common yellow ones. I think it probable that a longer residence at Dorey, a little farther in the interior, might show that several of the rarer kinds were found there, as I obtained a single female of the fine scale-breasted Ptiloris magnificus. I was told at Ternate of a bird that is certainly not yet known in Europe, a black King Paradise Bird, with the curled tail and beautiful side plumes of the common species, but all the rest of the plumage glossy black. The people of Dorey knew nothing about this, although they recognized by description most of the other species.

When the steamer left, I was suffering from a severe attack of fever. In about a week I got over this, but it was followed by such a soreness of the whole inside of the mouth, tongue, and gums, that for many days I could put nothing solid between my lips, but was obliged to subsist entirely on slops, although in other respects very well. At the same time two of my men again fell ill, one with fever, the other with dysentery, and both got very bad. I did what I could for them with my small stock of medicines, but they lingered on for some weeks, till on June 26th poor Jumaat died. He was about eighteen years of age, a native, I believe, of Bouton, and a quiet lad, not very active, but doing his work pretty steadily, and as well as he was able. As my men were all Mahometans, I let them bury him in their own fashion, giving them some new cotton cloth for a shroud.

On July 6th the steamer returned from the eastward. The weather was still terribly wet, when, according to rule, it should have been fine and dry. We had scarcely anything to eat, and were all of us ill. Fevers, colds, and dysentery were continually attacking us, and made me long to get away from New Guinea, as much as ever I had longed to come there. The captain of the *Etna* paid me a visit, and gave me a very interesting account of his trip. They had stayed at Humboldt Bay several days, and found it a much more beautiful and more interesting place than Dorey, as well as a better harbour. The natives were quite unsophisticated, being rarely visited except by stray whalers, and they were superior to the Dorey people, morally and physically. They went quite naked. Their houses were some in the water and some inland, and were all neatly and well built ; their fields were well cultivated, and the paths to them kept clear and open, in which respects Dorey is abominable. They were shy at first, and opposed the boats with hostile demonstrations, bending their bows, and intimating that they would shoot if an attempt was made to land. Very judiciously the captain gave way, but threw on shore a few presents, and after two or three trials they were permitted to land, and to go about and see the country, and were supplied with fruits and vegetables. All communication was carried on with them by signs—the Dorey interpreter, who accompanied the steamer, being unable to understand a word of their language. No new birds or animals were obtained, but in their ornaments the feathers of Paradise Birds were seen, showing, as might be expected, that these birds range far in this direction, and probably all over New Guinea.

It is curious that a rudimental love of art should co-exist with such a very low state of civilization. The people of Dorey are great carvers and painters. The outsides of the houses, wherever there is a plank, are covered with rude yet characteristic figures. The high-beaked prows of their boats are ornamented with masses of open filagree work, cut out of solid blocks of wood, and often of very tasteful design. As a figure-head, or pinnacle, there is often a human figure, with a head of cassowary feathers to imitate the Papuan "mop." The floats of their fishing-lines, the wooden beaters used in tempering the clay for their pottery, their tobacco-boxes, and other household articles, are covered with carving of tasteful and often elegant design. Did we not already know that such taste and skill are compatible with utter barbarism, we could hardly believe that the same people are, in other matters, entirely wanting in all sense of order, comfort, or decency. Yet such is the case. They live in the most miserable, crazy, and filthy hovels, which are utterly destitute of anything that can be called furniture ; not a stool, or bench, or board is seen in them; no brush seems to be known, and the clothes they wear are often filthy bark, or rags, or sacking. Along the paths where they daily pass to and from their provision grounds, not an over-hanging bough or straggling briar ever seems to be cut, so that you have to brush through a rank vegetation, creep under fallen trees and spiny creepers, and wade through pools of mud and mire, which cannot dry up because the sun is not allowed to penetrate. Their food is almost wholly roots and vegetables, with fish or game only as an occasional luxury, and they are consequently very subject to various skin diseases, the children especially being often miserable-looking objects, blotched all over with eruptions and sores. If these people are not savages, where shall we find any? Yet they have all a decided love for the fine arts, and spend their leisure time in executing works whose good taste and elegance would often be admired in our schools of design!

During the latter part of my stay in New Guinea the weather was very wet, my only shooter was ill, and birds became scarce, so that my only resource was insect-hunting. I worked very hard every hour of fine weather, and daily obtained a number of new species. Every dead tree and fallen log was searched and searched again; and among the dry and rotting leaves, which still hung on certain trees which had been cut down, I found an abundant harvest of minute Coleoptera. Although I never afterwards found so many large and handsome beetles as in Borneo, yet I obtained here a great variety of species. For the first two or three weeks, while I was searching out the best localities, I took about 30 different kinds of beetles a day, besides about half that number of butterflies, and a few of the other orders. But afterwards, up to the very last week, I averaged 49 species a day. On the 31st of May, I took 78 distinct sorts, a larger number than I had ever captured before, principally obtained among dead trees and under rotten bark. A good long walk on a fine day up the hill, and to the plantations of the natives, capturing everything not very common that came in my way, would produce about 60 species; but on the last day of June I brought home no less than 95 distinct kinds of beetles, a larger number than I ever obtained in one day before or since. It was a fine hot day, and I devoted it to a search among dead leaves, beating foliage,

CARVED TOOL FOR MAKING POTTERY.

and hunting under rotten bark, in all the best stations I had discovered during my walks. I was out from ten in the morning till three in the afternoon, and it took me six hours' work at home to pin and set out all the specimens, and to separate the species. Although I had already been working this spot daily for two months and a half, and had obtained over 800 species of Coleoptera, this day's work added 32 new ones. Among these were 4 Longicorns, 2 Carabidæ, 7 Staphylinidæ, 7 Curculionidæ, 2 Copridæ, 4 Chrysomelidæ, 3 Heteromera, 1 Elater, and 1 Buprestis. Even on the last day I went out, I obtained 16 new species; so that although I collected over a thousand distinct sorts of beetles in a space not much exceeding a square mile during the three months of my residence at Dorey, I cannot believe that this represents one half the species really inhabiting the same spot, or a fourth of what might be obtained in an area extending twenty miles in each direction.

On the 22nd of July the schooner *Hester Helena* arrived, and five days afterwards we bade adieu to Dorey, without much regret, for in no place which I have visited have I encountered more privations and annoyances. Continual rain, continual sickness, little wholesome food, with a plague of ants and flies, surpassing anything I had before met with, required all a naturalist's ardour to encounter; and when they were uncompensated by great success in collecting, became all the more insupportable. This long-thought-of and much-desired voyage to New Guinea had realized none of my expectations. Instead of being far better than the Aru Islands, it was in almost everything much worse. Instead of producing several of the rarer Paradise Birds, I had not even seen one of them, and had not obtained any one superlatively fine bird or insect. I cannot deny, however, that Dorey was very rich in ants. One small black kind was excessively abundant. Almost every shrub and tree was more or less infested with it, and its large papery nests were everywhere to be seen. They immediately took possession of my house, building a large nest in the roof, and forming papery tunnels down almost every post. They swarmed on my table as I was at work setting out my insects, carrying them off from under my very nose, and even tearing them from the cards on which they were gummed if I left them for an instant. They crawled continually over my hands and face, got into my hair, and roamed at will over my whole body, not producing much inconvenience till they began to bite, which they would do on meeting with any obstruction to their passage, and with a sharpness which made me jump again and rush to undress and turn out the offender. They visited my bed also, so that night brought no relief from their persecutions; and I verily believe that during my three and a half months' residence at Dorey I was never for a single hour entirely free from them. They were not nearly so voracious as many other kinds, but their numbers and ubiquity rendered it necessary to be constantly on guard against them.

The flies that troubled me most were a large kind of blue-bottle or blow-fly. These settled in swarms on my bird skins when first put out to dry, filling their plumage with masses of eggs, which, if neglected, the next day produced maggots. They would get under the wings or under the body where it rested on the drying-board, sometimes actually raising it up half an inch by the mass of eggs deposited in a few hours ; and every egg was so firmly glued to the fibres of the feathers, as to make it a work of much time and patience to get them off without in-juring the bird. In no other locality have I ever been troubled with such a plague as this.

On the 29th we left Dorey, and expected a quick voyage home, as it was the time of year when we ought to have had steady southerly and easterly winds. Instead of these, however, we had calms and westerly breezes, and it was seventeen days before we reached Ternate, a distance of five hundred miles only, which, with average winds, could have been done in five days. It was a great treat to me to find myself back again in my comfortable house, enjoying milk to my tea and coffee, fresh bread and butter, and fowl and fish daily for dinner. This New Guinea voyage had used us all up, and I determined to stay and recruit before I commenced any fresh expeditions. My succeeding journeys to Gilolo and Batchian have already been narrated, and it now only remains for me to give an account of my resi-dence in Waigiou, the last Papuan territory I visited in search of Birds of Paradise.

CHAPTER XXXV.

VOYAGE FROM CERAM TO WAIGIOU.

(JUNE AND JULY 1860.)

IN my twenty-fifth chapter I have described my arrival at Wahai, on my way to Mysol and Waigiou, islands which belong to the Papuan district, and the account of which naturally follows after that of my visit to the mainland of New Guinea. I now take up my narrative at my departure from Wahai, with the intention of carrying various necessary stores to my assistant, Mr. Allen, at Silinta, in Mysol, and then continuing my journey to Waigiou. It will be remembered that I was travelling in a small prau, which I had purchased and fitted up in Goram, and that, having been deserted by my crew on the coast of Ceram, I had obtained four men at Wahai, who, with my Amboynese hunter, constituted my crew.

Between Ceram and Mysol there are sixty miles of open sea,

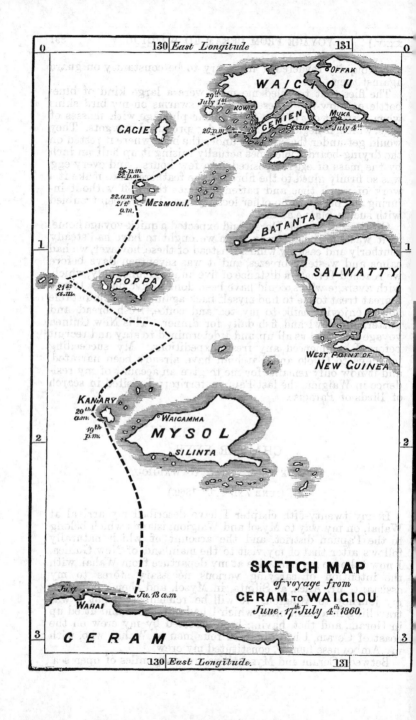

OFFAK

WAICIOU

July 1st

KOWIE

CEMIEN

MUKA

BESSIR

July 4th

CACIE

26. p.m.

25. p.m.
22. p.m.

22. a.m.
21st
p.m.

MESMON. I.

BATANTA

SALWATTY

1

21st
a.m.

POPPA

WEST POINT OF
NEW GUINEA

KANARY
20th
a.m.

19th
p.m.

WAICAMMA

MYSOL

SILINTA

2

SKETCH MAP
of voyage from
CERAM TO WAICIOU
June. 17th *July* 4th 1860.

Ju. 19

WAHAI

Ju. 18 a.m

3

CERAM

and along this wide channel the east monsoon blows strongly ; so that with native praus, which will not lay up to the wind, it requires some care in crossing. In order to give ourselves sufficient leeway, we sailed back from Wahai eastward, along the coast of Ceram, with the land-breeze ; but in the morning (June 18th) had not gone nearly so far as I expected. My pilot, an old and experienced sailor, named Gurulampoko, assured me there was a current setting to the eastward, and that we could easily lay across to Silinta, in Mysol. As we got out from the land the wind increased, and there was a considerable sea, which made my short little vessel plunge and roll about violently. By sunset we had not got halfway across, but could see Mysol distinctly. All night we went along uneasily, and at daybreak, on looking out anxiously, I found that we had fallen much to the westward during the night, owing, no doubt, to the pilot being sleepy and not keeping the boat sufficiently close to the wind. We could see the mountains distinctly, but it was clear we should not reach Silinta, and should have some difficulty in getting to the extreme westward point of the island. The sea was now very boisterous, and our prau was continually beaten to leeward by the waves, and after another weary day we found we could not get to Mysol at all, but might perhaps reach the island called Pulo Kanary, about ten miles to the north-west. Thence we might await a favourable wind to reach Waigamma, on the north side of the island, and visit Allen by means of a small boat.

About nine o'clock at night, greatly to my satisfaction, we got under the lee of this island, into quite smooth water—for I had been very sick and uncomfortable, and had eaten scarcely anything since the preceding morning. We were slowly nearing the shore, which the smooth dark water told us we could safely approach, and were congratulating ourselves on soon being at anchor, with the prospect of hot coffee, a good supper, and a sound sleep, when the wind completely dropped, and we had to get out the oars to row. We were not more than two hundred yards from the shore, when I noticed that we seemed to get no nearer although the men were rowing hard, but drifted to the westward ; and the prau would not obey the helm, but continually fell off, and gave us much trouble to bring her up again. Soon a loud ripple of water told us we were seized by one of those treacherous currents which so frequently frustrate all the efforts of the voyager in these seas ; the men threw down the oars in despair, and in a few minutes we drifted to leeward of the island fairly out to sea again, and lost our last chance of ever reaching Mysol ! Hoisting our jib, we lay to, and in the morning found ourselves only a few miles from the island, but with such a steady wind blowing from its direction as to render it impossible for us to get back to it.

We now made sail to the northward, hoping soon to get a more southerly wind. Towards noon the sea was much smoother, and with a S.S.E. wind we were laying in the direction of Salwatty, which I hoped to reach, as I could there easily get a boat to take provisions and stores to my companion in Mysol. This wind did not, however, last long, but died away into a calm ; and a light west wind springing up, with a dark bank of clouds, again gave us hopes of reaching Mysol. We were soon, however, again disappointed. The E.S.E. wind began to blow again with violence, and continued all night in irregular gusts, and with a short cross sea tossed us about unmercifully, and so continually took our sails aback, that we were at length forced to run before it with our jib only, to escape being swamped by our heavy mainsail. After another miserable and anxious night, we found that we had drifted westward of the island of Poppa, and the wind being again a little southerly, we made all sail in order to reach it. This we did not succeed in doing, passing to the north-west, when the wind again blew hard from the E.S.E., and our last hope of finding a refuge till better weather was frustrated. This was a very serious matter to me, as I could not tell how Charles Allen might act, if, after waiting in vain for me, he should return to Wahai, and find that I had left there long before, and had not since been heard of. Such an event as our missing an island forty miles long would hardly occur to him, and he would conclude either that our boat had foundered, or that my crew had murdered me and run away with her. However, as it was physically impossible now for me to reach him, the only thing to be done was to make the best of my way to Waigiou, and trust to our meeting some traders, who might convey to him the news of my safety.

Finding on my map a group of three small islands, twenty-five miles north of Poppa, I resolved, if possible, to rest there a day or two. We could lay our boat's head N.E. by N.; but a heavy sea from the eastward so continually beat us off our course, and we made so much leeway, that I found it would be as much as we could do to reach them. It was a delicate point to keep our head in the best direction, neither so close to the wind as to stop our way, or so free as to carry us too far to leeward. I continually directed the steersman myself, and by incessant vigilance succeeded, just at sunset, in bringing our boat to an anchor under the lee of the southern point of one of the islands. The anchorage was, however, by no means good, there being a fringing coral reef, dry at low water, beyond which, on a bottom strewn with masses of coral, we were obliged to anchor. We had now been incessantly tossing about for four days in our small undecked boat, with constant disappointments and anxiety, and it was a great comfort to have a night of quiet and comparative safety. My old pilot had never left the helm for more than an hour at a time, when one of the others would relieve him for a little sleep ; so I determined the next morning to look out for a secure and convenient harbour, and rest on shore for a day.

In the morning, finding it would be necessary for us to get round a rocky point, I wanted my men to go on shore and cut jungle-rope, by which to secure us from being again drifted away, as the wind was directly off shore. I unfortunately, however, allowed myself to be overruled by the pilot and crew, who all declared that it was the easiest thing possible, and that they would row the boat round the point in a few minutes. They accordingly got up the anchor, set the jib, and began rowing ; but, just as I had feared, we drifted rapidly off shore, and had to drop anchor again in deeper water, and much farther off. The two best men, a Papuan and a Malay, now swam on shore, each carrying a hatchet, and went into the jungle to seek creepers for rope. After about an hour our anchor loosed hold, and began to drag. This alarmed me greatly, and we let go our spare anchor, and, by running out all our cable, appeared tolerably secure again. We were now most anxious for the return of the men, and were going to fire our muskets to recall them, when we observed them on the beach, some way off, and almost immediately our anchors again slipped, and we drifted slowly away into deep water. We instantly seized the oars, but found we could not counteract the wind and current, and our frantic cries to the men were not heard till we had got a long way off, as they seemed to be hunting for shell-fish on the beach. Very soon, however, they stared at us, and in a few minutes seemed to comprehend their situation ; for they rushed down into the water, as if to swim off, but again returned on shore, as if afraid to make the attempt. We had drawn up our anchors at first not to check our rowing . but now, finding we could do nothing, we let them both hang down by the full length of the cables. This stopped our way very much, and we drifted from shore very slowly, and hoped the men would hastily form a raft, or cut down a soft-wood tree, and paddle out to us, as we were still not more than a third of a mile from shore. They seemed, however, to have half lost their senses, gesticulating wildly to us, running along the beach, then going into the forest ; and just when we thought they had prepared some mode of making an attempt to reach us, we saw the smoke of a fire they had made to cook their shell-fish ! They had evidently given up all idea of coming after us, and we were obliged to look to our own position.

We were now about a mile from shore, and midway between two of the islands, but we were slowly drifting out to sea to the westward, and our only chance of yet saving the men was to reach the opposite shore. We therefore set our jib and rowed hard ; but the wind failed, and we drifted out so rapidly that we had some difficulty in reaching the extreme westerly point of the island. Our only sailor left, then swam ashore with a rope, and helped to tow us round the point into a tolerably safe and secure anchorage, well sheltered from the wind, but exposed to a little swell which jerked our anchor and made us rather uneasy. We were now in a sad plight, having lost our two best men, and being doubtful if we had strength left to hoist our mainsail. We had only two days' water on board, and the small, rocky, volcanic island did not promise us much chance of finding any. The conduct of the men on shore was such as to render it doubtful if they would make any serious attempt to reach us, though they might easily do so, having two good choppers, with which in a day they could make a small out-rigger raft on which they could safely cross the two miles of smooth sea with the wind right aft, if they started from the east end of the island, so as to allow for the current. I could only hope they would be sensible enough to make the attempt, and determined to stay as long as I could to give them the chance.

We passed an anxious night, fearful of again breaking our anchor or rattan cable. In the morning (23rd), finding all secure, I waded on shore with my two men, leaving the old steersman and the cook on board, with a loaded musket to recall us if needed. We first walked along the beach, till stopped by the vertical cliffs at the east end of the island, find-ing a place where meat had been smoked, a turtle-shell still greasy, and some cut wood, the leaves of which were still green —showing that some boat had been here very recently. We then entered the jungle, cutting our way up to the top of the hill, but when we got there could see nothing, owing to the thickness of the forest. Returning, we cut some bamboos, and sharpened them to dig for water in a low spot where some sago-trees were growing ; when, just as we were going to begin, Hoi, the Wahai man, called out to say he had found water. It was a deep hole among the sago-trees, in stiff black clay, full of water, which was fresh, but smelt horribly from the quantity of dead leaves and sago refuse that had fallen in. Hastily con-cluding that it was a spring, or that the water had filtered in, we baled it all out as well as a dozen or twenty buckets of mud and rubbish, hoping by night to have a good supply of clean water. I then went on board to breakfast, leaving my two men to make a bamboo raft to carry us on shore and back without wading. I had scarcely finished when our cable broke, and we bumped against the rocks. Luckily it was smooth and calm, and no damage was done. We searched for and got up our anchor, and found that the cable had been cut by grating all night upon the coral. Had it given way in the night, we might have drifted out to sea without our anchor or been seriously damaged. In the evening we went to fetch water from the well, when, greatly to our dismay, we found nothing but a little liquid mud at the bottom, and it then became evident that the hole was one which had been made to collect rain water, and would never fill again as long as the present drought continued. As we did not know what we might suffer for want of water, we filled our jar with this muddy stuff so that it might settle. In the afternoon I crossed over to the other side of the island, and made a large fire, in order that our men might see we were still there.

The next day (24th) I determined to have another search for water; and when the tide was out rounded a rocky point and went to the extremity of the island without finding any sign of the smallest stream. On our way back, noticing a very small dry bed of a watercourse, I went up it to explore, although everything was so dry that my men loudly declared it was useless to expect water there; but a little way up I was rewarded by finding a few pints in a small pool. We searched higher up in every hole and channel where water marks appeared, but could find not a drop more. Sending one of my men for a large jar and teacup, we searched along the beach till we found signs of another dry watercourse, and on ascending this were so fortunate as to discover two deep sheltered rock-holes containing several gallons of water, enough to fill all our jars. When the cup came we enjoyed a good drink of the cool pure water, and before we left had carried away, I believe, every drop on the island.

In the evening a good-sized prau appeared in sight, making apparently for the island where our men were left, and we had some hopes they might be seen and picked up, but it passed along mid-channel, and did not notice the signals we tried to make. I was now, however, pretty easy as to the fate of the men. There was plenty of sago on our rocky island, and there would probably be some on the flat one they were left on. They had choppers, and could cut down a tree and make sago, and would most likely find sufficient water by digging. Shell-fish were abundant, and they would be able to manage very well till some boat should touch there, or till I could send and fetch them. The next day we devoted to cutting wood, filling up our jars with all the water we could find, and making ready to sail in the evening. I shot a small lory closely resembling a common species at Ternate, and a glossy starling which differed from the allied birds of Ceram and Matabello. Large wood-pigeons and crows were the only other birds I saw, but I did not obtain specimens.

About eight in the evening of June 25th we started, and found that with all hands at work we could just haul up our mainsail. We had a fair wind during the night and sailed north-east, finding ourselves in the morning about twenty miles west of the extremity of Waigiou with a number of islands intervening. About ten o'clock we ran full on to a coral reef, which alarmed us a good deal, but luckily got safe off again. About two in the afternoon we reached an extensive coral reef, and were sailing close alongside of it, when the wind suddenly dropped, and we drifted on to it before we could get in our heavy mainsail, which we were obliged to let run down and fall partly overboard. We had much difficulty in getting off, but at last got into deep water again, though with reefs and islands all around us. At night we did not know what to do, as no one on board could tell where we were or what dangers might surround us, the only one of our crew who was acquainted with the coast of Waigiou having been left on the island. We therefore took in all sail and allowed ourselves to drift, as we were some miles from the nearest land. A light breeze, however, sprang up, and about midnight we found ourselves again bumping over a coral reef. As it was very dark, and we knew nothing of our position, we could only guess how to get off again, and had there been a little more wind we might have been knocked to pieces. However, in about half an hour we did get off, and then thought it best to anchor on the edge of the reef till morning. Soon after daylight on the 27th, finding our prau had received no damage, we sailed on with uncertain winds and squalls, threading our way among islands and reefs, and guided only by a small map, which was very incorrect and almost useless, and by a general notion of the direction we ought to take. In the afternoon we found a tolerable anchorage under a small island and stayed for the night, and I shot a large fruit-pigeon new to me, which I have since named Carpophaga tumida. I also saw and shot at the rare white-headed kingfisher (Halcyon saurophaga), but did not kill it. The next morning we sailed on, and having a fair wind reached the shores of the large island of Waigiou. On rounding a point we again ran full on to a coral reef with our mainsail up, but luckily the wind had almost died away, and with a good deal of exertion we managed to get safely off.

We now had to search for the narrow channel among the islands, which we knew was somewhere hereabouts, and which leads to the villages on the south side of Waigiou. Entering a deep bay which looked promising, we got to the end of it, but it was then dusk, so we anchored for the night, and having just finished all our water could cook no rice for supper. Next morning early (29th) we went on shore among the mangroves, and a little way inland found some water, which relieved our anxiety considerably, and left us free to go along the coast in search of the opening, or of some one who could direct us to it. During the three days we had now been among the reefs and islands, we had only seen a single small canoe, which had approached pretty near to us, and then, notwithstanding, our signals went off in another direction. The shores seemed all desert; not a house, or boat, or human being, or a puff of smoke was to be seen; and as we could only go on the course that the ever-changing wind would allow us (our hands being too few to row any distance), our prospects of getting to our destination seemed rather remote and precarious. Having gone to the eastward extremity of the deep bay we had entered, without finding any sign of an opening, we turned westward; and towards evening were so fortunate as to find a small village of seven miserable houses built on piles in the water. Luckily the

Orang-kaya, or head man, could speak a little Malay, and informed us that the entrance to the strait was really in the bay we had examined, but that it was not to be seen except when close in-shore. He said the strait was often very narrow, and wound among lakes and rocks and islands, and that it would take two days to reach the large village of Muka, and three more to get to Waigiou. I succeeded in hiring two men to go with us to Muka, bringing a small boat in which to return; but we had to wait a day for our guides, so I took my gun and made a little excursion into the forest. The day was wet and drizzly, and I only succeeded in shooting two small birds, but I saw the great black cockatoo, and had a glimpse of one or two Birds of Paradise, whose loud screams we had heard on first approaching the coast.

Leaving the village the next morning (July 1st) with a light wind it took us all day to reach the entrance to the channel, which resembled a small river, and was concealed by a projecting point, so that it was no wonder we did not discover it amid the dense forest vegetation which everywhere covers these islands to the water's edge. A little way inside it becomes bounded by precipitous rocks, after winding among which for about two miles, we emerged into what seemed a lake, but which was in fact a deep gulf having a narrow entrance on the south coast. This gulf was studded along its shores with numbers of rocky islets, mostly mushroom shaped, from the water having worn away the lower part of the soluble coralline limestone, leaving them overhanging from ten to twenty feet. Every islet was covered with strange-looking shrubs and trees, and was generally crowned by lofty and elegant palms, which also studded the ridges of the mountainous shores, forming one of the most singular and picturesque landscapes I have ever seen. The current which had brought us through the narrow strait now ceased, and we were obliged to row, which with our short and heavy prau was slow work. I went on shore several times, but the rocks were so precipitous, sharp, and honeycombed, that I found it impossible to get through the tangled thickets with which they were everywhere clothed. It took us three days to get to the entrance of the gulf, and then the wind was such as to prevent our going any further, and we might have had to wait for days or weeks, when, much to my surprise and gratification, a boat arrived from Muka with one of the head men, who had in some mysterious manner heard I was on my way, and had come to my assistance, bringing a present of cocoa-nuts and vegetables. Being thoroughly acquainted with the coast, and having several extra men to assist us, he managed to get the prau along by rowing, poling, or sailing, and by night had brought us safely into harbour, a great relief after our tedious and unhappy voyage. We had been already eight days among the reefs, and islands of Waigiou, coming a distance of about fifty miles, and it was just forty days since we had sailed from Goram.

Immediately on our arrival at Muka, I engaged a small boat and three natives to go in search of my lost men, and sent one of my own men with them to make sure of their going to the right island. In ten days they returned, but to my great regret and disappointment, without the men. The weather had been very bad, and though they had reached an island within sight of that in which the men were, they could get no further. They had waited there six days for better weather, and then, having no more provisions, and the man I had sent with them being very ill and not expected to live, they returned. As they now knew the island I was determined they should make another trial and (by a liberal payment of knives, handkerchiefs, and tobacco, with plenty of provisions) persuaded them to start back immediately, and make another attempt. They did not return again till the 29th of July, having stayed a few days at their own village of Bessir on the way; but this time they had succeeded and brought with them my two lost men, in tolerable health, though thin and weak. They had lived exactly a month on the island; had found water, and had subsisted on the roots and tender flower-stalks of a species of Bromelia, on shell-fish, and on a few turtles' eggs. Having swum to the island, they had only a pair of trousers and a shirt between them, but had made a hut of palm-leaves, and had altogether got on very well. They saw that I waited for them three days at the opposite island, but had been afraid to cross, lest the current should have carried them out to sea, when they would have been inevitably lost. They had felt sure I would send for them on the first opportunity, and appeared more grateful than natives usually are for my having done so; while I felt much relieved that my voyage, though sufficiently unfortunate, had not involved loss of life.

CHAPTER XXXVI.

WAIGIOU.

(JULY TO SEPTEMBER 1860.)

THE village of Muka, on the south coast of Waigiou, consists of a number of poor huts, partly in the water and partly on shore, and scattered irregularly over a space of about half a mile in a shallow bay. Around it are a few cultivated patches, and a good deal of second-growth woody vegetation; while behind, at the distance of about half a mile, rises the virgin forest, through which are a few paths to some houses and plantations a mile or two inland. The country round is rather flat, and in places swampy, and there are one or two small streams which run behind the village into the sea below it. Finding that no house could be had suitable to my purpose, and having so often experienced the advantages of living close to or just within the forest, I obtained the assistance of half-a-dozen men ; and having selected a spot near the path and the stream, and close to a fine fig-tree, which stood just within the forest, we cleared the ground and set to building a house. As I did not expect to stay here so long as I had done at Dorey, I built a long, low, narrow shed, about seven feet high on one side and four on the other, which required but little wood, and was put up very rapidly. Our sails, with a few old attaps from a deserted hut in the village, formed the walls, and a quantity of "cadjans," or palm-leaf mats, covered in the roof. On the third day my house was finished, and all my things put in and comfortably arranged to begin work, and I was quite pleased at having got established so quickly and in such a nice situation.

It had been so far fine weather, but in the night it rained hard, and we found our mat roof would not keep out water. It first began to drop, and then to stream over everything. I had to get up in the middle of the night to secure my insect-boxes, rice, and other perishable articles, and to find a dry place to sleep in, for my bed was soaked. Fresh leaks kept forming as the rain continued, and we all passed a very miserable and sleepless night. In the morning the sun shone brightly, and everything was put out to dry. We tried to find out why the mats leaked, and thought we had discovered that they had been laid on upside down. Having shifted them all, and got every-thing dry and comfortable by the evening, we again went to bed, and before midnight were again awaked by torrents of rain and leaks streaming in upon us as bad as ever. There was no more sleep for us that night, and the next day our roof was again taken to pieces, and we came to the conclusion that the fault was a want of slope enough in the roof for mats, although it would be sufficient for the usual attap thatch. I therefore purchased a few new and some old attaps, and in the parts these would not cover we put the mats double, and then at last had the satisfaction of finding our roof tolerably water-tight.

I was now able to begin working at the natural history of the island. When I first arrived I was surprised at being told that there were no Paradise Birds at Muka, although there were plenty at Bessir, a place where the natives caught them and prepared the skins. I assured the people I had heard the cry of these birds close to the village, but they would not believe that I could know their cry. However, the very first time I went into the forest I not only heard but saw them, and was convinced there were plenty about ; but they were very shy, and it was some time before we got any. My hunter first shot a female, and I one day got very close to a fine male. He was, as I expected, the rare red species, Paradisea rubra, which alone inhabits this island, and is found nowhere else. He was quite low down, running along a bough searching for insects, almost like a woodpecker, and the long black riband-like filaments in his tail hung down in the most graceful double curve imaginable. I covered him with my gun, and was going to use the barrel which had a very small charge of powder and number eight shot, so as not to injure his plumage, but the gun missed fire, and he was off in an instant among the thickest jungle. Another day we saw no less than eight fine males at different times, and fired four times at them; but though other birds at the same distance almost always dropped, these all got away, and I began to think we were not to get this magnificent species. At length the fruit ripened on the fig-tree close by my house, and many birds came to feed on it; and one morning, as I was taking my coffee, a male Paradise Bird was seen to settle on its top. I seized my gun, ran under the tree, and, gazing up, could see it flying across from branch to branch, seizing a fruit here and another there, and then, before I could get a sufficient aim to shoot at such a height (for it was one of the loftiest trees of the tropics), it was away into the forest. They now visited the tree every morning; but they stayed so short a time, their motions were so rapid, and it was so difficult to see them, owing to the lower trees, which impeded the view, that it was only after several days' watching, and one or two misses, that I brought down my bird—a male in the most magnificent plumage.

This bird differs very much from the two large species which I had already obtained, and, although it wants the grace imparted by their long golden trains, is in many respects more remarkable and more beautiful. The head, back, and shoulders are clothed with a richer yellow, the deep metallic green colour of the throat extends further over the head, and the feathers are elongated on the forehead into two little erectile crests. The side plumes are shorter, but are of a rich red colour, terminating in delicate white points, and the middle tail-feathers are represented by two long rigid glossy ribands, which are black, thin, and semi-cylindrical, and droop gracefully in a spiral curve. Several other interesting birds were obtained, and about half-a-dozen quite new ones; but none of any remarkable beauty, except the lovely little dove, Ptilonopus pulchellus, which with several other pigeons I shot on the same fig-tree close to my house. It is of a beautiful green colour above, with a forehead of the richest crimson, while beneath it is ashy white and rich yellow, banded with violet red.

On the evening of our arrival at Muka I observed what appeared like a display of Aurora Borealis, though I could hardly believe that this was possible at a point a little south of the equator. The night was clear and calm, and the northern sky presented a diffused light, with a constant succession of faint vertical flashings or flickerings, exactly similar to an ordinary aurora in England. The next day was fine, but after

THE RED BIRD OF PARADISE. (*Paradisea rubra.*)

that the weather was unprecedentedly bad, considering that it ought to have been the dry monsoon. For near a month we had wet weather; the sun either not appearing at all, or only for an hour or two about noon. Morning and evening, as well as nearly all night, it rained or drizzled, and boisterous winds, with dark clouds, formed the daily programme. With the exception that it was never cold, it was just such weather as a very bad English November or February.

The people of Waigiou are not truly indigenes of the island, which possesses no "Alfuros," or aboriginal inhabitants.[1] They appear to be a mixed race, partly from Gilolo, partly from New Guinea. Malays and Alfuros from the former island have probably settled here, and many of them have taken Papuan wives from Salwatty or Dorey, while the influx of people from those places, and of slaves, has led to the formation of a tribe exhibiting almost all the transitions from a nearly pure Malayan to an entirely Papuan type. The language spoken by them is entirely Papuan, being that which is used on all the coasts of Mysol, Salwatty, the north-west of New Guinea, and the islands in the great Geelvink Bay—a fact which indicates the way in which the coast settlements have been formed. The fact that so many of the islands between New Guinea and the Moluccas —such as Waigiou, Guebé, Poppa, Obi, Batchian, as well as the south and east peninsulas of Gilolo—possess no aboriginal tribes, but are inhabited by people who are evidently mongrels and wanderers, is a remarkable corroborative proof of the distinctness of the Malayan and Papuan races, and the separation of the geographical areas they inhabit. If these two great races were direct modifications, the one of the other, we should expect to find in the intervening region some homogeneous indigenous race presenting intermediate characters. For example, between the whitest inhabitants of Europe and the black Klings of South India, there are in the intervening districts homogeneous races which form a gradual transition from one to the other ; while in America, although there is a perfect transition from the Anglo-Saxon to the negro, and from the Spaniard to the Indian, there is no homogeneous race forming a natural transition from one to the other. In the Malay Archipelago we have an excellent example of two absolutely distinct races, which appear to have approached each other, and intermingled in an unoccupied territory at a very recent epoch in the history of man ; and I feel satisfied that no unprejudiced person could study them on the spot without being convinced that this is the true solution of the problem, rather than the almost universally accepted view that they are but modifications of one and the same race.

The people of Muka live in that abject state of poverty that is almost always found where the sago-tree is abundant. Very few of them take the trouble to plant any vegetables or fruit, but live almost entirely on sago and fish, selling a little tripang or tortoiseshell to buy the scanty clothing they require. Almost all of them, however, possess one or more Papuan slaves, on whose labour they live in almost absolute idleness, just going out on little fishing or trading excursions, as an excitement in their monotonous existence. They are under the rule of the

[1] Dr. Guillemard met with some people who, he was told, were true indigenes. But it would be difficult to determine the point without a tolerably complete knowledge of all the surrounding islands and their languages.

Sultan of Tidore, and every year have to pay a small tribute of Paradise birds, tortoiseshell, or sago. To obtain these, they go in the fine season on a trading voyage to the mainland of New Guinea, and getting a few goods on credit from some Ceram or Bugis trader, make hard bargains with the natives, and gain enough to pay their tribute, and leave a little profit for themselves.

Such a country is not a very pleasant one to live in, for as there are no superfluities, there is nothing to sell ; and had it not been for a trader from Ceram who was residing there during my stay, who had a small vegetable garden, and whose men occasionally got a few spare fish, I should often have had nothing to eat. Fowls, fruit, and vegetables are luxuries very rarely to be purchased at Muka ; and even cocoa-nuts, so indispensable for Eastern cookery, are not to be obtained ; for though there are some hundreds of trees in the village, all the fruit is eaten green, to supply the place of the vegetables the people are too lazy to cultivate. Without eggs, cocoa-nuts, or plaintains, we had very short commons, and the boisterous weather being unpropitious for fishing, we had to live on what few eatable birds we could shoot, with an occasional cuscus, or eastern opossum, the only quadruped, except pigs, inhabiting the island.

I had only shot two male Paradiseas on my tree when they ceased visiting it, either owing to the fruit becoming scarce, or that they were wise enough to know there was danger. We continued to hear and see them in the forest, but after a month had not succeeded in shooting any more ; and as my chief object in visiting Waigiou was to get these birds, I determined to go to Bessir, where there are a number of Papuans who catch and preserve them. I hired a small outrigger boat for this journey, and left one of my men to guard my house and goods. We had to wait several days for fine weather, and at length started early one morning, and arrived late at night, after a rough and disagreeable passage. The village of Bessir was built in the water at the point of a small island. The chief food of the people was evidently shell-fish, since great heaps of the shells had accumulated in the shallow water between the houses and the land, forming a regular "kitchen-midden" for the exploration of some future archæologist. We spent the night in the chief's house, and the next morning went over to the mainland to look out for a place where I could reside. This part of Waigiou is really another island to the south of the narrow channel we had passed through in coming to Muka. It appears to consist almost entirely of raised coral, whereas the northern island contains hard crystalline rocks. The shores were a range of low limestone cliffs, worn out by the water, so that the upper part generally overhung. At distant intervals were little coves and openings, where small streams came down from the interior ; and in one of these we landed, pulling our

boat up on a patch of white sandy beach. Immediately above
was a large newly-made plantation of yams and plantains, and
a small hut, which the chief said we might have the use of if it
would do for me. It was quite a dwarf's house, just eight feet
square, raised on posts so that the floor was four and a half
feet above the ground, and the highest part of the ridge only
five feet above the floor. As I am six feet and an inch in my
stockings, I looked at this with some dismay ; but finding that
the other houses were much further from water, were dreadfully
dirty, and were crowded with people, I at once accepted the
little one, and determined to make the best of it. At first I
thought of taking out the floor, which would leave it high
enough to walk in and out without stooping ; but then there
would not be room enough, so I left it just as it was, had it

MY HOUSE AT BESSIR, IN WAIGIOU.

thoroughly cleaned out, and brought up my baggage. The
upper story I used for sleeping in, and for a store-room. In the
lower part (which was quite open all round) I fixed up a small
table, arranged my boxes, put up hanging-shelves, laid a mat
on the ground with my wicker-chair upon it, hung up another
mat on the windward side, and then found that, by bending
double and carefully creeping in, I could sit on my chair with
my head just clear of the ceiling. Here I lived pretty comfortably for six weeks, taking all my meals and doing all my
work at my little table, to and from which I had to creep in a
semi-horizontal position a dozen times a day ; and, after a few

severe knocks on the head by suddenly rising from my chair, learnt to accommodate myself to circumstances. We put up a little sloping cooking-hut outside, and a bench on which my lads could skin their birds. At night I went up to my little loft, they spread their mats on the floor below, and we none of us grumbled at our lodgings.

My first business was to send for the men who were accustomed to catch the Birds of Paradise. Several came, and I showed them my hatchets, beads, knives, and handkerchiefs ; and explained to them as well as I could by signs, the price I would give for fresh-killed specimens. It is the universal custom to pay for everything in advance ; but only one man ventured on this occasion to take goods to the value of two birds. The rest were suspicious, and wanted to see the result of the first bargain with the strange white man, the only one who had ever come to their island. After three days, my man brought me the first bird—a very fine specimen, and alive, but tied up in a small bag, and consequently its tail and wing feathers very much crushed and injured. I tried to explain to him, and to the others that came with him, that I wanted them as perfect as possible, and that they should either kill them, or keep them on a perch with a string to their leg. As they were now apparently satisfied that all was fair, and that I had no ulterior designs upon them, six others took away goods ; some for one bird, some for more, and one for as many as six. They said they had to go a long way for them, and that they would come back as soon as they caught any. At intervals of a few days or a week, some of them would return, bringing me one or more birds ; but though they did not bring any more in bags, there was not much improvement in their condition. As they caught them a long way off in the forest, they would scarcely ever come with one, but would tie it by the leg to a stick, and put it in their house till they caught another. The poor creature would make violent efforts to escape, would get among the ashes, or hang suspended by the leg till the limb was swollen and half-putrefied, and sometimes die of starvation and worry. One had its beautiful head all defiled by pitch from a dammar torch ; another had been so long dead that its stomach was turning green. Luckily, however, the skin and plumage of these birds is so firm and strong, that they bear washing and cleaning better than almost any other sort ; and I was generally able to clean them so well that they did not perceptibly differ from those I had shot myself.

Some few were brought me the same day they were caught, and I had an opportunity of examining them in all their beauty and vivacity. As soon as I found they were generally brought alive, I set one of my men to make a large bamboo cage with troughs for food and water, hoping to be able to keep some of them. I got the natives to bring me branches of a fruit they were very fond of, and I was pleased to find they ate it greedily, and would also take any number of live grasshoppers I gave them, stripping off the legs and wings, and then swallowing them. They drank plenty of water, and were in constant motion, jumping about the cage from perch to perch, clinging on the top and sides, and rarely resting a moment the first day till nightfall. The second day they were always less active, although they would eat as freely as before ; and on the morning of the third day they were almost always found dead at the bottom of the cage, without any apparent cause. Some of them ate boiled rice as well as fruit and insects ; but after trying many in succession, not one out of ten lived more than three days. The second or third day they would be dull, and in several cases they were seized with convulsions, and fell off the perch, dying a few hours afterwards. I tried immature as well as full-plumaged birds, but with no better success, and at length gave it up as a hopeless task, and confined my attention to preserving specimens in as good a condition as possible.

The Red Birds of Paradise are not shot with blunt arrows, as in the Aru Islands and some parts of New Guinea, but are snared in a very ingenious manner. A large climbing Arum bears a red reticulated fruit, of which the birds are very fond. The hunters fasten this fruit on a stout forked stick, and provide themselves with a fine but strong cord. They then seek out some tree in the forest on which these birds are accustomed to perch, and climbing up it fasten the stick to a branch and arrange the cord in a noose so ingeniously, that when the bird comes to eat the fruit its legs are caught, and by pulling the end of the cord, which hangs down to the ground, it comes free from the branch and brings down the bird. Sometimes, when food is abundant elsewhere, the hunter sits from morning till night under his tree with the cord in his hand, and even for two or three whole days in succession, without even getting a bite ; while, on the other hand, if very lucky, he may get two or three birds in a day. There are only eight or ten men at Bessir who practise this art, which is unknown anywhere else in the island. I determined, therefore, to stay as long as possible, as my only chance of getting a good series of specimens ; and although I was nearly starved, everything eatable by civilized man being scarce or altogether absent, I finally succeeded.

The vegetables and fruit in the plantations around us did not suffice for the wants of the inhabitants, and were almost always dug up or gathered before they were ripe. It was very rarely we could purchase a little fish ; fowls there were none ; and we were reduced to live upon tough pigeons and cockatoos, with our rice and sago, and sometimes we could not get these. Having been already eight months on this voyage, my stock of all condiments, spices and butter, was exhausted, and I found it impossible to eat sufficient of my tasteless and unpalatable food to support health. I got very thin and weak, and had a curious disease known (I have since heard) as brow-ague. Directly after breakfast every morning an intense pain set in on a small spot on the right temple. It was a severe burning ache, as bad as the worst toothache, and lasted about two hours, generally going off at noon. When this finally ceased, I had an attack of fever, which left me so weak and so unable to eat our regular food, that I feel sure my life was saved by a couple of tins of soup which I had long reserved for some such extremity. I used often to go out searching after vegetables, and found a great treasure in a lot of tomato plants run wild, and bearing little fruits about the size of gooseberries. I also boiled up the tops of pumpkin plants and of ferns, by way of greens, and occasionally got a few green papaws. The natives, when hard up for food, live upon a fleshy seaweed, which they boil till it is tender. I tried this also, but found it too salt and bitter to be endured.

Towards the end of September it became absolutely necessary for me to return, in order to make our homeward voyage before the end of the east monsoon. Most of the men who had taken payment from me had brought the birds they had agreed for. One poor fellow had been so unfortunate as not to get one, and he very honestly brought back the axe he had received in advance; another, who had agreed for six, brought me the fifth two days before I was to start, and went off immediately to the forest again to get the other. He did not return, however, and we loaded our boat, and were just on the point of starting, when he came running down after us holding up a bird, which he handed to me, saying with great satisfaction, "Now I owe you nothing." These were remarkable and quite unexpected instances of honesty among savages, where it would have been very easy for them to have been dishonest without fear of detection or punishment.

The country round about Bessir was very hilly and rugged, bristling with jagged and honey-combed coralline rocks, and with curious little chasms and ravines. The paths often passed through these rocky clefts, which in the depths of the forest were gloomy and dark in the extreme, and often full of fine-leaved herbaceous plants and curious blue-foliaged Lycopodiaceæ. It was in such places as these that I obtained many of my most beautiful small butterflies, such as Sospita statira and Taxila pulchra, the gorgeous blue Amblypodia hercules, and many others. On the skirts of the plantations I found the handsome blue Deudorix despœna, and in the shady woods the lovely Lycæna wallacei. Here, too, I obtained the beautiful Thyca aruna, of the richest orange on the upper side, while below it is intense crimson and glossy black; and a superb specimen of a green Ornithoptera, absolutely fresh and perfect, and one of the glories of my collection.

The birds obtained here, though not very rich in number of

species, were yet very interesting. I got another specimen of the rare New Guinea kite (Henicopernis longicauda), a large new goatsucker (Podargus superciliaris), and a most curious ground-pigeon of an entirely new genus, and remarkable for its long and powerful bill. It has been named Henicophaps albifrons. I was also much pleased to obtain a fine series of a large fruit-pigeon with a protuberance on the bill (Carpophaga tumida), and to ascertain that this was not, as had been hitherto supposed, a sexual character, but was found equally in male and female birds. I collected only seventy-three species of birds in Waigiou, but twelve of them were entirely new, and many others very rare ; and as I brought away with me twenty-four fine specimens of the Paradisea rubra, I did not regret my visit to the island, although it had by no means answered my expectations.

CHAPTER XXXVII.

VOYAGE FROM WAIGIOU TO TERNATE.

(SEPTEMBER 29 TO NOVEMBER 5 1860.)

I HAD left the old pilot at Waigiou to take care of my house and to get the prau into sailing order—to caulk her bottom, and to look after the upper works, thatch, and rigging. When I returned I found it nearly ready, and immediately began packing up and preparing for the voyage. Our mainsail had formed one side of our house, but the spanker and jib had been put away in the roof, and on opening them to see if any repairs were wanted, to our horror we found that some rats had made them their nest, and had gnawed through them in twenty places. We had therefore to buy matting and make new sails, and this delayed us till the 29th of September, when we at length left Waigiou.

It took us four days before we could get clear of the land, having to pass along narrow straits beset with reefs and shoals, and full of strong currents, so that an unfavourable wind stopped us altogether. One day, when nearly clear, a contrary tide and head-wind drove us ten miles back to our anchorage of the night before. This delay made us afraid of running short of water if we should be becalmed at sea, and we therefore determined, if possible, to touch at the island where our men had been lost, and which lay directly in our proper course. The wind was, however, as usual, contrary, being S.S.W. instead of S.S.E., as it should have been at this time of the year, and all we could do was to reach the island of Gagie, where we came to an anchor by moonlight under bare volcanic hills. In the morning we tried to enter a deep bay, at the head of which

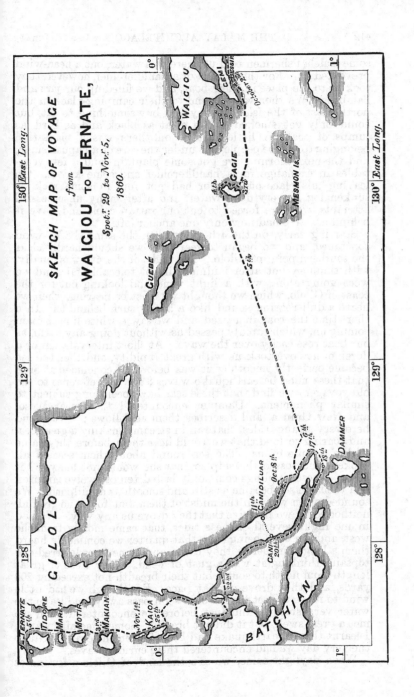

SKETCH MAP OF VOYAGE
from
WAIGIOU to TERNATE,
Sept.ʳ 29 to Nov.ʳ 5,
1860.

some Galela fishermen told us there was water, but a head-wind prevented us. For the reward of a handkerchief, however, they took us to the place in their boat, and we filled up our jars and bamboos. We then went round to their camping-place on the north coast of the island to try and buy something to eat, but could only get smoked turtle meat as black and as hard as lumps of coal. A little further on there was a plantation belonging to Guebe people, but under the care of a Papuan slave, and the next morning we got some plantains and a few vegetables in exchange for a handkerchief and some knives. On leaving this place our anchor had got foul in some rock or sunken log in very deep water, and after many unsuccessful attempts, we were forced to cut our rattan cable and leave it behind us. We had now only one anchor left.

Starting early, on the 4th of October, the same S.S.W. wind continued, and we began to fear that we should hardly clear the southern point of Gilolo. The night of the 5th was squally, with thunder, but after midnight it got tolerably fair, and we were going along with a light wind and looking out for the coast of Gilolo, which we thought we must be nearing, when we heard a dull roaring sound like a heavy surf, behind us. In a short time the roar increased, and we saw a white line of foam coming on, which rapidly passed us without doing any harm, as our boat rose easily over the wave. At short intervals, ten or a dozen others overtook us with great rapidity, and then the sea became perfectly smooth as it was before. I concluded at once that these must be earthquake waves; and on reference to the old voyagers we find that these seas have been long subject to similar phenomena. Dampier encountered them near Mysol and New Guinea, and describes them as follows : " We found here very strange tides, that ran in streams, making a great sea, and roaring so loud that we could hear them before they came within a mile of us. The sea round about them seemed all broken, and tossed the ship so that she would not answer her helm. These ripplings commonly lasted ten or twelve minutes, and then the sea became as still and smooth as a millpond. We sounded often when in the midst of them, but found no ground, neither could we perceive that they drove us any way. We had in one night several of these tides, that came mostly from the west, and the wind being from that quarter we commonly heard them a long time before they came, and sometimes lowered our topsails, thinking it was a gust of wind. They were of great length from north to south, but their breadth not exceeding 200 yards, and they drove a great pace. For though we had little wind to move us, yet these would soon pass away, and leave the water very smooth, and just before we encountered them we met a great swell, but it did not break." Some time afterwards, I learnt that an earthquake had been felt on the coast of Gilolo the very day we had encountered these curious waves.

When daylight came, we saw the land of Gilolo a few miles

off, but the point was unfortunately a little to windward of us. We tried to brace up all we could to round it, but as we approached the shore we got into a strong current setting northward, which carried us so rapidly with it that we found it necessary to stand off again, in order to get out of its influence. Sometimes we approached the point a little, and our hopes revived ; then the wind fell, and we drifted slowly away. Night found us in nearly the same position as we had occupied in the morning, so we hung down our anchor with about fifteen fathoms of cable to prevent drifting. On the morning of the 7th we were, however, a good way up the coast, and we now thought our only chance would be to get close in-shore, where there might be a return current, and we could then row. The prau was heavy, and my men very poor creatures for work, so that it took us six hours to get to the edge of the reef that fringed the shore ; and as the wind might at any moment blow on to it, our situation was a very dangerous one. Luckily, a short distance off, there was a sandy bay, where a small stream stopped the growth of the coral ; and by evening we reached this and anchored for the night. Here we found some Galela men shooting deer and pigs ; but they could not or would not speak Malay, and we could get little information from them. We found out that along shore the current changed with the tide, while about a mile out it was always one way, and against us ; and this gave us some hopes of getting back to the point, from which we were now distant twenty miles. Next morning we found that the Galela men had left before daylight, having perhaps some vague fear of our intentions, and very likely taking me for a pirate. During the morning a boat passed, and the people informed us that, at a short distance further towards the point, there was a much better harbour, where there were plenty of Galela men, from whom we might probably get some assistance.

At three in the afternoon, when the current turned, we started ; but having a head-wind made slow progress. At dusk we reached the entrance of the harbour, but an eddy and a gust of wind carried us away and out to sea. After sunset there was a land breeze, and we sailed a little to the south-east. It then became calm, and we hung down our anchor forty fathoms, to endeavour to counteract the current ; but it was of little avail, and in the morning we found ourselves a good way from shore. and just opposite our anchorage of the day before, which we again reached by hard rowing. I gave the men this day to rest and sleep ; and the next day (Oct. 10th) we again started at two in the morning with a land breeze. After I had set them to their oars, and given instructions to keep close in-shore, and on no account to get out to sea, I went below, being rather unwell. At daybreak I found, to my great astonishment, that we were again far off-shore, and was told that the wind had gradually turned more ahead, and had carried us out—none of them having the sense to take down the sail and row in-shore, or to call me. As soon as it was daylight, we saw that we had drifted back, and were again opposite our former anchorage, and, for the third time, had to row hard to get to it. As we approached the shore, I saw that the current was favourable to us, and we continued down the coast till we were close to the entrance to the lower harbour. Just as we were congratulating ourselves on having at last reached it, a strong south-east squall came on, blowing us back, and rendering it impossible for us to enter. Not liking the idea of again returning, I determined on trying to anchor, and succeeded in doing so, in very deep water and close to the reefs ; but the prevailing winds were such that, should we not hold, we should have no difficulty in getting out to sea. By the time the squall had passed, the current had turned against us, and we expected to have to wait till four in the afternoon, when we intended to enter the harbour.

Now, however, came the climax of our troubles. The swell produced by the squall made us jerk our cable a good deal, and it suddenly snapped low down in the water. We drifted out to sea, and immediately set our mainsail, but we were now without any anchor, and in a vessel so poorly manned that it could not be rowed against the most feeble current or the slightest wind, it would be madness to approach these dangerous shores except in the most perfect calm. We had also only three days' food left. It was therefore out of the question making any further attempts to get round the point without assistance, and I at once determined to run to the village of Gani-diluar, about ten miles further north, where we understood there was a good harbour, and where we might get provisions and a few more rowers. Hitherto winds and currents had invariably opposed our passage southward and we might have expected them to be favourable to us now we had turned our bowsprit in an opposite direction. But it immediately fell calm, and then after a time a westerly land breeze set in, which would not serve us, and we had to row again for hours, and when night came had not reached the village. We were so fortunate, however, as to find a deep sheltered cove where the water was quite smooth, and we constructed a temporary anchor by filling a sack with stones from our ballast, which being well secured by a network of rattans held us safely during the night. The next morning my men went on shore to cut wood suitable for making fresh anchors, and about noon, the current turning in our favour, we proceeded to the village, where we found an excellent and well-protected anchorage.

On inquiry, we found that the head men resided at the other Gani on the western side of the peninsula, and it was necessary to send messengers across (about half a day's journey) to inform them of my arrival, and to beg them to assist me. I then succeeded in buying a little sago, some dried deer-meat and cocoa-nuts, which at once relieved our immediate want of something to eat. At night we found our bag of stones still held us very well, and we slept tranquilly.

The next day (October 12th), my men set to work making anchors and oars. The native Malay anchor is ingeniously constructed of a piece of tough forked timber, the fluke being strengthened by twisted rattans binding it to the stem, while the cross-piece is formed of a long flat stone, secured in the same manner. These anchors, when well made, hold exceedingly firm, and, owing to the expense of iron, are still almost universally used on board the smaller praus. In the afternoon the head men arrived, and promised me as many rowers as I could put on the prau, and also brought me a few eggs and a little rice, which were very acceptable. On the 14th there was a north wind all day, which would have been invaluable to us a few days earlier, but which was now only tantalizing. On the 16th, all being ready, we started at daybreak with two new anchors and ten rowers, who understood their work. By evening we had

MALAY ANCHOR.

come more than half-way to the point, and anchored for the night in a small bay. At three the next morning I ordered the anchor up, but the rattan cable parted close to the bottom, having been chafed by rocks, and we then lost our third anchor on this unfortunate voyage. The day was calm, and by noon we passed the southern point of Gilolo, which had delayed us eleven days, whereas the whole voyage during this monsoon should not have occupied more than half that time. Having got round the point our course was exactly in the opposite direction to what it had been, and now, as usual, the wind changed accordingly, coming from the north and north-west—so that we still had to row every mile up to the village of Gani, which we did not reach till the evening of the 18th. A Bugis trader who was residing there, and the Senaji, or chief, were very kind ; the former assisting me with a spare anchor and a cable, and making me a present of some vegetables, and the latter baking fresh sago cakes for my men, and giving me a couple of fowls, a bottle of oil, and some pumpkins. As the weather was still very uncertain, I got four extra men to accompany me to Ternate, for which place we started on the afternoon of the 20th.

We had to keep rowing all night, the land breezes being too weak to enable us to sail against the current. During the afternoon of the 21st we had an hour's fair wind, which soon changed into a heavy squall with rain, and my clumsy men let the mainsail get taken aback and nearly upset us, tearing the sail, and, what was worse, losing an hour's fair wind. The night was calm, and we made little progress.

On the 22nd we had light head-winds. A little before noon we passed, with the assistance of our oars, the Paçiença Straits, the narrowest part of the channel between Batchian and Gilolo. These were well named by the early Portuguese navigators, as the currents are very strong, and there are so many eddies that even with a fair wind vessels are often quite unable to pass through them. In the afternoon a strong north wind (dead ahead) obliged us to anchor twice. At night it was calm, and we crept along slowly with our oars.

On the 23rd we still had the wind ahead, or calms. We then crossed over again to the mainland of Gilolo by the advice of our Gani men, who knew the coast well. Just as we got across we had another northerly squall with rain, and had to anchor on the edge of a coral reef for the night. I called up my men about three on the morning of the 24th, but there was no wind to help us, and we rowed along slowly. At daybreak there was a fair breeze from the south, but it lasted only an hour. All the rest of the day we had nothing but calms, light winds ahead, and squalls, and made very little progress.

On the 25th we drifted out to the middle of the channel, but made no progress onward. In the afternoon we sailed and rowed to the south end of Kaióa, and by midnight reached the village. I determined to stay here a few days to rest and recruit, and in hopes of getting better weather. I bought some onions and other vegetables, and plenty of eggs, and my men baked fresh sago cakes. I went daily to my old hunting-ground in search of insects, but with very poor success. It was now wet, squally weather, and there appeared a stagnation of insect life. We stayed five days, during which time twelve persons died in the village, mostly from simple intermittent fever, of the treatment of which the natives are quite ignorant. During the whole of this voyage I had suffered greatly from sun-burnt lips, owing to having exposed myself on deck all day to look after our safety among the shoals and reefs near Waigiou. The salt in the air so affected them that they would not heal, but became excessively painful, and bled at the slightest touch, and for a long time it was with great difficulty I could eat at all, being obliged to open my mouth very wide, and put in each mouthful with the greatest caution. I kept them constantly covered with ointment, which was itself very disagreeable, and they caused me almost constant pain for more than a month, as they did not get well till I had returned to Ternate, and was able to remain a week indoors.

A boat which left for Ternate the day after we arrived was obliged to return the next day, on account of bad weather. On the 31st we went out to the anchorage at the mouth of the harbour, so as to be ready to start at the first favourable opportunity.

On the 1st of November I called up my men at one in the morning, and we started with the tide in our favour. Hitherto it had usually been calm at night, but on this occasion we had a strong westerly squall with rain, which turned our prau broadside, and obliged us to anchor. When it had passed we went on rowing all night, but the wind ahead counteracted the current in our favour, and we advanced but little. Soon after sunrise the wind became stronger and more adverse, and as we had a dangerous lee shore which we could not clear, we had to put about and get an offing to the W.S.W. This series of contrary winds and bad weather ever since we started, not having had a single day of fair wind, was very remarkable. My men firmly believed there was something unlucky in the boat, and told me I ought to have had a certain ceremony gone through before starting, consisting of boring a hole in the bottom and pouring some kind of holy oil through it. It must be remembered that this was the season of the south-east monsoon, and yet we had not had even half a day's south-east wind since we left Waigiou. Contrary winds, squalls, and currents drifted us about the rest of the day at their pleasure. The night was equally squally and changeable, and kept us hard at work taking in and making sail, and rowing in the intervals.

Sunrise on the 2nd found us in the middle of the ten-mile channel between Kaióa and Makian. Squalls and showers succeeded each other during the morning. At noon there was a dead calm, after which a light westerly breeze enabled us to reach a village on Makian in the evening. Here I bought some pumelos (Citrus decumana), kanary-nuts, and coffee, and let my men have a night's sleep.

The morning of the 3rd was fine, and we rowed slowly along the coast of Makian. The captain of a small prau at anchor, seeing me on deck and guessing who I was, made signals for us to stop, and brought me a letter from Charles Allen, who informed me he had been at Ternate twenty days, and was anxiously waiting my arrival. This was good news, as I was equally anxious about him, and it cheered up my spirits. A light southerly wind now sprung up, and we thought we were going to have fine weather. It soon changed, however, to its old quarter, the west; dense clouds gathered over the sky, and in less than half an hour we had the severest squall we had experienced during our whole voyage. Luckily we got our great mainsail down in time, or the consequences might have been serious. It was a regular little hurricane, and my old Bugis

steersman began shouting out to "Allah! il Allah!" to preserve
us. We could only keep up our jib, which was almost blown to
rags, but by careful handling it kept us before the wind, and the
prau behaved very well. Our small boat (purchased at Gani)
was towing astern, and soon got full of water, so that it broke
away and we saw no more of it. In about an hour the fury of
the wind abated a little, and in two more we were able to hoist
our mainsail, reefed and half-mast high. Towards evening it
cleared up and fell calm, and the sea, which had been rather
high, soon went down. Not being much of a seamen myself I
had been considerably alarmed, and even the old steersman
assured me he had never been in a worse squall all his life. He
was now more than ever confirmed in his opinion of the un-
luckiness of the boat, and in the efficiency of the holy oil which
all Bugis praus had poured through their bottoms. As it was,
he imputed our safety and the quick termination of the squall
entirely to his own prayers, saying with a laugh, "Yes, that's
the way we always do on board our praus; when things are at
the worst we stand up and shout out our prayers as loud as we
can, and then Tuwan Allah helps us."

After this it took us two days more to reach Ternate, having
our usual calms, squalls, and head-winds to the very last; and
once having to return back to our anchorage owing to violent
gusts of wind just as we were close to the town. Looking at
my whole voyage in this vessel from the time when I left Goram
in May, it will appear that my experiences of travel in a native
prau have not been encouraging. My first crew ran away; two
men were lost for a month on a desert island; we were ten
times aground on coral reefs; we lost four anchors; the sails
were devoured by rats; the small boat was lost astern; we
were thirty-eight days on the voyage home, which should not
have taken twelve; we were many times short of food and
water; we had no compass-lamp, owing to there not being a
drop of oil in Waigiou when we left; and to crown all, during
the whole of our voyages from Goram by Ceram to Waigiou,
and from Waigiou to Ternate, occupying in all seventy-eight
days, or only twelve days short of three months (all in what
was supposed to be the favourable season), we had *not one single
day of fair wind*. We were always close braced up, always
struggling against wind, tide, and leeway, and in a vessel that
would scarcely sail nearer than eight points from the wind.
Every seamen will admit that my first voyage in my own boat
was a most unlucky one.

Charles Allen had obtained a tolerable collection of birds and
insects at Mysol, but far less than he would have done if I had
not been so unfortunate as to miss visiting him. After waiting
another week or two till he was nearly starved, he returned to
Wahai in Ceram, and heard, much to his surprise, that I had
left a fortnight before. He was delayed there more than a
month before he could get back to the north side of Mysol,

THE "KING" AND THE "TWELVE WIRED" BIRDS OF PARADISE.

which he found a much better locality, but it was not yet the season for the Paradise Birds ; and before he had obtained more than a few of the common sort, the last prau was ready to leave for Ternate, and he was obliged to take the opportunity, as he expected I would be waiting there for him.

This concludes the record of my wanderings. I next went to Timor, and afterwards to Bouru, Java, and Sumatra, which places have already been described. Charles Allen made a voyage to New Guinea, a short account of which will be given in my next chapter on the Birds of Paradise. On his return he went to the Sula Islands, and made a very interesting collection, which served to determine the limits of the zoological group of Celebes, as already explained in my chapter on the natural history of that island. His next journey was to Flores and Solor, where he obtained some valuable materials, which I have used in my chapter on the natural history of the Timor group. He afterwards went to Coti on the east coast of Borneo, from which place I was very anxious to obtain collections, as it is a quite new locality as far as possible from Saráwak, and I had heard very good accounts of it. On his return thence to Soura-baya in Java, he was to have gone to the entirely unknown Sumba or Sandal-wood Island. Most unfortunately, however, he was seized with a terrible fever on his arrival at Coti, and, after lying there some weeks, was taken to Singapore in a very bad condition, where he arrived after I had left for England. When he recovered he obtained employment in Singapore, and I lost his services as a collector.

The three concluding chapters of my work will treat of the Birds of Paradise, the Natural History of the Papuan Islands, and the Races of Man in the Malay Archipelago.

CHAPTER XXXVIII.

THE BIRDS OF PARADISE.

As many of my journeys were made with the express object of obtaining specimens of the Birds of Paradise, and learning something of their habits and distribution ; and being (as far as I am aware) the only Englishman who has seen these wonderful birds in their native forests, and obtained specimens of many of them, I propose to give here, in a connected form, the result of my observations and inquiries.

When the earliest European voyagers reached the Moluccas in search of cloves and nutmegs, which were then rare and precious spices, they were presented with the dried skins of birds so strange and beautiful as to excite the admiration even of those wealth-seeking rovers. The Malay traders gave them the name of "Manuk dewata," or God's birds ; and the Portuguese, finding that they had no feet or wings, and not being able to learn anything authentic about them, called them "Passaros de Sol," or Birds of the Sun ; while the learned Dutchmen, who wrote in Latin, called them "Avis paradiseus," or Paradise Bird. John van Linschoten gives these names in 1598, and tells us that no one has seen these birds alive, for they live in the air, always turning towards the sun, and never lighting on the earth till they die ; for they have neither feet nor wings, as, he adds, may be seen by the birds carried to India, and sometimes to Holland, but being very costly they were then rarely seen in Europe. More than a hundred years later Mr. William Funnel, who accompanied Dampier, and wrote an account of the voyage, saw specimens at Amboyna, and was told that they came to Banda to eat nutmegs, which intoxicated them and made them fall down senseless, when they were killed by ants. Down to 1760, when Linnæus named the largest species Paradisea apoda (the footless Paradise Bird), no perfect specimen had been seen in Europe, and absolutely nothing was known about them. And even now, a hundred years later, most books state that they migrate annually to Ternate, Banda, and Amboyna ; whereas the fact is, that they are as completely unknown in those islands in a wild state as they are in England. Linnæus was also acquainted with a small species, which he named Paradisea regia (the King Bird of Paradise), and since then nine or ten others have been named, all of which were first described from skins preserved by the savages of New Guinea, and generally more or less imperfect. These are now all known in the Malay Archipelago as "Burong mati," or dead birds, indicating that the Malay traders never saw them alive.

The Paradiseidæ are a group of moderate-sized birds, allied in their structure and habits to crows, starlings, and to the Australian honeysuckers ; but they are characterised by extraordinary developments of plumage, which are unequalled in any other family of birds. In several species large tufts of delicate bright-coloured feathers spring from each side of the body beneath the wings, forming trains, or fans, or shields ; and the middle feathers of the tail are often elongated into wires, twisted into fantastic shapes, or adorned with the most brilliant metallic tints. In another set of species these accessory plumes spring from the head, the back, or the shoulders ; while the intensity of colour and of metallic lustre displayed by their plumage, is not to be equalled by any other birds, except, perhaps, the humming-birds, and is not surpassed even by these. They have been usually classified under two distinct families, Paradiseidæ and Epimachidæ, the latter characterized by long and slender beaks, and supposed to be allied to the Hoopoes ; but the two groups are so closely allied in every essential point of structure and habits, that I shall consider

them as forming subdivisions of one family. I will now give
a short description of each of the known species, and then add
some general remarks on their natural history.

The Great Bird of Paradise (Paradisea apoda of Linnæus) is
the largest species known, being generally seventeen or eighteen
inches from the beak to the tip of the tail. The body, wings,
and tail are of a rich coffee-brown, which deepens on the breast
to a blackish-violet or purple-brown. The whole top of the head
and neck is of an exceedingly delicate straw-yellow, the feathers
being short and close set, so as to resemble plush or velvet ; the
lower part of the throat up to the eye is clothed with scaly
feathers of an emerald green colour, and with a rich metallic
gloss, and velvety plumes of a still deeper green extend in a
band across the forehead and chin as far as the eye, which is
bright yellow. The beak is pale lead blue ; and the feet, which
are rather large and very strong and well formed, are of a pale
ashy-pink. The two middle feathers of the tail have no webs,
except a very small one at the base and at the extreme tip,
forming wire-like cirrhi, which spread out in an elegant double
curve, and vary from twenty-four to thirty-four inches long.
From each side of the body, beneath the wings, springs a dense
tuft of long and delicate plumes, sometimes two feet in length,
of the most intense golden-orange colour and very glossy, but
changing towards the tips into a pale brown. This tuft of
plumage can be elevated and spread out at pleasure, so as
almost to conceal the body of the bird.

These splendid ornaments are entirely confined to the male
sex, while the female is really a very plain and ordinary-looking
bird of a uniform coffee-brown colour which never changes ;
neither does she possess the long tail wires, nor a single yellow
or green feather about the head. The young males of the first
year exactly resemble the females, so that they can only be
distinguished by dissection. The first change is the acquisition
of the yellow and green colour on the head and throat, and at
the same time the two middle tail feathers grow a few inches
longer than the rest, but remain webbed on both sides. At a
later period these feathers are replaced by the long bare shafts
of the full length, as in the adult bird ; but there is still no sign
of the magnificent orange side-plumes, which later still complete
the attire of the perfect male. To effect these changes there
must be at least three successive moultings ; and as the birds
were found by me in all the stages about the same time, it is
probable that they moult only once a year, and that the full
plumage is not acquired till the bird is four years old. It was
long thought that the fine train of feathers was assumed for a
short time only at the breeding season, but my own experience,
as well as the observation of birds of an allied species which I
brought home with me, and which lived two years in this
country, show that the complete plumage is retained during the

whole year, except during a short period of moulting as with most other birds.

The Great Bird of Paradise is very active and vigorous, and seems to be in constant motion all day long. It is very abundant, small flocks of females and young males being constantly met with ; and though the full-plumaged birds are less plentiful, their loud cries, which are heard daily, show that they also are very numerous. Their note is, "Wawk-wawk-wawk—Wŏk, wŏk-wŏk," and is so loud and shrill as to be heard a great distance, and to form the most prominent and characteristic animal sound in the Aru Islands. The mode of nidification is unknown ; but the natives told me that the nest was formed of leaves placed on an ant's nest, or on some projecting limb of a very lofty tree, and they believe that it contains only one young bird. The egg is quite unknown, and the natives declared they had never seen it ; and a very high reward offered for one by a Dutch official did not meet with success. They moult about January or February, and in May, when they are in full plumage, the males assemble early in the morning to exhibit themselves in the singular manner already described at p. 354. This habit enables the natives to obtain specimens with comparative ease. As soon as they find that the birds have fixed upon a tree on which to assemble, they build a little shelter of palm leaves in a convenient place among the branches, and the hunter ensconces himself in it before daylight, armed with his bow and a number of arrows terminating in a round knob. A boy waits at the foot of the tree, and when the birds come at sunrise, and a sufficient number have assembled, and have begun to dance, the hunter shoots with his blunt arrow so strongly as to stun the bird, which drops down, and is secured and killed by the boy without its plumage being injured by a drop of blood. The rest take no notice, and fall one after another till some of them take the alarm. (See Illustration facing p. 337.)

The native mode of preserving them is to cut off the wings and feet, and then skin the body up to the beak, taking out the skull. A stout stick is then run up through the specimen coming out at the mouth. Round this some leaves are stuffed, and the whole is wrapped up in a palm spathe and dried in the smoky hut. By this plan the head, which is really large, is shrunk up almost to nothing, the body is much reduced and shortened, and the greatest prominence is given to the flowing plumage. Some of these native skins are very clean, and often have wings and feet left on ; others are dreadfully stained with smoke, and all give a most erroneous idea of the proportions of the living bird.

The Paradisea apoda, as far as we have any certain knowledge, is confined to the mainland of the Aru Islands, never being found in the smaller islands which surround the central mass. It is certainly not found in any of the parts of New Guinea visited by the Malay and Bugis traders, nor in any of the other islands where Birds of Paradise are obtained. But this is by no means conclusive evidence, for it is only in certain localities that the natives prepare skins, and in other places the same birds may be abundant without ever becoming known. It is therefore quite possible that this species may inhabit the great southern mass of New Guinea, from which Aru has been separated; while its near ally, which I shall next describe, is confined to the north-western peninsula.

The Lesser Bird of Paradise (Paradisea papuana of Bechstein), "Le petit Emeraude" of French authors, is a much smaller bird than the preceding, although very similar to it. It differs in its lighter brown colour, not becoming darker or purpled on the breast; in the extension of the yellow colour all over the upper part of the back and on the wing coverts; in the lighter yellow of the side plumes, which have only a tinge of orange, and at the tips are nearly pure white; and in the comparative short-ness of the tail cirrhi. The female differs remarkably from the same sex in Paradisea apoda, by being entirely white on the under surface of the body, and is thus a much handsomer bird. The young males are similarly coloured, and as they grow older they change to brown, and go through the same stages in ac-quiring the perfect plumage as has already been described in the allied species. It is this bird which is most commonly used in ladies' head-dresses in this country, and also forms an im-portant article of commerce in the East.

The Paradisea papuana has a comparatively wide range, being the common species on the mainland of New Guinea, as well as on the islands of Mysol, Salwatty, Jobie, Biak and Sook. On the south coast of New Guinea, the Dutch naturalist, Muller, found it at the Oetanata river in longitude 136° E. I obtained it myself at Dorey; and the captain of the Dutch steamer *Etna* informed me that he had seen the feathers among the natives of Humboldt Bay, in 141° E. longitude. It is very prob-able, therefore, that it ranges over the whole of the mainland of New Guinea.

The true Paradise Birds are omnivorous, feeding on fruits and insects—of the former preferring the small figs; of the latter, grasshoppers, locusts, and phasmas, as well as cockroaches and caterpillars. When I returned home, in 1862, I was so fortunate as to find two adult males of this species in Singapore; and as they seemed healthy, and fed voraciously on rice, bananas, and cockroaches, I determined on giving the very high price asked for them—100*l*.—and to bring them to England by the overland route under my own care. On my way home I stayed a week at Bombay, to break the journey, and to lay in a fresh stock of bananas for my birds. I had great difficulty, however, in supplying them with insect food, for in the Peninsular and Oriental steamers cockroaches were scarce, and it was only by setting traps in the store-rooms, and by hunting an hour every night in the forecastle, that I could secure a few dozen of these

creatures—scarcely enough for a single meal. At Malta, where
I stayed a fortnight, I got plenty of cockroaches from a bake-
house, and when I left, took with me several biscuit-tins' full,
as provision for the voyage home. We came through the
Mediterranean in March, with a very cold wind ; and the only
place on board the mail-steamer where their large cage could be
accommodated was exposed to a strong current of air down a
hatchway which stood open day and night, yet the birds never
seemed to feel the cold. During the night journey from Mar-
seilles to Paris it was a sharp frost ; yet they arrived in London
in perfect health, and lived in the Zoological Gardens for one
and two years respectively, often displaying their beautiful
plumes to the admiration of the spectators. It is evident,
therefore, that the Paradise Birds are very hardy, and require
air and exercise rather than heat ; and I feel sure that if a good-
sized conservatory could be devoted to them, or if they could
be turned loose in the tropical department of the Crystal
Palace or the Great Palm House at Kew, they would live in this
country for many years.

The Red Bird of Paradise (Paradisea rubra of Viellot), though
allied to the two birds already described, is much more distinct
from them than they are from each other. It is about the same
size as Paradisea papuana (13 to 14 inches long), but differs
from it in many particulars. The side plumes, instead of being
yellow, are rich crimson, and only extend about three or four
inches beyond the end of the tail ; they are somewhat rigid,
and the ends are curved downwards and inwards, and are tipped
with white. The two middle tail feathers, instead of being
simply elongated and deprived of their webs, are transformed
into stiff black ribands, a quarter of an inch wide, but curved
like a split quill, and resembling thin hálf cylinders of horn or
whalebone. When a dead bird is laid on its back, it is seen
that these ribands take a curve or set, which brings them round
so as to meet in a double circle on the neck of the bird ; but
when they hang downwards during life they assume a spiral
twist, and form an exceedingly graceful double curve. They
are about twenty-two inches long, and always attract attention
as the most conspicuous and extraordinary feature of the
species. The rich metallic green colour of the throat extends
over the front half of the head to behind the eyes, and on the
forehead forms a little double crest of scaly feathers, which adds
much to the vivacity of the bird's aspect. The bill is gamboge
yellow, and the iris blackish olive.   (Figure at p. 403.)

The female of this species is of a tolerably uniform coffee-
brown colour, but has a blackish head, and the nape, neck, and
shoulders yellow, indicating the position of the brighter colours
of the male. The changes of plumage follow the same order of
succession as in the other species, the bright colours of the head
and neck being first developed, then the lengthened filaments

of the tail, and last of all, the red side plumes. I obtained a series of specimens, illustrating the manner in which the extraordinary black tail ribands are developed, which is very remarkable. They first appear as two ordinary feathers, rather shorter than the rest of the tail; the second stage would no doubt be that shown in a specimen of Paradisea apoda, in which the feathers are moderately lengthened, and with the web narrowed in the middle; the third stage is shown by a specimen which has part of the midrib bare, and terminated by a spatulate web; in another the bare midrib is a little dilated and semicylindrical, and the terminal web very small; in a fifth, the perfect black horny riband is formed, but it bears at its extremity a brown spatulate web; while in another specimen, part of the black riband itself bears, on one of its sides only, a narrow brown web. It is only after these changes are fully completed that the red side plumes begin to appear.

The successive stages of development of the colours and plumage of the Birds of Paradise are very interesting, from the striking manner in which they accord with the theory of their having been produced by the simple action of variation, and the cumulative power of selection by the females, of those male birds which were more than usually ornamental.[1] Variations of *colour* are of all others the most frequent and the most striking, and are most easily modified and accumulated by man's selection of them. We should expect, therefore, that the sexual differences of *colour* would be those most early accumulated and fixed, and would therefore appear soonest in the young birds; and this is exactly what occurs in the Paradise Birds. Of all variations in the *form* of birds' feathers, none are so frequent as those in the head and tail. These occur more or less in every family of birds, and are easily produced in many domesticated varieties, while unusual developments of the feathers of the body are rare in the whole class of birds, and have seldom or never occurred in domesticated species. In accordance with these facts, we find the scale-formed plumes of the throat, the crests of the head, and the long cirrhi of the tail, all fully developed before the plumes which spring from the side of the body begin to make their appearance. If, on the other hand, the male Paradise Birds have not acquired their distinctive plumage by successive variations, but have been as they are now from the moment they first appeared upon the earth, this succession becomes at the least unintelligible to us, for we can see no reason why the changes should not take place simultaneously, or in a reverse order to that in which they actually occur.

What is known of the habits of this bird, and the way in which it is captured by the natives, have already been described at page 408.

The Red Bird of Paradise offers a remarkable case of restricted

---

[1] I have since arrived at the conclusion that female selection is not the cause of the development of the ornamental plumes in the males. *See* my *Darwinism*, Chap. X.

range, being entirely confined to the small island of Waigiou, off the north-west extremity of New Guinea, where it replaces the allied species found in the other islands.[1]

The three birds just described form a well-marked group, agreeing in every point of general structure, in their comparatively large size, the brown colour of their bodies, wings, and tail, and in the peculiar character of the ornamental plumage which distinguishes the male bird. The group ranges nearly over the whole area inhabited by the family of the Paradiseidæ, but each of the species has its own limited region, and is never found in the same district with either of its close allies. To these three birds properly belongs the generic title Paradisea, or true Paradise Bird.[2]

The next species is the Paradisea regia of Linnæus, or King Bird of Paradise, which differs so much from the three preceding species as to deserve a distinct generic name, and it has accordingly been called Cicinnurus regius. By the Malays it is called "Burong rajah," or King Bird, and by the natives of the Aru Islands "Goby-goby."

This lovely little bird is only about six and a half inches long, partly owing to the very short tail, which does not surpass the somewhat square wings. The head, throat, and entire upper surface are of the richest glossy crimson red, shading to orange-crimson on the forehead, where the feathers extend beyond the nostrils more than half-way down the beak. The plumage is excessively brilliant, shining in certain lights with a metallic or glassy lustre. The breast and belly are pure silky white, between which colour and the red of the throat there is a broad band of rich metallic green, and there is a small spot of the same colour close above each eye. From each side of the body beneath the wing springs a tuft of broad delicate feathers about an inch and a half long, of an ashy colour, but tipped with a broad band of emerald green, bordered within by a narrow line of buff. These plumes are concealed beneath the wing, but when the bird pleases, can be raised and spread out so as to form an elegant semicircular fan on each shoulder. But another ornament still more extraordinary, and if possible more beautiful, adorns this little bird. The two middle tail feathers are modified into very slender wire-like shafts, nearly six inches long, each of which bears at the extremity, on the inner side only, a web of an emerald green colour, which is coiled up into a perfect spiral disc, and produces a most singular and charming effect. The bill is orange yellow, and the feet and legs of a fine cobalt blue. (See upper figure on the plate at the commencement of this chapter.)

---

[1] This species is said by Dr. Guillemard to be found also in the island of Batanta. (*Cruise of the Marchesa*, Vol. II., p. 225.)

[2] Three very distinct new species of Paradisea have since been discovered in Southeastern New Guinea, and two other less distinct local forms.

The female of this little gem is such a plainly coloured bird, that it can at first sight hardly be believed to belong to the same species. The upper surface is of a dull earthy brown, a slight tinge of orange red appearing only on the margins of the quills. Beneath, it is of a paler yellowish brown, scaled and banded with narrow dusky markings. The young males are exactly like the female, and they no doubt undergo a series of changes as singular as those of Paradisea rubra ; but, unfortunately, I was unable to obtain illustrative specimens.

This exquisite little creature frequents the smaller trees in the thickest parts of the forest, feeding on various fruits, often of a very large size for so small a bird. It is very active both on its wings and feet, and makes a whirring sound while flying, something like the South American manakins. It often flutters its wings and displays the beautiful fan which adorns its breast, while the star-bearing tail wires diverge in an elegant double curve. It is tolerably plentiful in the Aru Islands, which led to its being brought to Europe at an early period along with Paradisea apoda. It also occurs in the island of Mysol, and in every part of New Guinea which has been visited by naturalists.

We now come to the remarkable little bird called the " Magnificent," first figured by Buffon, and named Paradisea speciosa by Boddaert, which, with one allied species, has been formed into a separate genus by Prince Buonaparte, under the name of Diphyllodes, from the curious double mantle which clothes the back.

The head is covered with short brown velvety feathers, which advance on the back so as to cover the nostrils. From the nape springs a dense mass of feathers of a straw-yellow colour, and about one and a half inches long, forming a mantle over the upper part of the back. Beneath this, and forming a band about one-third of an inch beyond it, is a second mantle of rich, glossy, reddish-brown feathers. The rest of the back is orange-brown, the tail-coverts and tail dark bronzy, the wings light orange-buff. The whole under surface is covered with an abundance of plumage springing from the margins of the breast, and of a rich deep green colour, with changeable hues of purple. Down the middle of the breast is a broad band of scaly plumes of the same colour, while the chin and throat are of a rich metallic bronze. From the middle of the tail spring two narrow feathers of a rich steel blue, and about ten inches long. These are webbed on the inner side only, and curve outward, so as to form a double circle.

From what we know of the habits of allied species, we may be sure that the greatly developed plumage of this bird is erected and displayed in some remarkable manner. The mass of feathers on the under surface are probably expanded into a hemisphere, while the beautiful yellow mantle is no doubt elevated so as to give the bird a very different appearance from that which it presents in the dried and flattened skins of the natives, through which alone it is at present known. The feet appear to be dark blue.

THE MAGNIFICENT BIRD OF PARADISE. (*Diphyllodes speciosa.*)

This rare and elegant little bird is found only on the mainland of New Guinea, and in the island of Mysol.

A still more rare and beautiful species than the last is the Diphyllodes wilsoni, described by Mr. Cassin from a native skin in the rich museum of Philadelphia. The same bird was afterwards named "Diphyllodes respublica" by Prince Buonaparte, and still later, "Schlegelia calva," by Dr. Bernstein, who was so fortunate as to obtain fresh specimens in Waigiou.

In this species the upper mantle is sulphur yellow, the lower one and the wings pure red, the breast plumes dark green, and the lengthened middle tail feathers much shorter than in the allied species. The most curious difference is, however, that the top of the head is bald, the bare skin being of a rich cobalt blue, crossed by several lines of black velvety feathers.

It is about the same size as Diphyllodes speciosa, and is no doubt entirely confined to the island of Waigiou. The female, as figured and described by Dr. Bernstein, is very like that of Cicinnurus regius, being similarly banded beneath ; and we may therefore conclude that its near ally, the "Magnificent," is at least equally plain in this sex, of which specimens have not yet been obtained.

The Superb Bird of Paradise was first figured by Buffon, and was named by Boddaert, Paradisea atra, from the black ground colour of its plumage. It forms the genus Lophorina of Viellot, and is one of the rarest and most brilliant of the whole group, being only known from mutilated native skins. This bird is a little larger than the Magnificent. The ground colour of the plumage is intense black, but with beautiful bronze reflections on the neck, and the whole head scaled with feathers of brilliant metallic green and blue. Over its breast it bears a shield formed of narrow and rather stiff feathers, much elongated towards the sides, of a pure bluish-green colour, and with a satiny gloss. But a still more extraordinary ornament is that which springs from the back of the neck—a shield of a similar form to that on the breast, but much larger, and of a velvety black colour, glossed with bronze and purple. The outermost feathers of this shield

THE SUPERB BIRD OF PARADISE. (*Lophorina atra.*)

are half an inch longer than the wing, and when it is elevated it must, in conjunction with the breast shield, completely change the form and whole appearance of the bird. The bill is black, and the feet appear to be yellow.

This wonderful little bird inhabits the interior of the northern peninsula of New Guinea only. Neither I nor Mr. Allen could hear anything of it in any of the islands or on any part of the coast. It is true that it was obtained from the coast-natives by Lesson ; but when at Sorong in 1861, Mr. Allen learnt that it is only found three days' journey in the interior. Owing to these "Black Birds of Paradise," as they are called, not being so much valued as articles of merchandise, they now seem to be rarely preserved by the natives, and it thus happened that during several years spent on the coasts of New Guinea and in the

Moluccas I was never able to obtain a skin. We are therefore quite ignorant of the habits of this bird, and also of its female, though the latter is no doubt as plain and inconspicuous as in all the other species of this family.

The Golden, or Six-shafted Paradise Bird, is another rare species, first figured by Buffon, and never yet obtained in perfect condition. It was named by Boddaert, Paradisea sex-pennis, and forms the genus Parotia of Viellot. This wonderful bird is about the size of the female Paradisea rubra. The plumage appears at first sight black, but it glows in certain lights with bronze and deep purple. The throat and breast are scaled with broad flat feathers of an intense golden hue,

THE SIX-SHAFTED BIRD OF PARADISE. (*Parotia sexpennis.*)

changing to green and blue tints in certain lights. On the back of the head is a broad recurved band of feathers, whose brilliancy is indescribable, resembling the sheen of emerald and topaz rather than any organic substance. Over the forehead is a large patch of pure white feathers, which shine like satin ; and from the sides of the head spring the six wonderful feathers from which the bird receives its name. These are slender wires, six inches long, with a small oval web at the extremity. In addition to these ornaments, there is also an immense tuft of soft feathers on each side of the breast, which when elevated must entirely hide the wings, and give the bird an appearance of being double its real bulk. The bill is black, short, and rather compressed, with the feathers advancing over the nostrils, as in Cicinnurus regius. This singular and brilliant bird inhabits the same region as the Superb Bird of Paradise, and nothing whatever is known about it but what we can derive from an examination of the skins preserved by the natives of New Guinea.

The Standard Wing, named Semioptera wallacei by Mr. G. R. Gray, is an entirely new form of Bird of Paradise, discovered by myself in the island of Batchian, and especially distinguished by a pair of long narrow feathers of a white colour, which spring from among the short plumes which clothe the bend of the wing, and are capable of being erected at pleasure. The general colour of this bird is a delicate olive-brown, deepening to a kind of bronzy olive in the middle of the back, and changing to a delicate ashy violet with a metallic gloss, on the crown of the head. The feathers, which cover the nostrils and extend halfway down the beak, are loose and curved upwards. Beneath, it is much more beautiful. The scale-like feathers of the breast are margined with rich metallic blue-green, which colour entirely covers the throat and sides of the neck, as well as the long pointed plumes which spring from the sides of the breast, and extend nearly as far as the end of the wings. The most curious feature of the bird, however, and one altogether unique in the whole class, is found in the pair of long narrow delicate feathers which spring from each wing close to the bend. On lifting the wing-coverts they are seen to arise from two tubular horny sheaths, which diverge from near the point of junction of the carpal bones. As already described at p. 252, they are erectile, and when the bird is excited are spread out at right angles to the wing and slightly divergent. They are from six to six and a half inches long, the upper one slightly exceeding the lower. The total length of the bird is eleven inches. The bill is horny olive, the iris deep olive, and the feet bright orange.

The female bird is remarkably plain, being entirely of a dull pale earthy brown, with only a slight tinge of ashy violet on the head to relieve its general monotony; and the young males exactly resemble her. (See figures at p. 253.)

This bird frequents the lower trees of the forests, and, like most Paradise Birds, is in constant motion—flying from branch to branch, clinging to the twigs and even to the smooth and vertical trunks almost as easily as a woodpecker. It continually utters a harsh, creaking note, somewhat intermediate between that of Paradisea apoda and the more musical cry of Cicinnurus regius. The males at short intervals open and flutter their wings, erect the long shoulder feathers, and spread out the elegant green breast shields.

The Standard Wing is found in Gilolo as well as in Batchian, and all the specimens from the former island have the green breast shield rather longer, the crown of the head darker violet, and the lower parts of the body rather more strongly scaled with green. This is the only Paradise Bird yet found in the Moluccan district, all the others being confined to the Papuan Islands and North Australia.

We now come to the Epimachidæ, or Long-billed Birds of Paradise, which, as before stated, ought not to be separated from the Paradiseidæ by the intervention of any other birds. One of the most remarkable of these is the Twelve-wired Paradise Bird, Paradisea alba of Blumenbach, but now placed in the genus Seleucides of Lesson.

This bird is about twelve inches long, of which the compressed and curved beak occupies two inches. The colour of the breast and upper surface appears at first sight nearly black, but a close examination shows that no part of it is devoid of colour ; and by holding it in various lights, the most rich and glowing tints become visible. The head, covered with short velvety feathers, which advance on the chin much further than on the upper part of the beak, is of a purplish bronze colour ; the whole of the back and shoulders is rich bronzy green, while the closed wings and tail are of the most brilliant violet purple, all the plumage having a delicate silky gloss. The mass of feathers which cover the breast is really almost black, with faint glosses of green and purple, but their outer edges are margined with glittering bands of emerald green. The whole lower part of the body is rich buffy yellow, including the tuft of plumes which spring from the sides, and extend an inch and a half beyond the tail. When skins are exposed to the light the yellow fades into dull white, from which circumstance it derived its specific name. About six of the innermost of these plumes on each side have the midrib elongated into slender black wires, which bend at right angles, and curve somewhat backwards to a length of about ten inches, forming one of those extraordinary and fantastic ornaments with which this group of birds abounds. The bill is jet black, and the feet bright yellow. (See lower figure on the plate at the beginning of this chapter.)

The female, although not quite so plain a bird as in some other species, presents none of the gay colours or ornamental plumage of the male. The top of the head and back of the neck are black, the rest of the upper parts rich reddish brown ; while the under surface is entirely yellowish ashy, somewhat blackish on the breast, and crossed throughout with narrow blackish wavy bands.

The Seleucides alba is found in the island of Salwatty, and in the north-western parts of New Guinea, where it frequents flowering trees, especially sago-palms and pandani, sucking the flowers, round and beneath which its unusually large and powerful feet enable it to cling. Its motions are very rapid. It seldom rests more than a few moments on one tree, after which it flies straight off, and with great swiftness, to another. It has a loud shrill cry, to be heard a long way, consisting of "Cáh, cáh," repeated five or six times in a descending scale, and at the last note it generally flies away. The males are quite solitary in their habits, although, perhaps, they assemble at certain times like the true Paradise Birds. All the specimens shot and opened by my assistant Mr. Allen, who obtained this fine bird during his last voyage to New Guinea, had nothing in their stomachs but a brown sweet liquid, probably the nectar of the flowers on which they had been feeding. They certainly, however, eat both fruit and insects, for a specimen which I saw alive on board a Dutch steamer ate cockroaches and papaya fruit voraciously. This bird had the curious habit of resting at noon with the bill pointing vertically upwards. It died on the passage to Batavia, and I secured the body and formed a skeleton, which shows indisputably that it is really a Bird of Paradise. The tongue is very long and extensible, but flat and a little fibrous at the end, exactly like the true Paradiseas.

In the island of Salwatty, the natives search in the forests till they find the sleeping place of this bird, which they know by seeing its dung upon the ground. It is generally in a low bushy tree. At night they climb up the tree, and either shoot the birds with blunt arrows, or even catch them alive with a cloth. In New Guinea they are caught by placing snares on the trees frequented by them, in the same way as the Red Paradise Birds are caught in Waigiou, and which has already been described at page 408.

The great Epimaque, or Long-tailed Paradise Bird (Epimachus magnus) is another of these wonderful creatures, only known by the imperfect skins prepared by the natives. In its dark velvety plumage, glossed with bronze and purple, it resembles the Seleucides alba, but it bears a magnificent tail more than two feet long, glossed on the upper surface with the most intense opalescent blue. Its chief ornament, how-

THE LONG-TAILED BIRD OF PARADISE.
(*Epimachus magnus.*)

ever, consists in the group of broad plumes which spring from the sides of the breast, and which are dilated at the extremity, and banded with the most vivid metallic blue and green. The bill is long and curved, and the feet black, and similar to those of the allied forms. The total length of this fine bird is between three and four feet.

This splendid bird inhabits the mountains of New Guinea, in the same district with the Superb and the Six-shafted Paradise Birds, and I was informed is sometimes found in the ranges near the coast. I was several times assured by different natives that this bird makes its nest in a hole under ground, or under rocks, always choosing a place with two apertures, so that it may enter at one and go out at the other. This is very unlike what we should suppose to be the habits of the bird, but it is not easy to conceive how the story originated if it is not true ; and all travellers know that native accounts of the habits of animals, however strange they may seem, almost invariably turn out to be correct.

The Scale-breasted Paradise Bird (Epimachus magnificus of Cuvier) is now generally placed with the Australian Rifle birds in the genus Ptiloris. Though very beautiful, these birds are less strikingly decorated with accessory plumage than the other species we have been describing, their chief ornament being a more or less developed breastplate of stiff metallic green feathers, and a small tuft of somewhat hairy plumes on the sides of the breast. The back and wings of this species are of an intense velvety black, faintly glossed in certain lights with rich purple. The two broad middle tail feathers are opalescent green-blue with a velvety surface, and the top of the head is covered with feathers resembling scales of burnished steel. A large triangular space covering the chin, throat, and breast, is densely scaled with feathers, having a steel-blue or green lustre, and a silky feel. This is edged below with a narrow band of black, followed by shiny bronzy green, below which the body is covered with hairy feathers of a rich claret colour, deepening to black at the tail. The tufts of side plumes somewhat resemble those of the true Birds of Paradise, but are scanty, about as long as the tail, and of a black colour. The sides of the head are rich violet, and velvety feathers extend on each side of the beak over the nostrils.

I obtained at Dorey a young male of this bird, in a state of plumage which is no doubt that of the adult female, as is the case in all the allied species. The upper surface, wings, and tail are rich reddish brown, while the under surface is of a pale ashy colour, closely barred throughout with narrow wavy black bands. There is also a pale banded stripe over the eye, and a long dusky stripe from the gape down each side of the neck. This bird is fourteen inches long, whereas the native skins of the adult male are only about ten inches, owing to the way in which the tail is pushed in, so as to give as much prominence as possible to the ornamental plumage of the breast.

At Cape York, in North Australia, there is a closely allied species, Ptiloris alberti, the female of which is very similar to the young male bird here described. The beautiful Rifle Birds of Australia, which much resemble these Paradise Birds, are named Ptiloris paradiseus and Ptiloris victoriæ. The Scale-breasted Paradise Bird seems to be confined to the mainland of New Guinea, and is less rare than several of the other species.

There are three other New Guinea birds which are by some authors classed with the Birds of Paradise, and which, being almost equally remarkable for splendid plumage, deserve to be noticed here. The first is the Paradise pie (Astrapia nigra of Lesson), a bird of the size of Paradisea rubra, but with a very long tail, glossed above with intense violet. The back is bronzy black, the lower parts green, the throat and neck bordered with loose broad feathers of an intense coppery hue, while on the top of the head and neck they are glittering emerald green. All the plumage round the head is lengthened and erectile, and when spread out by the living bird must have an effect hardly surpassed by any of the true Paradise Birds. The bill is black and the feet yellow. The Astrapia seems to me to be somewhat intermediate between the Paradiseidæ and Epimachidæ.

There is an allied species, having a bare carunculated head, which has been called Paradigalla carunculata. It is believed to inhabit, with the preceding, the mountainous interior of New Guinea, but is exceedingly rare, the only known specimen being in the Philadelphia Museum.

The Paradise Oriole is another beautiful bird, which is now sometimes classed with the Birds of Paradise. It has been named Paradisea aurea and Oriolus aureus by the old naturalists, and is now generally placed in the same genus as the Regent Bird of Australia (Sericulus chrysocephalus). But the form of the bill and the character of the plumage seem to me to be so different that it will have to form a distinct genus. This bird is almost entirely yellow, with the exception of the throat, the tail, and part of the wings and back, which are black ; but it is chiefly characterized by a quantity of long feathers of an intense glossy orange colour, which cover its neck down to the middle of the back, almost like the hackles of a game-cock.

This beautiful bird inhabits the mainland of New Guinea, and is also found in Salwatty, but is so rare that I was only able to obtain one imperfect native skin, and nothing whatever is known of its habits.

I will now give a list of all the Birds of Paradise yet known, with the places they are believed to inhabit. Those printed in italics have been discovered since the first edition of my book was published.

1. PARADISEA APODA. The Great Paradise Bird. Aru Islands and Central New Guinea.

2. PARADISEA PAPUANA (or P. minor). The Lesser Bird of Paradise. North-West New Guinea, Mysol, and Jobie Islands.

3. PARADISEA RUBRA (or P. sanguinea). The Red Bird of Paradise. Waigiou and Batanta.

4. *Paradisea decora.* D'Entrecasteaux Islands. This beautiful species has red plumes rather more abundant than in P. rubra, having at the base a number of shorter plumes of a deeper red. The breast is of a soft lilac tint, the head and throat nearly as in P. papuana.

5. *Paradisea Raggiana.* South-East New Guinea. This species is something like P. apoda, but with red plumes.

6. *Paradisea Guilielmi II.* German New Guinea. The head, neck, and throat are green, the back yellow. Adult males not yet known.

7. *Paradisea novæguineæ.* A form of P. apoda inhabiting Southern New Guinea.

8. *Paradisea Finschi.* A form of P. papuana found in the south-east of New Guinea.

9. CICINNURUS REGIUS. The King Bird of Paradise. All New Guinea, Mysol, Aru Islands.

10. DIPHYLLODES SPECIOSA (or D. magnifica). The Magnificent Bird of Paradise. North-West New Guinea and Mysol.

11. DIPHYLLODES WILSONI (or D. respublica). The Red Magnificent. Island of Waigiou.

12. *Diphyllodes chrysoptera.* South-East New Guinea. Allied to the "Magnificent," but with richer and more varied colours.

13. *Diphyllodes Jobiensis.* An allied form from Jobie Island.

14. *Diphyllodes Hernsteini.* Another species, with red wings and brown crown, from the Horseshoe Mountains in South-East New Guinea.

15. *Diphyllodes Guilielmi III.* A gorgeous species with orange and red back and a green-tipped fan like the King bird, from the east of Waigiou.

16. LOPHORINA ATRA (or Lophorina superba). The Superb Bird of Paradise. Arfak Mountains, North-West New Guinea.

17. *Lophorina minor.* A smaller species from the Astrolabe Mountains in South-East New Guinea; differs in the form of the neck-collar, and somewhat in colour.

18. PAROTIA SEXPENNIS. The Golden Paradise Bird. The Arfak Mountains in North-West New Guinea.

19. *Parotia Lamesi.* A representative species from the Astrolabe Mountains in South-East New Guinea, differing slightly in colouration and in the form of the breast plumes.

20. SEMIOPTERA WALLACEI. Wallace's Standard Wing. The Islands of Batchian and Gilolo.

21. EPIMACHUS MAGNUS (or E. speciosus). The Long-tailed Paradise Bird. North-West New Guinea.

22. *Epimachus Macleayi.* A slightly different form, from the Astrolabe Mountains, South-East New Guinea.

23. *Epimachus Meyeri.* Another allied species from the Horseshoe Mountains, S.E. New Guinea. Female only known.

24. *Epimachus Elliotti.* A less brilliant species. Locality unknown.

25. SELEUCIDES ALBA (or S. nigricans). The Twelve-wired Paradise Bird. North-West to South-East New Guinea.

26. PTILORIS (CRASPEDOPHORA) MAGNIFICA. The Scale-breasted Paradise Bird. All New Guinea.

27. PTILORIS ALBERTI. Prince Albert's Paradise Bird. North Australia.

28. PTILORIS PARADISEA. The Rifle Bird. East Australia.
29. PTILORIS VICTORIÆ. Queen Victoria's Rifle Bird. North-East Australia.
30. *Ptiloris (Craspedophora) intercedens.* A species from South-East New Guinea, very closely allied to P. magnifica.
31. ASTRAPIA NIGRA. The Paradise Pie. Arfak Mountains, North-West New Guinea.
32. PARADIGALLA CARUNCULATA. The Carunculated Paradise Pie. Arfak Mountains.

The following species are so distinct from any previously known as to require to be placed in New Genera :—

33. *Drepanornis Albertisi.* ⎫ These are a group of rather small
34. *Drepanornis cervinicauda.* ⎬ and not highly ornamented birds from
35. *Drepanornis Bruijni.* ⎭ various parts of New Guinea.
36. *Astrarchia Stephaniæ.* A splendid bird from the Owen Stanley Mountains, allied to the Paradise Pie.
37. *Paradisornis Rudolphi.* Horseshoe Mountains in South-East New Guinea. It is small, but distinguished by its bright blue side-plumes, and lengthened middle tail feathers with small blue spatulate tips.

In my *Studies Scientific and Social,* Vol. I. Chap. xx., I have given figures of the last-named species, and of three other new Paradise-birds from the mountains of New Guinea. One of these, *Pteridophora Alberti,* is perhaps the most extraordinary of this wonderful family, having long fern-life appendages of a fine blue colour springing from the corner of the eye.

These, and other additions bring the number of the Birds of Paradise to fifty species, of which about forty are known to inhabit the great island of New Guinea. But if we consider those islands which are now united to New Guinea by a shallow sea to really form a part of it, we shall find that twenty-three of the Paradise Birds belong to that country, while three inhabit the northern and eastern parts of Australia, and one the Moluccas. All the more extraordinary and magnificent species are, however, entirely confined to the Papuan region.

Although I devoted so much time to a search after these wonderful birds, I only succeeded myself in obtaining five species during a residence of many months in the Aru Islands, New Guinea, and Waigiou. Mr. Allen's voyage to Mysol did not procure a single additional species, but we both heard of a place called Sorong, on the mainland of New Guinea, near Salwatty, where we were told that all the kinds we desired could be obtained. We therefore determined that he should visit this place, and endeavour to penetrate into the interior among the natives, who actually shoot and skin the Birds of Paradise. He went in the small prau I had fitted up at Goram, and through the kind assistance of the Dutch Resident at Ternate, a lieutenant and two soldiers were sent by the Sultan of Tidore to accompany and protect him, and to assist him in getting men and in visiting the interior.

Notwithstanding these precautions, Mr. Allen met with difficulties in this voyage which we had neither of us encountered before. To understand these, it is necessary to consider that the Birds of Paradise are an article of commerce, and are the monopoly of the chiefs of the coast villages, who obtain them at a low rate from the mountaineers, and sell them to the Bugis traders. A portion is also paid every year as tribute to the Sultan of Tidore. The natives are therefore very jealous of a stranger, especially a European, interfering in their trade, and above all of going into the interior to deal with the mountaineers themselves. They of course think he will raise the prices in the interior, and lessen the supply on the coast, greatly to their disadvantage ; they also think their tribute will be raised if a European takes back a quantity of the rare sorts ; and they have besides a vague and very natural dread of some ulterior object in a white man's coming at so much trouble and expense to their country only to get Birds of Paradise, of which they know he can buy plenty (of the common yellow ones which alone they value) at Ternate, Macassar, or Singapore.

It thus happened that when Mr. Allen arrived at Sorong, and explained his intention of going to seek Birds of Paradise in the interior, innumerable objections were raised. He was told it was three or four days' journey over swamps and mountains ; that the mountaineers were savages and cannibals, who would certainly kill him ; and, lastly, that not a man in the village could be found who dare go with him. After some days spent in these discussions, as he still persisted in making the attempt, and showed them his authority from the Sultan of Tidore to go where he pleased and receive every assistance, they at length provided him with a boat to go the first part of the journey up a river ; at the same time, however, they sent private orders to the interior villages to refuse to sell any provisions, so as to compel him to return. On arriving at the village where they were to leave the river and strike inland, the coast people returned, leaving Mr. Allen to get on as he could. Here he called on the Tidore lieutenant to assist him, and procure men as guides and to carry his baggage to the villages of the mountaineers. This, however, was not so easily done. A quarrel took place, and the natives, refusing to obey the imperious orders of the lieutenant, got out their knives and spears to attack him and his soldiers ; and Mr. Allen himself was obliged to interfere to protect those who had come to guard him. The respect due to a white man and the timely distribution of a few presents prevailed ; and, on showing the knives, hatchets, and beads he was willing to give to those who accompanied him, peace was restored, and the next day, travelling over a frightfully rugged country, they reached the villages of the mountaineers. Here Mr. Allen remained a month without any interpreter through whom he could understand a word or communicate a want. However, by signs and presents and a pretty liberal barter, he got on very well, some of them accompanying him every day in the forest to shoot, and receiving a small present when he was successful.

In the grand matter of the Paradise Birds, however, little was done. Only one additional species was found, the Seleucides alba, of which he had already obtained a specimen in Salwatty ; but he learnt that the other kinds, of which he showed them drawings, were found two or three days' journey farther in the interior. When I sent my men from Dorey to Amberbaki, they heard exactly the same story—that the rarer sorts were only found several days journey in the interior, among rugged mountains, and that the skins were prepared by savage tribes who had never even been seen by any of the coast people.

It seems as if Nature had taken precautions that these her choicest treasures should not be made too common, and thus be undervalued. This northern coast of New Guinea is exposed to the full swell of the Pacific Ocean, and is rugged and harbourless. The country is all rocky and mountainous, covered everywhere with dense forest, offering in its swamps and precipices and serrated ridges an almost impassable barrier to the unknown interior ; and the people are dangerous savages, in the very lowest stage of barbarism. In such a country, and among such a people, are found these wonderful productions of Nature, the Birds of Paradise, whose exquisite beauty of form and colour, and strange developments of plumage are calculated to excite the wonder and admiration of the most civilized and the most intellectual of mankind, and to furnish inexhaustible materials for study to the naturalist, and for speculation to the philosopher.

Thus ended my search after these beautiful birds. Five voyages to different parts of the district they inhabit, each occupying in its preparation and execution the larger part of a year, produced me only five species out of the fourteen known to exist in the New Guinea district. The kinds obtained are those that inhabit the coasts of New Guinea and its islands, the remainder seeming to be strictly confined to the central mountain-ranges of the northern peninsula ; and our researches at Dorey and Amberbaki, near one end of this peninsula, and at Salwatty and Sorong, near the other, enable me to decide with some certainty on the native country of these rare and lovely birds, good specimens of which have never yet been seen in Europe.

It must be considered as somewhat extraordinary that, during five years' residence and travel in Celebes, the Moluccas, and New Guinea, I should never have been able to purchase skins of half the species which Lesson, forty years ago, obtained during a few weeks in the same countries. I believe that all, except the common species of commerce, are now much more difficult to obtain than they were even twenty years ago ; and I impute it principally to them having been sought after by the Dutch officials through the Sultan of Tidore. The chiefs of the annual expeditions to collect tribute have had orders to get all the rare sorts of Paradise Birds ; and as they pay little or nothing for them (it being sufficient to say they are for the Sultan), the head men of the coast villages would for the future refuse to purchase them from the mountaineers, and confine themselves instead to the commoner species which are less sought after by amateurs, but are a more profitable merchandise. The same causes frequently lead the inhabitants of uncivilized countries to conceal minerals or other natural products with which they may become acquainted, from the fear of being obliged to pay increased tribute, or of bringing upon themselves a new and oppressive labour.

CHAPTER XXXIX.

THE NATURAL HISTORY OF THE PAPUAN ISLANDS.

NEW GUINEA, with the islands joined to it by a shallow sea, constitute the Papuan group, characterized by a very close resemblance in their peculiar forms of life. Having already, in my chapters on the Aru Islands and on the Birds of Paradise, given some details of the natural history of this district, I shall here confine myself to a general sketch of its animal productions, and of their relations to those of the rest of the world.

New Guinea is perhaps the largest island on the globe, being a little larger than Borneo. It is nearly fourteen hundred miles long, and in the widest part four hundred broad, and seems to be everywhere covered with luxuriant forests. Almost everything that is yet known of its natural productions comes from the north-western peninsula, and a few islands grouped around it.[1] These do not constitute a tenth part of the area of the whole island, and are so cut off from it, that their fauna may well be somewhat different ; yet they have produced (with a very partial exploration) no less than two hundred and fifty species of land birds, almost all unknown elsewhere, and comprising some of the most curious and most beautiful of the feathered tribes. It is needless to say how much interest attaches to the far larger unknown portion of this great island, the greatest *terra incognita* that still remains for the naturalist to explore, and the only region where altogether new and unimagined forms of life may perhaps be found. There is now, I am happy to say, some chance that this great country will no longer remain absolutely unknown to us. The Dutch Government have granted a well-equipped steamer to carry a naturalist (Mr. Rosenberg, already mentioned in this work) and assistants

---

[1] This is no longer true, very extensive collections having been made in German and British New Guinea (the south-eastern portion) which have more than doubled the number of species of birds.

to New Guinea, where they are to spend some years in circumnavigating the island, ascending its large rivers as far as possible into the interior, and making extensive collections of its natural productions.[1]

The Mammalia of New Guinea and the adjacent islands, yet discovered, are only seventeen in number. Two of these are bats, one is a pig of a peculiar species (Sus papuensis), and the rest are all marsupials. The bats are, no doubt, much more numerous, but there is every reason to believe that whatever new land Mammalia may be discovered will belong to the marsupial order. One of these is a true kangaroo, very similar to some of the middle-sized kangaroos of Australia, and it is remarkable as being the first animal of the kind ever seen by Europeans. It inhabits Mysol and the Aru Islands (an allied species being found in New Guinea), and was seen and described by Le Brun in 1714, from living specimens at Batavia. A much more extraordinary creature is the tree-kangaroo, two species of which are known from New Guinea. These animals do not differ very strikingly in form from the terrestrial kangaroos, and appear to be but imperfectly adapted to an arboreal life, as they move rather slowly, and do not seem to have a very secure footing on the limb of a tree. The leaping power of the muscular tail is lost, and powerful claws have been acquired to assist in climbing, but in other respects the animal seems better adapted to walk on *terra firma*. This imperfect adaptation may be due to the fact of there being no carnivora in New Guinea, and no enemies of any kind from which these animals have to escape by rapid climbing. Four species of Cuscus, and the small flying opossum, also inhabit New Guinea ; and there are five other smaller marsupials, one of which is the size of a rat, and takes its place by entering houses and devouring provisions.[2]

The birds of New Guinea offer the greatest possible contrast to the Mammalia, since they are more numerous, more beautiful, and afford more new, curious, and elegant forms than those of any other island on the globe. Besides the Birds of Paradise, which we have already sufficiently considered, it possesses a number of other curious birds, which in the eyes of the ornithologist almost serve to distinguish it as one of the primary divisions of the earth. Among its thirty species of parrots are the Great Black Cockatoo, and the little rigid-tailed Nasiterna, the giant and the dwarf of the whole tribe. The bare-headed Dasyptilus is one of the most singular parrots known ; while the beautiful little long-tailed Charmosyna, and the great variety of gorgeously-coloured lories, have no parallels elsewhere. Of pigeons it possesses about forty distinct species, among which are the magnificent crowned pigeons, now so well known

[1] The most important of the natural history travellers who have since visited New Guinea are the Italians Beccari and D'Albertis, the Germans Meyer and Finsch, Mr. H. O. Forbes, and several English and German collectors.
[2] Among the more interesting Mammalia since discovered are a species of Echidna, allied to the spiny ant-eater of Australia.

in our aviaries, and pre-eminent both for size and beauty; the curious Trugon terrestris, which approaches the still more strange Didunculus of Samoa; and a new genus (Henicophaps), discovered by myself, which possesses a very long and powerful bill, quite unlike that of any other pigeon.[1]   Among its sixteen kingfishers, it possesses the curious hook-billed Macrorhina, and a red and blue Tanysiptera, the most beautiful of that beautiful genus.   Among its perching birds are the fine genus of crow-like starlings, with brilliant plumage (Manucodia); the curious pale-coloured crow (Gymnocorvus senex); the abnormal red and black flycatcher (Peltops blainvillii); the curious little boat-billed flycatchers (Machærirhynchus); and the elegant blue flycatcher-wrens (Todopsis).

The naturalist will obtain a clearer idea of the variety and interest of the productions of this country, by the statement, that its land birds belong to 108 genera, of which 29 are exclusively characteristic of it; while 35 belong to that limited area which includes the Moluccas and North Australia, and whose species of these genera have been entirely derived from New Guinea. About one-half of the New Guinea genera are found also in Australia, about one-third in India and the Indo-Malay islands.

A very curious fact, not hitherto sufficiently noticed, is the appearance of a pure Malay element in the birds of New Guinea. We find two species of Eupetes, a curious Malayan genus allied to the forked-tail water-chats; two of Alcippe, an Indian and Malay wren-like form; an Arachnothera, quite resembling the spider-catching honeysuckers of Malacca; two species of Gracula, the Mynahs of India; and a curious little black Prionochilus, a saw-billed fruit-pecker, undoubtedly allied to the Malayan form, although perhaps a distinct genus. Now not one of these birds, or anything allied to them, occurs in the Moluccas, or (with one exception) in Celebes or Australia; and as they are most of them birds of short flight, it is very difficult to conceive how or when they could have crossed the space of more than a thousand miles, which now separates them from their nearest allies. Such facts point to changes of land and sea on a large scale, and at a rate which, measured by the time required for a change of species, must be termed rapid. By speculating on such changes we may easily see how partial waves of immigration may have entered New Guinea, and how all trace of their passage may have been obliterated by the subsequent disappearance of the intervening land.

There is nothing that the study of geology teaches us that is more certain or more impressive than the extreme instability of the earth's surface. Everywhere beneath our feet we find proofs that what is land has been sea, and that where seas now spread out has once been land; and that this change from sea

---

[1] Nearly ninety species of pigeons are now known to inhabit New Guinea and the adjacent Papuan Islands, while the parrot tribe in the same area has increased to about eighty species. Nearly 900 species of Papuan birds are now known.

to land, and from land to sea, has taken place not once or twice
only, but again and again, during countless ages of past time.
Now the study of the distribution of animal life upon the
present surface of the earth causes us to look upon this constant interchange of land and sea—this making and unmaking
of continents, this elevation and disappearance of islands—as
a potent reality, which has always and everywhere been in progress, and has been the main agent in determining the manner
in which living things are now grouped and scattered over the
earth's surface. And when we continually come upon such
little anomalies of distribution as that just now described, we
find the only rational explanation of them in those repeated
elevations and depressions which have left their record in
mysterious, but still intelligible characters on the face of
organic nature.

The insects of New Guinea are less known than the birds,
but they seem almost equally remarkable for fine forms and
brilliant colours. The magnificent green and yellow Ornithopteræ are abundant, and have most probably spread westward
from this point as far as India. Among the smaller butterflies
are several peculiar genera of Nymphalidæ and Lycænidæ, remarkable for their large size, singular markings, or brilliant
coloration. The largest and most beautiful of the clear-winged
moths (Cocytia d'urvillei) is found here, as well as the large and
handsome green moth (Nyctalemon orontes). The beetles
furnish us with many species of large size, and of the most
brilliant metallic lustre, among which the Tmesisternus mira-
bilis, a longicorn beetle of a golden green colour; the excessively brilliant rose-chafers, Lomaptera wallacei and Anacamp-
torhina fulgida; one of the handsomest of the Buprestidæ,
Calodema wallacei ; and several fine blue weevils of the genus
Eupholus, are perhaps the most conspicuous. Almost all the
other orders furnish us with large or extraordinary forms. The
curious horned flies have already been mentioned ; and among
the Orthoptera the great shielded grasshoppers are the most
remarkable. The species here figured (Megalodon ensifer) has
the thorax covered by a large triangular horny shield, two and
a half inches long, with serrated edges, a somewhat wavy,
hollow surface, and a faint median line, so as very closely to
resemble a leaf. The glossy wing-coverts (when fully expanded,
more than nine inches across) are of a fine green colour and so
beautifully veined as to imitate closely some of the large shining tropical leaves. The body is short, and terminated in the
female by a long curved sword-like ovipositor (not seen in the
cut), and the legs are all long and strongly-spined. These
insects are sluggish in their motions, depending for safety on
their resemblance to foliage, their horny shield and wing-
coverts, and their spiny legs.

The large islands to the east of New Guinea are very little

known, but the occurrence of crimson lories, which are quite absent from Australia, and of cockatoos allied to those of New Guinea and the Moluccas, shows that they belong to the Papuan group ; and we are thus able to define the Malay Archipelago as extending eastward to the Solomon's Islands. New Caledonia and the New Hebrides, on the other hand, seem more nearly allied to Australia ; and the rest of the islands of the Pacific, though very poor in all forms of life, possess a few peculiarities which compel us to class them as a separate group. Although as a matter of convenience I have always separated the Moluccas as a distinct zoological group from New Guinea, I have at the same time pointed out that its fauna was chiefly derived from that island, just as that of Timor was chiefly derived from Australia. If we were dividing the

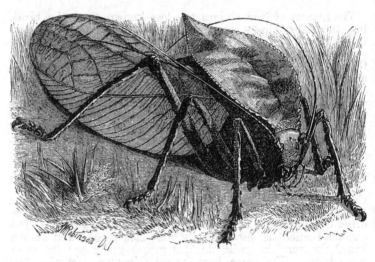

THE GREAT-SHIELDED GRASSHOPPER.

Australian region for zoological purposes alone, we should form three great groups : one comprising Australia, Timor, and Tasmania ; another New Guinea, with the islands from Bouru to the Solomon's group ; and the third comprising the greater part of the Pacific Islands.

The relation of the New Guinea fauna to that of Australia is very close. It is best marked in the Mammalia by the abundance of marsupials, and the almost complete absence of all other terrestrial forms. In birds it is less striking, although still very clear, for all the remarkable old-world forms which are absent from the one are equally so from the other, such as

Pheasants, Grouse, Vultures, and Woodpeckers ; while Cockatoos, Broad-tailed Parrots, Podargi, and the great families of the Honey-suckers and Brush-turkeys, with many others, comprising no less than twenty-four genera of land-birds, are common to both countries, and are entirely confined to them.

When we consider the wonderful dissimilarity of the two regions in all those physical conditions which were once supposed to determine the forms of life—Australia, with its open plains, stony deserts, dried up rivers, and changeable temperate climate ; New Guinea, with its luxuriant forests, uniformly hot, moist, and evergreen—this great similarity in their productions is almost astounding, and unmistakably points to a common origin. The resemblance is not nearly so strongly marked in insects, the reason obviously being, that this class of animals is much more immediately dependent on vegetation and climate than are the more highly organized birds and Mammalia. Insects also have far more effective means of distribution, and have spread widely into every district favourable to their development and increase. The giant Ornithopteræ have thus spread from New Guinea over the whole Archipelago, and as far as the base of the Himalayas ; while the elegant long-horned Anthribidæ have spread in the opposite direction, from Malacca to New Guinea, but owing to unfavourable conditions have not been able to establish themselves in Australia. That country, on the other hand, has developed a variety of flower-haunting Chafers and Buprestidæ, and numbers of large and curious terrestrial Weevils, scarcely any of which are adapted to the damp gloomy forests of New Guinea, where entirely different forms are to be found. There are, however, some groups of insects, constituting what appear to be the remains of the ancient population of the equatorial parts of the Australian region, which are still almost entirely confined to it. Such are the interesting sub-family of Longicorn coleoptera—Tmesisternitæ ; one of the best-marked genera of Buprestidæ—Cyphogastra ; and the beautiful weevils forming the genus Eupholus. Among butterflies we have the genera Mynes, Hypocista, and Elodina, and the curious eye-spotted Drusilla, of which last a single species is found in Java, but in no other of the western islands.

The facilities for the distribution of plants are still greater than they are for insects, and it is the opinion of eminent botanists that no such clearly-defined regions can be marked out in botany as in zoology. The causes which tend to diffusion are here most powerful, and have led to such intermingling of the floras of adjacent regions that none but broad and general divisions can now be detected. These remarks have an important bearing on the problem of dividing the surface of the earth into great regions, distinguished by the radical difference of their natural productions. Such difference we now know to be the direct result of long-continued separation by more or less

impassable barriers ; and as wide oceans and great contrasts of
temperature are the most complete barriers to the dispersal of
all terrestrial forms of life, the primary divisions of the earth
should in the main serve for all terrestrial organisms. However
various may be the effects of climate, however unequal the
means of distribution, these will never altogether obliterate the
radical effects of long-continued isolation ; and it is my firm
conviction, that when the botany and the entomology of New
Guinea and the surrounding islands become as well known as
are their mammals and birds, these departments of nature will
also plainly indicate the radical distinctions of the Indo-Malayan
and Austro-Malayan regions of the great Malay Archipelago.

CHAPTER XL.

THE RACES OF MAN IN THE MALAY ARCHIPELAGO.

I PROPOSE to conclude this account of my Eastern travels with
a short statement of my views as to the races of man which inhabit the various parts of the Archipelago, their chief physical
and mental characteristics, their affinities with each other and
with surrounding tribes, their migrations, and their probable
origin.

Two very strongly contrasted races inhabit the Archipelago
—the Malays, occupying almost exclusively the larger western
half of it, and the Papuans, whose headquarters are New Guinea
and several of the adjacent islands. Between these in locality
are found tribes who are also intermediate in their chief
characteristics, and it is sometimes a nice point to determine
whether they belong to one or the other race, or have been
formed by a mixture of the two.

The Malay is undoubtedly the most important of these two
races, as it is the one which is the most civilized, which has come
most into contact with Europeans, and which alone has any
place in history. What may be called the true Malay races, as
distinguished from others who have merely a Malay element in
their language, present a considerable uniformity of physical
and mental characteristics, while there are very great differences of civilization and of language. They consist of four
great, and a few minor semi-civilized tribes, and a number of
others who may be termed savages. The Malays proper inhabit
the Malay peninsula, and almost all the coast regions of Borneo
and Sumatra. They all speak the Malay language, or dialects
of it ; they write in the Arabic character, and are Mahometans
in religion. The Javanese inhabit Java, part of Sumatra,
Madura, Bali, and part of Lombock. They speak the Javanese
and Kawi languages, which they write in a native character.
They are now Mahometans in Java, but Brahmins in Bali and

Lombock. The Bugis are the inhabitants of the greater parts of Celebes, and there seems to be an allied people in Sumbawa. They speak the Bugis and Macassar languages, with dialects, and have two different native characters in which they write these. They are all Mahometans. The fourth great race is that of the Tagalas in the Philippine Islands, about whom, as I did not visit those islands, I shall say little. Many of them are now Christians, and speak Spanish as well as their native tongue, the Tagala. The Moluccan-Malays, who inhabit chiefly Ternate, Tidore, Batchian, and Amboyna, may be held to form a fifth division of semi-civilized Malays. They are all Mahometans, but they speak a variety of curious languages, which seem compounded of Bugis and Javanese with the languages of the savage tribes of the Moluccas.

The savage Malays are the Dyaks of Borneo ; the Battaks and other wild tribes of Sumatra ; the Jakuns of the Malay Peninsula ; the aborigines of Northern Celebes, of the Sula Islands, and of part of Bouru.

The colour of all these varied tribes is a light reddish brown, with more or less of an olive tinge, not varying in any important degree over an extent of country as large as all Southern Europe. The hair is equally constant, being invariably black and straight, and of a rather coarse texture, so that any lighter tint, or any wave or curl in it, is an almost certain proof of the admixture of some foreign blood. The face is nearly destitute of beard, and the breast and limbs are free from hair. The stature is tolerably equal, and is always considerably below that of the average European ; the body is robust, the breast well developed, the feet small, thick, and short, the hands small and rather delicate. The face is a little broad, and inclined to be flat ; the forehead is rather rounded, the brows low, the eyes black and very slightly, if at all, oblique ; the nose is rather small, not prominent, but straight and well-shaped, the apex a little rounded, the nostrils broad and slightly exposed ; the cheek-bones are rather prominent, the mouth large, the lips broad and well cut, but not protruding, the chin round and well-formed.

In this description there seems little to object to on the score of beauty, and yet on the whole the Malays are certainly not handsome. In youth, however, they are often very good-looking, and many of the boys and girls up to twelve or fifteen years of age are very pleasing, and some have countenances which are in their way almost perfect. I am inclined to think they lose much of their good looks by bad habits and irregular living. At a very early age they chew betel and tobacco almost incessantly ; they suffer much want and exposure in their fishing and other excursions ; their lives are often passed in alternate starvation and feasting, idleness and excessive labour—and this naturally produces premature old age and harshness of features.

In character the Malay is impassive. He exhibits a reserve, diffidence, and even bashfulness, which is in some degree attractive, and leads the observer to think that the ferocious and bloodthirsty character imputed to the race must be grossly exaggerated. He is not demonstrative. His feelings of surprise, admiration, or fear, are never openly manifested, and are probably not strongly felt. He is slow and deliberate in speech, and circuitous in introducing the subject he has come expressly to discuss. These are the main features of his moral nature, and exhibit themselves in every action of his life.

Children and women are timid, and scream and run at the unexpected sight of a European. In the company of men they are silent, and are generally quiet and obedient. When alone the Malay is taciturn; he neither talks nor sings to himself. When several are paddling in a canoe, they occasionally chant a monotonous and plaintive song. He is cautious of giving offence to his equals. He does not quarrel easily about money matters; dislikes asking too frequently even for payment of his just debts, and will often give them up altogether rather than quarrel with his debtor. Practical joking is utterly repugnant to his disposition; for he is particularly sensitive to breaches of etiquette, or any interference with the personal liberty of himself or another. As an example, I may mention that I have often found it very difficult to get one Malay servant to waken another. He will call as loud as he can, but will hardly touch, much less shake, his comrade. I have frequently had to waken a hard sleeper myself when on a land or sea journey.

The higher classes of Malays are exceedingly polite, and have all the quiet ease and dignity of the best-bred Europeans. Yet this is compatible with a reckless cruelty and contempt of human life, which is the dark side of their character. It is not to be wondered at, therefore, that different persons give totally opposite accounts of them—one praising them for their soberness, civility, and good-nature; another abusing them for their deceit, treachery, and cruelty. The old traveller, Nicolo Conti, writing in 1430, says: "The inhabitants of Java and Sumatra exceed every other people in cruelty. They regard killing a man as a mere jest; nor is any punishment allotted for such a deed. If any one purchase a new sword, and wish to try it, he will thrust it into the breast of the first person he meets. The passers-by examine the wound, and praise the skill of the person who inflicted it, if he thrust in the weapon direct." Yet Drake says of the south of Java: "The people (as are their kings) are a very loving, true, and just-dealing people;" and Mr. Crawfurd says that the Javanese, whom he knew thoroughly, are "a peaceable, docile, sober, simple, and industrious people." Barbosa, on the other hand, who saw them at Malacca about 1660, says: "They are a people of great ingenuity, very subtle in all their dealings; very malicious, great deceivers, seldom speaking the truth; prepared to do all manner of wickedness, and ready to sacrifice their lives."

The intellect of the Malay race seems rather deficient. They are incapable of anything beyond the simplest combinations of ideas, and have little taste or energy for the acquirement of knowledge. Their civilization, such as it is, does not seem to be indigenous, as it is entirely confined to those nations who have been converted to the Mahometan or Brahminical religions.

I will now give an equally brief sketch of the other great race of the Malay Archipelago, the Papuan.

The typical Papuan race is in many respects the very opposite of the Malay, and it has hitherto been very imperfectly described. The colour of the body is a deep sooty-brown or black, sometimes approaching, but never quite equalling, the jet-black of some negro races. It varies in tint, however, more than that of the Malay, and is sometimes a dusky brown. The hair is very peculiar, being harsh, dry, and frizzly, growing in little tufts or curls, which in youth are very short and compact, but afterwards grow out to a considerable length, forming the compact frizzled mop which is the Papuans' pride and glory. \ The face is adorned with a beard of the same frizzly nature as the hair of the head. The arms, legs, and breast are also more or less clothed with hair of a similar nature.

In stature the Papuan decidedly surpasses the Malay, and is perhaps equal, or even superior, to the average of Europeans The legs are long and thin, and the hands and feet larger than in the Malays. The face is somewhat elongated, the forehead flattish, the brows very prominent; the nose is large, rather arched and high, the base thick, the nostrils broad, with the aperture hidden, owing to the tip of the nose being elongated; the mouth is large, the lips thick and protuberant. The face has thus an altogether more European aspect than in the Malay, owing to the large nose; and the peculiar form of this organ, with the more prominent brows and the character of the hair on the head, face, and body, enable us at a glance to distinguish the two races. I have observed that most of these characteristic features are as distinctly visible in children of ten or twelve years old as in adults, and the peculiar form of the nose is always shown in the figures which they carve for ornaments to their houses, or as charms to wear round their necks.

PAPUAN CHARM.

The moral characteristics of the Papuan appear to me to separate him as distinctly from the Malay as do his form and features. He is impulsive and demonstrative in speech and action. His emotions and passions express themselves in shouts and laughter, in yells and frantic leapings. Women and children take their share in every discussion, and seem little alarmed at the sight of strangers and Europeans.

Of the intellect of this race it is very difficult to judge, but I am inclined to rate it somewhat higher than that of the Malays, notwithstanding the fact that the Papuans have never yet made any advance towards civilization. It must be remembered, however, that for centuries the Malays have been influenced by Hindoo, Chinese, and Arabic immigration, whereas the Papuan race has only been subjected to the very partial and local influence of Malay traders. The Papuan has much more vital energy, which would certainly greatly assist his intellectual development. Papuan slaves show no inferiority of intellect compared with Malays, but rather the contrary ; and in the Moluccas they are often promoted to places of considerable trust. The Papuan has a greater feeling for art than the Malay. He decorates his canoe, his house, and almost every domestic utensil with elaborate carving, a habit which is rarely found among tribes of the Malay race.

In the affections and moral sentiments, on the other hand, the Papuans seem very deficient. In the treatment of their children they are often violent and cruel ; whereas the Malays are almost invariably kind and gentle, hardly ever interfering at all with their children's pursuits and amusements, and giving them perfect liberty at whatever age they wish to claim it. But these very peaceful relations between parents and children are no doubt, in a great measure, due to the listless and apathetic character of the race, which never leads the younger members into serious opposition to the elders ; while the harsher discipline of the Papuans may be chiefly due to that greater vigour and energy of mind which always, sooner or later, leads to the rebellion of the weaker against the stronger—the people against their rulers, the slave against his master, or the child against its parent.

It appears, therefore, that, whether we consider their physical conformation, their moral characteristics, or their intellectual capacities, the Malay and Papuan races offer remarkable differences and striking contrasts. The Malay is of short stature, brown-skinned, straight-haired, beardless, and smooth-bodied. The Papuan is taller, is black-skinned, frizzly-haired, bearded, and hairy-bodied. The former is broad-faced, has a small nose, and flat eyebrows ; the latter is long-faced, has a large and prominent nose, and projecting eyebrows. The Malay is bashful, cold, undemonstrative, and quiet ; the Papuan is bold, impetuous, excitable, and noisy. The former is grave and seldom laughs ; the latter is joyous and laughter-loving—the one conceals his emotions, the other displays them.

Having thus described in some detail the great physical,

intellectual, and moral differences between the Malays and Papuans, we have to consider the inhabitants of the numerous islands which do not agree very closely with either of these races. The islands of Obi, Batchian, and the three southern peninsulas of Gilolo, possess no true indigenous population; but the northern peninsula is inhabited by a native race, the so-called Alfuros of Sahoe and Galela. These people are quite distinct from the Malays, and almost equally so from the Papuans. They are tall and well-made, with Papuan features, and curly hair; they are bearded and hairy-limbed, but quite as light in colour as the Malays. They are an industrious and enterprising race, cultivating rice and vegetables, and indefatigable in their search after game, fish, tripang, pearls, and tortoiseshell.

In the great island of Ceram there is also an indigenous race very similar to that of Northern Gilolo. Bouru seems to contain two distinct races—a shorter, round-faced people, with a Malay physiognomy, who may probably have come from Celebes by way of the Sula Islands; and a taller bearded race, resembling that of Ceram.

Far south of the Moluccas lies the island of Timor, inhabited by tribes much nearer to the true Papuan than those of the Moluccas.

The Timorese of the interior are dusky brown or blackish, with bushy frizzled hair, and the long Papuan nose. They are of medium height, and rather slender figures. The universal dress is a long cloth twisted round the waist, the fringed ends of which hang below the knee. The people are said to be great thieves, and the tribes are always at war with each other, but they are not very courageous or bloodthirsty. The custom of "tabu," called here "pomáli," is very general, fruit trees, houses, crops, and property of all kinds being protected from depredation by this ceremony, the reverence for which is very great. A palm branch stuck across an open door, showing that the house is tabooed, is a more effectual guard against robbery than any amount of locks and bars. The houses in Timor are different from those of most of the other islands; they seem all roof, the thatch overhanging the low walls and reaching the ground, except where it is cut away for an entrance. In some parts of the west end of Timor, and on the little island of Semau, the houses more resemble those of the Hottentots, being egg-shaped, very small, and with a door only about three feet high. These are built on the ground, while those of the eastern districts are raised a few feet on posts. In their excitable disposition, loud voices, and fearless demeanour, the Timorese closely resemble the people of New Guinea.

In the islands west of Timor, as far as Flores and Sandalwood Island, a very similar race is found, which also extends eastward to Timor-laut, where the true Papuan race begins to appear. The small islands of Savu and Rotti, however, to the west of Timor, are very remarkable in possessing a different and,

in some respects, peculiar race. These people are very handsome, with good features, resembling in many characteristics
the race produced by the mixture of the Hindoo or Arab with
the Malay. They are certainly distinct from the Timorese or
Papuan races, and must be classed in the western rather than
the eastern ethnological division of the Archipelago.

The whole of the great island of New Guinea, the Ké and
Aru Islands, with Mysol, Salwatty, and Waigiou, are inhabited
almost exclusively by the typical Papuans. I found no trace of
any other tribes inhabiting the interior of New Guinea, but the
coast people are in some places mixed with the browner races of
the Moluccas. The same Papuan race seems to extend over the
islands east of New Guinea as far as the Fijis.[1]

There remain to be noticed the black woolly-haired races of
the Philippines and the Malay peninsula, the former called
"Negritos," and the latter "Semangs." I have never seen these
people myself, but from the numerous accurate descriptions of
them that have been published, I have had no difficulty in
satisfying myself that they have little affinity or resemblance
to the Papuans, with which they have been hitherto associated.
In most important characters they differ more from the Papuan
than they do from the Malay. They are dwarfs in stature, only
averaging four feet six inches to four feet eight inches high, or
eight inches less than the Malays ; whereas the Papuans are
decidedly taller than the Malays. The nose is invariably represented as small, flattened, or turned up at the apex, whereas the
most universal character of the Papuan race is to have the nose
prominent and large, with the apex produced downwards, as it
is invariably represented in their own rude idols. The hair of
these dwarfish races agrees with that of the Papuans, but so it
does with that of the negroes of Africa. The Negritos and the
Semangs agree very closely in physical characteristics with
each other and with the Andaman islanders, while they differ
in a marked manner from every Papuan race.

A careful study of these varied races, comparing them with
those of Eastern Asia, the Pacific Islands, and Australia, has
led me to adopt a comparatively simple view as to their origin
and affinities.

If we draw a line (see Physical Map, p. 8), commencing to
the east of the Philippine Islands, thence along the western
coast of Gilolo, through the island of Bouru, and curving round
the west end of Flores, then bending back by Sandalwood
Island to take in Rotti, we shall divide the Archipelago into
two portions, the races of which have strongly marked distinctive peculiarities. This line will separate the Malayan and
all the Asiatic races from the Papuans and all that inhabit the
Pacific ; and though along the line of junction intermigration

---

[1] In the south-east peninsula of New Guinea are found some undoubtedly Polynesian
tribes, called Motu. These have probably settled here at an early period, and have mixed
more or less with the native Papuans.

and commixture have taken place, yet the division is on the whole almost as well defined and strongly contrasted, as is the corresponding zoological division of the Archipelago, into an Indo-Malayan and Austro-Malayan region.

I must briefly explain the reasons that have led me to consider this division of the Oceanic races to be a true and natural one. The Malayan race, as a whole, undoubtedly very closely resembles the East Asian populations, from Siam to Mandchouria. I was much struck with this, when in the island of Bali I saw Chinese traders who had adopted the costume of that country, and who could then hardly be distinguished from Malays ; and, on the other hand, I have seen natives of Java who, as far as physiognomy was concerned, would pass very well for Chinese. Then, again, we have the most typical of the Malayan tribes inhabiting a portion of the Asiatic continent itself, together with those great islands which, possessing the same species of large Mammalia with the adjacent parts of the continent, have in all probability formed a connected portion of Asia during the human period. The Negritos are, no doubt, quite a distinct race from the Malay ; but yet, as some of them inhabit a portion of the continent, and others the Andaman Islands in the Bay of Bengal, they must be considered to have had, in all probability, an Asiatic rather than a Polynesian origin.

Now, turning to the eastern parts of the Archipelago, I find, by comparing my own observations with those of the most trustworthy travellers and missionaries, that a race identical in all its chief features with the Papuan, is found in all the islands as far east as the Fijis ; beyond this the brown Polynesian race, or some intermediate type, is spread everywhere over the Pacific. The descriptions of these latter often agree exactly with the characters of the brown indigenes of Gilolo and Ceram.

It is to be especially remarked that the brown and the black Polynesian races closely resemble each other. Their features are almost identical, so that portraits of a New Zealander or Otaheitan will often serve accurately to represent a Papuan or Timorese, the darker colour and more frizzly hair of the latter being the only differences. They are both tall races. They agree in their love of art and the style of their decorations. They are energetic, demonstrative, joyous, and laughter-loving, and in all these particulars they differ widely from the Malay.

I believe, therefore, that the numerous intermediate forms that occur among the countless islands of the Pacific are not merely the result of a mixture of these races, but are, to some extent, truly intermediate or transitional ; and that the brown and the black, the Papuan, the natives of Gilolo and Ceram, the Fijian, the inhabitants of the Sandwich Islands and those of New Zealand, are all varying forms of one great Oceanic or Polynesian race.

It is, however, quite possible, and perhaps probable, that the brown Polynesians were originally the produce of a mixture of Malays, or some lighter coloured Mongol race with the dark Papuans; but if so, the intermingling took place at such a remote epoch, and has been so assisted by the continued influence of physical conditions and of natural selection, leading to the preservation of a special type suited to those conditions, that it has become a fixed and stable race with no signs of mongrelism, and showing such a decided preponderance of Papuan character, that it can best be classified as a modification of the Papuan type. The occurrence of a decided Malay element in the Polynesian languages has evidently nothing to do with any such ancient physical connexion. It is altogether a recent phenomenon, originating in the roaming habits of the chief Malay tribes; and this is proved by the fact that we find actual modern words of the Malay and Javanese languages in use in Polynesia, so little disguised by peculiarities of pronunciation as to be easily recognizable—not mere Malay roots only to be detected by the elaborate researches of the philologist, as would certainly have been the case had their introduction been as remote as the origin of a very distinct race—a race as different from the Malay in mental and moral, as it is in physical characters.

As bearing upon this question it is important to point out the harmony which exists between the line of separation of the human races of the Archipelago and that of the animal productions of the same country, which I have already so fully explained and illustrated. The dividing lines do not, it is true, exactly agree; but I think it is a remarkable fact, and something more than a mere coincidence, that they should traverse the same district and approach each other so closely as they do. If, however, I am right in my supposition that the region where the dividing line of the Indo-Malayan and Austro-Malayan regions of zoology can now be drawn, was formerly occupied by a much wider sea than at present, and if man existed on the earth at that period, we shall see good reason why the races inhabiting the Asiatic and Pacific areas should now meet and partially intermingle in the vicinity of that dividing line.

It has recently been maintained by Professor Huxley that the Papuans are more closely allied to the negroes of Africa than to any other race. The resemblance both in physical and mental characteristics had often struck myself, but the difficulties in the way of accepting it as probable or possible have hitherto prevented me from giving full weight to those resemblances. Geographical, zoological, and ethnological considerations render it almost certain that, if these two races ever had a common origin, it could only have been at a period far more remote than any which has yet been assigned to the antiquity of the human race. And even if their unity could be proved, it would in no way affect my argument for the close affinity of the Papuan and

Polynesian races, and the radical distinctness of both from the Malay.

Polynesia is pre-eminently an area of subsidence, and its great wide-spread groups of coral-reefs mark out the position of former lands and islands. The rich and varied, yet strangely isolated productions of Australia and New Guinea, also indicate an extensive land-area where such specialized forms were developed. The races of men now inhabiting these countries are, therefore, most probably the descendants of the races which inhabited these continents and islands. This is the most simple and natural supposition to make. And if we find any signs of direct affinity between the inhabitants of any other part of the world and those of Polynesia, it by no means follows that the latter were derived from the former. It is undoubtedly true that there are proofs of extensive migrations among the Pacific islands, which have led to community of language from the Sandwich group to New Zealand; but there are no proofs whatever of recent migration from any surrounding country to Polynesia, since there is no people to be found elsewhere sufficiently resembling the Polynesian race in their chief physical and mental characteristics.

If the past history of these varied races is obscure and uncertain, the future is no less so. The true Polynesians, inhabiting the farthest isles of the Pacific, are no doubt doomed to an early extinction. But the more numerous Malay race seems well adapted to survive as the cultivator of the soil, even when his country and government have passed into the hands of Europeans. If the tide of colonization should be turned to New Guinea, there can be little doubt of the early extinction of the Papuan race. A warlike and energetic people, who will not submit to national slavery or to domestic servitude, must disappear before the white man as surely as do the wolf and the tiger.

I have now concluded my task. I have given, in more or less detail, a sketch of my eight years' wanderings among the largest and the most luxuriant islands which adorn our earth's surface. I have endeavoured to convey my impressions of their scenery, their vegetation, their animal productions, and their human inhabitants. I have dwelt at some length on the varied and interesting problems they offer to the student of nature. Before bidding my readers farewell, I wish to make a few observations on a subject of yet higher interest and deeper importance, which the contemplation of savage life has suggested, and on which I believe that the civilized can learn something from the savage man.

We most of us believe that we, the higher races, have progressed and are progressing. If so, there must be some state of perfection, some ultimate goal, which we may never reach, but to which all true progress must bring us nearer. What is this ideally perfect social state towards which mankind ever has been, and still is tending ? Our best thinkers maintain that it is a state of individual freedom and self-government, rendered possible by the equal development and just balance of the intellectual, moral, and physical parts of our nature,—a state in which we shall each be so perfectly fitted for a social existence, by knowing what is right, and at the same time feeling an irresistible impulse to do what we know to be right, that all laws and all punishments shall be unnecessary. In such a state every man would have a sufficiently well-balanced intellectual organization to understand the moral law in all its details, and would require no other motive but the free impulses of his own nature to obey that law.

Now it is very remarkable that among people in a very low stage of civilization we find some approach to such a perfect social state. I have lived with communities of savages in South America and in the East, who have no laws or law courts but the public opinion of the village freely expressed. Each man scrupulously respects the rights of his fellow, and any infraction of those rights rarely or never takes place. In such a community, all are nearly equal. There are none of those wide distinctions, of education and ignorance, wealth and poverty, master and servant, which are the product of our civilization ; there is none of that wide-spread division of labour, which, while it increases wealth, produces also conflicting interests ; there is not that severe competition and struggle for existence, or for wealth, which the dense population of civilized countries inevitably creates. All incitements to great crimes are thus wanting, and petty ones are repressed, partly by the influence of public opinion, but chiefly by that natural sense of justice and of his neighbour's right which seems to be, in some degree, inherent in every race of man.

Now, although we have progressed vastly beyond the savage state in intellectual achievements, we have not advanced equally in morals. It is true that among those classes who have no wants that cannot be easily supplied, and among whom public opinion has great influence, the rights of others are fully respected. It is true, also, that we have vastly extended the sphere of those rights, and include within them all the brotherhood of man. But it is not too much to say, that the mass of our populations have not at all advanced beyond the savage code of morals, and have in many cases sunk below it. A deficient morality is the great blot of modern civilization, and the greatest hindrance to true progress.

During the last century, and especially in the last thirty years, our intellectual and material advancement has been too quickly achieved for us to reap the full benefit of it. Our mastery over the forces of nature has led to a rapid growth of population, and a vast accumulation of wealth ; but these have brought with them such an amount of poverty and crime, and have fostered the growth of so much sordid feeling and so many fierce passions, that it may well be questioned, whether the mental and moral status of our population has not on the average been lowered, and whether the evil has not overbalanced the good. Compared with our wondrous progress in physical science and its practical applications, our system of government, of administering justice, of national education, and our whole social and moral organization, remains in a state of barbarism.[1] And if we continue to devote our chief energies to the utilizing of our knowledge of the laws of nature with the view of still further extending our commerce and our wealth, the evils which necessarily accompany these when too eagerly pursued, may increase to such gigantic dimensions as to be beyond our power to alleviate.

We should now clearly recognize the fact, that the wealth and knowledge and culture of *the few* do not constitute civilization, and do not of themselves advance us towards the "perfect social state." Our vast manufacturing system, our gigantic commerce, our crowded towns and cities, support and continually renew a mass of human misery and crime *absolutely* greater than has ever existed before. They create and maintain in life-long labour an ever-increasing army, whose lot is the more hard to bear by contrast with the pleasures, the comforts, and the luxury which they see everywhere around them, but which they can never hope to enjoy; and who, in this respect, are worse off than the savage in the midst of his tribe.

This is not a result to boast of, or to be satisfied with; and, until there is a more general recognition of this failure of our civilization—resulting mainly from our neglect to train and develop more thoroughly the sympathetic feelings and moral faculties of our nature, and to allow them a larger share of influence in our legislation, our commerce, and our whole social organization—we shall never, as regards the whole community, attain to any real or important superiority over the better class of savages.

This is the lesson I have been taught by my observations of uncivilized man. I now bid my readers—Farewell!

NOTE.

THOSE who believe that our social condition approaches perfection will think the above word harsh and exaggerated, but it seems to me the only word that can be truly applied to us. We are the richest country in the world, and yet nearly one-twentieth of our population are parish paupers, and one-thirtieth known criminals. Add to these, the criminals who escape detection, and the poor who live mainly, or partly, on private charity (which, according to Dr. Hawkesley, expends seven millions sterling annually in London alone), and we may be sure that more than

[1] See note.

ONE-TENTH of our population are actually Paupers and Criminals. Both these classes we keep idle or at unproductive labour, and each criminal costs us annually in our prisons more than the wages of an honest agricultural labourer. We allow over a hundred thousand persons known to have no means of subsistence but by crime, to remain at large and prey upon the community, and many thousand children to grow up before our eyes in ignorance and vice, to supply trained criminals for the next generation. This, in a country which boasts of its rapid increase in wealth, of its enormous commerce and gigantic manufactures, of its mechanical skill and scientific knowledge, of its high civilization and its pure Christianity, —I can but term a state of social barbarism. We also boast of our love of justice, and that the law protects rich and poor alike, yet we retain money fines as a punishment, and make the very first steps to obtain justice a matter of expense—in both cases a barbarous injustice, or denial of justice to the poor. Again, our laws render it possible, that, by mere neglect of a legal form, and contrary to his own wish and intention, a man's property may all go to a stranger, and his own children be left destitute. Such cases have happened through the operation of the laws of inheritance of landed property; and that such unnatural injustice is possible among us, shows that we are in a state of social barbarism. One more example to justify my use of the term, and I have done. We permit absolute possession of the soil of our country, with no legal rights of existence on the soil to the vast majority who do not possess it. A great landholder may legally convert his whole property into a forest or a hunting-ground, and expel every human being who has hitherto lived upon it. In a thickly-populated country like England, where every acre has its owner and its occupier, this is a power of legally destroying his fellow-creatures; and that such a power should exist, and be exercised by individuals, in however small a degree indicates that, as regards true social science, we are still in a state of barbarism.

APPENDIX.

ON THE CRANIA AND THE LANGUAGES OF THE RACES OF MAN IN THE MALAY ARCHIPELAGO.

CRANIA.

A FEW years ago it was thought that the study of Crania offered the only sure basis of a classification of man. Immense collections have been formed; they have been measured, described, and figured; and now the opinion is beginning to gain ground, that for this special purpose they are of very little value. Professor Huxley has boldly stated his views to this effect; and in a proposed new classification of mankind has given scarcely any weight to characters derived from the cranium. It is certain, too, that though Cranioscopy has been assiduously studied for many years, it has produced no results at all comparable with the labour and research bestowed upon it. No approach to a theory of the excessive variations of the cranium has been put forth, and no intelligible classification of races has been founded upon it.

Dr. Joseph Barnard Davis, who has assiduously collected human crania for many years, has just published a remarkable work, entitled *Thesaurus Craniorum*. This is a catalogue of his collection (by far the most extensive in existence), classified according to countries and races, indicating the derivation and any special characteristics of each specimen; and by way of description, an elaborate series of measurements, nineteen in number when complete, by which accurate comparisons can be made, and the limits of variation determined.

This interesting and valuable work offered me the means of determining for myself whether the forms and dimensions of the crania of the Eastern races would in any way support or refute my classification of them. For the purposes of comparison, the whole series of nineteen measurements would have been far too cumbersome. I therefore selected three, which seem to me well adapted to test the capabilities of Cranioscopy for the purpose in view. These are :—1. The capacity of the cranium. 2. The proportion of the width to the length taken as 100. 3. The proportion of the height to the length taken as 100. These dimensions are given by Dr. Davis in almost every case, and have furnished me with ample materials. I first took the "*means*" of groups of crania of the same race from distinct localities, as given by Dr. Davis himself, and thought I could detect differences characteristic of the great divisions of the Malayans and Papuans ; but some anomalies induced me to look at the amount of individual variation, and this was so enormous that I became at once convinced that even this large collection could furnish no trustworthy average. I will now give a few examples of these variations, using the terms—Capacity, W: L, H: L, for the three dimensions compared. In the Capacity, I always compare only male crania, so as not to introduce the sexual difference of size. In the other proportionate dimensions, I use both sexes to get a larger average, as I find these proportions do not vary definitely according to sex, the two extremes often occurring in the series of male specimens only.

MALAYS.—Thirteen male Sumatra crania had :—Capacity, from 61·5 to 87 ounces of sand ; W : L, ·71 to ·86 ; H : L, ·73 to ·85. Ten male Celebes crania varied thus :—Capacity, from 67 to 83 ; W : L, ·73 to ·92 ; H : L, ·76 to ·90.

In the whole series of eighty-six Malay skulls from Sumatra, Java, Madura, Borneo, and Celebes, the variation is enormous. Capacity (66 skulls), 60 to 91 ounces of sand ; W : L, ·70 to ·92 ; H : L, ·72 to ·90. And these extremes are not isolated abnormal specimens, but there is a regular gradation up to them, which always becomes more perfect the larger the number of specimens compared. Thus, besides the extreme Dolicocephalic skull (70) in the supposed Brachycephalic Malay group, there are others which have W : L, ·71, ·72 and ·73, so that we have every reason to believe that with more specimens we should get a still narrower form of skull. So the very large cranium, 91 ounces, is led up to by others of 87 and 88.

The largest, in an extensive series of English, Scotch, and Irish crania, was only 92·5 ounces.

PAPUANS.—There are only four true Papuan crania in the collection, and these vary considerably (W : L, ·72 to ·83). Taking, however, the natives of the Solomon Islands, New Caledonia, New Hebrides, and the Fijis as being all decidedly of Papuan race, we have a series of 28 crania (23 male), and these give us :—Capacity, 66 to 80 ; W : L, ·65 to ·85 ; H : L, ·71 to ·85 ; so nearly identical with some of the Malayan groups as to offer no clear points of difference.

The Polynesians, the Australians, and the African negroes offer equally wide ranges of variation, as will be seen by the following summary of the dimensions of the crania of these races and the preceding :—

Number of Crania.	CAPACITY.	W : L.	H : L.
83. Malays (66 male) .	60 to 91	·70 to ·92	·72 to ·90
28. Papuans (23 m.) .	66 ,, 80	·65 ,, ·85	·71 ,, ·85
156. Polynesians (90 m.)	62 ,, 91	·69 ,, ·90	·68 ,, ·88
23. Australians (16 m.)	59 ,, 86	·57 ,, ·80	·64 ,, ·80
72. Negroes (38 m.). .	66 ,, 87	·64 ,, ·83	·65 ,, ·81

The only conclusions that we can draw from this table are, that the Australians have the smallest crania, and the Polynesians the largest ; the Negroes, the Malays, and Papuans not differing perceptibly in size. And this accords very well with what we know of their mental activity and capacity for civilization.

The Australians have the *longest* skulls ; after which come the Negroes ; then the Papuans, the Polynesians, and the Malays.

The Australians have also the *lowest* skulls ; then the Negroes ; the Polynesians and Papuans considerably higher and equal, and the Malay the highest.

It seems probable, therefore, that if we had a much more extensive series of crania the averages might furnish tolerably reliable race-characters, although, owing to the large amount of individual variation, they would never be of any use in single examples, or even when moderate numbers only could be compared.

So far as this series goes, it seems to agree well with the conclusions I have arrived at, from physical and mental characters observed by myself. These conclusions briefly are : that the Malays and Papuans are radically distinct races ; and that the Polynesians are most nearly allied to the latter, although they have probably some admixture of Malayan or Mongolian blood.

LANGUAGES.

During my travels among the islands of the Archipelago, I collected a considerable number of vocabularies, in districts hitherto little visited. These represent about fifty-seven distinct languages (not including the common Malay and Javanese), more than half of which I believe are quite unknown to philologists, while only a few scattered words have been recorded of some others. Unfortunately, nearly half the number have been lost. Some years ago I lent the whole series to the late Mr. John Crawford, and having neglected to apply for them for some months, I found that he had in the meantime changed his residence, and that the books, containing twenty-five of the vocabularies, had been mislaid ; and they have never since been recovered. Being merely old and much battered copy-books, they probably found their way to the dust-heap along with other waste paper. I had previously copied out nine common words in the whole series of languages, and these are here given, as well as the remaining thirty-one vocabularies in full.

Having before had experience of the difficulty of satisfactorily determining any words but nouns and a few of the commonest adjectives, where the people are complete savages and the language of communication but imperfectly known, I selected about a hundred and twenty words, and have adhered to them throughout as far as practicable. After the English, I give the Malay word for comparison with the other languages. In orthography I adopt generally the continental mode of sounding the vowels, with a few modifications, thus :—

English .... a e i *or* ie ei o ŭ ū
Sounded .... ah a ee i o é *or* éh oo

These sounds come out most prominently at the end of a syllable ; when followed by a consonant the sounds are very little different from the usual pronunciation. Thus, "Api" is pronounced *Appee*, while "Minta" is pronounced *Mintah*. The short ŭ is pronounced like *er* in English, but without any trace of the guttural. Long, short, and accented syllables are marked in the usual way. The languages are grouped geographically, passing from west to east; those from the same or adjacent islands being as much as possible kept together.

I profess to be able to draw very few conclusions from these vocabularies. I believe that the languages have been so much modified by long intercommunication among the islands, that resemblances of words are no proof of affinity of the people who use those words. Many of the wide-spread similarities can be traced to organic onomatopœia. Such are the prevalence of *g* (hard), *ng*, *ni*, in words meaning "tooth"; of *l* and *m* in those for "tongue"; of *nge*, *ung*, *sno*, in those for "nose." Others are plainly commercial words, as "salaka" and "ringgit" (the Malay word for dollar) for silver, and "mas" for gold. The Papuan group of languages appear to be distinguished by harsher combinations of letters, and by monosyllabic words ending in a consonant, which are rarely or never found in the Malay group. Some of the tribes who are decidedly of Malay race, as the people of Ternate, Tidore, and Batchian, speak languages which are as decidedly of a Papuan type; and this, I believe, arises from their having originally immigrated to these islands in small numbers, and by marrying native women acquired a considerable portion of their language, which later arrivals of Malays were obliged to learn and adopt if they settled in the country. As I have hardly mentioned in my narrative some of the names of the tribes whose languages are here given, I will now give a list of them, with such explanatory remarks as I may think useful to the ethnologist, and then leave the vocabularies to speak for themselves.

LIST OF VOCABULARIES COLLECTED.

*Those marked * are lost.*

1. **Malay.**—The common colloquial Malay as spoken in Singapore; written in the Arabic character.

2. **Javanese.**—Low or colloquial Javanese as spoken in Java; written in a native character.

*3. **Sassak.**—Spoken by the indigenes of Lombock, who are Mahometans, and of a pure Malay race.

*4. **Macassar.**—Spoken in the district of Southern Celebes, near Macassar; written in a native character. Mahometans.

*5. **Bugis.**—Spoken over a large part of Southern Celebes; written in a native character distinct from that of Macassar. Mahometans.

6. **Bouton.**—Spoken in Boutong, a large island south of Celebes. Mahometans.

7. **Salayer.**—Spoken in Salayer, a smaller island south of Celebes. Mahometans.

*8. **Tomore.**—Spoken in the eastern peninsula of Celebes, and in Batchian, by emigrants who have settled there. Pagans.

Note.—The people who speak these five languages of Celebes are of pure Malayan type, and (all but the last) are equal in civilization to the true Malays.

*9. **Tomohon;** *10. **Langowen.**—Villages on the plateau of Minahasa.

*11. **Ratahan;** *12. **Belang.**—Villages near the south-east coast of Minahasa. *13. **Tanawanko.**—On the west coast. *14. **Kema.**—On the east coast. *15. **Bantek.**—A suburb of Menado.

16. **Menado.**—The chief town. 17. **Bolang-hitam.**—A village on the north-west coast, between Menado and Licoupang.

These nine languages, with many others, are spoken in the north-west peninsula of Celebes, by the people called Alfuros, who are of Malay race, and seem to have affinities with the Tagalas of the Philippines through the Sanguir islanders. These languages are falling into disuse, and Malay is becoming the universal means of communication. Most of the people are being converted to Christianity.

18. **Sanguir Islands** and **Siau**—Two groups of islands between Celebes and the Philippines. The inhabitants wear a peculiar costume, consisting of a loose cotton gown hanging from the neck nearly to the feet. They resemble, physically, the people of Menado.

19. **Salibabo Islands,** also called **Talaut.**—This vocabulary was given me from memory by Captain Van der Beck. See page 269.

20. **Sula Islands.**—These are situated east of Celebes, and their inhabitants seem to be Malays of the Moluccan type, and are Mahometans.

21. **Cajeli;** 22. **Wayapo;** 23. **Massaratty.**—These are three villages on the eastern side of Bouru. The people are allied to the natives of Ceram. Those of Cajeli itself are Mahometans.

24. **Amblau.**—An island a little south-east of Bouru. Mahometans.

*25. **Ternate.**—The northernmost island of the Moluccas. The inhabitants are Mahometans of Malay race, but somewhat mixed with the indigenes of Gilolo.

26. **Tidore.**—The next island of the Moluccas. The inhabitants are undistinguishable from those of Ternate.

*27. **Kaioa Islands.**—A small group north of Batchian.

*28. **Batchian.**—Inhabitants like the preceding. Mahometans, and of a similar Malay type.

29. **Gani.**—A village on the south peninsula of Gilolo. Inhabitants, Moluccan-Malays, and Mahometans.

*30. **Sahoe**; 31. **Galela.**—Villages of Northern Gilolo. The inhabitants are called Alfuros. They are indigenes of Polynesian type, with brown skins, but Papuan hair and features. Pagans.

32. **Liang.**—A village on the north coast of Amboyna. Several other villages near speak the same language. They are Mahometans, or Christians, and seem to be of mixed Malay and Polynesian type.

33. **Morella** and **Mamalla.**—Villages in North-West Amboyna. The inhabitants are Mahometans.

34. **Batu-merah.**—A suburb of Amboyna. Inhabitants Mahometans, and of Moluccan-Malay type.

35. **Lariki, Asilulu, Wakasiho.**—Villages in West Amboyna inhabited by Mahometans, who are reported to have come originally from Ternate.

36. **Saparua.**—An island east of Amboyna. Inhabitants of the brown Polynesian type, and speaking the same language as those on the coast of Ceram opposite.

37. **Awaiya**; 38. **Camarian.**—Villages on the south coast of Ceram. Inhabitants indigenes of Polynesian type, now Christians.

39. **Teluti** and **Hoya**; 40. **Ahtiago** and **Tobo.**—Villages on the south coast of Ceram. Inhabitants Mahometans, of mixed brown Papuan or Polynesian and Malay type.

41. **Ahtiago.**—Alfuros or indigenes inland from this village. Pagans, of Polynesian or brown Papuan type.

42. **Gah.**—Alfuros of East Ceram.

43. **Wahai.**—Inhabitants of much of the north coast of Ceram. Mahometans of mixed race. Speak several dialects of this language.

*44. **Goram.**—Small islands east of Ceram. Inhabitants of mixed race, and Mahometans.

45. **Matabello.**—Small islands south-east of Goram. Inhabitants of brown Papuan or Polynesian type. Pagans.

46. **Teor.**—A small island south-east of Matabello. Inhabitants a tall race of brown Papuans. Pagans.

*47. **Ké Islands.**—A small group west of the Aru Islands. Inhabitants true black Papuans. Pagans.

*48. **Aru Islands.**—A group west of New Guinea. Inhabitants true Papuans. Pagans.

49. **Mysol** (coast).—An island north of Ceram. Inhabitants Papuans with mixture of Moluccan Malays. Semi-civilized.

50. **Mysol** (interior).—Inhabitants true Papuans. Savages.

*51. **Dorey.**—North coast of New Guinea. Inhabitants true Papuans. Pagans.

*52. **Teto**; *53. **Vaiqueno, East Timor**; *54. **Brissi, West Timor.**—Inhabitants somewhat intermediate between the true and the brown Papuans. Pagans.

*55. **Savu**; *56. **Rotti.**—Islands west of Timor. Inhabitants of mixed race, with apparently much of the Hindoo type.

*57. **Allor**; *58. **Solor.**—Islands between Flores and Timor. Inhabitants of dark Papuan type.

59. **Bajau,** or **Sea Gipsies.**—A roaming tribe of fishermen of Malayan type, to be met with in all parts of the Archipelago.

NINE WORDS IN FIFTY-NINE LANGUAGES

	English	Group	BLACK.	FIRE.	LARGE.	NOSE.
1	Malay		Itam	Api	Bŭsar	Idong
2.	Javanese		Iran	Gūni	Gedé	Irong
3.	Sasak (Lombock)		Bidan	Api	Ble	Idong
4.	Macassar	S. Celebes.	Leling	Pepi	Lompo	Kamūrong
5.	Bugis	S. Celebes.	Malotong	Api	Marája	Ingok
6.	Bouton	S. Celebes.	Amáita	Whá	Monghi	Oánu
7.	Salayer	S. Celebes.	Hitam	Api	Bakéh	Kumor
8.	Tomóre	S. Celebes.	Moito	Api	Owhosi	Hengénto
9.	Tomohon	North Celebes.	Rūmdum	Api	Tuwón	Ngerun
10.	Langowan	North Celebes.	Wūlin	Api	Wanko	Ngilung
11.	Ratahan	North Celebes.	Mahítum	Pūtong	Loben	Irun
12.	Belang	North Celebes.	Mūhónde	Sūlu	Musolah	Niyun
13.	Tanawanko	North Celebes.	Rūmdum	Api	Súla	Ngerun
14.	Kema	North Celebes.	Hirun	Api	Súla	Ngerun
15.	Bantek	North Celebes.	Maitung	Pūtung	Ramoh	Idung
16.	Menado	North Celebes.	Maitung	Pūtung	Raboh	Idong
17.	Bolang Itam	North Celebes.	Moitomo	Pūro	Morokaro	Djunga
18.	Sanguir Is.		Maítum	Pūtun	Labo	Hirong
19.	Salibabo Is.		Maitu	Pūton	Bagewa	
20.	Sula Is.		Miti	Api	Ea	Ne
21.	Cajeli	Bouru	Metan	Ahú	Lehai	Nem
22.	Wayapo	Bouru	Miti	Bána	Bagut	Nien
23.	Massaratty	Bouru	Miti	Bána	Haat	Nieni
24.	Amblaw		Kameichei	Afu	Plaré	Neinya téha
25.	Ternate		Kokotu	Uku	Lamu lamu	Nunu
26.	Tidore		Kokótu	Uku	Lamu	Un
27.	Kaióa Is.		Kūda•	Lūtan	Lol	Usnod
28.	Batchian		Ngóa	Api	Rá	Hidom
29.	Gani	Gilolo	Kitkudu	Lūtan	Talalólo	Usnut
30.	Sahoe	Gilolo	Kokótu	Uhuh	Lamu	Ngūnu
31.	Galela	Gilolo	Tatataro	Uku	Elamo	Ngūno
32.	Liang	Amboyna.	Méte	Aów	Nila	Hiruka
33.	Morella	Amboyna.	Méte	Aów	Hella	Iuka
34.	Batu-merah	Amboyna.	Meteni	Aow	Enda-á	Ninura
35.	Lariki, &c.	Amboyna.	Méte	Aow	Era	Iru
36.	Saparua		Meteh	Háo	Ilahil	Iri
37.	Awaiya	Ceram.	Meténi	Aousa	Iláhe	Nua-mo
38.	Camarian	Ceram.	Méti	Hao	Eräámei	Hili-mo
39.	Teluti	Ceram.	Méte	Yafo	Elau	Olicolo
40.	Ahtiago (Mah.)	Ceram.	Memétan	Yaf	Aíyuk	Iiin
41.	Ahtiago (Alf.)	Ceram.	Meten	Wahum	Poten	Ilnum
42.	Gah	Ceram.	Miatan	Aif	Bobuk	Sonina
43.	Wahai	Ceram.	Meten	Aow	Maína	Inóre
44.	Goram		Meta metan	Hai	Bobok	Suwera
45.	Matabello		Meten	Efi	Leleh	Wiramáni
46.	Teor		Miten	Yaf	Lēn	Gilinkani
47.	Ké Is.		Metan	Youf	Lih	Nirun
48.	Aru Is.		Būré	Ow	Jinny	Djurul
49.	Mysol (Coast)		Mūlmetan	Lap	Sala	Shong gulu
50.	Do. (Interior)		Bit	Yap	Klen	Mot mobi
51.	Dorey		Paisim	Voor	Iba	Snori
52.	Teto, E.	Timor	Metan	Hahi	Bot	Inur
53.	Vaiqueno, E.	Timor	Meta	Hai	Naiki	Inu
54.	Brissi, W.	Timor	Metan	Ai	Naaik-Bena	Panan
55.	Savu		Meddi	Ai	Moneái	Hewonga
56.	Rotti		Ngéo	Hai	Matua, Malóa	Idun
57.	Allor		Mité	Api	Bé	Niru
58.	Solor		Mitang	Api	Belang	Irung
59.	Bájau (Sea Gipsies)		Lawon	Api	Basar	Uroh

OF THE MALAY ARCHIPELAGO.

	SMALL.	TONGUE.	TOOTH.	WATER.	WHITE.
1.	Kíchil	Lídah	Gígi	Ayer	Pūtih.
2.	Chili	Ilat	Untu	Banyu	Pūteh.
3.	Bri	Ellah	Gigi	Aie	Pūtih.
4.	Chadi	Lelah	Gigi	Yéni	Kebo.
5.	Becho	Líla	Isi	Uwál	Mapūte.
6.	Kidikidi	Lílah	Nichi	Mânu	Mapūti.
7.	Kedi	Lilah	Gigi	Aer	Pūtih.
8.	Odidi	Elunto	Nisinto	Mánu	Mopūtih.
9.	Koki	Lilah	Baan	Rano	Kuloh.
10.	Toyáan	Lilah '	Ipan	Rano	Kuloh.
11.	Iok	Rilah	Isi	Aki	Mawuroh.
12.	Mohintek	Lilah	Mopon	Tivi	Pūtih.
13.	Koki	Lilah	Wään	Rano	Kūloh.
14.	Koki	Dilah	Waang.	Dorr	Pūtih.
15.	Kokonio	Dilrah	Isy	Akéi	Mabida.
16.	Dodío	Lilah	Ngísi	Akéi	Mabida.
17.	Moisiko	Dila	Dongito	Sarugo	Mopotiho.
18.	Anióu	Lilah	Isi	Aki	Mawérah.
19.	Kadodo			Wai	Mawiralt.
20.	Mahé	Maki	Nihi	Wai	Bóti.
21.	Koi	Mahmo	Nisini	Waili	Umpoti.
22.	Roit	Maän	Nisi	Wai	Boti.
23.	Roi	Maanen	Nisinen	Wai	Boti.
24.	Bakoti	Munartea	Nisnya-teha	Wai	Purini.
25.	Ichi ichi	Aki	Ingin	Namo	Bobūdo.
26.	Kéni	Aki	Ing	Aki	Bubulo.
27.	Kūtu	Mod	Hahlo	Woya	Bulam.
28.	Díkit	Lidah	Gigi	Paisu	Putih.
29.	Wai-waio	Imōd	Afod	Waiyr	Wūlan.
30.	Cheka	Yeidi	Ngedi	Namo	Būdo.
31.	Dechéki	Nangaládi	Ini	Aki	Daari.
32.	Koi	Meka	Niki	Wehr	Pūtih.
33.	Ahuntai	Meka	Nikin	Wehl	Pūtih.
34.	Ana-á	Numawa	Nindíwa	Weyl	Pūtih.
35.	Koi	Méh'	Niki	Weyl	Pūtih.
36.	Ihihil	Mé	Nio	Wai	Pūtil.
37.	Olihil	Mei	Nisi-mo	Waëli	Pūtile.
38.	Kokanéii	Meëm	Nikim	Waeli	Pūtih.
39.	Anan	Mecolo	Lilico	Welo	Pūtih.
40.	Nelak	Melin	Nifan	Wai	Babut.
41.	Anaanin	Ninúm	Nesnim	Waín	Pūtih.
42.	Wota wota	Lemukonína	Nisikonina	Arr	Maphutu.
43.	Kiiti	Mé	Lesin	Tólun	Pūteh.
44.	Tutúin	Kelo	Nisium	Arr	Mehūti.
45.	Enéna	Tumoma	Nifoa	Arr	Maphūti.
46.	Fek	Mën	Nifin	Wehr	Sélūp.
47.	Kot	Nefan	Oin	Wehr	Neah.
48.	Sie	Gigi	Mulu	Wehr	Eren.
49.	Gūnam	Aran	Kalifin	Wayr	Būs.
50.	Senpoh	Aran	Kelif		Boo.
51.	Besarbamba	Kaprendi	Nasi	Waar	Piūper.
52.	Lüik	Nañal	Nian	Vé	Mūty.
53.	Anâ	Iemal	Nissy	Hoi	Mūty.
54.	Ana	Man	Nissin	Oü	Mūty.
55.	Anaíki	Weo	Ngútu	Uilóko	Pūdi.
56.	Anoána, Loaána	Máan	Nissi	Oée	Fūla.
57.	Kaái	Wewelli	Ulo	Wé	Būráka.
58.		Ewel	Ipa	Wai	Būrang.
59.	Didiki	Délah	Gigi	Boi	Potih.

One Hundred and Seventeen Words in Thirty-three

English	ANT.	ASHES.	BAD.	BANANA.
1. Malay	Sŭmut	Hábū	Jáhat	Pisang
2. Javanese	Sūmut	A'vu	Ollo	Gudang
6. Bouton } S. Celebes.	Oséa	Orápu	Madúki	Olóka
7. Salayer }	Kalihara	Umbo	Seki	Loka
16. Menado }				
17. Bolang- } N. Celebes.	Singeh	Abū	Dalruy	Lénsa
hitam }	Tohomo	Awu	Moiatu	Pagie
18. Sanguir, Sian	Kiáso	Henáni	Lai	Busa
19. Salibabo		Reoh		
20. Sula Is	Kokoi	Aftúha	Busár	Fía
21. Cajeli }	Mosisin	Aptai	Nakié	Umpúlue
22. Wayapo } Bouru.	Fosisin	Aptai	Dabóho	Fūat
23. Massaratty }	Misisin	Ogotīn	Dabóho	Fúati
24. Amblaw	Kakai	Lávu	Behei	Biyeh
26. Tidore	Bifi	Fíka	Jíra	Koi
29. Gani } Gilolo.	Laim	Tapin	Lekat	Lókka
31. Galela }	Golúdo	Kapok	Atoró	Bóle
32. Liang }	Umu	Awmáti	Ahia	Kula
33. Morella } Amboyna.	Oön	Armatei	Ahia	Kula
34. Batumerah }	Manisiá	Howaluxi	Akahia	Iáni
35. Lariki }	Aten	Aow matei	Ahia	Kōra
36. Saparua	Sumakow	Hamatanyo	Ahía	Kúla
37. Awaiya }	Tumúe	Ahwotoí	Ahia	Wūri
38. Camarian }	Sümukáo	Hao matei	Ahié	U'ki
39. Teluti } Ceram.	Phóino	Yafow matán	Ahia	Peléwa
40. Ahtiago and Tobo }	Fóin	Laftaín	A'vet	Fūd
41. Ahtiago (Alfuros) }		Laf teinim	Kafetáia	Phitim
42. Gah }	Niéfer	Aif tai	Nungalótuk	Fúdia
43. Wahai }	Isalema	Tókar	Aháti	Uri
45. Matabello	Otúma	Aow lómi	Ráhat	Phúdi
46. Teor	Singa singat	Yaf leit	Yat	Mūk
49. Mysol	Kamili	Gelap	Lek	Talah
50. Mysol	Kumlih	Geni	Leak	Máh
59. Baju	Sumut	Habu	Ráhat	Pisang

Languages of the Malay Archipelago.—*Continued.*

	BELLY.	BIRD.	BLACK.	BLOOD.	BLUE.	BOAT.
1.	Prút............Búrung........Itam			......Dárah	......Bíru	........Praū.
2.	Wūtan........Manok.........Iran			......Gŭte	......Biru	........Prau.
6.	Kompo.........Manumanu...Amaíta			...Oráh	........Ijan	........Búnka.
7.	Pompon ......Burung ......Hitam			...Rara	........Láo	........Lopi.
16.	Tijan ....Mánu	........Maitung...Daha		........Mabidu	......Sakaen.	
17.	Teo ....Manoko ......Moitomo		Dugu	........Morono	......Bolato.	
18.	Tian............Manu ....Maitun		...Daha	........Biru	........Sakaen.	
19.	................Manu urarutang Ma-itu			....	Biru	........Kasáneh.
20.	Téna ....Mánu ....Miti		.....Póha	........Biru	........Lótu.	
21.	Tihumo ......Manúi........Métan		...Lála............	Biru ......Waä.		
22.	Tihen ....Manúti........Miti		......Raha	........Biru	........Wága.	
23.	Fukanen ......Mánúti........Miti		......Ráha	........Biru	........Waga.	
24.	Remnati kuroiManúe........Kame icheiHahanatéa ...Biroi				........Waa.	
26.	Yóru ....Namo bangowKokótu		...Yán............	Rúru	........O'ti.	
29.	Tutut ....Manik ....Kitkúdu...Sislor			........Biru	........Wōg.	
31.	Poko ....Namo ....Tatatáro...LarahnangowBiru				........Déru.	
32.	Hétuáka ......Tuwi ....Méte		......Lala	........Mala	........Haka.	
33.	Tiáka ....Mano ....Méte		......Lala	........Mala	........Haka.	
34.	Tiáva ....Burung ......Meténi		...Lalaí	........Amála	......Háka.	
35.	Tia ....Mano ....Méte		......Lala	........Mála	........Sepó.	
36.	Teho...........Mano ....Meteh......Lalah			........Lala	........Tala.	
37.	Tia ....Manúe........Meténi		...Lalah	........Meteni	......Siko.	
38.	Tiámo ....Mánu ....Méti		......Lála	........Lála	........Tála.	
39.	Teocólo ......Manúo........Méte		......Láia	........Lala	........Yalopei.	
40.	Tian............Nióva ....Memétan Láwa			........Biru	........Wáha.	
41.	Tapura ....Manuwan ...Meten......Lahim........MasounaniniWaim.					
42.	Toniña ....Manok ....Miatan		...Lalai	........Biri	........Wúna.	
43.	Tiare...........Malok ....Meten......Lasin			........Marah	........Polútu.	
45.	Abúda ....Mánok.........Meten......Lárah			........Biru	........Sóa.	
46.	Kabin ....Manok.........Miten......Larah			........Biru	........Hól.	
49.	Nan ....Mulmetan Lomos........Melah				......Owé.	
50.	Mot ni ....Bít ....Lemoh			....	Owáwi.	
59.	Bútah ....Mano ......Lawön......Lahah........Lawu				........Bido.	

ONE HUNDRED AND SEVENTEEN WORDS IN THIRTY-THREE

English	BODY.	BONE.	BOW.	BOX.
1. Malay	Bádan	Túlang	Pánah	Púti
2. Javanese	Awah	Bálong	Panah	Krobak
6. Bouton } S. Celebes..	Karóko	Obúku	Opána	Buéti
7. Salayer	Kaleh	Boko	Panah	Puti
16. Menado				
17. Bolang- } N.Celebes.	Dokoku, Aoh. Duhy			Mabida
hitam	Botanga	Tula		
18. Sanguir, Sian	Badan	Buko		Bantali
19. Salibabo			Papite	
20. Sula Is.	Kóli	Hoi	Djūb	Burúa
21. Cajeli	Batum	Lolimo	Panah	Bueti
22. Wayapo } Bouru.	Fatan	Rohin		Buéti
23. Massaratty.	Fatanin	Rohin	Pánat	Buéti
24. Amblaw	Nanau	Koknatéa	Busu	Poroso
26. Tidore	Róhi	Yóbo	Jobi jobi	Barúa
29. Gani	Badan	Momud	Pusi	Barúa
31. Galela. } Gilolo.	Nangaróhi	Kovo	Ngámi	Barúa
32. Liang	Nanáka	Ruri	Husur	Buéti
33. Morella	Dada	Luli	Husul	Buéti
34. Batumerah.	Anáro	Lulivá	Apúsu	Saüpa
35. Lariki	Anána	Ruri	Husur	Buèti
36. Saparua	Inawallah	Riri	Husu	Ruūwai
37. Awaiya	Sanawála	Lila	Husúli	Pūéti
38. Camarian	Patani	Nili	Husúli	Buéti
39. Teluti	Hatáko	Toicólo	Osio	Huéti
40. Ahtiago and Tobo	Whátan	Lúin	Bánah	Kúnchi
41. Ahtiago (Alfuros)	Nufátanim	Lūim	Husūūm	Husum
42. Gah	Rísi	Lului	Usulah	Kuincha
43. Wahai	Hatare	Luni	Helu	Kapai
45. Matabello	Watan	Lúru	Lóburr	Udiss
46. Teor	Telimin	Urut	Fun	Fud
49. Mysol	Badan	Kaboom	Fean	Bus
50. Mysol	Padan	Mot bom	Aan	Boo
59. Baju	Badan	Bákas	Panah	Puti

LANGUAGES OF THE MALAY ARCHIPELAGO.—*Continued.*

	BUTTERFLY.	CAT.	CHILD.	CHOPPER.	COCOA-NUT.	COLD.
1.	Kūpūkūpū...	Kūching..	A'nak	Párang	Klápa ...	..Dingin,Tijok.
2.	Kūpu.	Kuching..	Anak	Parang	Krambil	...A'dam.
6.	Kumberá.....	Ombutá...	Oánana	Kapuru	Kalimbúngo Magári.	
7.	Kolikoti	Miaò	Anak	Berang	Nyóroh	Dingin.
16.	Karinboto....	Tusa	Dodio	Kompilang..	Bángoh	Madadun.
17.	Wieto	Ngeäu.....	Anako.	Boroko	Bongo	Motimpia.
18.	Kalibumbong	Miau	Anak	Pedah	Bángu.	Matuno.
19.		Miau	Pigi-neneh	.Galéleh	Nyu.	
20.	Maápa.	Nāo	Ninána	Péda	Núi	Bagóa.
21.	Lahen	Sika	A'nai	Tolie	Niwi	Numniri.
22.	Lahei.	Sika	Nánat	Tódo	Niwi	Damóti.
23.	Tapalápat...	Māo	Naánati	Katúen	Niwi	Dabridi.
24.	Koláfi	Mau	Emlúmo	...Laiey	Niwi	Komoriti.
26.	Kopa kopa...	Túsa	Ngófa	Péda	Igo	Góga.
29.	Kalibobo.....	Tusa	Untúna	Barakas	Níwitwan...	Makufin.
31.	Mimáliki	...Bóki	Mangópa	...Taíto	Igo	Damála.
32.	Kakópi	Túsa	Niana	Lobo	Nier	Periki.
33.	Pepeül.	Sie	Wana	Lopho	Niwil	Periki.
34.	Kupo kupo...	Temai	Opoliána	...Ikíti	Niwéli	Mutí.
35.	Lowar lowar.	Sía	Wári	Lopo	Nimil	Periki.
36.	Kokohan	....Siah	Anahei	Lopo	Muöllo	Puriki.
37.	Korūli.	Maōw	Wána	Aáti	Liwéli	Pepéta.
38.		Sía	Ana	Lopo	Niwéli	Maríki.
39.	Tutupúno	...Sia	Anan	Lopo	Nuélo	Pilikéko.
40.	Bubúmái	...Sikar	Iniának	Béda	Núa	Bäidik.
41		Láfim	Anavim	Tafim	Nuim	Makáriki.
42.	Kowa kowa..	Shika	Dúia	Péde	Niūla.	Lifie.
43.	Koháti	Sika	A'la	Tulumaina..	Lúen	Mariri.
45.	Obaóba	Odára	Enéna	Béda	Dar	Arídin.
46.	Kokop	Sika	Anìk	Funén.	Nōr	Giridin.
49.	Kalabubun...	Mar	Kachun	Keío	Nea	Kabluji.
50.		Miau	Wai	Yeu	Nen	Pátoh.
59.	Titúe	Miau	Anáko	Bádi	Salóka.	Jérnih,

One Hundred and Seventeen Words in Thirty-three

English	COME.	DAY.	DEER.	DOG.
1. Malay	Mári	A'ri (Siang.)	Rūsa	A'ujing ...
2. Javanese	Marein	Aivan	Rusa	Asu ....
6. Bouton ⎫ S. Celebes..	Maivé	Héo	Orúsa	Muntóa ...
7. Salayer ⎭	Maika	Allo	Rusa	Asu ....
16. Menado ⎫	Simépu	Roū	Rusa	Kapuna ...
17. Bolang- ⎬ N. Celebes.. hitam ⎭	Aripa	Unuveno	Rusa	Ungu ......
18. Sanguir, Sian	Dumahi	Rókadi	Rusa	Kapúna ...
19. Salibabo	Maranih			Assu ......
20. Sula Is	Mái	Dawíka	Munjangan	Asu ....
21. Cajeli ⎫	Omai	Gáwak	Mūnjángan	Aso.........
22. Wayapo ⎬ Bouru.	Ikomai	Dówa	Mūnjángan	Asu ....
23. Massaratty... ⎭	Gumáhi	Liar	Munjangan	Asu ....
24. Amblaw	Buoma	Laei	Munjaráni	Asu ....
26. Tidore	Ino keré	Wellusita	Munjangan	Káso ......
29. Gani ⎫ Gilolo.	Mai	Balanto	Munjangan	Iyór ....
31. Galela ⎭	Nehíno	Taginíta	Munjangan	Gáso ......
32. Liang ⎫	Uimai	Kikir	Munjangan	Asu.........
33. Morella ⎬ Amboyna.	Oimai	Alowata	Munjangan	Asu ......
34. Batumerah ⎪	Omai	Watiëla	Munjangan	Asu ......
35. Lariki ⎭	Mai	Aoaaóa	Munjangan	Asu ....
36. Saparua	Mai	Kai	Rusa	Asu ....
37. Awaiya ⎫	Alowei	Apaláwe	Maiyáni	A'su.......
38. Camarian ⎪	Mai		Maiyánani	Asúa ......
39. Teluti ⎬ Ceram.	Mai	Kíla	Meisakano	Wasu ......
40. Ahtiago and Tobo ⎪	Kulé	Matalima	Rúsa	Yás.........
41. Ahtiago (Alfuros) ⎪	Dak Lápar	Pília	Tusim	Nawang...
42. Gah ⎪	Mai	Malal	Rusa	Kafúni ...
43. Wahai ⎭	Mai	Kaseiella	Mairáran	Asu ......
45. Matabello	Gomári	Larnumwáə	Rúsa	Afúna .....
46. Teor	Yef man	Liléw	Rusa	How ......
49. Mysol	Jog mah	Seasan	Mengangan	Yes.........
50. Mysol	Bo mun	Kluh	Menjangan	Yem ......
59. Baju	Paituco	Lau	Paiów	Asu ....

LANGUAGES OF THE MALAY ARCHIPELAGO.—*Continued.*

	DOOR.	EAR.	EGG.	EYE.	FACE.	FATHER.
1.	Píntu ......Telínga	.....Túlor .....Máta			........Múka ....Bápa.	
2.	Lawang ...Kúping	.....U'ndok ...Móto			........Ral ....Baba.	
6.	Obámba ...Talinga	.....Ontólo ...Máta			........Oroku ....Amana.	
7.	Pintu ......Toli	Tanar ......Mata			........Rupa ....Ama.	
16.	Raroangen.Túri............	Natu .....Mata			........Duhn ....Jama.	
17.	Pintu ......Boronga ......Natu	.....Mata			........Paio............Kiamat.	
18.	Pintu ......Toli ....Tuloi	.....Mata			........Gáti............Yaman.	
19.	....					
20.	Yamáta ...Telinga	.....Metélo ...Háma			........Lúgi............Nibaba.	
21.	Lilolono ...Telilan.........Telon	.....Lamūmo	.....Uhamo		......A'mam.	
22.	Káren ......Telingan	.....Télo	.....Raman.........Pupan		........Náma.	
23.	Henóloni ...Linganani ...Telo	.....Ramani			......Pupan lalin...Náama.	
24.	Sowéni......Herenatia	...Rchöi .....Lumatibukói.Ufnati lareni.Amao.				
26.	Móra.........Ngan	.......Gósi	.....Lau		............Gái ....Baba.	
29.	Nára.........Tingēt	.........Toli	.....Umtowt	......Gonaga	.....Bápa.	
31.	Ngóra ......Nangów	.....Magosi	...Láko		.........Nangabío ...Nambába.	
32.	Metenúre...Terina	........Muntiro...Máta			........Hihika ......Ama.	
33.	Metenulu...Telina	.......Mantirhui.Mata			........Uwaka.........A'ma.	
34.	Lamáta......Telinawa......Munteloá.Matava			.....Uwaro........Kopapa.		
35.	Metoüru ...Terina	.......Momatíro.Mata			........U'wa ......Ama.	
36.	Metoro......Teréna.........Tero	.....Mata			........Wáni ....Ama.	
37.	Aleáni ......Terína mo ...Telúli......Mata mo			.....Wámu mo ...Ama.		
38.	Metanorúi..Terinam ......Terúni	...Máta			........Wamo ....Ama.	
39.	Untaniyún.Tinacóno......Tin	.........Matacolo	.....Facólo		........Amacolo.	
40.	Lolamatan.Líkan ....Tólin	.....Mátan			........U'fan ....Iáman.	
41.	Motūlnim..Telikeinlúim.Tolnim ...Mátara.........Uhúnam			......Amái.		
42.	Yebúteh ...Tanomulino..Tolor	.....Matanina	...Funonína		...Mama.	
43.	Olamatan...Teninare......Latun	.....Mata			........Matalalin ...Ama.	
45.	Fidin ......Tilgár ....Atulú.....Matáda			.....Omomanía ...Ieí.		
46.	Remátin ...Karin ....Telli	.....Matin			........Matinóin...A'ma.	
49.	Batal ......Tenaan ......Tolo	.....Tūn			.........Tunah ....Mám.	
50.	Bata ....Mot na.........Tolo	.....Mut morobu..Mutino			......Mām.	
59.	Boláwah ...Telinga	.....Untello ...Mata			.........Rúa ....Uáh,	

One Hundred and Seventeen Words in Thirty-three

English	FEATHER.	FINGER.	FIRE.	FISH.
1. Malay	Būlū	Jári	A'pi	...Ikan
2. Javanese	Wūlu	Jári	Gúni	...Iwa
6. Bouton } S. Celebes..	Owhú	Saranga	Whá	...Ikáni
7. Salayer }	Bulu	Karaami	Api	...Jugo
16. Menado }	Mombulru.Talrimido		Pūtung.Maranigan.	
17. Bolang- } N. Celebes.	Burato	Sagowari	Puro	...Sea
hitam }				
18. Sanguir, Sian	Doköl	Limado	Putún..Kina	
19. Salibabo			Puton..Inásah	
20. Sula Is	Nifóa	Kokowana	Api	...Kéna
21. Cajeli }	Bolon	Limam kokon	Ahū	...Iáni
22. Wayapo } Bouru.	Fulun	Wangan	Bána	...Ikan
23. Massaratty.. }	Folun	Wangan	Bána	...Ikan
24. Amblaw	Boloi	Lemnati kokoli	Afu	...Ikiani
26. Tidore	Gógo	Gia marága	U'ku	...Nýan
29. Gani } Gilolo.	Lonko	Odeso	Lútan..Ian	
31. Galela }	Ló	Raríga	Uku	...Náu
32. Liang	Huru	Rimaka hatu	Aow	...Iyan
33. Morella	Manuhrui..Limaka hatui		Aow	...Iyan
34. Batumerah	Hulúni	...Limáwa kukualima..Aow		...Iáni
35. Lariki	Manhúru	...Lima hato	Aow	...Ian
36. Saparua	Huruni	Uūn	Hao	...Ian
37. Awaiya	Hulúe	Saäti	Aoúsa..Iáni	
38. Camarian	Phulúi	Tarüni	Haō	...Iáni
39. Teluti	Wicolo	Limaco hunilo	Yáfo	...Yáno
40. Ahtiago and Tobo	Fulin	Uin	Yāf	...I'an
41. Ahtiago (Alfuros)	Toholim	...Tai-ímara likéluni...Wáham I'em		
42. Gah	Veolührr	...Numonin tutulo	Aif	...Ikan
43. Wahai	Hulun	Kukur	Aow	...Ian
45. Matabello	Alolú	Taga tagan	Efi	...I'an
46. Teor	Phulin	Limin tagin	Yaf	...Ikan
49. Mysol	Guf	Kanin ko	Lap	...Ein
40. Mysol	Gan	Kanin ko	Yap	...Ein
59. Baju	Bolo	Eríke	Api	...Déiah

The bracket labels: 32–35 Amboyna; 37–43 Ceram.

LANGUAGES OF THE MALAY ARCHIPELAGO.—*Continued.*

	FLESH.	FLOWER.	FLY.	FOOT.	FOWL.	FRUIT.
1.	Dáging......Búnga	........Lálah	...Káki	........A'yam	......Búa.	
2.	Dáging......Kembang	...Lálah	...Síkil	........Pitek	......Wowóan.	
6.	U'ntok......Obúnga	.....Oráli	...Oei	...........Mánu	......Bakena.	
7.	Asi ....Bunga	........Katinali.	Bunkin	......Jangan	...Bua.	
16.	Gisini ......Burány	......Ralngoh.	Raédai	........Mánu	......Bua.	
17.	Sapu.........Wringonea	...Rango	...Teoro	........Mano	......Bunganea.	
18.	Gusi.........Lelun	........Lango	...Laidi	........Manu	......Buani.	
19.					Manu	......Buwah.
20.	Ni'ihi ......Saía	...........Kafini	...Yiéi	...........Mánu	......Kao fua.	
21.	Isim........Mnúrū	........Bena	...Bitim	........Tehúi	......Búan.	
22.	Isin ....Tatan	........Féna	...Kadan	........Téput	......Fūan.	
23.	Isinini ......Kao tutun	...Féna	...Fitinen	......Téputi	......Fuan.	
24.	Isnatéa......Kakali	........Béna	...Beernyáti atani.	Rufúa	......Buani.	
26.	Róhe.........Hatimoöto s:ya.	Gúphu...	Yóhu	........Toko	......Hatimoöto sopho	
29.	Woknu......Bunga	........Búbal	...Wed	...........Manik	......Sapu.	
31.	Nangaláki..Mabúnga	......Gúpu	...Nandóhu	......Tóko	......Masópo.	
32.	Isi...........Powta	........Lari	......Aika	........Mano	......Húa.	
33.	Isi...........Powti	........Lali	......Aika	........Manu	......Hua.	
34.	Isíva.........Kahuka	......Henai	...Aíva	...........Máno	......Aihuwána.	
35.	Isi...........Kupang	......Pénah	...Ai	...........Mano	......Ai hua.	
36.	Isini.........Kupar	........Upenah..	Ai	...........Mano hena.	Hwányo.	
37.	Waoúti......Lahówy	......Pepénah.	Aì	...........Manulúma.	Huváiy.	
38.	..............Kupáni	......Upéna	...Ai	...........Mánu	......Huwái.	
39.	Isicolo ......Tifin	........Upéna	...Yaicólo	......Manuo	......Huan.	
40.	Isin ....Futin	.........Lákar	...Yái	...........Tóñ	........Vúan.	
41.	Isnum ......Eiheitnum	...Phenem.	Wáira	........Towim	...Eifuanum.	
42.	Sesiún ......Fuis	...........Langar...	Kaieniña	......Manok	...Woya.	
43.	Héla.........Loen	........Mumun..	Ai	...........Malok	......Huan.	
45.	Ahí ....Ai wöi	.........Wéger	...Owéda	........Manok	...Woi imotta.	
46.	Henin ......Pus	...........Omiss	...Yain	........Manok	...Phuin.	
49.	Wamut ...Gáp heu	................Kanin pap	...Kakep	......Gapeah.		
50.	Mot nut ...Ioh	.........Kelang...	Mat wey	......Tekayap	...I'po.	
59.	Isi...........Bunga	........Langow..	Nai	...........Mano	......Bua.	

ONE HUNDRED AND SEVENTEEN WORDS IN THIRTY-THREE

English	GO.	GOLD.	GOOD.	HAIR.
1. Malay	Púrgi	Mās	Baik	Rámbut
2. Javanese	Lungo	Mas	Butje	Rambut
6. Bouton ⎰ S. Celebes..	Lipano	Huláwa	Marápe	Bulwa
7. Salayer ⎱	Lampa	Bulain	Baji	Uhu
16. Menado ⎱ N. Celebes.	Máko	Bolraong	Sahenie	Uta
17. Bolang- ⎰	Korunu	Bora	Mopia	Woöko
hitam				
18. Sanguir, Sian	Dako	Mas	Mapiah, Ma holi.	Utan
19. Salibabo	Ma puréteh.	Bulawang..	Mapyia	
20. Sula Is	Láka	Famaká	Pía	O'ga
21. Cajeli ⎱	Oweho	Blawan	Ungano	Buloni
22. Wayapo ⎰ Bouru.	Iko	Balówan	Dagósa	Folo
23. Massaratty..	Wíko	Hawan	Dagósa	Olofólo
24. Amblaw	Buoh	Bulówa	Parei	Olnáti
26. Tidore	Tagi	Guráchi	Láha	Hútu
29. Gani ⎱ Gilolo.	Tahn	Omas	Fiar	Iklct
31. Galela ⎰	Notági	Gurachi	Talóha	Hútu
32. Liang ⎱	Oï	Halowan	Ia	Kaiola
33. Morella	Oi	Halowan	Ia	Keiúle
34. Batumerah	Awái	Halowani..	Amaísi	Huá
35. Lariki ⎰	Oi	Halowan	Mai	Keö
36. Saparua	Ai	Halowan	Malopi	Uwóhoh
37. Awaiya ⎱	Aeó	Halowáni..	Aólo	Uwoleíha mo
38. Camarian	Aeo	Halowani..	Mái	Keóri
39. Teluti	Itái	Hulawano.	Fia	Këülo
40. Ahtiago and Tobo	Akó	Masa	Komúin	Ulvú
41. Ahtiago (Alfuros)	Teták	Masen	Komia	Ulufúim
42. Gah	Ketángo	Mas	Guphïn	Uka
43. Wahai ⎰	Aou	Hulaän	Ia	Húe
45. Matabello	Fanów	Mása	Fía	U'a
46. Teor	Takek	Mas	Phien	Wultáfun
49. Mysol	Jog	Plehan	Fci	Peleah
50. Mysol	Bo	Phean	Ti	Mutlen
59. Baju	Molch	Mas	Alla	Buli tokolo...

LANGUAGES OF THE MALAY ARCHIPELAGO.—*Continued.*

	HAND.	HARD.	HEAD.	HONEY.	HOT.	HOUSE.
1.	Tángan	Kras...:....Kapála	Mádu	Pánas	Rúmah.
2.	Tángan	Kras	U'ndass	Mádu	Páuas	Umah.
6.	Olima	Tobo	Obaku	Ogora	Mopáni	Bánna.
7.	Lima	Teras	Ulu	Ngongnou	Bumbung..Sapu.
16.	Rilma	Maketihy..Timbónang	Madu	Matéti	Balry.
17.	Rima	Murugoso.Urie	Teoka	Mopaso	Bore.
18.	Lima	Makúti	Tumbo	Matúti	Bali.
19.						Bareh.
20.	Lima	Kadiga	Náp	Baháha	U'ma.
21.	Limámo	Namkana..Olum	Madu	Poton	Lúma.
22.	Fahan	Lumé	Ulun fatu	Dapóto	Húma.
23.	Fahan	Digíwi	Olun	Dapótoni	Húma.
24.	Lemnatia	Unkiweh...Olimbukói.Násu	Umpána	Lúmah.
26.	Gia	Futúro	Defólo	Sasáhu	Fola.
29.	Komud	Maséti	Poi	San	U'm.
31.	Gia	Daputúro..Nangasáhi.Mangópa	Dasáho	Táhu.
32.	Rimak	Makána-...Uruka	Niri	Putu	Rumah.
33.	Limaka	Makana	Uruka	Keret	Loto	Lumah.
34.	Limáwa	Amakana..Ulúra	Aputu	Lumá.
35.	Lima	Makána	Uru	Miropenah	Pútu	Rumah.
36.	Rimah	Makanah..Uru	Madu	Kuno	Rumah.
37.	A'la	Uru	Ulu mo	Helímah	Maoúso	Lúūma.
38.	Limamo	Makána	Ulu	Násu	Pútu	Luma.
39.	Limacolo	Unté	Oyúko	Penanûn	Pútu	Uma.
40.	Niman	Kakówan...Yúlin	Músa	Bafánat	Umah.
41.	Tai-ímara	Mocolá...Ulukátim..Lukaras	Asála	Feióm.
42.	Numoniña	Kaforat	Luníni	Nasu musun.Mofánas	Lúme.
43.	Mimare	Mukola	Ulure	Kinsumi	Mulai	Luman.
45.	Dumada lomia..Máitan	Alúda	Limlimur	Ahúan	Orúma.
46.	Limin	Keherr	Ulin		Horip	Sarin.
49.	Kanin	Umtoo	Kahutu	Fool	Benis	Kom.
50.	Mot mor	Net	Mullud	Fool	Pelah	Dé.
59.	Tangan	Kras	Tikolo		Panas	Rumah.

480

ONE HUNDRED AND SEVENTEEN WORDS IN THIRTY-THREE

English	HUSBAND.	IRON.	ISLAND.	KNIFE.
1 Malay	Láki	Bŭsi	Pūlo	Písau
2. Javanese	Bedjo	Wusi	...Pulo	Lading
6. Bouton ⎱ S. Celebes..	Obawinena	A'sé	Liwúto	Pisau
7. Salayer ⎰	Burani	Busi	Pulo	Pisau
16. Menado ⎱	Gagijannee	Wasey	...Mapuroh	...Pahegy
17. Bolang- ⎱ N. Celebes. hitam ⎰	Taroraki	Oäse	Riwuto	Piso
18. Sanguir, Sian	Kapopungi	Wasi	...Toadi	Pisau
19. Salibabo	Essah		Taranusa	...Lari
20. Sula Is	Túa	Mŭm	...Pássi	Kóbi
21. Cajeli ⎫	Umlanei	Awin	...Núsa	Iliti
22. Wayapo ⎬ Bouru.	Mori	Kawil	...Núsa	Irit
23. Massaratty.. ⎭	Gebhá	Momul	...Nusa	Katánan ...
24. Amblaw	Emanow	Awi	Nusa	Kamarasi ...
26. Tidore	Nau	Búsi	Gurumongópho	Dari
29. Gani ⎱ Gilolo.	Mondemapin	...Busi	Wāf	Kobit
31. Galela ⎰	Maróka	Dodiódo..	Gurongópa	...Díha
32. Laing ⎫	Mahinatima malona	Taä	Nusa	Seë
33. Morella ⎬ Amboyna.	Amolono	Ta	Nusa	Seëti
34. Batumerah ⎪	Mundai	Saëi	Nusa	Opiso
35. Lariki ⎭	Malona	Mamōr	...Nusa	Séi
36. Saparua	Manowa	Mamōlo..	Nusa	Seit
37. Awaiya ⎫	Manowai	Mamóle..	Mísa	Amasáli
38. Camarian ⎪	Malóna	Mamóle..	Nusa	Seíti
39. Teluti ⎪ Ceram.	Ihina manowa	...Momollo.	Nusa	Sëito
40. Ahtiago and Tobo ⎬	Imyóna	Momŭm .Túbil		Tuána
41. Ahtiago (Alfuros) ⎪	Ifnéinin sawanim	Momolin.	Tuplim	Macouosim .
42. Gah ⎪	Bulana	Momŭmi.	Tubur	Tuka
43. Wahai ⎭	Pulahan	Héta	Lusan	Tuluangan..
45. Matabello	Helameranna	...Momŭmo	Tobūr	Mirass
46. Teor	Wehoin	Momúm..	Lowánik	...Isowa
49. Mysol	Man	Seti	Yef	Cheni
50. Mysol	Mo man	Leti	Ef	Yeaói
59. Baju	Ndáko	Bisi	Pulow	Pisau

LANGUAGES OF THE MALAY ARCHIPELAGO.—*Continued*.

	LARGE.	LEAF.	LITTLE.	LOUSE.	MAN.	MAT.
1.	Būsar......Daūn		Kíchil	Kūtū ...	Orang lákilaki.	Tíkar.
2.	Gedé......Godong		Chili........	Kūtu ...	Wong lanan ...	Klosso.
6.	Moughí ...Tawána		Kidikidi ...	Okútu..	Omani	Kiwaru.
7.	Bákeh ...Taha		Kédi	Kutu ...	Tau	Tupur.
16.	Raboh ...Daun		Dodio	Kutu ...	Taumata esen..	Sapie.
17.	Morokaro.Lungianea ...		....Moisiko......	Kutu ...	Roraki	Boraru.
18.	Labo ......Decaluni		....Aníou ...	Kutú ...	Manesh	Sapieh.
19.	Bagewa ...		....Kadodo		Tomatá	Bilátah.
20.	Eá ....Kao hósa		Mahé........	Kóta ...	Maona	Saváta.
21.	Léhai......Atétun		Köi	Olta ...	Umlanai	A'pine.
22.	Bágut......Kroman		Roit	Kóto ...	Gemana........	A'tin.
23.	Haat ......Kóman		Roi...........	Koto ...	Anamhána ...	Kátini.
24.	Plaré ......Lai obawai		Bakoti	Uru ...	Remau	Arimi.
26.	Lámu......Hatimoöto merow.		Kéni ...	Túma...	Nonán	Junúito.
29.	Talalólo ...Nilonko		Waiwáio ...	Kútu ...	Mon	Kalása.
31.	Elámo ...Misóka ...		....Dechéki......	Gáni ...	Anów...........	Jungúto.
32.	Nila ......Ailow		Koi..........	Utu ...	Malona	Päi.
33.	Hella ......Ailow		Ahúntai ...	Utu ...	Malono	Hilil.
34.	Enda-a ...Aitéti		Aná-á ...	Utu ...	Mundai	Towai.
35.	Ira ....Ai rawi		Koi	Kutu ...	Malona	Pafl.
36.	Ilahil ......Laun		Ihíhil	Utu ...	Tumata........	Pai.
37.	Iláhe ......Laíni		Olíhil........	U'tu ...	Tumata........	Kaili.
38.	Eräämei...Airówi		Kokaneii ...	Utúa ...	Tumata........	Paílí.
39.	Elau ......Daun		Anan........	Utu ...	Manusia	Pai-ilo.
40.	Aíyuk ...Lan		Nélak	Tínan...	Muána	Láb.
41.	Poten......Eilúnim		....Anaanin ...	Kutim..	Muruleinum ...	Lapim.
42.	Bobuk ...Lino		Wota wota ..	Kutu ...	Beláne	Kiël.
43.	Mäina......Totun ...		....Kiiti	Utun ...	Ala híeiti	Kihu.
45.	Leléh ... . Arehín		Enena	U'tu ...	Marananna ...	I'ra.
46.	Lën........Chafen		Fek	Hut ...	Meránna	Fira.
49.	Sala ......Kaluin		Gunam	Ut	Motu	Tin.
50.	Klen ......Idun		Senpoh	Uti......	Mot	Tin.
59.	Basar ......Daun		Didiki	Kutu ...	Lélah...........	Tepoh.

ONE HUNDRED AND SEVENTEEN WORDS IN THIRTY-THREE

English	MONKEY.	MOON.	MOSQUITO.	MOTHER.
1. Malay	Münyeet	Būlan	Nyámok	Ma
2. Javanese	Budéss	Wulan	Nyámok	Mbo
6. Bouton } S. Celebes.	Róke	Búla	Burótok	Inaná
7. Salayer }	Dáre	Bulan	Kasisili	Undo
16. Menado } N. Celebes.	Bohen	Bulrang	Tenie	Inany
17. Bolang-hitam	Kurango	Wura	Kongito	Leyto
18. Sanguir, Sian	Babah	Buran	Túni	Inúngi
19. Salibabo		Burang		
20. Sula Is.	Mía	Fasina	Samábu	Nieía
21. Cajeli }	Kessi	Būlani	Suti	Inámo
22. Wayapo } Bouru.	Kess	Fhūlan	Múmun	Neína
23. Massaratty }		Fhulan	Seúgeti	Neína
24. Amblaw	Kess	Bular	Sphúre	Ina
26. Tidore	Mía	O'ra	Sisi	Yaíya
29. Gani } Gilolo.	Nok	Pai	Nini	Mamo
31. Galela }	Mía	O'sa	Gumóma	Maówa
32. Liang }	Sia	Hulanita	Séne	Ina
33. Morella	Aruka	Hoolan	Sisil	Inaö
34. Batumerah } Amboyna.	Késs	Huláni	Sisili	Inao
35. Lariki	Rúa	Haran	Sūn	Ina
36. Saparua	Rua	Phulan	Sonot	Ina
37. Awaiya	Kesi	Phuláni	Manisíe	Ina
38. Camarian	Kesi	Wuláni	Senóto	Ina
39. Teluti	Lúka	Hiáno	Sumóto	Inaú
40. Ahtiago and Tobo } Ceram.	Lūkar	Phúlan	Minís	Aína
41. Ahtiago (Alfuros)	Meiram	Melim	Manis	Inái
42. Gah	Lēk	Wúan	Umiss	Nina
43. Wahai	Yakiss	Hulan	U'muti	Ina
45. Matabello	Léhi	Wúlan	U'muss	Nína
46. Teor	Lek	Phulan	Rophun	I'na
49. Mysol		Pet	Kamumus	Nin
50. Mysol		Náh	Owei	Nin
59. Baju	Mondo	Bulan	Sisil	Máko

LANGUAGES OF THE MALAY ARCHIPELAGO.—*Continued.*

	MOUTH.	NAIL(FINGER).	NIGHT.	NOSE.	OIL.	PIG.
1.	Múlūt	Kúkū............	Málam......	Idong	...Mínyak	...Bábi.
2.	Sánkum......	Kūku............	Bungi	Irong	...Lūngo	...Chilong.
6.	Nánga	Kuku............	Maromó	...Oánu	Mínak	...Abáwhu.
7.	Bawa.........	Kanuko.........	Bungi	Kumor.....	...Minyak	...Bahi.
16.	Mohong......	Kanuku	Máhri	Hidong	Rana	Babi.
17.	Nganga	Kamiku.........	Gubie	Jjunga.........	Rana	Rioko.
18.	Mohon	Kanuko.........	Hubbi-......	Hirong	Lana	Bawi.
19.						Bawi.
20.	Beióni	Kowóri	Bohúwi	...Né	Wági	Faũ.
21.	Nūūm	Uloimo	Petū.........	Nem	Nielwíne..	Babúe.
22.	Muen	Utlobin.........	Béto.........	Nïen	Newiyn...	Fafu.
23.	Naónen	Logini	Béto.........	Nieni	Newiny...	Fafú.
24.	Numátéa	...Hernenyati	...Pirue	Neínya téha	Nivehöi...	Bawu.
26.	Móda	Gulichiũ	Sophúto	...Ŭn	Guróho	...Sóho.
29.	Sumut	Kuyut	Becómo	...Usnut	Nimósu	...Boh.
31.	Nangúru	...Gitipi...........	Daputo	...Ngúno.........	Gosóso	...Titi.
32.	Hihika	Terëina	Hatóru	...Hirúka	Neerwiyn.	Hahow.
33.	Soöka	Tereiti	Hatolu.....	Iúka............	Neerliyn..	Hahu.
34.	Suara	Kuku............	Hulaniti	...Ninúra	Wakéli	...Hahu.
35.	Ihi	Terein	Halometi...	I'ru	Nimimein	Hahu.
36.	Nuku	Teri	Potu	Iri	Warisini..	Hahul.
37.	Ihi mo	Talü	Müte	Nua mo	Wailasini.	Hāhu.
38.	So	....	Améti	Hilimo	Wailisini .	Hāwhúa.
39.	Hihico	Talicólo.........	Humoloi	...Olicolo.........	Fofótu	...Hahu.
40.	Vudin	Selíki...........	Matabūt	...I'lin	Kūl	Wār.
41.	Tafurnum ...	....	Potūūn	...I'lnum.........	Félim	...Fafuim.
42.	Lonina	Wuku	Garagaran	Sonina.........	Gúa	Bóia.
43.	Siurure	Talahikun......	Manemi	...Inore	Héli	Hahu.
45.	Ilida	Asiliggir	Olawáha	...Werámani	...Gúla	Boör.
46.	Huin	Limin kukin...	Pogaragara	Gilinkani	...Hīp	Faf.
49.	Gulan.........	Kasebo	Maléh	Shong gulu...	Majulu	...Boh.
50.	Mot po	Kok nesib......	Mau.........	Mot mobi	..Menik	...Boh.
59.	Boah	Kuku............	Sangan	...Uroh	Mánge	...Góh.

One Hundred and Seventeen Words in Thirty-three

English	POST.	PRAWN.	RAIN.	RAT.
1. Malay	Tíeng	Udong	Hūjan	Tíkus
2. Javanese	Soko	..Uran	Hudan	Tikus
6. Bouton ⎱ S. Celebes..	Otúko	Meláma	Waó	Bokóti
7. Salayer ⎰	Palayaran	...Doön	Bosi	Blaha
16. Menado ⎱	Dihi	Udong	Tahíty	Barano
17. Bolang- ⎬ N.Celebes..				
hitam ⎰	Panterno	Ujango	Oha	Borabu
18. Sanguir, Sian	Dihi	Udong	Tahiti	Balango
19. Salibabo	Pari-arang		Urong	
20. Sula Is.	H'ii	U'ha	Húya	Saáfa
21. Cajeli ⎞	Ateoni	Ulai	U'lani	Boti
22. Wayapo .... ⎬ Bouru.	Katehan	Uran	Dekat	Boti
23. Massaratty . ⎠	Katéheni	Uran	Dekati	Tíkuti
24. Amblaw	Hampowne	...Ulai	Ulah	Púe
26. Tidore	Ngasu	Búrowi	Béssar	Múti
29. Gani ⎱ Gilolo.	Li	Níke	Ulan	Lūf
31. Galela ⎰	Golingáso	...Dódi	Húra	Lúpu
32. Liang	Riri	Méter	Hulan	Maláha
33. Morella	Lili	Metar	Hulan	Malaha
34. Batumerah	Lili	Metáli	Huláni	Puéni
35. Lariki	Leilein	Mítal	Haran	Maláha
36. Saparua	Riri	Mital	Tiah	Mulahah
37. Awaiya	Lili	Mitáli	Uláne	Maláha
38. Camarian	Lili	Mitali	Uláni	Maláha
39. Teluti	Hili	Mutáyo	Gia	Maiyáha
40. Ahtiago and Tobo	Fólan	Filúan	U'lan	Meláva
41. Ahtiago (Alfuros)	Faolnim	Hoim	Roim	Sikim
42. Gah	Usa	Gurun	U'an	Karúfei
43. Wahai	Hinin	Bokoti	Ulan	Mulahan
45. Matabello	Faléra	Gúrun	Udáma	Arófa
46. Teor	Pelérr	Gurun	Hurani	Fudarúa
49. Mysol	Fehan	Kasána	Golim	Keluf
50. Mysól	Felian	Kasana	Golim	Quóh
59. Baju	Tikala	Dóah	Huran	Tikus

LANGUAGES OF THE MALAY ARCHIPELAGO.—*Continued.*

	RED.	RICE.	RIVER.	ROAD.	ROOT.	SALIVA.
1.	Mérah	Brās	Sūngei	Jálan	A'kar	Lúdah.
2.	Abang	Bras	Sungei	Malaku	...Oyok	I'du.
6.	Merüí	Bai	Uvé	Dára	Koleséna	...Ovilu.
7.	Eja	Biras	Balang	Lalan	Akar	Pedro.
16.	Mahamu	Bogáseh	...Raríou	Dalren	Hámu	Edu.
17.	Mopoha	Bugasa	...Ongagu	Lora	Wakatia	...Due.
18.	Hamu	Bowáseh	...Sawán	Dalin	...Pungenni	..Udu.
19.	Maramutah	...Boras				
20.	Mia	Bíra	Sungei	A'ya	Kao akar	...Bihú.
21.	Unmíla	Hálai	Wai lé	Lalani	Alamúti	...Bulai.
22.	Míha	Hála	Wai fatan	...Tuhun		Púhah.
23.	Miha	Pála	Wai	Tóhoni	...Kao lahin	..Fúhah.
24.	Meháni	Fála	Waibatang	...Lahuléa	...Owáti	Rnbunatéa.
26.	Kohóri	Bira	Wai	Lolinga	...Hatimoöto.	Gidi.
29.	Mecoit	Samasi	Waiyr	Lolan	Niwolo	...Iput.
31.	Desoélla	...Itámo	Siléra	Néko		Kiví.
32.	Kao	Allar	Weyr	Lahan	Waäta	Tehula.
33.	Kao	Allar	Weyl hatei	...Lalan	...Eiwaäti	...Tehula.
34.	Awow	..Allá
i		Laláni	Ai	Tohulá.
35.	Kao	Hála	Wai hatei	...Lalan	...Ai waat	...Tohural.
36.	Kao	Hálal	Walil	Lalano?	...Aiwaári	...Tohulah.
37.	Meranáte	Hála	Waliláhe	Laláni	Lamúti	...Tohulah.
38.	Kaõ	Hála	Waliráhi	Lalani	Haiwaári	...Tohúlah.
39.	Kao	Fála	Wailolún	Latína	Yai	Apícolo.
40.	Dadow	Fála	Wailálan	Lólan	(Ai) waht	..Béber.
41.	Lahanín	Hálim	Wailanim	...Lalim	Ai liléham.	Píto.
42.	Merah	Faasi	Arr lehn	Lään	Akar	Gunísia.
43.	Mosina	Allan	Tolo maina	...Olamatan	..Tamun	...Aito.
45.	Ulúli	...Fáha	Arr sūasūa	...Laran	Ai áha	Ananihi.
46.	Fulifúli	Paser	Wehr fofowt.	.Lagain	Woki	Munini.
49.	Mamé	Fās	Wayr	Lelin	Gaka watu.	Clif.
50.	Shei	Fās	Weyoh	Má	Aikówa	...Tefoo.
59.	Merah	Buas	Ngusor	Lalan		Lijah.

One Hundred and Seventeen Words in Thirty-three

English	SALT.	SEA.	SILVER.	SKIN.
1. Malay	Gáram	Laut	Pérak	Kúlit
2. Javanese	Uyah	Segóro	Perak	Kūlit
6. Bouton } S. Celebes.	Gára	Andal	Riáli	Okulit
7. Salayer	Sela	Laut	Salaka	Balulan
16. Menado } N. Celebes.	Asing	Sási	Salraka	Pisy
17. Bolang-hitam	Simuto	Borango	Ringit	Kurito
18. Sanguir, Sian	Asing	Laudi	Perak	Pisi
19. Salibabo	Tagaroang	Salaba	Timokah
20. Sula Is	Gási	Mahi	Salaka	Koli
21. Cajeli } Bouru.	Sasi	Olat	Siláka	Usum
22. Wayapo	Sasi	Olat	Siláka	Usam
23. Massaratty..	Sasi	Masi	Silaka	Okonen
24. Amblaw	Sasieh	Lanti	Siláka	Tinyau
26. Tidore	Gási	Nólo	Saláka	A'hi
29. Gani } Gilolo.	Gási	Wólat	Salaka	Kakutut
31. Galela	Gási	Teow	Salaka	Makáhi
32. Liang } Amboyna.	Tasi	Mit	Pisiputi	Urita
33. Morella	Tasi	Met	Salaka	Uliti
34. Batumerah	Tási	Lauti	Salaka	Asáva
35. Lariki	Tasi	Lautan	Salaka	U'sa
36. Saparua	Tasi	Sawah	Salaka	Kutai
37. Awaiya } Ceram.	Tasíe	Lauhaha	Salaka	Lelutini
38. Camarian	Tasíe	Lauhaha	Salaka	Wehúi
39. Teluti	Lósa	Toweín	Salák	Lilicolo
40. Ahtiago and Tobo	Másin	Tási	Salaka	Ikulit
41. Ahtiago (Alfuros)	Teisim	Taisin	Salaka	
42. Gah	Síle	Tasok	Salak	Likito
43. Wahai	Tasi	Laut	Seláka	Unin
45. Matabello	Síra	Táhi	Saláha	Aliti
46. Teor	Siren	Hoak	Silaka	Holit
49. Mysol	Lesin	Sol	Sulūp	Kine
50. Mysol	Garam	Belot	Salup	Mot kehin
59. Baju	Garam	Medilaut	Salaka	Kulit

LANGUAGES OF THE MALAY ARCHIPELAGO.—*Continued*.

	SMOKE.	SNAKE.	SOFT.	SOUR.	SPEAR.	STAR.
1. A'sap	Ŭ'lar	..Lúmbūt	....Másam	.....Tómbak	...Bíntang.
2. Kukos	Ulo	....Gárno	.....A'sam	.....Tombak	...Lintang.
6. Ombu	Sávha	...Marobá	.....Amopára	...Pandáno	...Kalipopo.
7. Minta	Saa	....Lumut	....Kusi	.....Poki	Bintang.
16. Pūpūsy	....Katoün..Marobo	....Maresing	...Budiak	...Bitūy.	
17. Obora	Noso	...Murumpito.Morosomo		Matitie.	
18.	Katóan..Musikomi	...Naloso	.....Malehan	...Bitúin.	
19.				Kanumpitah.	
20. Apfé	Túi	...Maóma	.....Maníli	.....Pedwihi	...Fatúi.
21. Melūn	Nehei	...Namlomo	...Numnino	...Tombak	...Tūlin.
22. Fénen	Níha	...Lómo	......Dumílo	.....Néro	.....Tūlu.
23. Fenen	Wao	....Lumlóba	...Dumwilo	...Nero	.....Tólóti.
24. Mipéli	Nife	....Maloh	.....Numliloh	...Tuwáki	...Maralai.
26. Munyépho	...Yéya	...Bóleh	....Logi	....Sagu-sagu..Ngóma.	
29. Iáso	Bow	....Iklūt	......Manil	Sagu-sagu..Betól.	
31. Odópo	Inhíar	...Damúdo	...Dakíopi	....Tombak	...Ngóma.
32. Kunu	Nia	....Apoka	....Marino	.....Taha	.....Marin.
33. Aowaht	Nía	....Polo	......Marino	.....Túpa	.....Marin.
34. Asaha	Niéi	....Maluta	.....Amokinino..Sapolo	....Alanmatána.	
35. Aow pōt	Niar	....Máro	......Marino	.....Topar	..... Mari.
36. Poho	Niar	....Maru	......Marimo	....Kalēi	.....Mareh.
37. Weíli	Tepéli	...Mamouni	...Maalino	.....Soláni	.....Oōna.
38. Poöti	Nía	....Máru	.....Maaríno	.....Sanóko	...Umáli.
39. Yafoin	Nifar	...Málu	.....Malim	.....Tupa	.....Meléno.
40. Numi	Búfin	...Mamálin	...Manil	....Túba	.....Tói.
41. Waham rapoi.Koioim	..Mulisním	...Kounim	.....Leis-ánum.Kohim.		
42. Kobun	Tekoss	...Malúis	.....Mateïbi	.....Oika	.....Tilassa.
43. Honin	Tipolum.Mulumu	...Manino	.....Tite	....Teën.	
45. Ef ubun	Tofágin..Malúis	.....Matílū	.....Galla galla.Tóin.		
46. Yaf mein	Urubai	...Máfon	.....Metiloi	.....Gala gala...Tokun.	
49. Las	Pok	....Umblo	.....Embisin	...Chei	.......Toen.
50. Yap hoi	Pok	....Rum	......Pep	.....Dei	.........Náh.
59. Umbo	Ular	....Lúmah	...Gúsuh	.....Wijah	.....Kúliginta.

ONE HUNDRED AND SEVENTEEN WORDS IN THIRTY-THREE

English	SUN.	SWEET.	TONGUE.	TOOTH.
1. Malay	Máta-ári	Mánis	Lídah	Gígi
2. Javanese	Sungingi	Lūgi	I'lat	U'ntu
6. Bouton } S. Celebes.	Soremo	Maméko	Lilah	Nichi
7. Salayer }	Mata-alo	Tuni	Lilah	Gigi
16. Menado }	Mata roú	Manisy	Lilah	Ngisi
17. Bolang- } N. Celebes.				
hitam }	Unu	Mogingo	Dila	Dongito
18. Sanguir, Sian	Kaliha	Mawangi	Lilah	Isi
19. Salibabo	Allo			
20. Sula Is	Léa	Mína	Máki	Níhi
21. Cajeli }	Léhei	Emmínei	Mahmo	Nisim
22. Wayapo } Bouru.	Hangat	Dumína	Maan	Nisi
23. Massaratty.. }	Lia	Durianaa	Maanen	Nisinen
24. Amblaw	Laei	Mina	Munartéa	Nisnyatéa
26. Tidore	Wángi	Mámi	Aki	Ing
29. Gani } Gilolo.	Fowé	Gamis	Imōd	Afod
31. Galela }	Wangi	Damúti	Nangaládi	Ini
32. Liang	Riamata	Masusu	Meka	Niki
33. Morella	Liamátei	Masusu	Méka	Nikin
34. Batumerah	Limatáni	Kaséli	Numáwa	Nindiwa
35. Lariki	Liamáta	Masúma	Méh	Niki
36. Saparua	Riamatani	Mosuma	Me	Nio
37. Awaiya	Líamatei	Emási	Méi	Nisi mo
38. Camarian	Liamatei	Masóma	Meëm	Nikim
39. Teluti	Liamatan	Sunsúma	Mecólo	Lilico
40. Ahtiago and Tobo	Liamátan	Merasan	Mélin	Nifan
41. Ahtiago (Alfuros)	Léum		Nínum	Nesnim
42. Gah	Woleh	Masárat	Lemukonina	Nisikonina
43. Wahai	Leän	Moleli	Me	Lesin
45. Matabello	Olēr	Mateltelátan	Tumomá	Nifóa
46. Teor	Lew	Minek	Mēn	Nifin
49. Mysol	Seasan	Krismis	Aran	Kalifin
50. Mysol	Kluh	Mis	Aran	Kelif
59. Baju	Matalon	Manis	Délah	Gigi

Languages of the Malay Archipelago.—*Continued*.

	WATER.	WAX.	WHITE.	WIFE.	WING.	WOMAN.
1. A'yer	...Lílin	...Pūtih ..	...Bíni	...........Sayap	......Purumpuan.
2. Banyu	...Lílin	...Puté........	Seng wedo	...Sewíwi	.....Wong wedo.
6. Mánu	...Taru......	Mapúti	...Orakenana	...Opáni	......Bawíne.
7. Aer	......Pantis	...Putih	......Baini	........Kapi	......Baini.
16. Akéi	......Tadu	...Mabida	...Gagijan	......Panidey	...Taumatababiney.
17. Sarúgo	...Tajo......	Mopotiho..	Wure	........Poripikia...	Bibo.
18. Aki	......Lilin	...Mawirah	...Sawa	........Tula........	Mahoweni.
19. Wai	..................Mawirah	...Babineh	....................Babineh.
20. Wai	......Tócha	...Boti........	Nifáta	........Sóba........	Fina.
21. Wäili	...Lilin	....Umpóti	...Sówom	......Ahiti	......Umbinei.
22. Wai	..................Bóti........	Gefína	........Ahit........	Gefíneh.
23. Wai	..................Bóti........	Fínha	........Panin	......Fíneh.
24. Wai	......Lilin	...Purini	......Elwinyo	......Aféti	... . Remau elwinyo.
26. Aki	......Tóeha	...Bubúlo	...Foyá	.......Fila fila	...Fofoyá.
29. Waiyr	...Tócha	...Wulan......	Mapīn	........Nifako...	...Mapīn.
31. Aki	......Tócha	...Daári	......Mapidéka	...Gulupúpo..	Opedéka.
32. Weyr	...Kina......	Putih	......Mahina	......Aïna........	Mahina.
33. Weyl	...Lilin	...Putih	......Mahina	.....Ihóti	......Mahina.
34. Weyl	.............Putih	......Mahinai	......Kihoá	......Mainai
35. Weyl	...Lilin	...Putih	......Mahina	.....I'ho	........Mahina.
36. Wai	......Riruiah..	Putil	......Pipina	........Ihol	........Pipina.
37. Wäéli	...Lilin	...Putíle	......Mumahéna	...Teyhóli	...Mahína.
38. Wäéli	...Lilin	...Putih	......Nímahína	...Ihóri	...Mahína.
39. Wélo	...Nínio	...Putih	......Nihina	.....Hihóno	...Ihina.
40. Wai	.....Lilin	...Babút	......Invína........	Yeón	......Vína.
41. Wai-im	..............Putih	......Ifnéinin	....................Ifnéinin.
42. Arr	......Lilin	...Maphutu...	Bina............	Wákul.....	Binei.
43. Tólun	...Lilin	...Puteh	......Pinan ...	......Keheil......	Pina híeti.
45. Arr	......Lilin	...Maphúti	..Ahéhwá	......Olilífi	......Felelára.
46. Wehr	....Sélup	......Wewina	......Fanik	......Mewina.
49. Wayr	...Telilin...	Bus ..	......Pin	............Kufeu	...Pin.
50.	.......................Boo	.........Ji yu	........Fieh........	...Mot yu.
59. Boi	..................Potih	......Lako	........Kapéna	...Dindah.

One Hundred and Seventeen Words in Thirty-three

English		WOOD.	YELLOW.	ONE.	TWO.
1. Malay		Káyū	Kūning	Sátu	Dúa
2. Javanese		Kayu	Kuning	Sa, Sawiji	Loro
6. Bouton	S. Celebes.	Okao	Mákuni	Saangu	Ruano
7. Salayer	S. Celebes.	Kaju	Didi	Sedri	Rua
16. Menado	N. Celebes.	Kalun	Madidihey	Esa	Dudua
17. Bolang-hitam	N. Celebes.	Kayu	Morohago	Soboto	Dia
18. Sanguir, Sian		Kalu	Ridihi	Kusa	Dua
19. Sulibabo		Kalu	Maririkah	Sembäow	Dua
20. Sula Is		Kaō	Kuning	Hía	Gahú
21. Cajeli	Bouru.	Aow	Umpóro	Silei	Lua
22. Wayapo	Bouru.	Kaō	Konin	Umsiun	Rua
23. Massaratty	Bouru.	Kaō	Koni	Nosiúni	Rua
24. Amblaw		Ow	Umpotoi	Sabi	Lua
26. Tidore		Lúto	Kuráchi	Remoi	Malófo
29. Gani	Gilolo.	Gagi	Madímal	Lepso	Leplu
31. Galela	Gilolo.	Góta	Decokuráti	Moi	Sinuto
32. Liang	Amboyna.	Ayer	Poko	Sa	Rua
33. Morella	Amboyna.	Ai	Poko	Sa	Lua
34. Batumerah	Amboyna.	Ai	Apoo	Wása	Luá
35. Lariki	Amboyna.	Ai	Poko	Isa	Dua
36. Saparua		Ai	Pocu	Esa	Rua
37. Awaiya	Ceram.	Ai	Poporólo	Lai-isa	Lūūa
38. Camarian	Ceram.	Ai	Pocu	Isái	Lúa
39. Teluti	Ceram.	Lyeií	Poko	San	Lua
40. Ahtiago and Tobo	Ceram.	A'i	Ununing	San	Lua
41. Ahtiago (Alfuros)	Ceram.	Ai-im	Uninim	Esá	Elúa
42. Gah	Ceram.	Kaya	Kunukunu	So	Lotu
43. Wahai	Ceram.	Ai	Masikuni	Sali	Lua
45. Matabello		A'i	Wuliwulan	Sa	Rua
46. Teor		Kai	Kúni	Kayée	Rúa
49. Mysol		Gáh	Kumenis	Katim	Lu
50. Mysol		Ei	Flo	K'tim	Lu
59. Baju		Kayu	Kuning	Sa	Dua

LANGUAGES OF THE MALAY ARCHIPELAGO.—*Continued.*

	THREE.	FOUR.	FIVE.	SIX.	SEVEN.	EIGHT.
1.	Tíga......A'mpat	......Líma	........A'nam	.....Tújoh	......Delápan.
2.	Talu .....Papat	........Lima	........Nanam	......Pitu ....Wola.
6.	Taruáno..Patánu	.....Limánu	......Namano	...Pituáno......Veluáno.
7.	Tello.....Ampat	........Lima	........Unam	......Tujoh ......Karna.
16.	Tateru...Pa ...	........Rima	........Num ....Pitu ....Walru.
17.	Toro......Opato	.... Rima	........Onomo ......Pitu ....Waro.
18.	Tellon ...Kopa	........Lima	.. ..Kanum ......Kapitu ......Walu.
19.	Tetálu....Apátah	......Delima	......Annuh ......Pitu ....Waru.
20.	Gatíl......Garíha	........Lima	........Gané ....Gapítu ......Gatahúa.
21.	Tello......Há	........Lima	........Ne ....Hito ....Walo.
22.	Tello......Pá	.... Lima	........Né ....Pito ....Etrúa.
23.	Tello......Pa	........Lima	........Né ....Pito ....Trúa.
24.	Relu......Faä	........Lima	........Noh ....Pitu ....Walu.
26.	Rangi ...Ráha	........Runtóha ......Rora ....Tumodí......Tuíkángi.
29.	Leptol ...Lepfoht	......Leplim	......Lepwonan...Lepfit ......Lepwal.
31.	Sängi ...Iha	.....Matóha	......Butánga ...Tumidingi.. Itupangi.
32.	Tero......Hani	........Rima	........Nena ....Itu...........Waru.
33.	Telo......Hata	........Lima	........Nena ....Itu...........Waru.
34.	Telua ...Atá	........Limá	........Nená ....Ituá ....Walúa.
35.	Toro......Aha	........Rima	........Nöo ....Itu...........Waru.
36.	Toru......Haä	........Rima	........Noöh ....Hitu ....Waru.
37.	Te-elu ...Aäta	........Lima	........Nöme ......Witu ......Walu.
38.	Tello......A'ä	........Lima	........Nöme ......Itu...........Walu.
39.	Toi ......Fai	........Lima	........Noi ....Fitu ....Wagu.
40.	Töl ......Fet	........Lima	........Num ....Fit ....Wal.
41.	E'ntol ...Enháta	......Enlima	......Ennói ......Enhit........Enwol.
42.	Tolo......Faat	........Lim ....Wonen ......Fiti ....Alu.
43.	Tolo......Ati	........Nima	........Lomi ....Itu...........Alu.
45.	Tolu......Fata	........Rima	........Onam ......Fitu ....Allu.
46.	Tel ......Faht	........Lima	........Nem ....Fit ....Wal.
49.	Tol ......Fut	........Lim ....Onum ......Fit ....Wal.
50.	Tol ......Fut	........Lim ....Onum ......Tit .. ..Wal.
59.	Tiga......Ampat	........Lima	........Nam ....Tujoh........Dolapan.

One Hundred and Seventeen Words in Thirty-three

English	NINE.	TEN.	ELEVEN.
1. Malay	Sambílan	Sapúloh	Sapúloh sátu
2. Javanese	Sanga	Pulah	Swalas
6. Bouton } S. Celebes..	Sioánu	Sapúloh	Sapúloh sano
7. Salayer }	Kasa	Sapuloh	Sapuloh sedrú
16. Ménado } N. Celebes.	Sio	Mapulroh	
17. Bolang-hitam }	Sio	Mopuru	
18. Sanguir, Sian	Kasiow	Kapuroh	Mapurosa
19. Salibabo	Sioh	Mapuroh	Ressa
20. Sula Is.	Gatasía	Póha	Poha di ha
21. Cajeli	Siwa	Boto	Boto lesile
22. Wayapo } Bouru.	Eshía	Polo	Polo geren en sium.
23. Massaratty.. }	Chía	Polo	Polo tem sia
24. Amblaw	Siwa	Buro	Buro lani sebi
26. Tidore	Sio	Nigimói	Nigimói seremoi ..
29. Gani } Gilolo.	Lepsiu	Yagimso	Yagimso lepso
31. Galela }	Sio	Megió	Megió demoi
32. Liang } Amboyna.	Sia	Husa	Huséla
33. Morella	Siwa	Husá	Huselali
34. Batumerah	Siwá	Husa	Husalaisa
35. Lariki	Siwa	Husa	Husaelel
36. Saparua	Siwa	Husani	Husani lani
37. Awaiya } Ceram.	Siwa	Hutūsa	Sinleūsa
38. Camarian	Siwa	Tineín	Salaise
39. Teluti	Siwa	Hútu	Mesileë
40. Ahtiago and Tobo	Siwa	Vūta	Vut säilan
41. Ahtiago (Alfuros)	Ensiwa	Fotusa	Fotusa elése
42. Gah	Sia	Ocha	Ocha le se
43. Wahai	Sia	Husa	Husa lesa
45. Matabello	Sia	Sow	Terwahei
46. Teor	Siwer	Hutá	Ocha kilu
49. Mysol	Si	Lafu	Lafu kutim
50. Mysol	Sin	Yah	Yah tem metim
59. Baju	Sambilan	Sapuloh	

LANGUAGES OF THE MALAY ARCHIPELAGO.—*Continued.*

	TWELVE.	TWENTY.	THIRTY.	ONE HUNDRED.
1.	Sapúloh dúa ....Dúa pūloh.........Tiga pūloh ......Sarátus.			
2.	Rolas ....Rongpuluh ......Talupuluh.........Atus.			
6.	Sapúlohruano ....Ruapulo ... .Tellopulo ....Sáatu.			
7.	Sapuloh rua ....Ruampuloh ... ..Tellumpuloh......Sabilangan.			
16.	....Mahasu.			
17.	....Gosoto.			
18.	Mapuro dua ....Duampuloh ......Tellumpulo ......Mahásu.			
19.	Ressa dua ....Dua puroh ......Tetalu puroh ..Ma rasu.			
20.	Poha di gahú ....Poha gahú ......Poha gatíl........O'ta.			
21.	Betele dua ....Botlua ....Bot telo............Bot ha.			
22.	Polo geren rua ....Porúa...............Potéllo ....U'tun.			
23.	Polo tem rua............Porúa...............Potello ....U'tun.			
24.	Bŏr lan lua ....Borolua... ....Borélo.............Uruni.			
26.	Nigimói semolopho ...Negimelopho ...Negerangi ....Ratumoi.			
29.	Yagimso leplu ....Yofalu ....Yofatol ....Utinso.			
31.	Megió desinoto.........Menohallo........Muruangi ....Rátumoi.			
32.	Husa lua ....Huturúa ....Hutáro ....Hutúna.			
33.	Husa lua ....Huturua ....Hutatilo ....Hutūn.			
34.	Husalaisa lua ....Hotulua ....Hotelo ....Hutunsá.			
35.	Husendua .. ..Hutorua ....Hutóro ....Hutūn.			
36.	Husani elarua ....Huturua ....Hutoro ....Utúni.			
37.	Sinlūa ....Hutulúa ....Hututēlo ....Utúni.			
38.	Salalua ....Hutulua ....Hututello ... .Hutunérs.			
39.	Hutulelúa ....Hutulúa ....Hututoi ....Hutún.			
40.	Vut sailan lūa ....Vut lua...........Vut tol ....Utin.			
41.	Elelúa...................Fotulúa ....Fotol ....Hutnisá.			
42.	Husa la lua .... Otoru ....Otólu ....Lutcho.			
43.	Ocha siloti..............Hutu a ....Hutu tololu .....Utun.			
45.	Ternorua ....Teranrua ....Terantolo ....Rátua.			
46.	Arúa ....Oturúa ....Otil ....Rása.			
49.	Fufu lu ....Lufu lu ....Lufu tol............Uton.			
50.	Yah mulu ....Ya luh ....Yatol ....Toon.			
59.	....Datus.			

INDEX

INDEX.

A.

Abel, Dr. Clarke, his account of a mias, 48

Acacia, in the Archipelago, 6

Acarus, bites of the, 274

Æschynanthus, climbing plants in Borneo, 62

African negroes, on the crania and languages of the, 459, 460

Ahtiago, village of, 276

Ahtiago and Tobo vocabularies, 466

Alcedo dea, 349

Alfuros, the true indigenes of Gilolo, 241, 243; of Papuan race, the predominant type in Ceram, 280

Ali, the author's attendant boy, 240, 241, 244, 252; the author's head man, 312

Allen, Charles, the author's assistant, 36; sent with the collections to Saráwak, 49; finds employment, and leaves the author for four years, 156; rejoins the author, 232; news of, 240, 244, 288, 292, 394; letter received from, 417; his collections, 418; his difficulties, 419; his wanderings, 419; finally obtains employment in Singapore, 419; his voyage to Sorong, and his difficulties, 437 *et seq.*

Allor vocabulary, 467

Amahay, bay of, 270; visit to, 274

Amberbaki, visit to, 385

Amblau vocabulary, 465

Amboyna, island of, 4; voyage to, from Banda, 223; map of, 224; the town of, 224; volcanoes in remote times, 224 (*see* Water, limpid); the author's cottage in, 226 (*see* Interior); general character of the people, 229; habits and customs, 230, 231 (*see* Shells), 231–233; clove cultivation established at, 237; departure from, 267; map of, 268

Amboyna lory, 273

Ampanam, 117, 118; birds of, 118; cause of the tremendous surf at, 125

Anchors of the Malays, 414, 415

Andaman Islands, in the Bay of Bengal, 453

Animal life, luxuriance and beauty of, in the Moluccas, 308

Animals, distribution of, the key to facts in the past history of the earth, 111–115, 155–162; geographical distribution of, 372, 373

Anonaceous trees in Borneo, 62

Anthribidæ, species of, 248

Ants, noxious, 357; at Dorey, tormented by, 390, 391

Ape, the Siamang, 103

Arabs in Singapore, 16

Archipelago, Malay, physical geography of, 1; productions of, in some cases unknown elsewhere, 1 (*see* Islands); extent of, 2; natural division of, into two parts, 7 (*see* Austro-Malayan, and Natural productions); shallow waters of, 9–13 (*see* Races)

Architectural remains in Java, 77; ruined temples, 79-81

Arfaks, of New Guinea, 381, 382, 386

Arjuna Mount, 76

Arndt, M., a German resident in Coupang, 142

Arrack, demand for, 351

Art, rudimental love of, among barbarians, 389

Aru Islands, 6; voyage to, from Macassar in a native prau, 308 *et seq.*; diary of the voyage, 312

N.

O.

Medals awarded me.

Royal Medal of the Royal Society — 1868

Gold Medal of the Société de Geographie
                          of Paris — 1870

Darwin Medal of the Royal Society — 1890

Gold Medal of the Roy. Geog. Society — 1892

Gold Medal of the Linnean Society — 1892

Elected Fellow of the Royal Society 1893.

Extracts about Mias (8½ pp. 31-40) in "International
Library of Literature" – American. Rec.d fee of 2 guineas for it.